Robust Control

Wiley-ASME Press Series

Robust Control: Youla Parameterization Approach
Farhad Assadian, Kevin R. Mallon

Fabrication of Process Equipment
Owen Greulich, Maan H. Jawad

Practice with Oilfield and Drilling Applications
Donald W. Dareing

Flow-Induced Vibration Handbook for Nuclear and Process Equipment
Michel J. Pettigrew, Colette E. Taylor, Nigel J. Fisher

Vibrations of Linear Piezostructures
Andrew J. Kurdila, Pablo A. Tarazaga

Bearing Dynamic Coefficients in Rotordynamics: Computation Methods and Practical Applications
Lukasz Brenkacz

Advanced Multifunctional Lightweight Aerostructures: Design, Development, and Implementation
Kamran Behdinan, Rasool Moradi-Dastjerdi

Vibration Assisted Machining: Theory, Modelling and Applications
Li-Rong Zheng, Dr. Wanqun Chen, Dehong Huo

Two-Phase Heat Transfer
Mirza Mohammed Shah

Computer Vision for Structural Dynamics and Health Monitoring
Dongming Feng, Maria Q Feng

Theory of Solid-Propellant Nonsteady Combustion
Vasily B. Novozhilov, Boris V. Novozhilov

Introduction to Plastics Engineering
Vijay K. Stokes

Fundamentals of Heat Engines: Reciprocating and Gas Turbine Internal Combustion Engines
Jamil Ghojel

Offshore Compliant Platforms: Analysis, Design, and Experimental Studies
Srinivasan Chandrasekaran, R. Nagavinothini

Computer Aided Design and Manufacturing
Zhuming Bi, Xiaoqin Wang

Pumps and Compressors
Marc Borremans

Corrosion and Materials in Hydrocarbon Production: A Compendium of Operational and Engineering Aspects
Bijan Kermani and Don Harrop

Design and Analysis of Centrifugal Compressors
Rene Van den Braembussche

Case Studies in Fluid Mechanics with Sensitivities to Governing Variables
M. Kemal Atesmen

The Monte Carlo Ray-Trace Method in Radiation Heat Transfer and Applied Optics
J. Robert Mahan

Dynamics of Particles and Rigid Bodies: A Self-Learning Approach
Mohammed F. Daqaq

Primer on Engineering Standards, Expanded Textbook Edition
Maan H. Jawad and Owen R. Greulich

Engineering Optimization: Applications, Methods and Analysis
R. Russell Rhinehart

Compact Heat Exchangers: Analysis, Design and Optimization using FEM and CFD Approach
C. Ranganayakulu and Kankanhalli N. Seetharamu

Robust Adaptive Control for Fractional-Order Systems with Disturbance and Saturation
Mou Chen, Shuyi Shao, and Peng Shi

Robot Manipulator Redundancy Resolution
Yunong Zhang and Long Jin

Stress in ASME Pressure Vessels, Boilers, and Nuclear Components
Maan H. Jawad

Combined Cooling, Heating, and Power Systems: Modeling, Optimization, and Operation
Yang Shi, Mingxi Liu, and Fang Fang

Applications of Mathematical Heat Transfer and Fluid Flow Models in Engineering and Medicine
Abram S. Dorfman

Bioprocessing Piping and Equipment Design: A Companion Guide for the ASME BPE Standard
William M. (Bill) Huitt

Nonlinear Regression Modeling for Engineering Applications: Modeling, Model Validation, and Enabling Design of Experiments
R. Russell Rhinehart

Geothermal Heat Pump and Heat Engine Systems: Theory and Practice
Andrew D. Chiasson

Fundamentals of Mechanical Vibrations
Liang-Wu Cai

Introduction to Dynamics and Control in Mechanical Engineering Systems
Cho W.S. To

Robust Control

Youla Parameterization Approach

Farhad Assadian and Kevin R. Mallon
University of California, Davis, USA

This Work is a co-publication between John Wiley & Sons Ltd and ASME Press.

ASME PRESS

WILEY

This Work is a co-publication between John Wiley & Sons Ltd and ASME Press.

Registered Offices
John Wiley & Sons, Inc., 111 River Street, Hoboken, NJ 07030, USA
John Wiley & Sons Ltd., The Atrium, Southern Gate, Chichester, West Sussex, PO19 8SQ, UK

Editorial Office
The Atrium, Southern Gate, Chichester, West Sussex, PO19 8SQ, UK

For details of our global editorial offices, customer services, and more information about Wiley products visit us at www.wiley.com.

Wiley also publishes its books in a variety of electronic formats and by print-on-demand. Some content that appears in standard print versions of this book may not be available in other formats.

Library of Congress Cataloging-in-Publication Data applied for
[ISBN: 9781119500360 (hardback)]

Cover Design: Wiley
Cover Image: © VikaSuh/Getty Images; Image by Farhad Assadian and Kevin R. Mallon

Set in 9.5/12.5pt STIXTwoText by Straive, Chennai, India
Printed and bound by CPI Group (UK) Ltd, Croydon, CR0 4YY

C9781119500360_170122

To Professor John Brewer

Contents

Preface

In the past several decades, there have been many published control system design books in classical (PID), modern (state-space), and neoclassical and optimal robust control. Then, why is it necessary to publish yet another control system design book?

To clearly answer this question, I need to say a few words about my own background. Before becoming a Professor in System Dynamics and Control, I have been a control engineer with both energy and automotive industries for over 30 years. In these industries, feedback is still a novelty and majority of feedback control technique is around designing PID controllers. Control theory has drastically advanced and yet with these advancements, the gap between industrial application of controls and theoretical controls has been increased. Majority of the current control publications are very much math heavy and make the assumptions that all practitioners and students have the right background to fully appreciate these theoretical development. Most importantly, in all these developments, the beauty of control design at times is not clearly addressed and hence, large number of the students and the practitioners are not still have a good understanding of why feedback control is needed.

In this book, we try to close this gap by illustrating that feedback control could be elegantly designed in the frequency domain using Youla parameterization approach. It is the system bandwidth which elegantly defines a trade-off between performance and robustness. The father of quantitative feedback theory (QFT), Issac Horowitz, said that the bandwidth is like an open window during a hot summer night. If the window is kept widely open, then one will let the flies in and if the window is kept shut, then one will limit the summer night breeze. This simple and elegant metaphor defines the trade-off or balance between performance (speed) and noise rejection capability in feedback control design.

Youla parameterization technique is not new; the technique was first developed in the 1970s [1] and addressed in several other existing textbooks, for instance [2, 3]. However, in this book, we will provide deeper insights with many practical applications in utilizing this technique in both single input single output (SISO) and multiple input multiple output (MIMO) frequency domain feedback control design. In addition, an estimation technique using Youla parameterization and controller output observer both for SISO and MIMO plants, for the first time, is introduced. Although, the emphasis of these two parts is on frequency domain, we will discuss optimal robust control approaches such as H_2/LQG and H_∞ using state-space approaches in Chapter 5 of this book. We will provide comparative results through practical examples between the optimal robust control methods and the Youla parameterization approach.

We have to apologize for the level of mathematical rigor provided in this book. It goes without saying that control system design is a subject which requires mastery of basic mathematical concepts. We have tried hard to avoid theorem and proof format when introducing mathematical concepts and provided the level mathematics to give the readers the necessary background for understanding and appreciating the control theoretical concepts discussed in this book.

This book is divided into two parts: Part I covers SISO control and estimation design using the Youla parameterization approach, while Part II covers MIMO control and estimation design using Youla parameterization approach.

Part I of this book includes eight chapters:

In Chapter 1, a review of the Laplace transform is given. In this chapter, we tried to avoid many details in introducing the Laplace transform and concentrated on providing the properties of the Laplace transform. It is these properties will be utilized throughout this book for developing the main concepts.

In Chapter 2, the response of linear time-invariant (LTI) dynamic systems is provided. A review of stability of LTI dynamic systems is given and frequency response including Bode and Nyquist diagrams are discussed.

In Chapter 3, we provide a discussion of feedback principals, a review of well-posedness and internal stability is given, and the role of Youla parameter and interpolation conditions in assuring internal stability is provided. Complementary sensitivity transfer function T and sensitivity transfer function S are derived, and modeling uncertainty is discussed.

In Chapter 4, we provide the fundamentals of feedback design using Youla parameterization. We discuss algebraic and analytical constraints in feedback design and we reveal how Youla parameter could be utilized in designing a feedback control by considering the constraints and trade-offs in SISO control design.

In Chapter 5, a review of signal and system size (norms) is provided. linear fractional transformation (LFT) platform and its use in control design are discussed including uncertainty modeling with the use of this platform. A brief review of state-space method including concepts such as controllability and observability are discussed.

In Chapter 6, we discuss the Mixed Sensitivity Method and H_∞ method for SISO robust control design. In this chapter, we provide insights to the frequency domain solution of H_∞ control and selection of the weighting filters in shaping complementary sensitivity, sensitivity, and Youla transfer functions.

In Chapter 7, we discuss how Youla parameterization could be utilized in deriving a state and input estimation method. In this chapter, an introduction to Luenberger observer and Kalman filtering is provided including a discussion about the Controlled Output Observer approach. A step-by-step transformation approach from Luenberger observer to Youla parameterization estimation using the Controller Output Observer framework is developed.

In Chapter 8, we present two automotive examples, one for control example and another for estimation. The control example demonstrates yaw stability control with an active differential. The estimation example consists of vehicle yaw rate and sideslip estimation using the Youla controller output observer (YCOO) method. In the control example, we reveal the differences between using SISO robust control design using the standard Matlab toolbox and Youla parameterization hand computation technique. In the estimation example, we compare the result of YCOO to both linear and nonlinear Kalman filter method.

Part II of this book includes nine chapters.

In Chapter 9, we provide an overview of different multivariable feedback control methods and the importance of Youla parameterization approach in designing multivariable feedback control system. In the same chapter, we introduce multivariable transfer function matrix, Schur complement, and Rosenbrock's system matrix including its application for gaining insight into poles and zeros of a transfer function matrix.

In Chapter 10, Matrix Fractional Description, unimodular matrices, and the concept of polynomial matrices' equivalency with several insightful examples are introduced. A discussion on Smith canonical form and Smith-McMillan form of a transfer function matrix including several examples is provided. The idea of coprimeness and several approaches for transforming a transfer function matrix to its equivalent stat-space form are discussed.

In Chapter 11, we discuss the eigenvalue problem and matrix diagonalization using eigenvectors and eigenvalues. The diagonalization issues using eigenvalues and eigenvectors are discussed, and the use of singular value decomposition as a general method for transfer function matrices diagonalization is reviewed.

In Chapter 12, we discuss MIMO feedback principals. We introduce MIMO return ratio and return difference, complementary sensitivity and sensitivity transfer function matrices, and Youla transfer function matrix. Concepts such as Nyquist stability, well-posedness, and internal stability for MIMO feedback systems are discussed. Some of the properties of singular value decomposition are reviewed and their implications on the limit of MIMO feedback performance are revealed.

In Chapter 13, we introduce the concept of MIMO feedback control design using Youla parameterization. We illustrate the simplicity of this technique by showing that for low-order transfer function matrices, the MIMO control design could be easily performed by hand.

In Chapter 14, a review of vector signal and MIMO system size (norms) is provided. Linear fractional transformation (LFT) platform is revisited, and the uncertainty modeling for MIMO feedback systems with the use of this platform is discussed.

In Chapter 15, we utilize the norms discussed in the previous chapter to derive MIMO controllers. We discuss optimal control methods. We review the derivation of linear quadratic regulator (LQR) and linear quadratic Gaussian (LQG) and discuss the equivalency between H_2 and LQG optimal control. We provide the state-space solutions to H_2 and H_∞ optimal robust control.

In Chapter 16, we revisit the system from Chapter 7 and discuss how Youla parameterization could be utilized in deriving state and input estimation method for MIMO systems. A step-by-step design approach for MIMO Youla parameterization estimation using the Controller Output Observer framework is developed.

In Chapter 17, we present three MIMO control examples and one MIMO estimation example. The control examples include active suspension control, advanced engine speed control for hybrid vehicles, and robust control for the powered descent of a multibody lunar landing system. The estimation example revisits the vehicle yaw rate and sideslip estimation example of Chapter 8, transforming the SISO problem into a MIMO one through the introduction of rear wheel steering. In the control examples, we reveal the differences between using MIMO robust control design using the standard Matlab toolbox and MIMO YCOO hand computation technique. In the estimation example, we compare the result of MIMO YCOO with both linear and nonlinear Kalman filter method.

University of California, Davis
July, 2017

Farhad Assadian
Kevin R. Mallon

Acknowledgments

My sincere thanks and appreciation to Professor John Brewer who introduced me to the concept of Youla parameterization in 1990s. Unfortunately, Professor Brewer passed way on 17 October 2003 at the age of 65. His enthusiasms and passion for teaching control system design to his students was exemplary. He shall always be remembered as one of the best teachers by his students.

F.A.

Introduction

"It seems that without electronic, controls nothing works."

I.1 Why Feedback Control?

Control systems are increasingly becoming an integral part of every piece of equipment that we humans use. Without control systems our lives would drastically change. From aircrafts to household appliances, control systems play a vital role in the operation of these machines. In the automotive domain, nowadays, the majority of vehicle attributes are software and control-system based. It is therefore important for everyone, especially machine operators and technical managers, to have a basic understanding of control system operation and their impact. It is also becoming significantly important that all engineers, from every specialty, should be required to take at least one good control system design class throughout their academic education.

When it comes to control system design, our emphasis in this book is on feedback control system design, particularly model-based feedback control design. We will describe what we mean by model based in the next few paragraphs. But, first let's clarify the need for feedback controls. It is interesting, especially in industry, that there is still not a good understanding about why feedback control is necessary.

Let's take a look at Figure I.1a, where a person without eyesight is trying to pick up a book, which is located on a table. From this person's point of view, there is an uncertainty about the exact position of this book. To reduce this uncertainty, the person would have to find the book by examining different locations on the table with their hands. Hence, it normally takes more effort or energy to overcome the uncertainty and locate this book. Now, imagine a person with eyesight, as in Figure I.1b. For this person, the exact location of the book is found by making a measurement using the person's eyesight and feeding back this information to the brain. Therefore, uncertainty about the position of the book is reduced by this feedback measurement. Interestingly, the effort and energy required to locate the book is also reduced when compared to the person without eyesight. We can say then that the fundamental reason for feedback in control system design is simply to reduce **uncertainty**. We will see that this uncertainty exist in many sources including the act of the measurement itself. Amazingly, the same uncertainty that gives rise to the use of feedback control itself and can result in its demise!

Another important takeaway from this discussion is that reducing uncertainty requires effort and energy and the source of this effort and energy is the actuator. We will relate the actuator effort and energy to system robustness when discussing robust control development in the following chapters.

There are several different methods that could be utilized in designing feedback controllers. Our intent in this book is to illustrate that the use of Youla parameterization technique is one of the most straightforward methods in designing model-based feedback control system, especially for single-input, single-output (SISO) control system design. We will show that control system design becomes procedural with minimal calibration effort when implementing these controllers on the actual systems. Reducing controller calibration is significantly important to industry, as reducing calibration effort directly reduces development costs. We will argue that the elegance of Youla parameterization is its simplicity and in reducing controller calibration overhead.

Imagine designing a control system for a very complex system which could be mathematically described by nonlinear partial differential equations (PDEs). What are the choices to design a controller for this complex system?

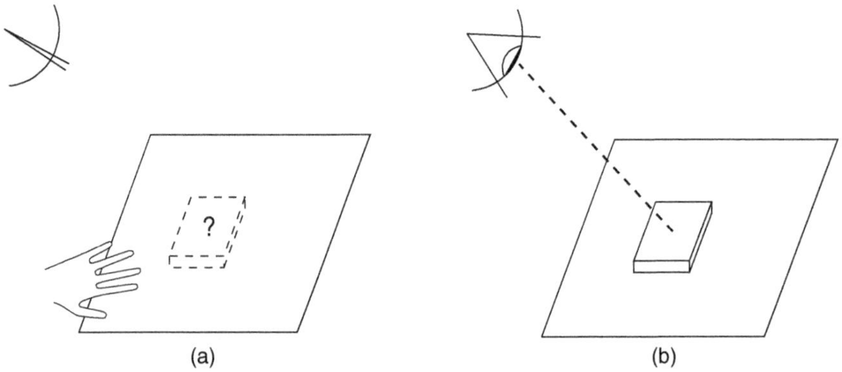

Figure I.1 There is more uncertainty about the book's location with (a) closed eyes, compared to (b) open eyes.

- **Direct Control of Nonlinear PDEs**: As far as the authors of this book are aware, there are no methods which directly address the control design of nonlinear PDEs, especially the open loop unstable kinds. In addition, if there were control methodologies to address nonlinear PDEs directly, quantifying model errors and uncertainties would have been an extremely difficult task. The quote by George Box "Essentially, all models are wrong, but some are useful" should be kept in mind when complex systems are modeled using nonlinear PDEs.
- **Indirect Control by Model Reduction of the Nonlinear PDEs to linear time-invariant (LTI) PDEs**: There are methods which address the control of LTI PDEs, such as boundary control of LTI PDEs using backstepping control method [4], or robust control of infinite dimensional LTI using H_∞ control [5].

It should be noted that at times, linearization of particular nonlinear systems results in loss of controllability. We will not address these systems in this book.

We will use model reduction techniques tp go one step further than indirect control by reducing the original complex model to LTI ordinary differential equations (ODEs). Here stands the trade-off between complexity and uncertainty: As the complex model is reduced to a simpler model, naturally the complexity is reduced. However, uncertainty is increased due to loss of information. Therefore, the entire idea of robust control is to quantify these uncertainties in the control design phase.

There are generally two objectives in designing a feedback control system: performance and robustness. "Performance" indicates the ability of the control system to respond quickly and accurately under ideal conditions. "Robustness" indicates the ability of the control system to respond accurately in the face of disturbances, noise, or other sources of uncertainty. Generally (but not always), there is a trade-off between robustness and performance: Speeding up the response requires sacrificing robustness, while increasing robustness requires sacrificing the nominal performance. This concept – the trade-off between robustness and performance – is illustrated in Figure I.2.

More specifically, the conflicting measures are:

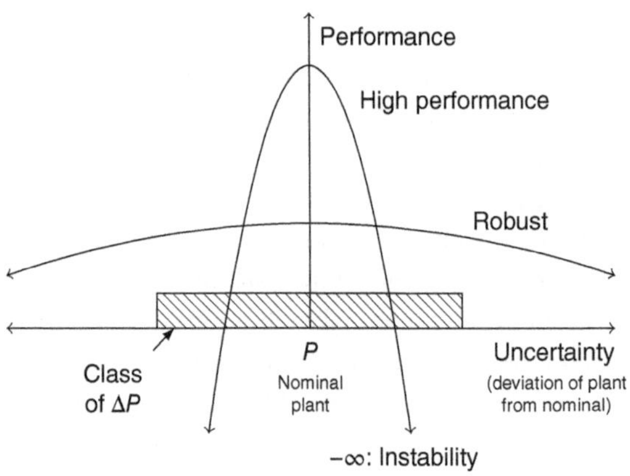

Figure I.2 Robustness and performance.

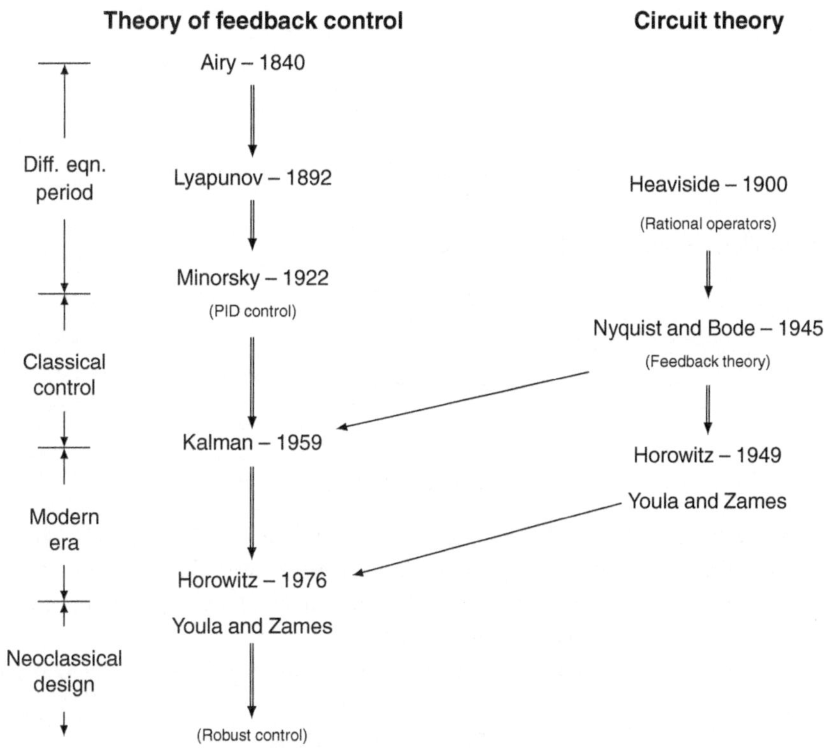

Figure I.3 Brief history of feedback control.

- Stability (is the controller robustly stable?)
- Command tracking and disturbance rejection
- Sensor noise rejection
- Actuator effort and saturation.

The history of control systems theory and the path to robust control is briefly outlined in Figure I.3.

I.2 Why Youla Parameterization?

In the prior section, we briefly discussed the benefits of feedback control and various options for model-based control development. In the next few paragraph, we will elaborate more on the benefits of feedback control and the utility of Youla parameterization in feedback control design. This section presumes some knowledge of the Laplace transform, transfer functions, and block diagram algebra.

When dealing with model-based controls, the starting point is a mathematical description of the system. Consider the simple system shown in Figure I.4, where P is the plant model and Q is the control.,

In order to enforce perfect target following, where $y(s) = r(s)$ and s is the Laplace operator, we have to select the controller $Q(s) = P^{-1}(s)$. This very simple example illustrates that plan invertability is very important in control system design. We will elaborate on this point when designing feedback controllers using Youla parameterization method. Let's assume for a moment that this plant is invertible. What happens when we implement this simple controller on an actual plant?

In order to capture the reality a bit better, we added a disturbance signal at the output of the plant, as illustrated in Figure I.5. This disturbance signal could, for example, model a wind gust acting on a vehicle while driving. Based on the

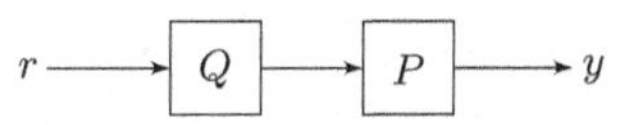

Figure I.4 System model.

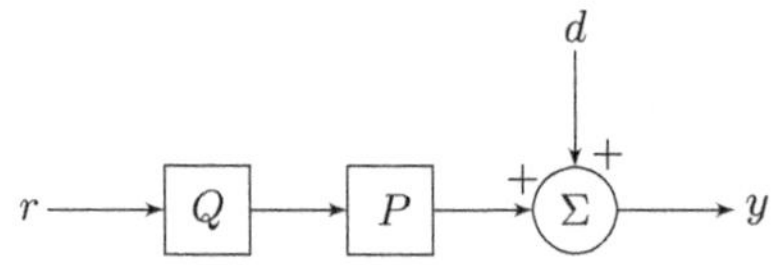

Figure I.5 System model with output disturbance.

previous feedforward controller, we can write the following equations.

$$y(s) = d(s) + P(s)Q(s)r(s) \tag{1}$$

but

$$Q(s) = P^{-1}(s) \tag{2}$$

so therefore,

$$y(s) = d(s) + r(s) \tag{3}$$

So, we are here stating the obvious point that if there is a disturbance or a source of uncertainty, the feedforward controller is not capable of coping with this. As a result, the target following in a feedforward loop deteriorates when compared to the situation when perfect knowledge of the plant was available. The output y no longer follows the commanded input r.

Let's now add a feedback to Figure I.5 as illustrated in Figure I.6.

The output $y(s)$ can be written as a function of the target signal, $r(s)$, and disturbance signal, $d(s)$, as follows.

$$y(s) = \frac{P(s)Q(s)}{1 + P(s)Q(s)} r(s) + \frac{1}{1 + P(s)Q(s)} d(s) \tag{4}$$

Two important points should be evident from this equation. First, we cannot simply set $Q(s) = P^{-1}(s)$, as we did in the feedforward case, if we are designing for good target following and disturbance rejection. Second, if we set $P(s)Q(s) \gg 1$, then we will recover the target following of the feedforward controller by rejecting or minimizing the disturbance or uncertainty effect. There are other possible disturbances and uncertainties to consider, but the value of feedback controls in the presence of uncertainties has been made clear.

In the next few paragraphs, we provide a very basic introduction to Youla parameterization. The feedback control loop is revisited in Figure I.7. However, this time, we replaced the dynamic model of the plant $P(s)$ with a generalized symbol $G_p(s)$ and replaced the feedforward controller, $Q(s)$ with a generalized symbol for feedback controller $G_c(s)$.

The closed-loop transfer function or the complementary sensitivity transfer function, $T(s)$, can be written as a function of $G_p(s)$ and $G_c(s)$ as follows:

$$T(s) = \frac{G_p(s)G_c(s)}{1 + G_p(s)G_c(s)} \tag{5}$$

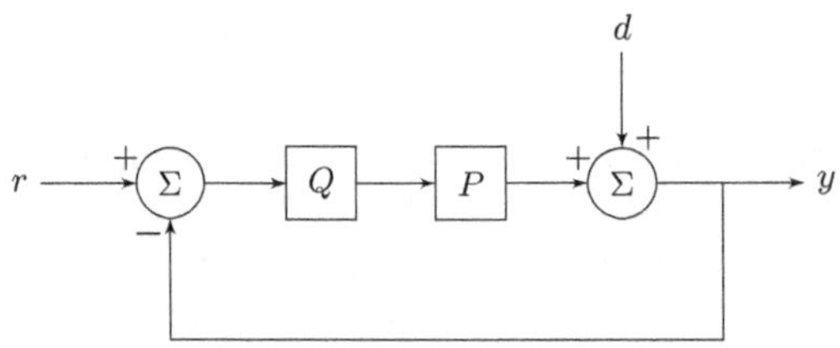

Figure I.6 System model.

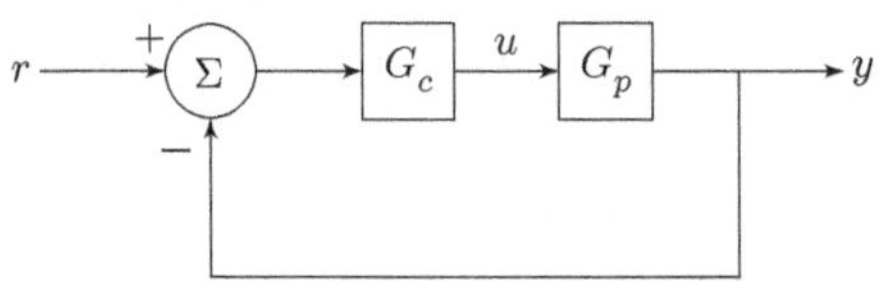

Figure I.7 Closed-loop feedback system.

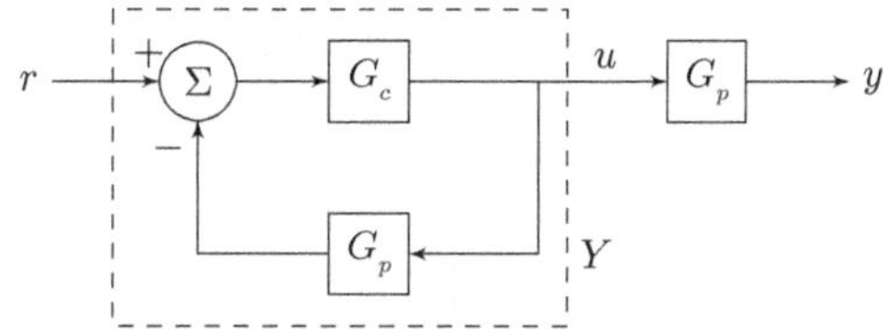

Figure I.8 Block diagram manipulation of system.

Equivalently, we define the Youla transfer function $Y(s)$ to be the transfer function from the input target, $r(s)$ to the actuator input, $u(s)$,

$$Y(s) = \frac{G_c(s)}{1 + G_p(s)G_c(s)} \tag{6}$$

At the first glance, defining the Youla transfer function seems to be trivial and not useful. However, as we discuss below and in the following chapters, this simple mapping will have important consequences both in SISO and multi-input, multi-output (MIMO) control and estimation design.

The feedback control loop presented in Figure I.8 can be easily manipulated to be drawn as a function of $Y(s)$ as follows:

It can be seen, from this manipulation, that the complementary transfer function $T(s)$ can be written as a function of $Y(s)$ as follows:

$$T(s) = Y(s)G_p(s) \tag{7}$$

The transfer function $Y(s)$ is simply equal to the original feedforward controller, $Q(s)$. In Eq. (I.7), the linearity in the free stable transfer function $Y(s)$ is what gives its practicality in hand computing or optimizing for a SISO or MIMO controller. Given a particular plant transfer function $G_p(s)$, one typically fixes a preferred closed-loop transfer function $T(s)$ as described in the following chapters, and then computes a proper and stable $Y(s)$ using (I.7).

From Eq. (I.6), the feedback controller, G_c, can be computed as follows:

$$G_c(s) = \frac{Y(s)}{1 - G_p(s)Y(s)} \tag{8}$$

The details for addressing the feedback loop internal stability are addressed in the following chapters.

Before finishing this introduction, we illustrate the utility of using $Y(s)$ and the Youla parameterization method through a very simple classical example.

Example I.1 Consider a plant with an unstable pole

$$G_p(s) = \frac{1}{s - 1} \tag{9}$$

and assume the controller

$$G_c(s) = K \tag{10}$$

where K is a simple gain (proportional type) controller, as shown in Figure I.9,

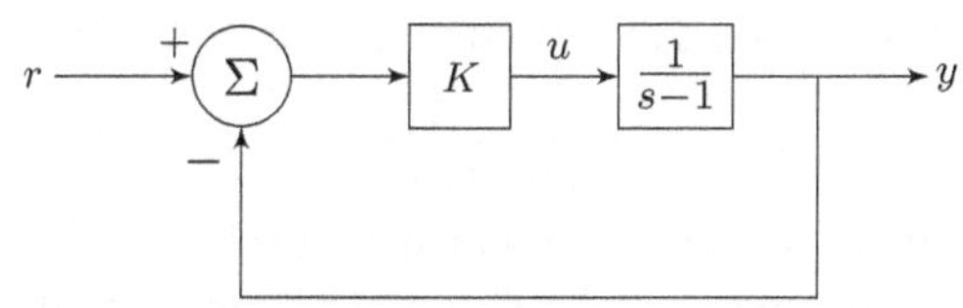

Figure I.9 Closed-loop feedback system.

The closed-loop transfer function, $T(s)$, can be written as

$$T(s) = \frac{K}{s - 1 + K} \tag{11}$$

The simple gain can be computed to stabilize the feedback control loop, by selecting the controller K to be $1 < K < \infty$. The root locus is illustrated in Figure I.10,

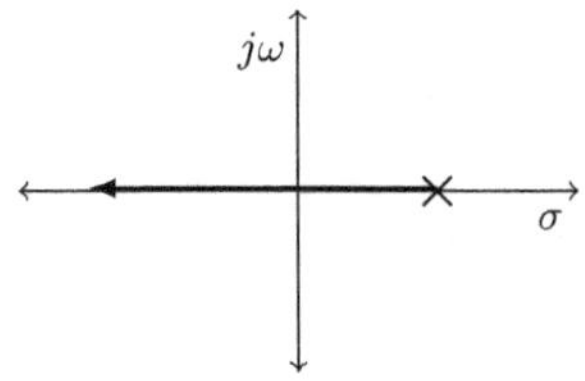

Figure I.10 Plant s-domain.

Stabilizing this plant is a simple and straightforward task. However, students who are given this simple problem, traditionally, are not given the proper reasons as of why,

1. The gain, K, could not be chosen freely without additional consequences, e.g. actuator saturation.
2. A simple gain can stabilize the plant but can not accomplish an appropriate target following, a performance measure.

For this simple example, let's compute the $Y(s)$ transfer function, which is equal to the ratio of the actuator input to the target input, $\frac{u(s)}{r(s)}$.

$$Y(s) = \frac{K(s-1)}{s-1+K} \tag{12}$$

It is a straightforward matter to obtain a Bode plot of this transfer function for various gains, K, as shown in Figure I.11,

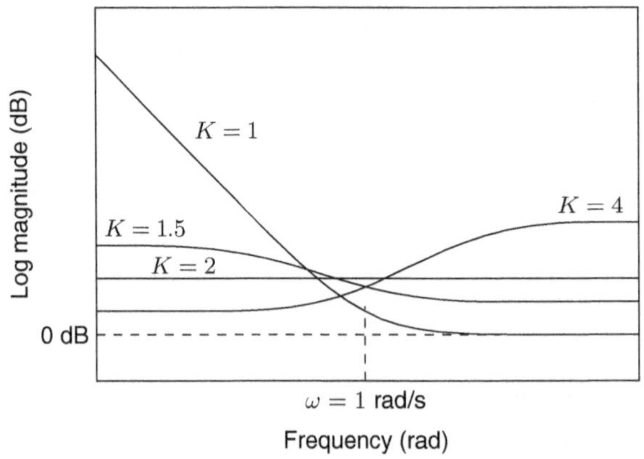

Figure I.11 Frequency response of $Y(s)$ for various gains K.

Now, it should be obvious, by looking at this plot, as we increase the gain, K, we are increasing the high-frequency demand on the actuator. More importantly, by increasing this gain, we are introducing noise to our system, as the $Y(s)$ transfer function is a high pass filter with a simple gain controller type, K.

Now, we use the Youla approach: we compute a stabilizing controller for this plant. According to this method, first, we need to select a closed-loop transfer function, $T(s)$. For this example, we select $T(s)$ to be equal to

$$T(s) = \frac{K}{\tau s+1} \tag{13}$$

where K is some gain and is not equal to the previously computed controller gain. Preferably, we would like to have $K = 1$, to ensure good target following as the closed-loop transfer function becomes equal to 1 at steady state with this selection for K . But as we shall soon see, this won't be possible as internal stability will be compromised if we select $K = 1$, so we cannot simultaneously satisfy both target following and stability with this choice of controller, a simple gain.

Using Eq. (I.7), we can compute the $Y(s)$ transfer function as $Y(s) = \frac{T(s)}{G_p(s)}$,

$$Y(s) = \frac{K(s-1)}{\tau s+1} \tag{14}$$

Now using Eq. (I.8), we can compute the controller, G_c,

$$G_c(s) = \frac{K(s-1)}{\tau s + (1-K)} \tag{15}$$

We are free to select K and τ controller coefficients. However, as we know from classical control, the controller G_c must not cancel the unstable pole of the plant as in this case, the internal stability of the feedback control loop will be compromised. Therefore, the coefficients K and τ must be selected as such to cancel $s-1$ polynomial in the numerator of G_c. There is only one unique solution for the coefficients K and τ, to see this, let's rearrange Eq. (I.15) and write it as

$$G_c(s) = \frac{K}{\tau}\frac{s-1}{s+\frac{1-K}{\tau}} \tag{16}$$

By looking at this Equation, it is evident that we must select $\frac{1-K}{\tau} = -1$, to cancel the numerator polynomial, with this selection, the controller G_c can be written as $G_c = \frac{K}{\tau}$ where $K = \tau + 1$, hence,

$$G_c(s) = 1 + \frac{1}{\tau} \tag{17}$$

Again this controller is a pure gain controller, but this time, this controller is a function of τ, the first-order time constant. So, if we make τ small or the closed-loop time response fast, according to our selection of closed-loop transfer function, $T(s)$, we have a higher gain controller. In addition, when we computed this controller using the proposed method, we saw that we are unable to simultaneously satisfy both target following and stability requirements using a simple gain controller. It should be evident that the Youla method exposed several important feedback control concepts along the way, which were not obvious using the classical approach. We will discuss these points and more about the utility of the Youla approach in the next chapters.

About the Companion Website

Robust Control: Youla Parameterization Approach is accompanied by a companion website:

www.wiley.com/go/assadian/robust

The website includes:

- Matlab Code
- Instructor Slides

Part I

Control Design Using Youla Parameterization: Single Input Single Output (SISO)

1

Review of the Laplace Transform

Our emphasis in this book is on constant parameter linear time invariant dynamic systems. Laplace transform theory is most useful when applied to analysis of linear systems, and this transformation allows the analyst to reduce differential equation problems to algebraic problems. The theory discussed later will be utilized in both the dynamic system analysis and control system design portions of the text. It is demonstrated in this chapter that linear, time-invariant dynamic equations can be transformed into rational functions. Many important input (test) signals can be transformed in this way as well.

1.1 The Laplace Transform Concept

One of the most important problems in dynamic systems analysis is to predict the outputs of a physical system when the inputs are known. We can then ask the following question, Does a transformation scheme exist to help in the tasks of dynamic system analysis? The answer is yes. In fact, there are several transformations that could be used in addressing this problem. For example, the Fourier transformation can be used; however, the domain for the Fourier transformation is "limited." That is, too many important types of inputs do not belong to the domain of the Fourier transform. On the other hand, the Laplace transform has a very broad domain, and the existence of the Laplace transform tables aid the execution of the steps in the indirect path shown in Figure 1.1.

1.2 Singularity Functions

Before we can take up the definition of the Laplace transform, it is necessary to consider those functions of time which have a discontinuity at some instants. Especially important is some understanding of what we mean by an impulse function or the so called Dirac-δ function.

Our first definitions provide a terminology for values of a discontinuous function at either side of the instant of discontinuity.

Definition 1.1 The initial condition, denoted $y(0^-)$, is the limit from the left at $t = 0$:

$$y(0^-) = \lim_{t \to 0^-} y(t) \tag{1.1}$$

Definition 1.2 The initial value, denoted $y(0^+)$, is the limit from the right at $t = 0$:

$$y(0^+) = \lim_{t \to 0^+} y(t) \tag{1.2}$$

Notice that the restrictions are strict inequalities. In other words, the initial condition is the limiting value of the function just before a possible jump discontinuity at $t = 0$ and the initial value is the limiting value of the function just after a possible jump discontinuity at $t = 0$.

Robust Control: Youla Parameterization Approach, First Edition. Farhad Assadian and Kevin R. Mallon.

Companion Website: www.wiley.com/go/assadian/robust

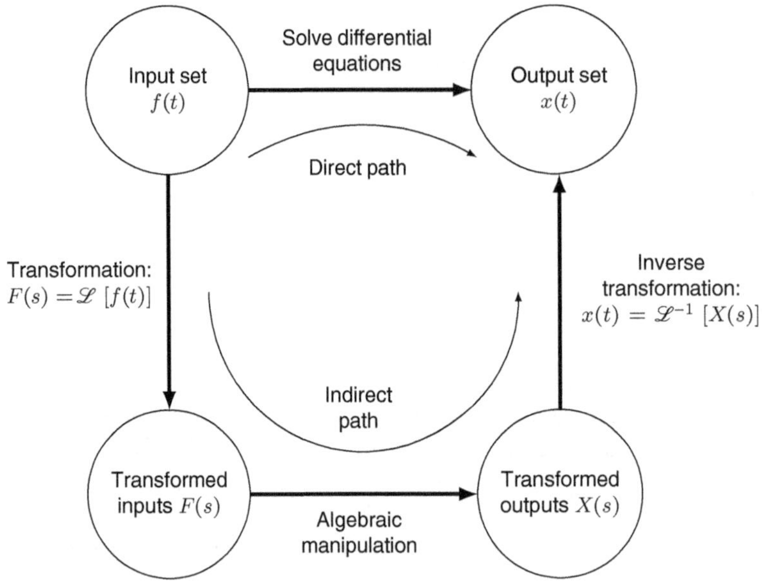

Figure 1.1 Laplace transform process.

Example 1.1 We consider the initial condition and initial value of a unit step function,

$$u(t) = 1(t) \tag{1.3}$$

$$u(0^-) = 0,\ u(0^+) = 1,\ \mathscr{L}[u(t)] = \frac{1}{s} \tag{1.4}$$

1.2.1 Definition of the Impulse Function

The impulse (or Dirac Delta) function $\delta(t)$ is defined such that

$$\delta(t) = 0 \text{ for } t \neq 0 \tag{1.5}$$

$$\int_{t_0}^{t_f} \delta(t-\tau)y(t)\mathrm{d}t = \begin{cases} y(\tau), & t_0 < \tau \le t_f \\ 0, & \text{otherwise} \end{cases} \quad \text{for any integrable } y(t) \tag{1.6}$$

An important question is why the limits of this integration are chosen such that $t_0 < \tau \le t_f$. Suppose we write things more symmetrically: This means that the boundary on τ is $t_0 \le \tau \le t_f$. Then

$$\begin{aligned} y(\tau) &= \int_{t_0}^{t_f} \delta(t-\tau)y(t)\mathrm{d}t \\ &= \int_{t_0}^{\tau} \delta(t-\tau)y(t)\mathrm{d}t + \int_{\tau}^{t_f} \delta(t-\tau)y(t)\mathrm{d}t \\ y(\tau) &= y(\tau) + y(\tau) = 2y(\tau) \end{aligned} \tag{1.7}$$

A contradiction! Similarly, suppose the boundary on τ is $t_0 < \tau < t_f$. Then

$$\begin{aligned} y(\tau) &= \int_{t_0}^{t_f} \delta(t-\tau)y(t)\mathrm{d}t \\ &= \int_{t_0}^{\tau} \delta(t-\tau)y(t)\mathrm{d}t + \int_{\tau}^{t_f} \delta(t-\tau)y(t)\mathrm{d}t \\ y(\tau) &= 0 + 0 = 0 \end{aligned} \tag{1.8}$$

So, we must have $t_0 < \tau \le t_f$ or $t_0 \le \tau < t_f$.

Another important point is even with the right definition of the boundary on τ, the definition of the impulse function given in Eq. (1.6) is inconsistent with the definition of the Riemann integral, the definition used in introduction to calculus. Without covering the details of this inconsistency (as this is outside the scope of this text), this issue is briefly discussed in Section 1.2.2.

1.2.2 The Impulse Function and the Riemann Integral

The impulse function is inconsistent with the definition of the Riemann integral (illustrated in Figures 1.2 and 1.3).

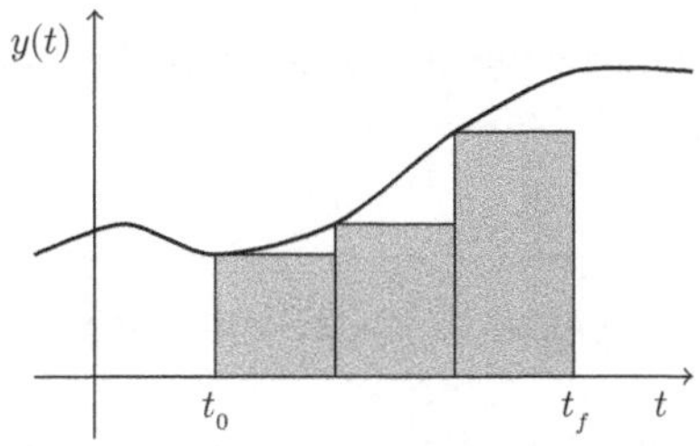

Figure 1.2 Riemann integral lower sum, S_L.

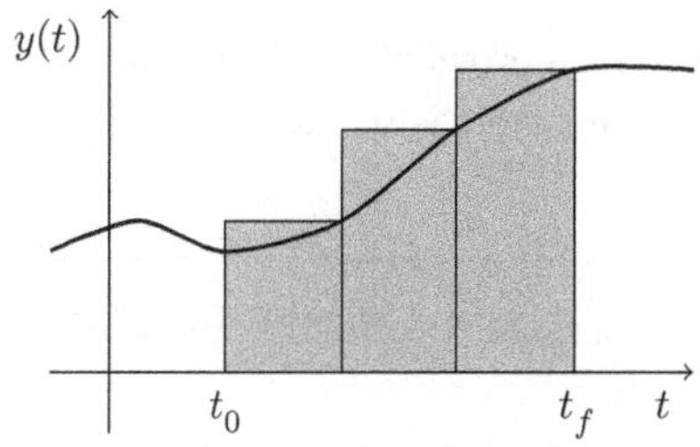

Figure 1.3 Riemann integral upper sum, S_U.

The Riemann integral (with lower sum S_L and upper sum S_U) is the common limit as the partitioning is refined. For the impulse function with $t_0 < \tau \leq t_f$, however, the lower limit is defined as $S_L \equiv 0$ while the upper limit is nonzero. The Riemann integral simply does not work. There are several ways to resolve this mathematical difficulty, the simplest is to use the Stieltjes definition of the integral [6].

1.2.3 The General Definition of Singularity Functions

The kth order derivative of singularity function $\delta^{(k)}(t)$ is defined such that:

$$\delta^{(k)}(t) \text{ for } t \neq 0 \tag{1.9}$$

$$\int_{t_0}^{t_f} \delta^{(k)}(t-\tau)y(t)\mathrm{d}t = \begin{cases} (-1)^k \frac{\mathrm{d}^k y(\tau)}{\mathrm{d}t^k}, & t_0 < \tau \leq t_f \\ 0, & \text{otherwise} \end{cases} \tag{1.10}$$

Once again, it can be shown that this definition is inconsistent with Riemann's definition of integral.

1.2.3.1 "Graphs" of Some Singularity Functions

The main reason for the persistent use of the singularity functions in the control-system literature, in spite of the mathematical problems, is the simple fact that they are useful theoretical devices. In addition, we are able to manufacture signals in the laboratory or on the computer which seem to be endowed with all of the theoretical properties of the singularity functions. A good way to think of all this is that the physical notion of the impulse is not what creates the difficulty; rather, it is the notion of what we mean by the integral that lies at the root of the theoretical problems.

The laboratory impulse is illustrated as dash lines overlaid on the function, $\rho(t)$ in Figure 1.4. (Notice that this pulse-like function ends at $t = \tau$.) An apparent mathematical justification for this claim derives from use of the mean-value theorem provided by the following equation,

$$\int_a^b y(t)\mathrm{d}t = y(\zeta)(b-a),\ a \leq \zeta \leq b \tag{1.11}$$

That is, imagine multiplying this pulse by $y(t)$ and integrating. It then seems to follow from the mean-value theorem,

$$\int_{t_0}^{t_f} \rho(t)y(t)\mathrm{d}t = \int_{\tau-\epsilon}^{\tau} \rho(t)y(t)\mathrm{d}t = \frac{1}{\epsilon}\cdot y(\zeta)\cdot\epsilon = y(\zeta),\ \tau-\epsilon \leq \zeta \leq \tau. \tag{1.12}$$

If we continue this line of reasoning, we can imagine laboratory versions of the higher order singularity functions. For example, a finite difference approximation for the first singularity function. To "construct" $\delta^{(1)}(\tau) \Rightarrow -\frac{\mathrm{d}y}{\mathrm{d}t}(\tau)$, refer to the finite difference approximation:

$$\frac{\mathrm{d}y(\tau)}{\mathrm{d}t} \approx \frac{y(\tau)-y(\tau-\epsilon)}{\epsilon} \tag{1.13}$$

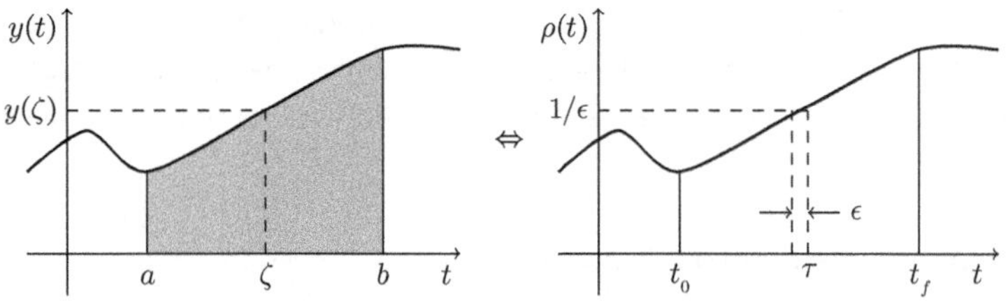

Figure 1.4 Mean value theorem development of singularity functions.

The following example provides an illustration of an impulse function and a finite difference approximation for the first singularity function and their associated dynamic responses.

Example 1.2 We compare the impulse and doublet response for a moving mass. Consider applying a force $F(t)$ to the mass shown in Figure 1.5.

First, let $F(t)$ be an approximate impulse or doublet function, as shown in Figures 1.6 and 1.7. In these cases, ϵ is considered to be a very small (but finite) number.

The approximate impulse function is zero except for a length of time ϵ starting at $t = -\epsilon$, where the magnitude is $1/\epsilon$. The velocity response is a ramp between $t = -\epsilon$ and $t = 0$, ending at a magnitude of 1. In turn, the displacement response is a parabola between $t = -\epsilon$ and $t = 0$ and a ramp with slope equal to 1 for $t > 0$.

The approximate doublet is zero except for a length of time 2ϵ starting at $t = -2\epsilon$. For the first half of the duration, the approximate doublet has a magnitude of $1/\epsilon^2$. For the second half, it has a magnitude of $-1/\epsilon^2$. The velocity response is a positive ramp from $t = -2\epsilon$ to $t = -\epsilon$ and a negative ramp from $t = -\epsilon$ to $t = 0$, ending with a magnitude of $1/\epsilon$. Further integration gives the displacement response, a parabolic increase from $t = -2\epsilon$ to $t = -\epsilon$ and parabolic decrease from $t = -\epsilon$ to $t = 0$, such that the magnitude at $t = -\epsilon$ is $1/2$ and the magnitude at $t = 0$ is 1.

Next, consider the exact response to an impulse or doublet force input, as shown in Figures 1.8 and 1.9. As $\epsilon \to 0$, the approximate impulse and doublet functions approach true impulse and doublet functions and integrate into the more familiar step and ramp functions.

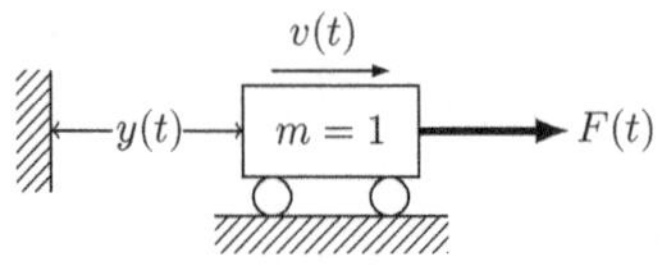

Figure 1.5 An mass with an applied force.

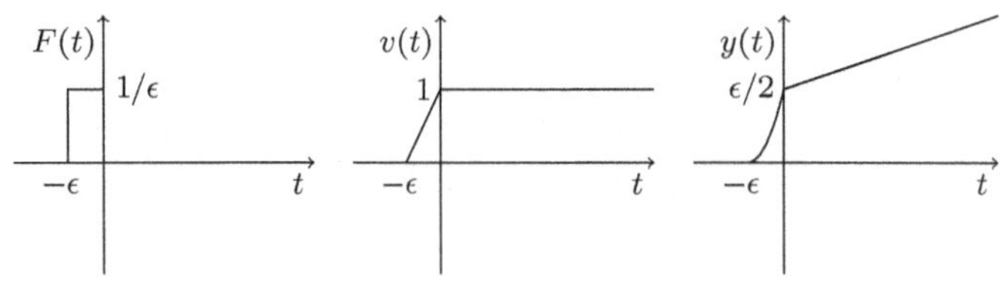

Figure 1.6 Impulse approximate response.

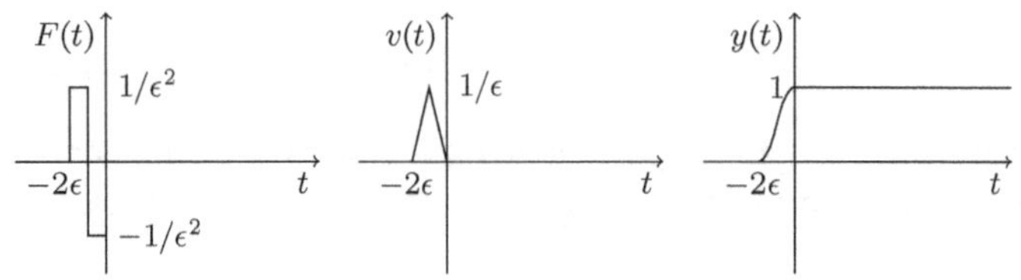

Figure 1.7 Doublet approximate response.

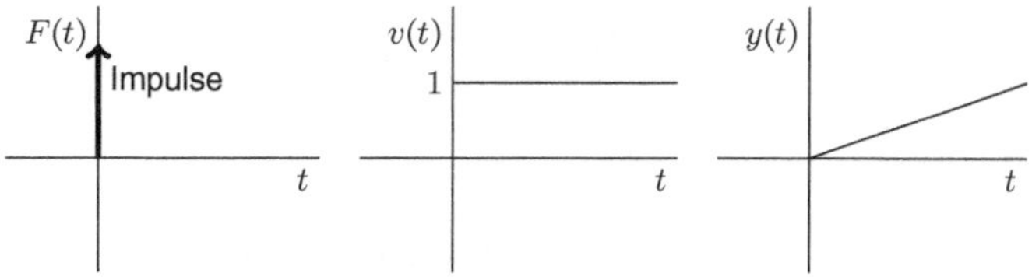

Figure 1.8 Impulse exact response ($\epsilon \to 0$).

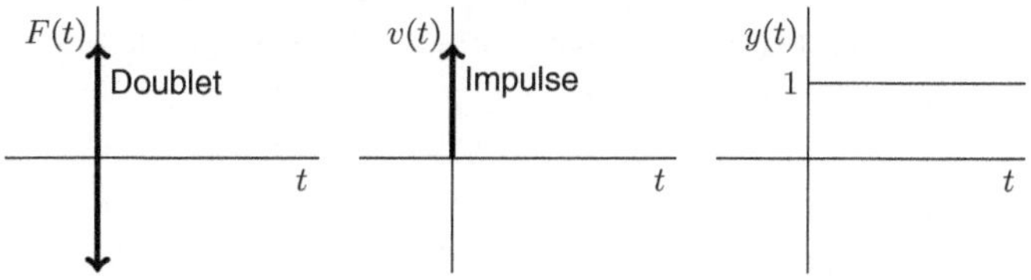

Figure 1.9 Doublet exact response ($\epsilon \to 0$).

Similarly, we can construct a finite difference approximation for the second derivative and then infer a laboratory version of the third singularity function. In a similar way, we can approximate all the singularity functions.

1.3 The Laplace Transform

1.3.1 Definition of the Laplace Transform

The Laplace transform converts a function from the time domain with variable t, to the Laplace domain with variable s (the Laplace domain is sometimes referred to as the s-domain). The Laplace variable s is a complex number and can be written in terms of its real part σ and imaginary part ω.

$$s = \sigma + j\omega \tag{1.14}$$

The choice for s is arbitrary but proves to be very convenient in the application of dynamic systems theory [7]. The Laplace transform of $y(t)$ is denoted by $\hat{y}(s)$ or $\mathscr{L}[y(t)]$ and is defined as [8]

$$\mathscr{L}[y(t)] = \int_{0^-}^{\infty} e^{-st} y(t)dt \tag{1.15}$$

Notice the lower limit of 0^-, which reads "slightly before zero." The Fourier transform is defined by

$$\mathscr{F}[y(t)] = \int_{-\infty}^{\infty} e^{-j\omega t} y(t)dt \tag{1.16}$$

The Laplace transformation domain [9] is given as follows: for a function to be Laplace transformable, $f(t)$ must be

1. Sectionally continuous; i.e. $f(t)$ may have only a finite number of discontinuities in any finite interval; and
2. Of exponential order; i.e. for some real $\sigma_0, A > 0$, and $t_1 > 0, Ae^{-\sigma_0 t} \geq |f(t)|$ for $t_1 > t$. The region of the s plane for which $Re(s) > \sigma_0$ is called the "region of convergence."

These conditions are sufficient conditions for the existence of the integral in Eq. (1.15).

Example 1.3 We consider the Laplace transform of the unit step function, illustrated in Figure 1.10.

$$u(t) = \begin{cases} 1, & t \geq 0 \\ 0, & t < 0 \end{cases} \tag{1.17}$$

The unit step function, $u(t)$, can be written as $1(t)$.

$$\hat{u}(s) = \int_{0^-}^{\infty} e^{-st} 1(t)dt$$

$$= \int_0^{\infty} e^{-st} dt$$
$$= \left. \frac{-e^{-st}}{s} \right|_0^{\infty}$$
$$= +\frac{1}{s} \text{ (if } Re\ s > 0, \text{ where } Re\ s = \sigma) \tag{1.18}$$

Plotting $\hat{u}(s)$ vs. s cannot be done in a very convenient manner, as was done with $u(t)$ vs. t, because s is complex. However, it is very useful to plot the critical points of $\hat{u}(s)$, i.e. points where $\hat{u}(s)$ is infinite (called poles) and is zero (called zeros), on the s plane. For the unit step function, $\hat{u}(s)$ has a single pole at origin as illustrated in pole–zero diagram of Figure 1.11.

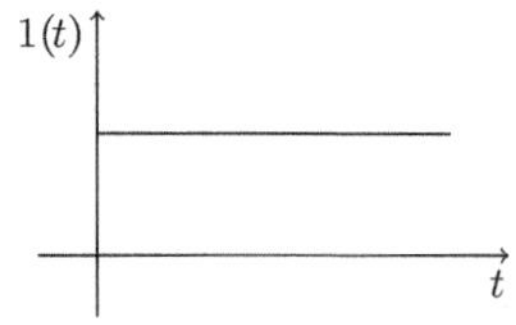

Figure 1.10 Unit step time domain.

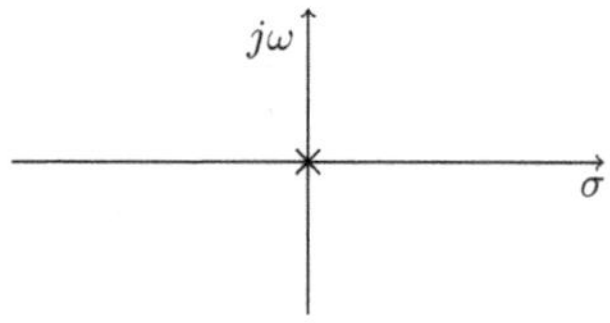

Figure 1.11 Unit step s-domain.

1.3.2 Laplace Transform Properties

In this section, we discuss the properties of the Laplace transform by series of propositions with their associated proofs. This should aid in quickly deriving Laplace transforms for sets of functions.

Proposition 1.1 *The Laplace transform is linear, i.e.*

$$\mathcal{L}[c_1 y(t) + c_2 g(t)] = c_1 \mathcal{L}[y(t)] + c_2 \mathcal{L}[g(t)] \tag{1.19}$$

for any c_1, c_2, $y(t)$, $g(t)$.

Proof: Linearity of the integration.

$$\int [c_1 f(t) + c_2 h(t)] dt = c_1 \int f(t) dt + c_2 \int h(t) dt,$$

Then, per the definition of the Laplace transform in Eq. (1.15), the Laplace transform is linear. ∎

Proposition 1.2 *The Laplace transform is nonmultiplicative, i.e.*

$$\mathcal{L}[y(t)g(t)] \neq \mathcal{L}[y(t)]\mathcal{L}[g(t)] \tag{1.20}$$

due to the properties of integration.

Proof:

$$\int f(t)h(t) dt \neq \int f(t) dt \int h(t) dt$$

Therefore, per the definition of the Laplace transform in Eq. (1.15), the Laplace transform cannot be multiplicative. ∎

1.3.3 Shifting the Laplace Transform

The definition of the Laplace transform is manipulated to yield results that are useful for obtaining the transforms of an important set of functions and for manipulating and interpreting mathematical models of physical systems.

Proposition 1.3 (Complex Shift of the Laplace Transform): *If* $\hat{y}(s) = \mathscr{L}[y(t)]$, *then*

$$\mathscr{L}[e^{-\sigma t}y(t)] = \hat{y}(s+\sigma) \tag{1.21}$$

Proof:

$$\begin{aligned}\mathscr{L}[e^{-\sigma t}y(t)] &= \int_{0^-}^{\infty} e^{-st}e^{-\sigma t}y(t)dt \\ &= \int_{0^-}^{\infty} e^{-(s+\sigma)t}y(t)dt \\ &= \hat{y}(s+\sigma)\end{aligned}$$

Therefore, multiplying $y(t)$ by $e^{-\sigma t}$ shifts $\hat{y}(s)$ by σ. ∎

This result shares a symmetry with the following proposition.

Proposition 1.4 (Time Shift of the Laplace Transform): *If* $\hat{y}(s) = \mathscr{L}[y(t)]$ *and* $y(t) \equiv 0$ *for* $t < 0^-$, *then*

$$\mathscr{L}[y(t-\tau)] = e^{-\tau s}\hat{y}(s) \tag{1.22}$$

Proof: Let $\zeta = t - \tau$, so

$$\begin{aligned}\mathscr{L}[y(t-\tau)] &= \int_{0^-}^{\infty} e^{-st}y(t-\tau)dt \\ &= \int_{-\tau^-}^{\infty} e^{-s(\zeta+\tau)}y(\zeta)d\zeta \\ &= e^{-s\tau}\int_{-\tau^-}^{\infty} e^{-s\zeta}y(\zeta)d\zeta \\ &\quad -\tau^- = 0^- \text{ because } y(t) \equiv 0 \text{ for } t < 0^-, \text{ so} \\ &= e^{-s\tau}\int_{0^-}^{\infty} e^{-s\zeta}y(\zeta)d\zeta \\ &= e^{-s\tau}\hat{y}(s)\end{aligned}$$

Therefore, multiplying $\hat{y}(s)$ by $e^{-\tau s}$ shifts $y(t)$ by τ. ∎

Example 1.4 Consider the function illustrated in Figure 1.12

$$y(t) = e^{-\sigma t}1(t) \tag{1.23}$$

The critical points of this function are plotted in Figure 1.13.

$$\mathscr{L}[1(t)] = \frac{1}{s}, \text{ so } \mathscr{L}[e^{-\sigma t}1(t)] = \frac{1}{s+\sigma} \tag{1.24}$$

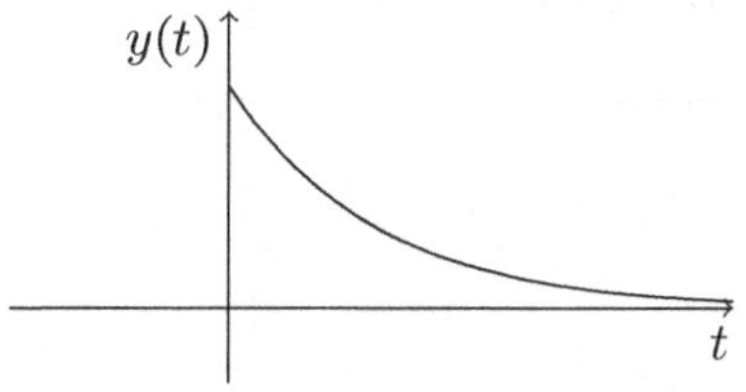

Figure 1.12 Decaying exponential time response.

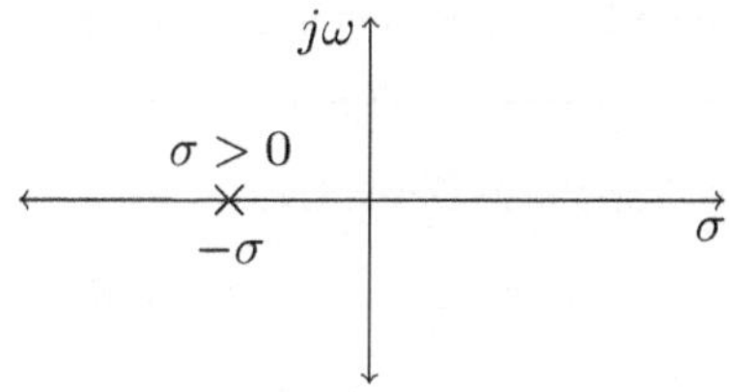

Figure 1.13 Decaying exponential time domain.

Example 1.5 Consider the function

$$y(t) = \cos(\omega t)1(t) \tag{1.25}$$

As illustrated in Figure 1.14, if z is complex, then by definition $e^z = 1 + z + \frac{z^2}{2!} + \frac{z^3}{3!} + \cdots$. Suppose $z = j\theta$:

$$e^{j\theta} = 1 + j\theta - \frac{\theta^2}{2!} - \frac{j\theta^3}{3!} + \frac{\theta^4}{4!} + \frac{j\theta^5}{5!} + \cdots \tag{1.26}$$

$$e^{j\theta} = \cos(\theta) + j\sin(\theta),\ e^{-j\theta} = \cos(\theta) - j\sin(\theta) \tag{1.27}$$

$$\cos(\theta) = \frac{e^{j\theta} + e^{-j\theta}}{2} \tag{1.28}$$

Therefore,

$$\begin{aligned}\mathscr{L}[\cos(\omega t)1(t)] &= \frac{1}{2}\mathscr{L}[e^{j\omega t}1(t)] + \frac{1}{2}\mathscr{L}[e^{-j\omega t}1(t)] \\ &= \frac{1}{2}\frac{1}{s - j\omega} + \frac{1}{2}\frac{1}{s + j\omega} \\ &= \frac{s}{s^2 + \omega^2}\end{aligned} \tag{1.29}$$

The critical points of this function are plotted in Figure 1.15.

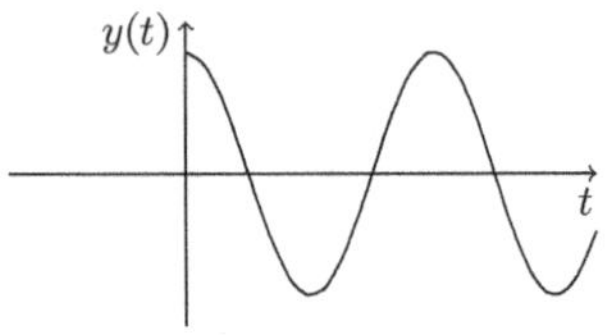

Figure 1.14 Cosine function time response.

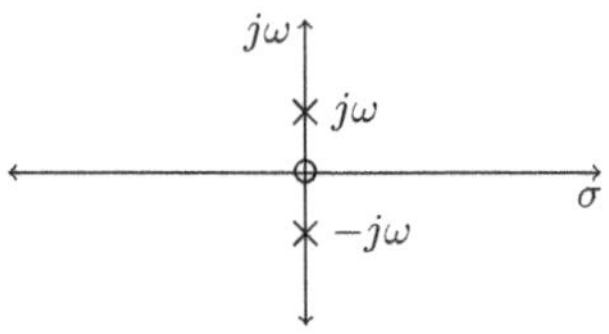

Figure 1.15 Cosine function s-domain.

Example 1.6 Consider the function

$$y(t) = e^{-\sigma t}\cos(\omega t)1(t) \tag{1.30}$$

As illustrated in Figure 1.16,

$$\mathscr{L}[e^{-\sigma t}\cos(\omega t)1(t)] = \frac{s + \sigma}{(s + \sigma)^2 + \omega^2} \tag{1.31}$$

The critical points of this function are plotted in Figure 1.17.

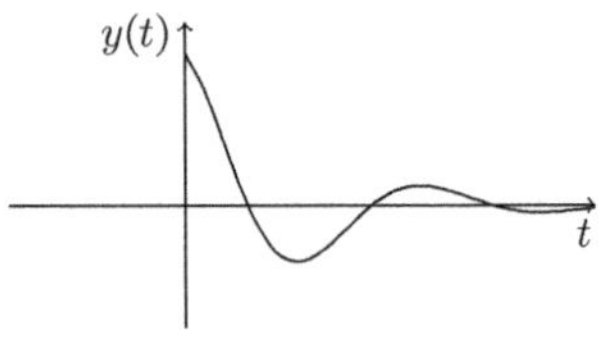

Figure 1.16 Decaying cosine time response.

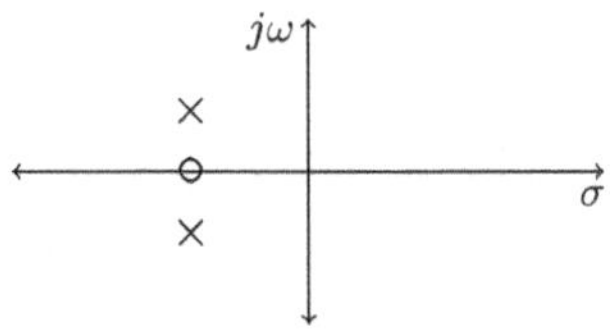

Figure 1.17 Decaying cosine s-domain.

1.3.4 Laplace Transform Derivatives

The next two propositions are also symmetrical.

Proposition 1.5 (Derivative of a Transformed Function): *If $\hat{y}(s) = \mathscr{L}[y(t)]$, then*

$$\mathscr{L}[ty(t)] = -\frac{d}{ds}\hat{y}(s) \tag{1.32}$$

Proof:

$$-\frac{\mathrm{d}}{\mathrm{d}s}\hat{y}(s) = -\frac{\mathrm{d}}{\mathrm{d}s}\int_{0^-}^{\infty} e^{-st}y(t)\mathrm{d}t$$
$$= \int_{0^-}^{\infty} e^{-st}(ty(t))\mathrm{d}t = \mathscr{L}[ty(t)]$$

Therefore, when a function $y(t)$ is multiplied by t, the Laplace transform of the resulting output is the negative derivative in s of $\mathscr{L}[y(t)]$. ∎

Example 1.7 Consider the ramp function, $t(1(t))$, as shown in Figure 1.18.

$$\mathscr{L}[t(1(t))] = -\frac{\mathrm{d}}{\mathrm{d}s}\left[\frac{1}{s}\right] = \frac{1}{s^2} \tag{1.33}$$

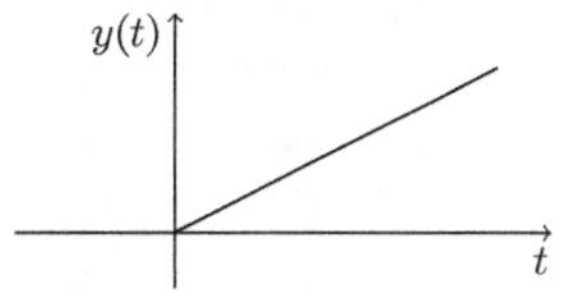

Figure 1.18 Ramp input.

Apply this proposition in an iterative fashion to establish

$$\mathscr{L}[t^n y(t)] = (-1)^n \frac{\mathrm{d}^n}{\mathrm{d}s^n}\hat{y}(s) \tag{1.34}$$

Proposition 1.6 (Transformation of a Derivative): *If $\hat{y}(s) = \mathscr{L}[y(t)]$, then the Laplace transform of the derivative is related to the initial condition by*

$$\mathscr{L}\left[\frac{\mathrm{d}y}{\mathrm{d}t}\right] = s\hat{y}(s) - y(0^-) \tag{1.35}$$

Proof: Reminder: Integration by parts.

$$\int u\ \mathrm{d}v = uv - \int v\ \mathrm{d}u \tag{1.36}$$

Then,

$$\mathscr{L}\left[\frac{\mathrm{d}y}{\mathrm{d}t}\right] = \int_{0^-}^{\infty} e^{-st}\frac{\mathrm{d}y}{\mathrm{d}t}\mathrm{d}t$$

Integration by parts:

$$u = e^{-st},\ \mathrm{d}u = -se^{-st}\mathrm{d}t,\ v = y(t), \mathrm{d}v = \left(\frac{\mathrm{d}y(t)}{\mathrm{d}t}\right)\mathrm{d}t$$

$$= e^{-st}y(t)|_{0^-}^{\infty} - \int_{0^-}^{\infty} -se^{-st}y(t)\mathrm{d}t$$

knowing that

$$e^{-st}y(t)|_{0^-}^{\infty} = 0 - y(0^-)$$

so

$$\mathscr{L}\left[\frac{\mathrm{d}y}{\mathrm{d}t}\right] = s\hat{y}(s) - y(0^-)$$

Therefore, when a function $y(t)$ is differentiated with respect to time, the Laplace transform of the resulting output is s times $\mathscr{L}[y(t)]$ minus the initial condition $y(0^-)$. ∎

Applying this result in an iterative way establishes a corollary to this proposition.

Corollary 1.1

$$\mathcal{L}\left[\frac{\mathrm{d}^n y(t)}{\mathrm{d}t^n}\right] = s^n\hat{y}(s) - \sum_{k=1}^{n} s^{n-k}\frac{\mathrm{d}^{k-1}}{\mathrm{d}t^{k-1}}y(0^-) \tag{1.37}$$

Example 1.8 Consider a force applied to a mass, as shown in Figure 1.19.

Newton's law: $m\frac{\mathrm{d}^2y}{\mathrm{d}t^2} = u(t)$, where $u(t) = F(t)$

$$\mathcal{L}\left[\frac{\mathrm{d}^2y}{\mathrm{d}t^2}\right] = s^2\hat{y}(s) - sy(0^-) - \frac{\mathrm{d}}{\mathrm{d}t}(y(0^-)) \tag{1.38}$$

$$m[s^2\hat{y}(s) - sy(0^-) - \frac{\mathrm{d}y}{\mathrm{d}t}(0^-)] = \hat{u}(s) \tag{1.39}$$

$$\hat{y}(s) = \underbrace{\frac{sy(0^-) + \frac{\mathrm{d}y}{\mathrm{d}t}(0^-)}{s^2}}_{\text{I.C. response}} + \underbrace{\frac{1}{ms^2}\hat{u}(s)}_{\text{forced response}} \tag{1.40}$$

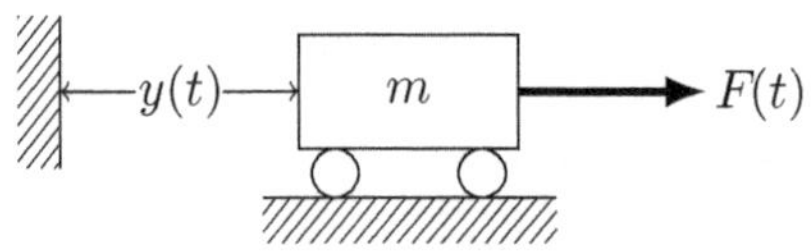

Figure 1.19 A mass with an applied force.

When all initial conditions are zero, we are left with the transfer function, $G(s)$, from input $\hat{u}$ to output $\hat{y}$.

$$G(s) = \frac{\hat{y}}{\hat{u}}(s) = \frac{1}{ms^2} \tag{1.41}$$

The block diagram and critical points of this system are illustrated in Figures 1.20 and 1.21, respectively.

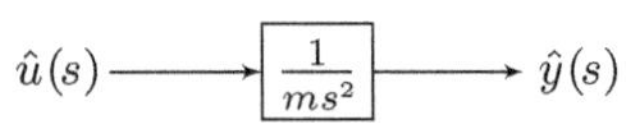

Figure 1.20 Forced mass block diagram.

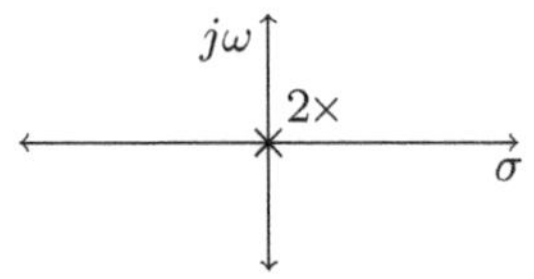

Figure 1.21 Poles of the forced mass.

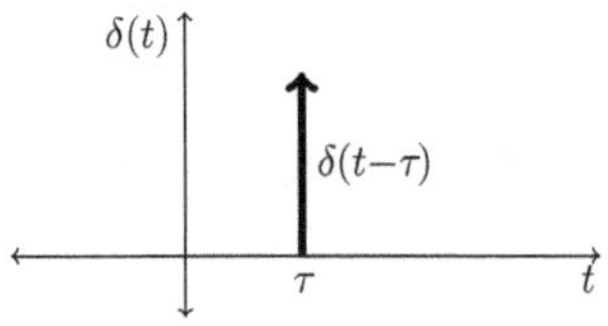

Figure 1.22 Impulse function.

1.3.5 Transforms of Singularity Functions

To this point, our example transforms have been rational functions. Here, we determine the transforms of the singularity functions that are rational functions (polynomials actually) for the special case of $\tau = 0$. As it turns out, it is this value of

τ that is used most often in our discussions. A simple application of Eq. (1.15) and the definition of an impulse function of the last section yields the transform

$$\mathscr{L}[\delta(t-\tau)] = \int_{0^-}^{\infty} \delta(t-\tau)e^{-st}dt = e^{-s\tau} \tag{1.42}$$

where the impulse $\delta(t-\tau)$ is illustrated in Figure 1.22. Note: If $\tau = 0$, then $\mathscr{L}[\delta(t)] = 1$. By linearity,

$$\mathscr{L}^{-1}[R_0] = R_0\delta(t) \tag{1.43}$$

More generally, the application of Eq. (1.15) and the previous definition of an impulse function of the last section results in

$$\mathscr{L}[\delta^{(k)}(t-\tau)] = s^k e^{-s\tau}, \quad \mathscr{L}[\delta^{(k)}(t)] = s^k \tag{1.44}$$

1.4 Inverse Laplace Transform

At this point, it is clear that the differential equations of linear, time invariant, systems can be converted to polynomials and rational functions where the Laplace transform is effected; the Laplace transform can be used to solve differential equations by purely algebraic means. Nearly all that is needed to complete the solution of ordinary differential equations with constant coefficients by Laplace transformation is contained in the propositions and examples of previous sections. The final ingredient is a method for obtaining the inverse transform, $y(t) = \mathscr{L}^{-1}[\hat{y}(s)]$. One can certainly utilize the Cauchy integral formula, which is discussed in Appendix A, for this purpose. Other, more mundane but useful methods are developed in this section.

1.4.1 Inverse Laplace Transformation by Heaviside Expansion

In this section, we shall use the simpler procedure of partial fraction or Heaviside expansion to obtain an inverse transformation. That is to say, we exploit the properties of rational functions as follows:

Suppose we have a transfer function, $G(s)$ and want $g(t)$. Also suppose that $G(s)$ is strictly proper, where the number of poles, m, is greater than number of zeros, n, $m > n$. Write any strictly proper transfer function as

$$G(s) = \frac{q_{n-1}s^{n-1} + q_{n-2}s^{n-2} + \cdots + q_1 s + q_0}{(s+p_1)(s+p_2)\cdots(s+p_m)} \tag{1.45}$$

The problem of converting this form with a Heaviside expansion is divided into subcases according to whether or not the poles, p_k, are repeated.

1.4.1.1 Distinct Poles

Where the p_k are distinct, it is possible to make the Heaviside expansion as

$$G(s) = \frac{R_1}{s+p_1} + \frac{R_2}{s+p_2} + \cdots + \frac{R_n}{s+p_n} \tag{1.46}$$

The R_k are called residues. Multiply this by $(s + p_k)$ and note that result must apply to all values of s. In particular, we can set $s = -p_k$ to obtain the formula

$$R_k = (s+p_k)G(s)\big|_{s=-p_k} \tag{1.47}$$

The residuals can then be independently inverted.

$$g(t) = [R_1e^{-p_1t} + R_2e^{-p_2t} + \cdots + R_ne^{-p_nt}]1(t) \tag{1.48}$$

1.4.1.2 Distinct Poles with $G(s)$ Being Proper

When $G(s)$ is proper, the degree of the numerator is equal to the degree of the denominator. If $G(s)$ is proper, then it can be written as

$$G(s) = R_0 + \frac{R_1}{s+p_1} + \frac{R_2}{s+p_2} + \cdots + \frac{R_n}{s+p_n}, \text{ where } R_0 = G(\infty) \tag{1.49}$$

$$g(t) = R_0\delta(t) + [R_1e^{-p_1t} + R_2e^{-p_2t} + \cdots + R_ne^{-p_nt}]1(t) \tag{1.50}$$

Remember that $g(t) = \mathscr{L}^{-1}[G(s)]$.

1.4.1.3 Repeated Poles

Suppose one of the poles of $G(s)$, call it $-p_1$, is repeated a times,

$$G(s) = \frac{q_{n-1}s^{n-1} + q_{n-2}s^{n-2} + \cdots + q_1 s + q_0}{(s+p_1)^a(s+p_2)\cdots(s+p_{n-a+1})} \tag{1.51}$$

The Heaviside expansion is modified in the following way,

$$G(s) = \frac{R_{11}}{s+p_1} + \frac{R_{12}}{(s+p_1)^2} + \cdots + \frac{R_{1a}}{(s+p_1)^a} + \frac{R_2}{s+p_2} + \cdots + \frac{R_{n-a+1}}{s+p_{n-a+1}} \tag{1.52}$$

It is easy to verify that the residues associated with linear denominator terms are calculated as in Eq. (1.47). For a single repeated pole ($a = 2$), it is easily verified that

$$R_{12} = \frac{\mathrm{d}}{\mathrm{d}s}[(s+p_1)^a G(s)]\bigg|_{s=-p_1} \tag{1.53}$$

In addition, for values of $a - l \geq 2$, we find that

$$R_{1l} = \frac{1}{(a-l)!}\frac{\mathrm{d}^{a-l}}{\mathrm{d}s^{a-l}}[(s+p_1)^a G(s)]\bigg|_{s=-p_1} \tag{1.54}$$

Then $g(t)$ can be written as

$$\begin{aligned} g(t) = \big[&R_{11}e^{-p_1 t} + R_{12}te^{-p_1 t} + \cdots \\ &+ R_{1a}t^a e^{-p_1 t} + R_2 e^{p_2 t} + \cdots \\ &+ R_{n-a+1}e^{-p(n-a+1)t}\big]\, 1(t) \end{aligned} \tag{1.55}$$

The examples below illustrate the utility of Heaviside expansion in inverse Laplace transformation.

Example 1.9 Consider the transfer function (distinct poles – proper)

$$G(s) = \frac{s^2+5s+6}{s^2+3s+2} \tag{1.56}$$

This is the case where $G(s)$ is proper. First, carry out any pole/zero cancellations.

$$G(s) = \frac{\cancel{(s+2)}(s+3)}{(s+1)\cancel{(s+2)}} = \frac{s+3}{s+1} \tag{1.57}$$

Therefore,

$$G(s) = \frac{s+3}{s+1} = R_0 + \frac{R_1}{s+p_1} = R_0 + \frac{R_1}{s+1} \tag{1.58}$$

According to Eq. (1.49),

$$R_0 = G(\infty) = 1 \tag{1.59}$$

And according to Eq. (1.47),

$$R_1 = (s+1)G(s)|_{s=-1} \tag{1.60}$$

yields $R_1 = 2$. Therefore, the partial fraction expansion of $G(s)$ is

$$G(s) = 1 + \frac{2}{s+1} \tag{1.61}$$

and therefore,

$$g(t) = \delta(t) + [2e^{-t}]1(t) \tag{1.62}$$

Example 1.10 Consider the transfer function (distinct poles, strictly proper)

$$G(s) = \frac{4s^2 + 13s + 12}{s(s^2 + 5s + 6)} \tag{1.63}$$

First, factor $G(s)$.

$$G(s) = \frac{4s^2 + 13s + 12}{s(s + 2)(s + 3)} \tag{1.64}$$

Therefore, the partial fraction expansion is of the form

$$G(s) = \frac{R_1}{s + p_1} + \frac{R_2}{s + p_2} + \frac{R_3}{s + p_3} = \frac{R_1}{s} + \frac{R_2}{s + 2} + \frac{R_3}{s + 3} \tag{1.65}$$

According to Eq. (1.47)

$$R_k = (s + p_k)G(s)|_{s=-p_k} \tag{1.66}$$

This yields

$$\begin{aligned} R_1 &= sG(s)|_{s=0} = 2 \\ R_2 &= (s + 2)G(s)|_{s=-2} = -1 \\ R_3 &= (s + 3)G(s)|_{s=-3} = 3 \end{aligned} \tag{1.67}$$

Therefore, the partial fraction expansion of $G(s)$ is

$$G(s) = \frac{2}{s} - \frac{1}{s + 2} + \frac{3}{s + 3} \tag{1.68}$$

and

$$g(t) = [2 - e^{-2t} + 3e^{-3t}]1(t) \tag{1.69}$$

Example 1.11 Consider the transfer function (repeated poles)

$$G(s) = \frac{4s^3 + 16s^2 + 23s + 13}{(s + 1)^3(s + 2)} \tag{1.70}$$

The partial fraction expansion is of the form

$$G(s) = \frac{R_{11}}{(s + 1)} + \frac{R_{12}}{(s + 1)^2} + \frac{R_{13}}{(s + 1)^3} + \frac{R_2}{(s + 2)} \tag{1.71}$$

According to Eqs. (1.53) and (1.54)

$$\begin{aligned} R_{11} &= \frac{1}{2!}\frac{d^2}{ds^2}[(s + 1)^3G(s)]\bigg|_{s=-1} = 3 \\ R_{12} &= \frac{d}{ds}[(s + 1)^3G(s)]\bigg|_{s=-1} = 1 \\ R_{13} &= [(s + 1)^3G(s)]\bigg|_{s=-1} = 2 \\ R_2 &= [(s + 2)G(s)]|_{s=-2} = 1 \end{aligned} \tag{1.72}$$

Therefore, the partial fraction expansion of $G(s)$ is

$$G(s) = \frac{3}{(s + 1)} + \frac{1}{(s + 1)^2} + \frac{2}{(s + 1)^3} + \frac{1}{(s + 2)} \tag{1.73}$$

and

$$g(t) = [(3 + t + 2t^2)e^{-t} + e^{-2t}]1(t) \tag{1.74}$$

1.5 The Transfer Function and the State Space Representations (State Equations)

The true value of the Laplace transform is that it provides an important, mathematical characterization of a dynamic system, the transfer function. In this text, for control system design with Youla parameterization, we discover that the transfer function is more valuable than the underlying state space representation. Of course the state equations are the usual product of dynamic system analysis and as discussed later, the transfer function is obtained from the state equations. The state equations tell us much that is important about the dynamics of any system and are used here to draw-out crucial mathematical properties of the transfer function and the transfer function approach to feedback design.

1.5.1 The Transfer Function

Our picture of the dynamic system is that of Figure 1.23. We shall use "u" to denote input and "y" to denote output throughout the text. In continuous-time application, the transfer function is the ratio of the transformed output to the transformed input $G(s) = \frac{\hat{y}}{\hat{u}}$ and all initial conditions are zero.

One can obtain this ratio anywhere the transforms exist. In some cases, the term "transfer function" is utilized to mean some input–output characteristic with direct physical measures rather than the "s-plane" transform. In these cases, the transfer function depends upon the particular input. In the case of the linear, time-invariant (LTI) system, the transfer function is completely specified by the parameters of the system and is independent of the input.

This independence of the transfer function for the input, for the LTI systems, is easily established if one considers the differential equation describing a dynamic system. Since such an equation transforms to a polynomial operator times the output equal to a polynomial times the input, the ratio of output to input is a rational function. Clearly, this rational function is independent of the particular input.

We have shown through several examples, one can always derive the equation of motion (the differential equation) and then write the transfer function by inspection. That is almost the way things go, but not quite.

1.5.2 The State Equations

Rather than a single, high-order equation, it is usually more convenient to write a system of simultaneous equations to describe the dynamics of a physical system. In mechanical engineering, for example, Hamilton's principle leads to a system of first-order equations. In general, the dynamics of a system can be written in the matrix form.[1]

$$\begin{aligned} \underline{\dot{x}} &= A\underline{x} + \underline{b}u \\ \underline{y} &= \underline{c}^T\underline{x} + D\underline{u} \end{aligned} \tag{1.75}$$

The first equation is called the dynamic equation, while the second is referred to as the message equation. The combination is called the *state variable model.*[2] A great deal of computer software is based upon the assumption that the state equations are an important part of the user's natural language.

The vector $\underline{x}$ represents a set of intermediate variables (momenta and generalized displacements in the context of Hamilton's principle) needed to complete the description of the relation of input to output. These intermediates are called *state variables.* If one takes the input–output point-of-view, as is often the case in control studies, the identities of the state variables are immaterial.

The state equations are represented in the block diagram of Figure 1.24.

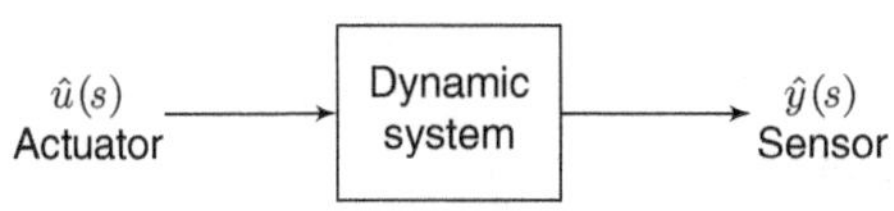

Figure 1.23 Block diagram of a dynamic system.

1 The matrix notation used here is conventional: column matrices ("vectors") are represented by lower case and rectangular matrices are represented by upper case. The "T" superscript represents the matrix transpose so c^T represents a row matrix.

2 Kalman called the combination a state-variable realization. His explanation is that an "analog" computer circuit could be directly wired from these equations and such a circuit is a physical synthesis or "realization" of the equations. The analog computer is antiquated at this date so the modern student may have difficulty interpreting these ideas.

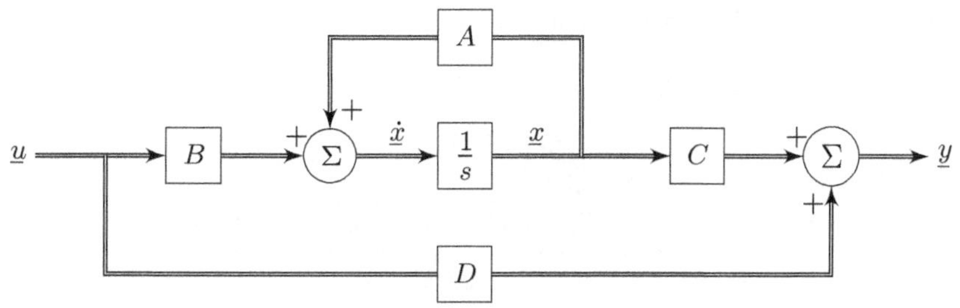

Figure 1.24 State-space block diagram.

The transfer function is related to state variable models after some preparation. For one thing, we must decide what we mean by the Laplace transform of a vector or a matrix. We let the Laplace transform of a vector be a vector of Laplace transforms.

$$\mathscr{L}[\underline{x}(t)] = \mathscr{L}\begin{bmatrix} x_1(t) \\ x_2(t) \\ \vdots \\ x_n(t) \end{bmatrix} \triangleq \begin{bmatrix} \mathscr{L}[x_1(t)] \\ \mathscr{L}[x_2(t)] \\ \vdots \\ \mathscr{L}[x_n(t)] \end{bmatrix} \tag{1.76}$$

The kth component of the Laplace transform of the vector is the Laplace transform of the kth component. The Laplace transform of a matrix is defined in a similar fashion. The linearity property has implications for the Laplace transform of matrix products. Notice that by linearity, $\mathscr{L}[A\underline{x}(t)] = A\mathscr{L}[\underline{x}(t)]$. In fact, we can use the linearity property to obtain many matrix-transform relations similar to the scalar counterparts.

The state variable model of (1.75) can be transformed into the Laplace domain.

$$\begin{aligned} s\hat{\underline{x}} - \underline{x}(0^-) &= A\hat{\underline{x}} + \underline{b}\hat{\underline{u}} \\ \hat{\underline{y}} &= \underline{c}^T\hat{\underline{x}} + D\hat{\underline{u}} \end{aligned} \tag{1.77}$$

These equations are manipulated to obtain

$$\hat{\underline{y}} = \left[\underline{c}^T(sI - A)^{-1}\right]\hat{\underline{x}}(0^-) + \left[\underline{c}^T(sI - A)^{-1}\underline{b} + D\right]\hat{\underline{u}} \tag{1.78}$$

where for any variable v, $\hat{v} = \mathscr{L}[v(t)]$.

It is here that we can justify the name "state." In order to obtain the output, $\hat{y}(s)$, we must know the input, $\hat{u}(s)$ (that is, $u(t)$ for $t \geq 0$) and the initial conditions, $\underline{x}(0^-)$. These initial conditions contain all the information that is required about what had happened before the input was applied; they completely describe the current state of the system.

In order to obtain a connection with the transfer function, set $\underline{x}(0^-) = \underline{0}$ and knowing that $G(s) = \frac{\hat{y}}{\hat{u}}$, we obtain

$$G(s) = \underline{c}^T(sI - A)^{-1}\underline{b} + D \tag{1.79}$$

1.5.3 Transfer Function Properties

Consider a rational function

$$G(s) = \frac{\hat{y}}{\hat{u}} = \frac{\text{num}}{\text{den}} \tag{1.80}$$

Definition 1.3 $G(s)$ is proper if deg(num) = deg(den).

Definition 1.4 $G(s)$ is strictly proper if deg(num) < deg(den).

Definition 1.5 $G(s)$ is improper otherwise (i.e. if deg(num) > deg(den)).

Recall the formula for matrix inversion.

$$M^{-1} = \frac{\text{adj } M}{\det M} \tag{1.81}$$

We can then write

$$\underset{n \times n}{(sI - A)^{-1}} = \frac{\text{adj }(sI - A)}{\det (sI - A)} = \frac{\text{poly. with deg} \leq n - 1}{\text{poly. with deg} = n} \tag{1.82}$$

where n is the dimension of matrix A. Then we can write $G(s)$ as

$$\begin{aligned}\frac{\hat{y}}{\hat{u}} = G(s) &= \underline{c}^T(sI - A)^{-1}\underline{b} + D \\ &= \frac{\underline{c}^T \text{adj}\,(sI - A)\underline{b}}{\det(sI - A)} + D \\ &= \frac{\underline{c}^T \text{adj}\,(sI - A)\underline{b} + D\det(sI - A)}{\det(sI - A)}\end{aligned} \tag{1.83}$$

An element of the adjugate matrix, adj $(sI - A)$, is formed by crossing out a single row and a single column and applying the determinant operation. This means that the elements of adj $(sI - A)$ are polynomials of degree no greater than $n - 1$. Using this observation, we can say the followings about $G(s)$:

- Strictly proper if $D = 0$.
- Proper if $D \neq 0$.
- Never improper.

The question arises whether any transfer function is improper: Is there another formulation of the dynamic equations such that such a thing is possible? One does see such formulations from time to time, but one should be skeptical on physical grounds. The Paley–Wiener theorem [10] establishes that an improper transfer function is physically impossible. In Section 1.5.4 of this chapter, we argue in a more heuristic way that the transfer function of a physical system is not improper.

1.5.4 Poles and Zeros of a Transfer Function

Given the state equations below, the transfer function $G(s)$ can be written as

$$\begin{aligned}\dot{\underline{x}} &= A\underline{x} + \underline{b}\underline{u} \\ \underline{y} &= \underline{c}^T\underline{x} + D\underline{u}\end{aligned} \tag{1.84}$$

$$G(s) = \frac{\underline{c}^T \text{adj}\,(sI - A)\underline{b} + D\det(sI - A)}{\det(sI - A)} = \frac{\eta(s)}{\Delta(s)}, \quad \deg \eta \le \deg \Delta \tag{1.85}$$

Poles of $G(s)$ are roots of $\det(sI - A) = 0$ and depend only upon the internal connections of the system; that is, the poles depend only upon the A-matrix. The zeros of $G(s)$ are the roots of $\underline{c}^T\text{adj}\,(sI - A)\underline{b} + D\det(sI - A) = 0$. In contrast to poles, zeros also depend upon the way that the system connected to its environment (characterized by $\underline{c}^T$ and $\underline{b}$). The physical interpretation of these state matrices is provided in Table 1.1.

Let $G(s)$ be the transfer function from $\hat{u}(s)$ to $\hat{y}(s)$ without any repeated poles,

$$G(s) = \frac{\eta(s)}{\Delta(s)} = \frac{(s + z_1)(s + z_2)\cdots(s + z_m)}{(s + p_1)(s + p_2)\cdots(s + p_n)}, \quad m \le n \tag{1.86}$$

then, as shown in the previous sections, the output response $\hat{y}(s)$ is

$$\hat{y}(s) = G(s)\hat{u}(s) = \cdots + \frac{R_k}{s + p_k} + \cdots \tag{1.87}$$

Table 1.1 Matrix form parameters.

Matrix	Function
A	State interaction
B	Connection to actuator (“environment”)
C^T	Connection to sensor (“environment”)
D	Direct connection from actuator to sensor

where,

$$\mathscr{L}\left[\frac{R_k}{s+p_k}\right]^{-1} = R_k e^{-p_k t} \tag{1.88}$$

In this context, a pole and zero cancellation could be problematic in feedback design, and it will be discussed in the following sections. The example below introduces the idea of a pole and zero cancellation.

Example 1.12 Pole cancellation. Consider a system with dynamics

$$\hat{y}(s) = \frac{(s+1)}{(s+1)(s+2)}\hat{u}(s) \tag{1.89}$$

Partial fraction expansion without carrying out pole–zero cancellation will result in a term with denominator $(s+1)$ and numerator 0.

$$\hat{y}(s) = \frac{(s+1)}{(s+1)(s+2)}\hat{u}(s) = \cdots + \frac{0}{s+1} + \cdots \tag{1.90}$$

Inverse Laplace and retransform:

$$\ddot{y} + 3\dot{y} + 2 = \dot{u} + u \tag{1.91}$$

$$s^2\hat{y}(s) - \dot{y}(0^-) - sy(0^-) + 3s\hat{y}(s) - 3y(0^-) + 2\hat{y}(s) = s\hat{u} + 1\hat{u} \tag{1.92}$$

$$(s^2 + 3s + 2)\hat{y}(s) = (s+1)\hat{u}(s) + \dot{y}(0^-) + sy(0^-) + 3y(0^-) \tag{1.93}$$

$$\hat{y}(s) = \frac{(s+1)}{(s+1)(s+2)}\hat{u} + \frac{[y(0^-)]s + [3y(0^-) + \dot{y}(0^-)]}{(s+1)(s+2)} \tag{1.94}$$

$$\hat{y}(s) = \cdots + \frac{R}{s+1} + \cdots, \quad \text{where } R \neq 0 \tag{1.95}$$

The pole is canceled from the forced response, but not the initial condition response. This is called a "hidden mode."

1.5.5 Physical Realizability

It is important to obtain results that indicate the range of possibilities for real, physical transfer functions. The most satisfactory results are obtained by imposing a constraint of "cause and effect": the output of a causal physical system cannot anticipate the input. We claim that these causal physical systems are proper, and we ask the following question, Can we build a system for which $G(s)$ is improper? Is an improper transfer function realizable?

The simplest improper $G(s)$ is illustrated in Figure 1.25, this system is a simple differentiator.

Let $u(0^-) = 0, \hat{y}(s) = s\hat{u}(s), G(s) = \frac{\hat{y}}{\hat{u}}(s)$. More generally, suppose $G(s) = \frac{\eta(s)}{\Delta(s)}$, where $\deg \eta \geq \deg \Delta + 1$. We can expand $G(s)$ such that

$$G(s) = R_0 + R_{-1}s + \cdots + \underbrace{G'(s)}_{\text{strictly proper}} \tag{1.96}$$

as illustrated in Figure 1.26.

But, what is the definition of a derivative? The derivative at time t is the common limit (if it exists) of the slope from the right and left.

$$\frac{dy(t)}{dt} = \lim_{\Delta t \to 0^-} \frac{y(t) - y(t-\Delta t)}{\Delta t} = \lim_{\Delta t \to 0^+} \frac{y(t+\Delta t) - y(t)}{\Delta t} \tag{1.97}$$

This is shown in Figure 1.27.

We can determine the limit from the left, $\Delta t \to 0^-$ (from past values), but it is impossible to measure the limit from the right $\Delta t \to 0^+$ (and therefore the common limit), because no device cannot predict the future.

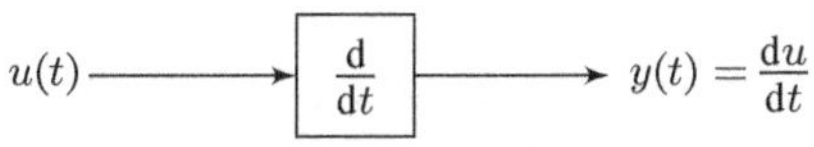

Figure 1.25 An improper system.

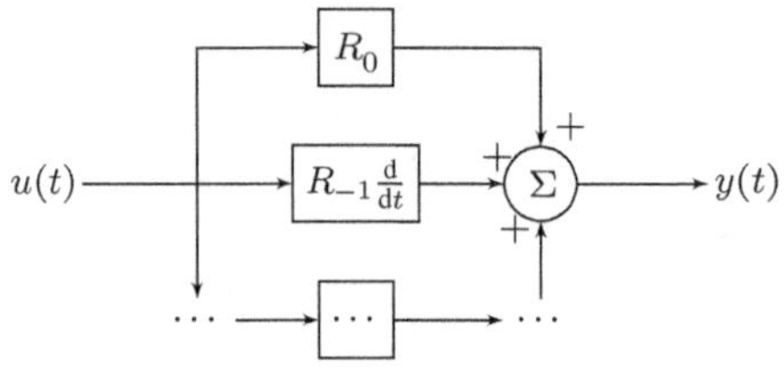

Figure 1.26 An expansion of $G(s)$ into proper and improper components.

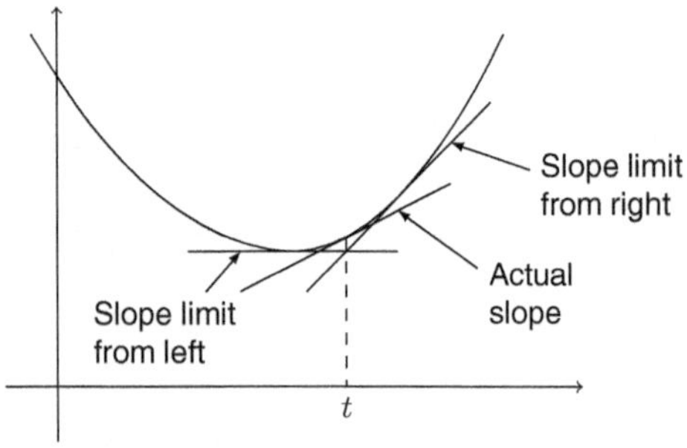

Figure 1.27 Derivative common limit.

Instead, we can implement some delay, e.g. $\frac{s}{\tau s+1}$, (making the $G(s)$ proper), in order to approximately measure the derivative. This delay results in using only the past values to approximate the computation of a differential operator.

It should be noted that some limit processes, such as integration, do not require a device that can foresee the future, as discussed later.

One can ask, is integration physically realizable? Integration, as shown in Figure 1.28, is the common limit (if it exists) as partitioning is refined Figure 1.29, shown in Figure 1.29, is refined. Information about future states is not required for integration, so integration is realizable.

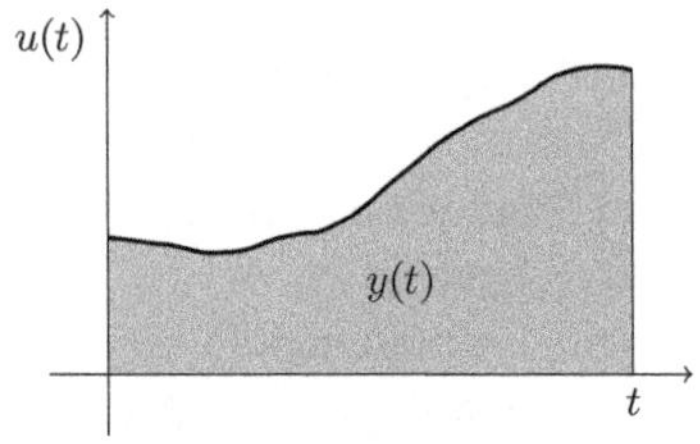

Figure 1.28 Integral of $u(t)$.

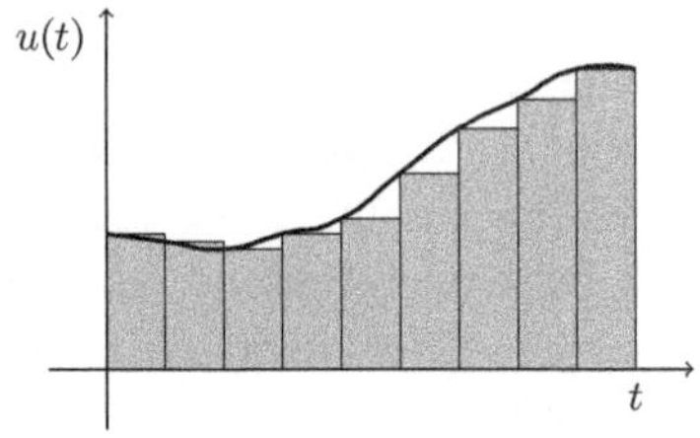

Figure 1.29 The integral is the common limit of the upper and lower sum.

The integral operator is shown in Figure 1.30.

The Laplace transform of this operator is given by

$$\frac{dy}{dt}(t) = u(t) \tag{1.98}$$

$$\text{Transform: } s\hat{y}(s) = \hat{u}(s) \tag{1.99}$$

$$G(s) = \frac{\hat{y}}{\hat{u}} = \frac{1}{s} \tag{1.100}$$

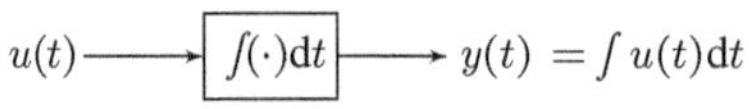

Figure 1.30 Integration block diagram.

A consideration of frequency response allows one to anticipate the more rigorous result laid down by the Paley–Wiener theorem [10], which is outside the scope of this text. Intuitively, it seems impossible for a physical system to have increasing gain with increasing frequency: If this sort of thing were possible, the power requirements would be enormous. This leads to speculate that *the degree of the numerator of a transfer function of a physically realizable dynamic system is never greater than the degree of the denominator*.

Notice that we satisfy this constraint wherever formula (1.79) is used: State equations (1.75) meet expectations for physical realizability. Something negative can be said about those formulations of the state equations from which improper transfer functions are calculated.

The same sort of reasoning leads some to argue that a physical system must be strictly proper (similar to our integral operator), where the D-matrix is zero, since it is hard to imagine that a real system can pass a very high frequency input without some attenuation. That is fair enough, but this conclusion is not supported by the more rigorous realizability theory based upon causality principles and complex analysis.

1.6 Problems

1.1 Find the Laplace transform of the two functions of time illustrated in Figure 1.31a,b. Hint: You can write an expression for these signals using step and delayed steps.

1.2 Obtain the Laplace transform of

$$y(t) = te^{-t}\cos(2t)$$

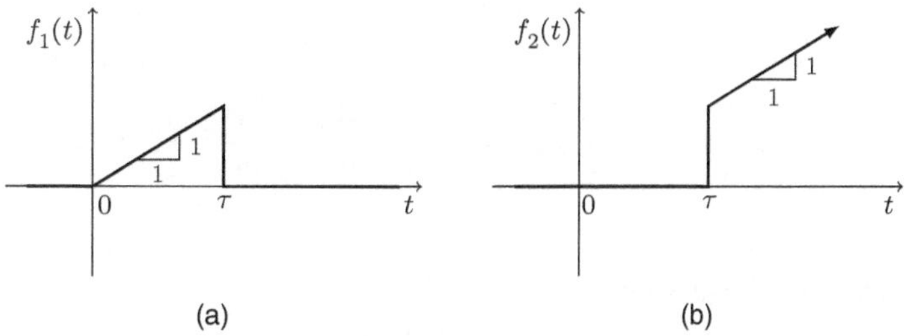

Figure 1.31 Time domain functions for Problem 1.1. (a) Truncated ramp. (b) Delayed ramp.

1.3 Find the inverse Laplace transforms of

(a) $\frac{7s-6}{s^2-s-6}$

(b) $\frac{2s^2+5}{s^2+3s+2}$

(c) $\frac{6(s+34)}{s(s^2+10s+34)}$

(d) $\frac{8s+10}{(s+1)(s+2)^3}$

1.4 Consider the following state-space equation,

$$\dot{x} = Ax + \underline{b}u$$

$$y = \underline{c}^T x$$

where

$$A = \begin{pmatrix} 0 & 1/J \\ 0 & -b/J \end{pmatrix} \quad \underline{b} = \begin{pmatrix} 0 \\ 1 \end{pmatrix}$$

$$\underline{c}^T = \begin{pmatrix} 1 & 0 \end{pmatrix} \quad D = 0$$

Find the transfer function $G(s)$ where $G(s) = \frac{\hat{y}(s)}{\hat{u}(s)}$.

1.5 For the system shown in Figure 1.32, the elements are designed so that

$$\frac{dc}{dt} = r(t) - c(t-\tau)$$

The unavoidable delay τ occurs because the measuring device must be placed downstream of the roller. Use the time-shifting property to find the transfer function $G(s) = \frac{C(s)}{R(s)}$.

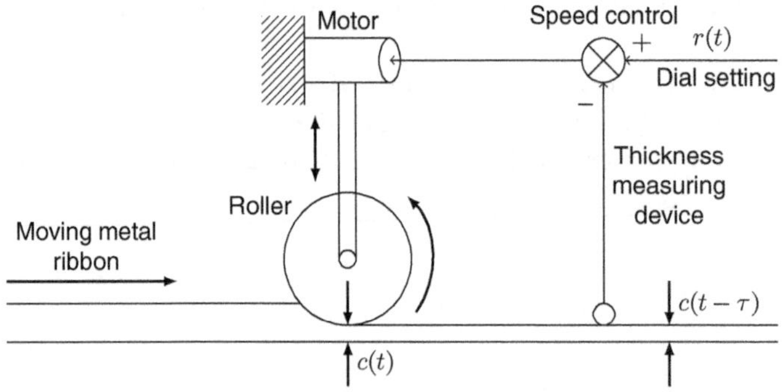

Figure 1.32 Controlled metal-rolling device.

1.6 Apply the relevant propositions and proofs discussed in the class in a iterative fashion to establish the following corollaries:

(a)

$$\mathscr{L}[t^n y(t)] = (-1)^n \frac{d^n}{ds^n}\hat{y}(s)$$

(b)

$$\mathscr{L}\left[\frac{d^n}{dt^n}y(t)\right] = s^n\hat{y}(s) - \sum_{k=1}^{n} s^{n-k}\frac{d^{k-1}}{dt^{k-1}}y(0^-)$$

1.7 Consider the following state-space equation,

$$\dot{x} = Ax + \underline{b}u_1$$

$$y = \underline{c}^T x$$

where

$$A = \begin{bmatrix} 0 & \frac{1}{m} \\ 0 & -\frac{b}{m} \end{bmatrix}, \quad B = \begin{bmatrix} \frac{b}{m} \\ -\frac{b^2}{m} \end{bmatrix}, \quad \underline{c}^T = [1 \;\; 0]$$

Find the transfer function $G(s)$, where

$$G(s) = \frac{\hat{y}}{\hat{u}_1}(s)$$

Explain how u_1 has to change so that the transfer function $G(s)$ will be equal to

$$G(s) = \frac{-\frac{1}{m}}{s\left(s+\frac{b}{m}\right)}$$

1.8 Newton's second law for the submersible shown is

$$m\frac{d^2y}{dt^2} = F \tag{1.101}$$

where F is regulated by pumping water in or out of tanks. The relationship between F and the pumping rate, $u(t)$, is

$$\frac{dF}{dt} = k_B u \tag{1.102}$$

(a) Find the transfer function $\frac{\hat{y}}{\hat{u}}(s)$. Indicate the pole–zero diagram.

(b) Actually, if the water is expelled or brought in from the bottom of the craft, then we get a little added thrust, so the second law becomes

$$m\frac{d^2y}{dt^2} = F + k_T u \tag{1.103}$$

What is the transfer function $\frac{\hat{y}}{\hat{u}}(s)$ in this case? Indicate the pole–zero diagram.

(c) If, instead, the water is expelled or brought in from the top of the craft, then the second law becomes

$$m\frac{d^2y}{dt^2} = F - k_T u \tag{1.104}$$

What is the transfer function $\frac{\hat{y}}{\hat{u}}(s)$ in this case? Indicate the pole–zero diagram.

2

The Response of Linear, Time-Invariant Dynamic Systems

This chapter is a review of the basics of dynamic system response. In particular, we wish to relate the parameters of the transfer function or the state equations to the quantitative features of the time response of forced and unforced response. Special emphasis is placed upon the steady-state frequency response. The frequency response relationships derived in this chapter are fundamental to the main issues of feedback design developed in Chapter 3.

Transient response is addressed as well. This is appropriate because improved dynamic, transient response can be thought of as an important, albeit a secondary, aspect of control system design. It stands to reason that one can improve dynamic response with feedback, so the designer might as well deal with these issues while tending to the more important matters, as will be discussed in the following chapters.

2.1 The Time Response of Dynamic Systems

Begin with the elementary transfer function relation where $\hat{y}(s) = G(s)\hat{u}(s)$, as indicated in Figure 2.1.

Consider $G(s)$ to be a dynamic system as indicated in Figure 2.2. Now consider an impulse input, where $\hat{u}(s) = \delta(t)$, so that $\hat{y}(s) = G(s)$. Then, $y(t) = g(t)$, where $g(t) = \mathscr{L}^{-1}[G(s)]$ is called the **impulse response of the dynamic system.**

Using an impulse response input, we will generalize the response of a dynamic system to any test input. However, before accomplishing this, we will go over three fundamental theorems.

2.1.1 Final Value Theorem

For a linear time-invariant system, the final value theorem can be used to find the value of $y(t)$ as $t \to \infty$.

Theorem 2.1 (Final Value Theorem) *If $\mathscr{L}[y(t)] = \hat{y}(s)$ and poles of $s\hat{y}(s)$ lie strictly in the LHP, then the final value theorem states*

$$\lim_{t\to\infty} y(t) = \lim_{s\to 0} s\hat{y}(s) \tag{2.1}$$

Proof:

$$\mathscr{L}\left[\frac{dy}{dt}\right] = s\hat{y}(s) - y(0^-)$$

$$\begin{aligned}
s\hat{y}(s) &= y(0^-) + \mathscr{L}\left[\frac{dy}{dt}\right] \\
&= y(0^-) + \int_{0^-}^{\infty} e^{-st}\frac{dy}{dt}(t)dt \\
\lim_{s\to 0} s\hat{y}(s) &= y(0^-) + \int_{0^-}^{\infty} \frac{dy}{dt}(t)dt \\
&= y(0^-) + \lim_{t\to\infty} y(t) - y(0^-) \\
&= \lim_{t\to\infty} y(t)
\end{aligned}$$

Therefore, the final value of $y(t)$ can be obtained from the Laplace transform of $y(t)$ using the final value theorem. ■

Robust Control: Youla Parameterization Approach, First Edition. Farhad Assadian and Kevin R. Mallon.

Companion Website: www.wiley.com/go/assadian/robust

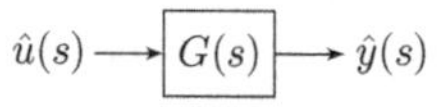

Figure 2.1 Dynamic system.

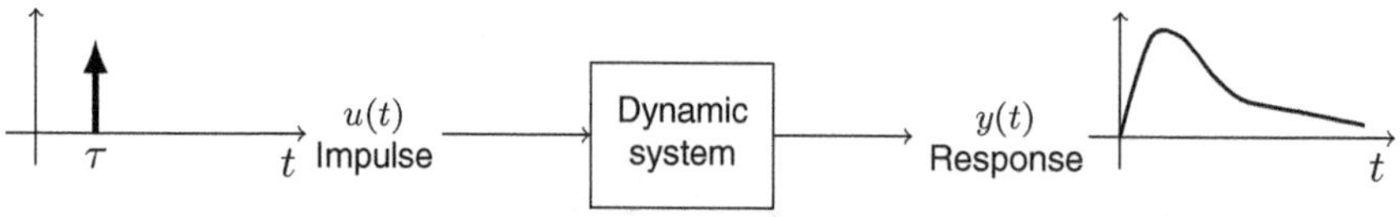

Figure 2.2 Response of a dynamic system.

2.1.2 Initial Value Theorem

For a linear time-invariant system, the initial value theorem can be used to find the initial value of $y(t)$ at the instant some input $u(t)$ is applied.

Theorem 2.2 (Initial Value Theorem) *If $\mathscr{L}[y] = \hat{y}$, then the initial value theorem states*

$$\lim_{s\to\infty} s\hat{y}(s) = y(0^+) \tag{2.2}$$

Proof:

$$\begin{aligned}
s\hat{y}(s) &= y(0^-) + \int_{0^-}^{\infty} e^{-st}\frac{dy}{dt}(t)dt \\
&= y(0^-) + \int_{0^-}^{0^+} e^{-st}\frac{dy}{dt}(t)dt + \int_{0^+}^{\infty} e^{-st}\frac{dy}{dt}(t)dt \\
\lim_{s\to\infty} s\hat{y}(s) &= y(0^-) + \int_{0^-}^{0^+} \underbrace{e^{-s\cdot 0}}_{=1}\frac{dy}{dt}(t)dt + \int_{0^+}^{\infty} \underbrace{e^{-st}}_{\to 0 \text{ as } s\to\infty}\frac{dy}{dt}(t)dt \\
&= y(0^-) + \int_{0^-}^{0^+} \frac{dy}{dt}(t)dt + 0 \\
&= y(0^-) + y(0^+) - y(0^-) \\
&= y(0^+)
\end{aligned}$$

Therefore, the initial value of $y(t)$ can be obtained from the Laplace transform of $y(t)$ using the initial value theorem. ■

Example 2.1 Initial and final values for a step and ramp response.

$$G(s) = \frac{s+1}{s+2},\quad u(t) = 1(t),\quad \hat{u}(s) = \frac{1}{s} \tag{2.3}$$

Given an initial condition $y(0^-) = 0$,

$$\hat{y}(s) = \frac{s+1}{s(s+2)} \implies y(\infty) = \lim_{s\to 0} s\hat{y}(s) = \frac{1}{2},\quad y(0^+) = \lim_{s\to\infty} s\hat{y}(s) = 1 \tag{2.4}$$

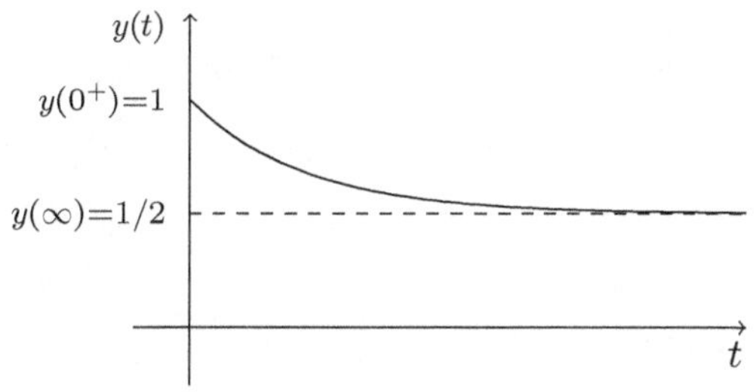

Figure 2.3 Response to step input.

The initial and final values are illustrated in Figure 2.3. Now suppose $u(t)$ is a ramp input.

$$u(t) = t1(t), \ \hat{u}(s) = \frac{1}{s^2} \tag{2.5}$$

$$\hat{y}(s) = \frac{s+1}{s^2(s+2)} \implies y(\infty) = \infty, \ y(0^+) = 0 \tag{2.6}$$

Example 2.2 Initial and final values for a step response.

$$G(s) = \frac{s+1}{s^2+s+1}, \ u(t) = 1(t), \ \hat{u}(s) = \frac{1}{s} \tag{2.7}$$

$$\hat{y}(s) = \frac{s+1}{s(s^2+s+1)}, \text{given that } y(0^-) = 0, \ \dot{y}(0^-) = 0 \tag{2.8}$$

$$y(\infty) = \lim_{s\to 0} s\hat{y}(s) = 1 \tag{2.9}$$

$$y(0^+) = \lim_{s\to \infty} s\hat{y}(s) = 0 \tag{2.10}$$

$$\dot{y}(0^+) = \lim_{s\to \infty} s^2\hat{y}(s) = 1 \tag{2.11}$$

Note that there is a discontinuity in the slope: $\dot{y}(0^-) = 0$ and $\dot{y}(0^+) = 1$.

2.1.3 Convolution and the Laplace Transform

Consider a dynamic system $G(s)$ as indicated in Figure 2.1.

$$\hat{y}(s) = G(s)\hat{u}(s) \tag{2.12}$$

Theorem 2.3 **(Convolution Theorem)** *If for* $0^- \leq \zeta \leq t \leq \infty$, $\mathscr{L}[u(t)] = \hat{u}(s)$ *and* $\mathscr{L}[g(t)] = G(s)$, *then*

$$\mathscr{L}\left[\int_{0^-}^{t} g(t-\zeta)u(\zeta)\mathrm{d}\zeta\right] = G(s)\hat{u}(s) = \mathscr{L}\left[\int_{0^-}^{t} g(\zeta)u(t-\zeta)\mathrm{d}\zeta\right] \tag{2.13}$$

Convolution can be written as a short-hand notation using the "$*$" operator, i.e.

$$[g(t) * u(t)] = \left[\int_{0^-}^{t} g(t-\zeta)u(\zeta)\mathrm{d}\zeta\right] \tag{2.14}$$

Equation (2.14) is illustrated in Figure 2.4.

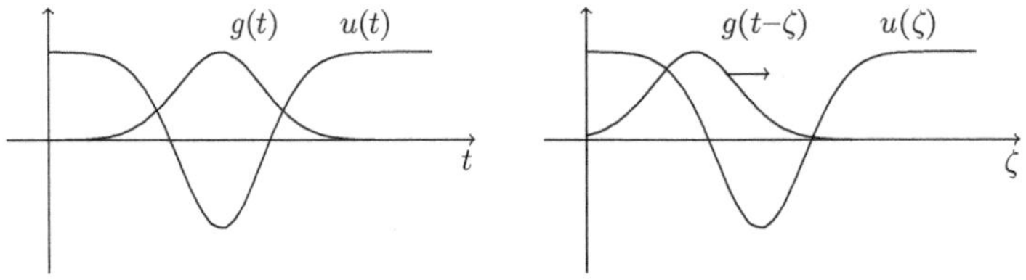

Figure 2.4 Convolution.

Proof: From the definition of the transform

$$\mathcal{L}\left[g(t) * u(t)\right] = \int_{0^-}^{\infty} e^{-st} \int_{0^-}^{t} g(t-\zeta)u(\zeta)d\zeta \; dt$$

Interchange the order of integration but preserve the ordering $0^- \leq \zeta \leq t \leq \infty$ so that

$$\mathcal{L}\left[g(t) * u(t)\right] = \int_{0^-}^{\infty} e^{-s\zeta}u(\zeta) \int_{\zeta^-}^{\infty} e^{-s(t-\zeta)}g(t-\zeta)dt \; d\zeta$$

Let $\eta = t - \zeta$ in the inner integration so $dt = d\eta$. Then,

$$\begin{aligned}\mathcal{L}\left[g(t) * u(t)\right] &= \int_{0^-}^{\infty} e^{-s\zeta}u(\zeta) \int_{0^-}^{\infty} e^{-s\eta}g(\eta)d\eta \; d\zeta \\ &= G(s)\int_{0^-}^{\infty} e^{-s\zeta}u(\zeta)d\zeta \end{aligned} \tag{2.15}$$

Therefore, $\mathcal{L}[g(t) * u(t)] = G(s)\hat{u}(s)$. ∎

This means that for a dynamic system as shown in Figure 2.1,

$$y(t) = \int_{0^-}^{t} g(t-\zeta)u(\zeta)d\zeta \tag{2.16}$$

We are now in a position to write the general time response of a LTI dynamic system. From Section 1.5, we know that

$$\underline{\hat{y}} = \left[\underline{c}^T(sI-A)^{-1}\right]\underline{x}(0^-) + \left[\underline{c}^T(sI-A)^{-1}\underline{b} + D\right]\underline{\hat{u}} \tag{2.17}$$

So,

$$\begin{aligned} y(t) &= \underline{c}^T\Phi(t)\underline{x}(0^-) + g(t) * u(t) \\ &= \underline{c}^T\Phi(t)\underline{x}(0^-) + \int_{0^-}^{t} g(t-\zeta)u(\zeta)d\zeta \end{aligned} \tag{2.18}$$

then we can write

$$y(t) = \underline{c}^T\Phi(t)\underline{x}(0^-) + \underline{c}^T\left[\int_{0^-}^{t}\Phi(t-\zeta)u(\zeta)d\zeta\right]\underline{b} + Du(t) \tag{2.19}$$

where $\Phi(t) = \mathcal{L}^{-1}[\underline{\Phi}(s)] = \mathcal{L}^{-1}[(sI-A)^{-1}]$ is known as the *State Transition Matrix* and $g(t) = \mathcal{L}^{-1}[G(s)]$.

There is a useful interpretation of $g(t)$ as the impulse response of the dynamic system. One can imagine a procedure where the impulse response is measured and the output to any other input predicted with the convolution (2.18).

The first term in (2.18) is called the *initial condition response* and the second term is called the *forced response*. The initial condition response will be a transient response for stable systems and the forced response makes some contribution to the transient response as well as to the steady-state response. The third term in (2.19) is called *direct transmission or static gain* response and is most obvious where things are in terms of the state equations. We reason that there must be a similar response component in (2.18) but it is obscured in the convolution integral.

The stability concept requires some elaboration, which will be addressed in this section.

Proposition 2.1 *The step response of a dynamic system is the integral of the impulse response:*

$$y(t) = \int_{0}^{t} g(\zeta)d\zeta \tag{2.20}$$

Proof: From the simple fact that $\hat{y}(s) = G(s) \cdot 1/s$ and $G(s) = \mathcal{L}[g(t)]$. ∎

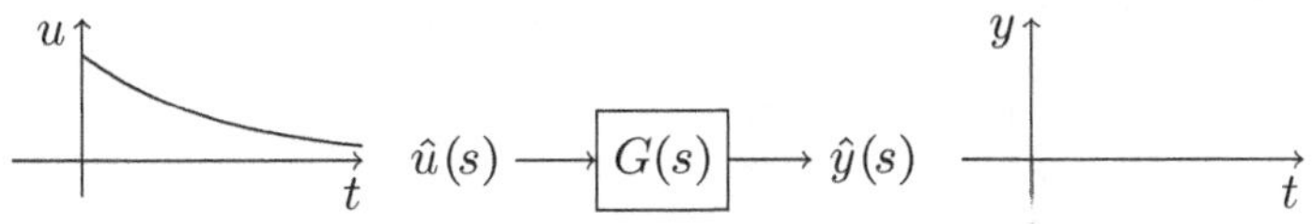

Figure 2.5 Transmission blocking response.

2.1.4 Transmission Blocking Response

In this subsection, we provide an interpretation of a transfer function zero. We propose the following definitions to distinguish between zeros and zeros with transmission blocking response properties.

Definition 2.1 A transfer function, $G(s)$, has a transmission zero at $s = z_k$ if there exists an input such that $u(t) = u_k e^{z_k t} 1(t)$ results in the output $y(t)$ not containing $e^{z_k t}$ terms.

Definition 2.2 A transfer function, $G(s)$, has a transmission zero at $s = z_k$ with the *transmission blocking response* property if there exists an input such that $u(t) = u_k e^{z_k t} 1(t)$ results in exactly zero output $y(t) \equiv 0 \quad \forall t$. These zeros are called transmission blocking zeros.

We do not make a distinction between zeros of a transfer function or transmission zeros. We will address this point in more detail in the MIMO part of this text.

Now, suppose $u(t) = e^{-\mu t} 1(t)$. Is it possible to design $G(s)$ so that $y(t)$ has exactly zero response to $u(t)$, as shown in Figure 2.5? The answer to this question is *no* if all initial conditions are precisely equal to zero. The best we could do is to eliminate this particular input from the output response, but we are unable to force the output $y(t)$ to be exactly equal to zero for all times.

We then wonder, is it possible to apply a nonzero input to a system with a nonzero gain and find that the output is exactly zero? The answer to this question is "yes" if both the *input* and *the initial conditions* are just right. The particular input we have in mind is the exponential $u(t) = e^{-\mu t} 1(t)$, so that $\hat{u}(s) = 1/(s + \mu)$ and the output is a superposition of forced and initial condition responses. Denote the transfer function $G(s) = \eta(s)/\Delta(s)$ so that

$$\hat{y}(s) = G(s)\hat{u}(s) + \text{I.C.} = \frac{\eta(s)}{\Delta(s)} \cdot \frac{1}{s + \mu} + \frac{q(s)}{\Delta(s)} \equiv 0 \tag{2.21}$$

where $q(s)$ is some polynomial due to initial conditions. In order to cancel the dynamics of $u(t)$, we require that $\eta(s)$ has $(s + \mu)$ as a factor, $(s + \mu)$ must be a zero of $G(s)$ (Definition 2.1). However, appropriate dynamics from initial conditions are still required to fully block a response.

$$\frac{\eta(s)}{\Delta(s)} \cdot \frac{1}{s + \mu} + \frac{q(s)}{\Delta(s)} = 0 \quad \Longrightarrow \quad q(s) = \frac{-\eta(s)}{s + \mu} \tag{2.22}$$

Knowing, the procedure for selecting the right combination of the input and initial conditions for transmission blocking response, we can then ask, are poles and zeros ever really canceled? We can factor out the $(s + \mu)$ term from $G(s)$ such that

$$G(s) = (s + \mu)G'(s) \tag{2.23}$$

Then, Eq. (2.21) can be written as

$$\hat{y}(s) = G'(s) + I.C. \tag{2.24}$$

From Eq. (2.24), we can state that even if we cancel the unwanted pole by a properly located zero in $G(s)$, this unwanted pole may still result in an output response if the initial conditions are not zero. The two examples below shed further light to the concepts of *transmission blocking response* and *pole–zero cancellation*.

Example 2.3 Suppose we assume a step input, $u(t) = 1(t)$ to the following system:

$$G(s) = \frac{s}{s + 1}, \; u(t) = 1(t), \; \hat{u}(s) = \frac{1}{s} \tag{2.25}$$

Then we ask, what should be the initial condition to generate an output, $y(t) \equiv 0 \quad \forall t$,

$$\frac{s}{s+1} \cdot \frac{1}{s} + \frac{q(s)}{s+1} = 0 \quad \Longrightarrow \quad q(s) = -1 \tag{2.26}$$

For simulating this simple system, it is easier to write the output response $\hat{y}(s)$ in the following form:

$$\underline{\hat{y}} = \left[\underline{c}^T(sI - A)^{-1}\right]\underline{x}(0^-) + \left[\underline{c}^T(sI - A)^{-1}\underline{b} + D\right]\underline{\hat{u}} \tag{2.27}$$

As will be discussed later, the state matrices A, $\underline{b}$, $\underline{c}^T$, and D can be computed from $G(s)$. Knowing $G(s) = \frac{s}{s+1}$ then $A = -1$, $\underline{b} = 1$, $\underline{c}^T = -1$, and $D = 1$. Comparing Eqs. (2.27) and (2.26), it is not difficult to see that $q(s) = \underline{c}^T \cdot \underline{x}(0^-)$. Therefore, knowing that $q(s) = -1$ and $\underline{c}^T = -1$, then $\underline{x}(0^-) = 1$. We can then simulate the system shown generally in Figure 2.6.

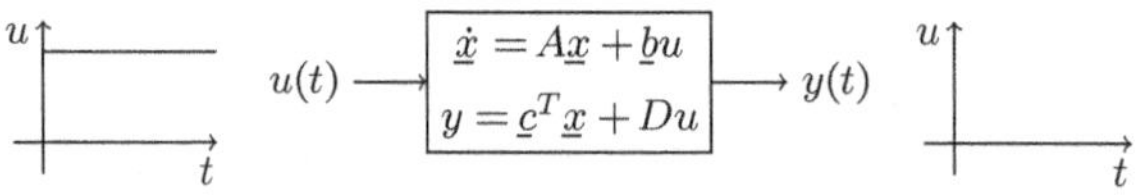

Figure 2.6 Transmission blocking of a unit step input.

We then observe that the output $y(t) \equiv 0 \quad \forall t$.

Example 2.4 Given the following system:

$$\hat{y} = \frac{(s+1)}{(s+1)(s+2)}\hat{u} = \cdots + \frac{0}{s+1} + \cdots \tag{2.28}$$

if we do not cancel the common pole–zero and utilize the Heaviside expansion to compute the residuals, as discussed in Section 1.4, we will find that the residual associated with the $s + 1$ is equal to zero. we can equivalently write the system equations as:

$$(s^2 + 3s + 2)\hat{y} = (s+1)\hat{u} \tag{2.29}$$

now, utilizing the inverse Laplace transform, we obtain

$$\ddot{y} + 3\dot{y} + 2 = \dot{u} + u \tag{2.30}$$

and re-transform,

$$s^2\hat{y} - \dot{y}(0^-) - sy(0^-) + 3s\hat{y} - 3y(0^-) + 2\hat{y} = s\hat{u} + \hat{u} \tag{2.31}$$

The output can be written in terms of the input and the initial conditions:

$$\hat{y} = \frac{(s+1)}{(s+1)(s+2)}\hat{u} + \frac{sy(0^-) + 3y(0^-) + \dot{y}(0^-)}{(s+1)(s+2)} \tag{2.32}$$

again, utilizing the Heaviside expansion,

$$\hat{y} = \cdots + \frac{R}{s+1} + \cdots, \quad \text{where} \quad R \neq 0 \tag{2.33}$$

The pole is canceled by a common zero from the forced response, but not the initial condition response. We will study the impact of this type of pole cancellation from forced but not from the initial condition response in feedback control design in the following chapter.

2.1.5 Stability

In order to discuss the intuitive and important notion of stability, one must describe stability or, at least, instability in precise terms. There are some subtleties here, and these are related to *Inverse Laplace Transformation by Heaviside Expansion* discussed in Chapter 1. It is clear that $y(t)$ will grow without bound where a pole exists in the right half-plane or where *multiple poles exist on the imaginary axis.* In the first case, the *unbounded growth* is exponential while in the latter case the growth is some polynomial in t. Unbounded growth is certainly something we want to call *unstable.*

It is important to notice that the system poles directly define the components in the initial condition response (known also as unforced or homogeneous response) part of $y(t)$. This observation leads to the following working descriptions of stability.

Definition 2.3 If the initial condition response is unbounded, we call the dynamic system *unstable*, otherwise we call the system *stable.* We call a system *asymptotically stable* if the initial condition response $\rightarrow 0$ as time $\rightarrow \infty$.

Clearly, a **stable** system is one for which the poles of the transfer function are **(i) in the left half-plane**, and **(ii) any pole on the imaginary axis is not a repeated pole**. Clearly, a system is **asymptotically stable** if and only if the poles of the transfer function are **strictly in the left half-plane**.

Example 2.5 Consider the simple case of an "integrator":

$$G(s) = \frac{1}{s} \tag{2.34}$$

The state equations are

$$\begin{aligned} A &= 0, \quad b = 1 \\ c^T &= 1, \quad D = 0 \end{aligned} \tag{2.35}$$

Since the impulse response is $g(t) = \mathscr{L}^{-1}[1/s] = 1(t)$, the output response $y(t)$ can be written as:

$$y(t) = y(0^-) + \int_0^t u(\zeta)\mathrm{d}\zeta \tag{2.36}$$

where $y(0^-)$ is the initial value on the integrator. First of all, suppose that $u \equiv 0$. In this case, it is clear that $y(t)$ does not approach zero at $t \rightarrow \infty$; the system is stable but not asymptotically stable. This is a good point to state something about physical devices that we hope will serve as an integrator: clearly, the true integral of the input,

$$\int_0^t u(\zeta)\mathrm{d}\zeta$$

is achieved by our "integrator" only where

$$y(0^-) = 0$$

That is, in the general case, there is an unfortunate offset from the true integral!

The stability discussions so far are around "initial condition response." Still, there is another important way to view the stability condition by considering the effects of "inputs."

Definition 2.4 **Bounded-input, bounded-output (BIBO)** stability means that the output remains bounded for any input that is bounded.

If a transfer function is improper, the Heaviside expansion of the output will have terms in positive powers of "s" as we saw in Chapter 1. This will be true even when the input is bounded. However, s^k for $k \geq 0$ transforms to a singularity function, an unbounded output! Thus, the implication is that *even if an improper transfer function were realizable, it would be BIBO unstable.* Another implication of the definition is with regard to a transfer function with poles on the imaginary axis; for example, if there is a pole at the origin and the input is a step input (bounded), the Heaviside expansion will contain a term proportional to $1/s^2$, an output term proportional to $t^2/2$ (unbounded). That is, a system characterized by a transfer

Table 2.1 Stability definitions.

Definition	Criterion
Asymp. stable	Poles strictly in LHP
BIBO stable	Poles strictly in LHP
BIBO unstable	Poles in RHP or on the $j\omega$-axis

function with a pole at the origin is BIBO unstable. Similarly, if the transfer function has imaginary poles on the imaginary axis and the input is a sine or cosine function at the resonant frequency, the response is unbounded.

Thus, we have the next proposition:

Proposition 2.2 *A dynamic system is BIBO stable only if it is characterized by a proper transfer function. When a transfer function is proper, a necessary and sufficient condition for both BIBO stability and asymptotic stability is that all poles of the transfer function must lie strictly in the left half-plane.*

A necessary and sufficient condition for BIBO stability, of a proper transfer function, is the same as for asymptotic stability; all poles of the transfer function must lie strictly within the left half-plane.

We summarize our stability discussions by providing a table with stability definitions and several examples below. Consider the transfer function $G(s)$.

$$G(s) = \frac{\eta(s)}{\Delta(s)} \tag{2.37}$$

In general, the output $\hat{y}(s)$ has a forced response and an initial condition response.

$$\hat{y}(s) = \underbrace{G(s)\hat{u}(s)}_{\text{forced response}} + \overbrace{\frac{q(s)}{\Delta(s)}}^{\text{I.C. response}} = \frac{R_1}{s+p_1} + \frac{R_2}{s+p_2} + \cdots \tag{2.38}$$

There are fundamentally three definitions of stability for LTI systems as given in Table 2.1. Stability in classical dynamics is based on the initial condition response, **asymptotic stability**. For controls, however, it is more appropriate to look at **BIBO** stability based on the forced response. ■

Example 2.6 The definitive, mechanical engineering example is

$$G(s) = 1/(ms^2) \tag{2.39}$$

which corresponds to

$$m\frac{\mathrm{d}^2y}{\mathrm{d}t^2} = u(t) \tag{2.40}$$

Suppose that we apply the bounded input:

$$u(t) = A\sin(\omega t)$$

It is a simple matter to demonstrate that the solution to the differential equation is

$$y(t) = -\frac{A}{m\omega^2}\sin(\omega t) + \left[\frac{A}{m\omega} + \frac{\mathrm{d}y(0^-)}{\mathrm{d}t}\right]t + y(0^-) \tag{2.41}$$

The second term is unbounded and nonzero if $A \neq 0$ where $\mathrm{d}y(0^-)/\mathrm{d}t = 0$ (BIBO unstable) and where $A = 0$ and $\mathrm{d}y(0^-)/\mathrm{d}t \neq 0$ (asymptotically unstable). Don't miss the point that even though the input force is bounded and has zero mean, the output displacement is unbounded and has a nonzero mean.

Example 2.7 Consider the dynamics:

$$\dot{y}(t) + \alpha y(t) = u(t) \tag{2.42}$$

with initial condition $y(0) = y_0$. Then,

$$s\hat{y}(s) - y(0) + \alpha\hat{y}(s) = \hat{u}(s) \tag{2.43}$$

$$\hat{y}(s) = \frac{1}{s+\alpha}\left(\hat{u}(s) + y_0\right) \tag{2.44}$$

Asymptotic stability is found by setting $\hat{u}(s) = 0$ and checking the initial condition response:

$$y(t) = \mathcal{L}^{-1}\left[\hat{y}(s)|_{\hat{u}(s)=0}\right] = \mathcal{L}^{-1}\left[\frac{y_0}{s+\alpha}\right] = y_0 e^{-\alpha t} \tag{2.45}$$

For $\alpha > 0$, $y(t) \to 0$ as $t \to \infty$. Therefore, the system is asymptotically stable for $\alpha > 0$.

BIBO stability is found by setting the initial conditions to zero and considering the response to bounded inputs $\hat{u}(s)$. Letting

$$\hat{g}(s) = \frac{1}{s+\alpha} \tag{2.46}$$

and

$$g(t) = \mathcal{L}^{-1}[\hat{g}(s)] \tag{2.47}$$

Then,

$$y(t) = \mathcal{L}^{-1}\left[\hat{y}(s)|_{y_0=0}\right] = \mathcal{L}^{-1}\left[\hat{g}(s)\hat{u}(s)\right] \tag{2.48}$$

$$= \int_{0^-}^{\infty} g(\tau)u(t-\tau)\mathrm{d}\tau \tag{2.49}$$

$$\therefore |y(t)| \leq \int_{0^-}^{\infty} |g(\tau)|\,|u(t-\tau)|\,\mathrm{d}\tau \tag{2.50}$$

If $u(t)$ is bounded, then

$$|u(t-\tau)| < K_1 < \infty \tag{2.51}$$

Then,

$$|y(t)| \leq K_1 \int_{0^-}^{\infty} |g(\tau)|\,\mathrm{d}\tau \tag{2.52}$$

Hence for BIBO stability,

$$\int_{0^-}^{\infty} |g(\tau)|\,\mathrm{d}\tau < \infty \tag{2.53}$$

In other words, if the impulse response is absolutely integrable, then the system is BIBO stable. This is true if and only if the poles of $\hat{g}(s)$ are in the left-hand plane (that is, $\alpha > 0$).

Example 2.8 Consider the dynamics

$$\ddot{y}(t) + y(t) = u(t) \tag{2.54}$$

with initial condition $y(0) = y_0$ and $\dot{y}(0) = v_0$. Then,

$$s^2\hat{y}(s) - \dot{y}(0) - sy(0) + \hat{y}(s) = \hat{u}(s) \tag{2.55}$$

$$\hat{y}(s) = \frac{1}{s^2+1}\left(\hat{u}(s) + v_0 + sy_0\right) \tag{2.56}$$

Asymptotic stability is found by setting $\hat{u}(s) = 0$ and checking the initial condition response.

$$y(t) = \mathcal{L}^{-1}\left[\hat{y}(s)|_{\hat{u}(s)=0}\right] = \mathcal{L}^{-1}\left[\frac{1}{s^2+1}(v_0 + sy_0)\right] \tag{2.57}$$

$$= y_0\cos(t) + v_0\sin(t) \tag{2.58}$$

For any nonzero initial conditions, $y(t)$ will not go to zero. It is therefore not asymptotically stable.

BIBO stability is found by setting the initial conditions to zero and considering the response to bounded inputs $\hat{u}(s)$. Following the same procedure as previously, letting

$$\hat{g}(s) = \frac{1}{s^2+1} \tag{2.59}$$

and

$$g(t) = \mathscr{L}^{-1}[\hat{g}(s)] \tag{2.60}$$

Then,

$$y(t) = \mathscr{L}^{-1}\left[\hat{y}(s)|_{y_0=0}\right] = \mathscr{L}^{-1}\left[\hat{g}(s)\hat{u}(s)\right] \tag{2.61}$$

$$= \int_{0^-}^{\infty} g(\tau)u(t-\tau)d\tau \tag{2.62}$$

$$\therefore |y(t)| \leq \int_{0^-}^{\infty} |g(\tau)|\,|u(t-\tau)|\,d\tau \tag{2.63}$$

If $u(t)$ is bounded, then

$$|u(t-\tau)| < K_1 < \infty \tag{2.64}$$

Then,

$$|y(t)| \leq K_1 \int_{0^-}^{\infty} |g(\tau)|\,d\tau \tag{2.65}$$

Hence, for BIBO stability,

$$\int_{0^-}^{\infty} |g(\tau)|\,d\tau < \infty \tag{2.66}$$

In other words, if the impulse response is absolutely integrable, then the system is BIBO stable. In this case, we know that this is not true and Eq. (2.66) is not finite. Therefore, the system is not BIBO stable.

Example 2.9 BIBO stability (sine input).

$$G(s) = \frac{1}{s^2+\omega^2} = \frac{1}{(s+j\omega)(s-j\omega)} \tag{2.67}$$

Critical points are shown in Figure 2.7. Apply a sinusoidal input:

$$\hat{u}(s) = \frac{\omega}{s^2+\omega^2} \Longrightarrow \hat{y}(s) = \frac{\omega}{s^2+\omega^2}\cdot\frac{1}{s^2+\omega^2} \tag{2.68}$$

Therefore, similar to the last example, a bounded input (sine wave) generates an unbounded response, as shown in Figure 2.8. Therefore, the system is BIBO unstable.

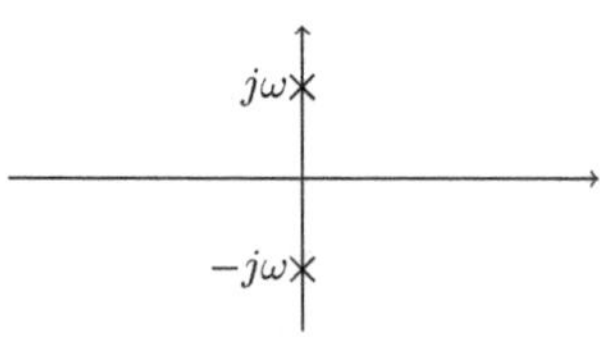

Figure 2.7 $G(s)$ roots.

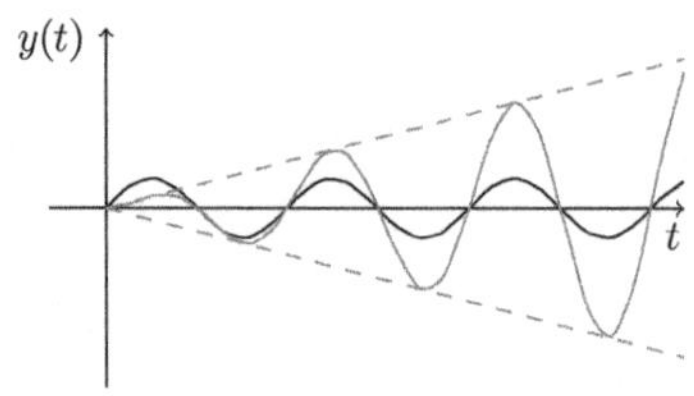

Figure 2.8 Unbounded response.

Example 2.10 BIBO stability (step input).

$$G(s) = \frac{1}{s} \tag{2.69}$$

Suppose $u(t)$ is a step input.

$$\hat{u}(s) = \frac{1}{s} \implies \hat{y}(s) = \frac{1}{s^2} \tag{2.70}$$

$$y(t) = \mathscr{L}^{-1}[\hat{y}(s)] = t1(t) \tag{2.71}$$

$y(t)$ has an unbounded (ramp) response to a bounded (step) input, therefore BIBO unstable.

Example 2.11 BIBO stability (improper transfer function).

$$G(s) = \frac{s^2}{s+1} \tag{2.72}$$

$G(s)$ can be expanded into an improper residual and a proper transfer function.

$$G(s) = Rs + \underbrace{G'(s)}_{\text{proper}} = s + \frac{-s}{s+1} \tag{2.73}$$

Suppose $\hat{u}(s) = \frac{1}{s}$ (step input).

$$\hat{y}(s) = 1 + \frac{1}{s} \cdot \frac{-s}{s+1} = 1 - \frac{1}{s+1} \tag{2.74}$$

$$y(t) = \delta(t) - e^{-t} \tag{2.75}$$

A bounded input (step function) results in an unbounded output (impulse function). Therefore improper transfer functions are BIBO unstable.

2.1.6 Initial Values and Reverse Action

Much can be learned from an examination of the record of the response of a dynamic system. In this subsection and the next, we examine the interpretation of observables at two extremes, $t = 0^+$ and $t \to \infty$. In order to be specific, we consider the response to step inputs:

$$u(t) = 1(t) \tag{2.76}$$

and we take n as the degree of the denominator of $G(s)$ and take m as the degree of the numerator. Taken together, this means,

$$\hat{y}(s) = G(s)/s \tag{2.77}$$

Also, we write

$$G(s) = \frac{\eta_m s^m + \eta_{m-1} s^{m-1} + \cdots + \eta_1 s + \eta_0}{\delta_n s^n + \delta_{n-1} s^{n-1} + \cdots + \delta_1 s + \delta_0} \tag{2.78}$$

Theorem 2.4 (Derivative Discontinuity Theorem) *Where all initial conditions are zero, a discontinuity occurs in the $(n-m)$th derivative of the step response of a dynamic system:*

$$\frac{\mathrm{d}^{n-m}}{\mathrm{d}t^{n-m}} y(0^+) = \frac{\eta_k}{\delta_k} \tag{2.79}$$

where k is the coefficient of the highest order term in either the numerator or denominator of Eq. (2.78).

Proof: Left as an exercise to the reader. Hint: Use the initial value theorem and (2.77). ∎

The point is that where a discontinuity can be observed, the difference in degree can be determined empirically. In practice, only discontinuities in y and $\mathrm{d}y/\mathrm{d}t$ can perceived easily so that only small differences in n and m are inferred in this way. Refer to Example 2.2.

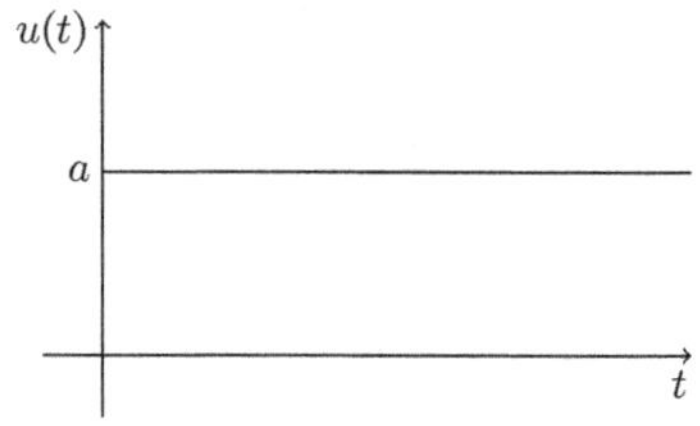

Figure 2.9 Step input.

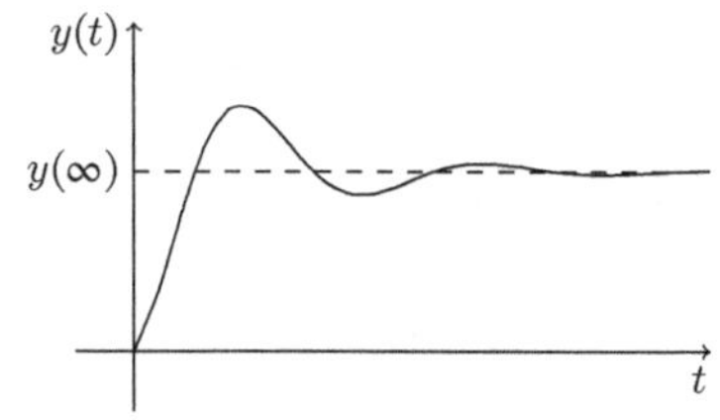

Figure 2.10 Step response.

In particular, where the transfer function is proper, $n = m$,

$$y(0^+) = \frac{\eta_k}{\delta_k} \tag{2.80}$$

In this case, a discontinuity occurs in the output.

Some systems display reverse action, which means that the initial response is in the direction opposite to that of the steady-state response. If a system is stable and strictly proper, and if $\eta_0 \geq 0$, the system displays this unusual behavior only where an odd number of zeros of the transfer function are in the right half-plane. This assertion is an immediate consequence of (2.80) because it is this case where

$$\eta_k < 0 \tag{2.81}$$

Generally, we have reverse action only where the system is non-minimum phase. Example 2.12 shows the step response of a system with non-minimum phase zero.

2.1.7 Final Values and Static Gain

The theories of automatic control laid down in this text are strongly related to the way in which systems magnify signals. The quantification of this magnification is the **gain** of the system. We begin the study of the simplest form of gain in this subsection.

Knowing the final value theorem and $G(s)$ in (2.78), we can write that the final value of a response to a step input of magnitude u_0 as:

$$y(\infty) = \frac{\eta_0}{\delta_0} u_0 \tag{2.82}$$

or

$$y(\infty) = G(0)u_0 \tag{2.83}$$

Definition 2.5 A system is said to be of type ℓ if δ_ℓ is the denominator coefficient of lowest degree which is nonzero, the ratio:

$$K = \frac{\eta_0}{\delta_\ell} \tag{2.84}$$

is called the *error constant*. If $\ell = 0$, K is called **the static gain**, if $\ell = 1$, K is called **the velocity error coefficient**, and if $\ell = 2$, K is called **the acceleration error coefficient**.

This is consistent with (2.82) since for a type-0 system and for Figures 2.9 and 2.10 below, we can write

$$\hat{y}(s) = G(s)\hat{u}(s), \ \ \hat{u}(s) = \frac{a}{s} \tag{2.85}$$

$$y(\infty) = \lim_{s\to 0} s\hat{y}(s) = \lim_{s\to 0} sG(s)\frac{a}{s} = a \cdot G(0) \tag{2.86}$$

$$K = \frac{y(\infty)}{a} = G(0) \tag{2.87}$$

Example 2.12 Static gain for minimum and non-minimum phase transfer functions:

$$\text{Minimum phase: } G(s) = \frac{s+1}{s^2+s+1} \Longrightarrow G(0) = 1 \tag{2.88}$$

$$\text{Non-minimum phase: } G(s) = \frac{1-s}{s^2+s+1} \Longrightarrow G(0) = 1 \tag{2.89}$$

$$\text{Non-minimum phase: } G(s) = \frac{s-1}{s^2+s+1} \Longrightarrow G(0) = -1 \tag{2.90}$$

The critical points are drawn for the minimum phase system in Figure 2.11 and for the non-minimum phase system in Figure 2.12.

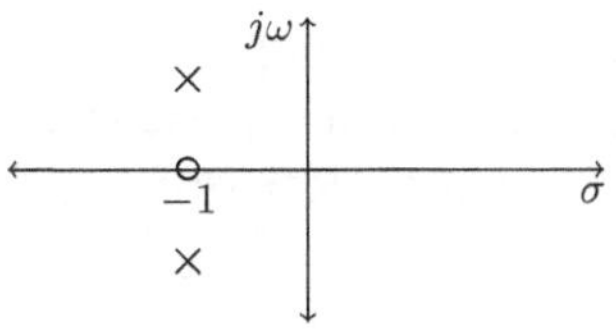

Figure 2.11 Minimum phase.

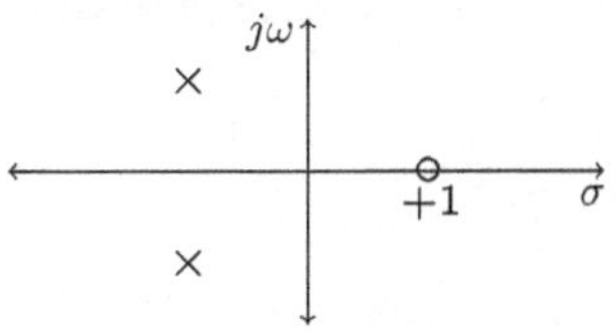

Figure 2.12 Non-minimum phase.

Consider a unit step ($a = 1$):

$G(s)$	$\frac{s+1}{s^2+s+1}$	$\frac{1-s}{s^2+s+1}$	$\frac{s-1}{s^2+s+1}$
$y(\infty)$	+1	+1	−1
$y(0^+)$	0	0	0
$\dot{y}(0^+)$	1	−1	1

The time response is drawn for the minimum phase system in Figure 2.13 and for the non-minimum phase systems in Figure 2.14.

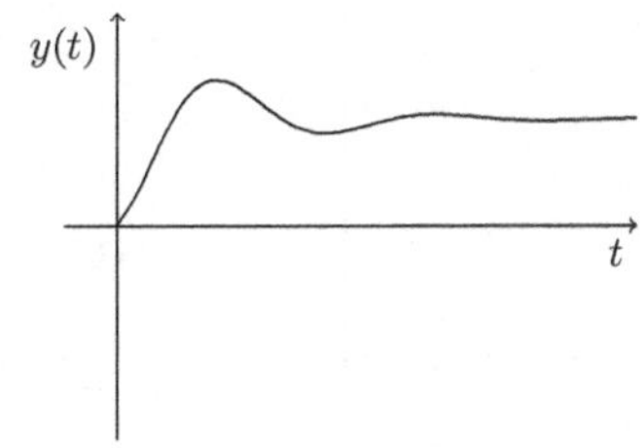

Figure 2.13 Minimum phase response.

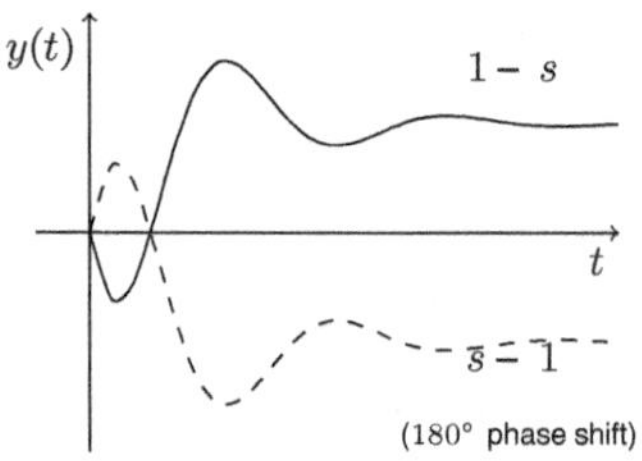

Figure 2.14 Non-minimum phase response.

Non-minimum phase transfer functions experience reverse action: the response initially moves *away* from the steady-state value. Negative static gain indicates a negative response.

2.1.8 Time Response Metrics

The performance of a dynamic system is often measured with time-domain response metrics. The responses of first- and second-order systems are described in this section. At first, it will seem that too much effort is expended in this way. However, where we lay down the details of feedback design by "loop shaping," it is clear that the developments in this section are indispensable.

2.1.8.1 First-Order System (Single-Pole Response)

Consider the transfer function $G(s)$, where K is the static gain and τ is the transfer function time constant.

$$G(s) = \frac{K}{\tau s + 1} \Longrightarrow \text{static gain } G(0) = K \tag{2.91}$$

The pole–zero diagram for $G(s)$ is show in Figure 2.15.

The system's step response can be found as follows:

$$\hat{y}(s) = G(s)\frac{1}{s} = \frac{K}{s(\tau s + 1)} \tag{2.92}$$

It is easy to verify that if the initial condition is zero, the time response to a unit step input is

$$y(t) = K(1 - e^{-t/\tau})1(t) \tag{2.93}$$

This classic response is illustrated in Figure 2.16. Notice the graphical interpretations of the time constant, τ. The initial value of the slope of this response is

$$\frac{\mathrm{d}y}{\mathrm{d}t}(t) = \frac{K}{\tau}e^{-t/\tau} \Longrightarrow \frac{\mathrm{d}y}{\mathrm{d}t}(0) = \frac{K}{\tau} \tag{2.94}$$

$$y(\tau) = K(1 - e^{-1}) \approx 0.632 \cdot K \tag{2.95}$$

So, the step response will have an initial slope K/τ, a final value of K, and will rise to 63.2% of the final value at $t = \tau$.

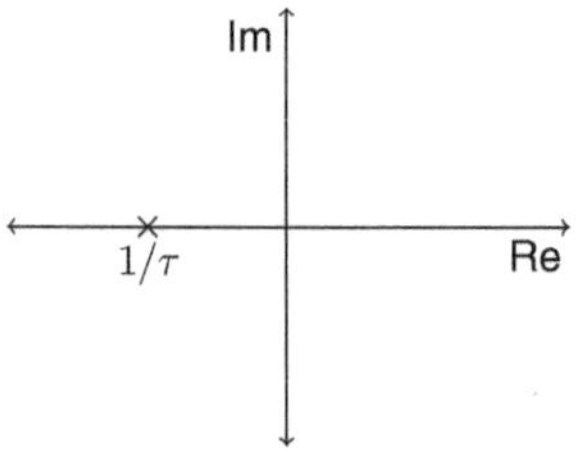

Figure 2.15 Pole–zero diagram of a first order system.

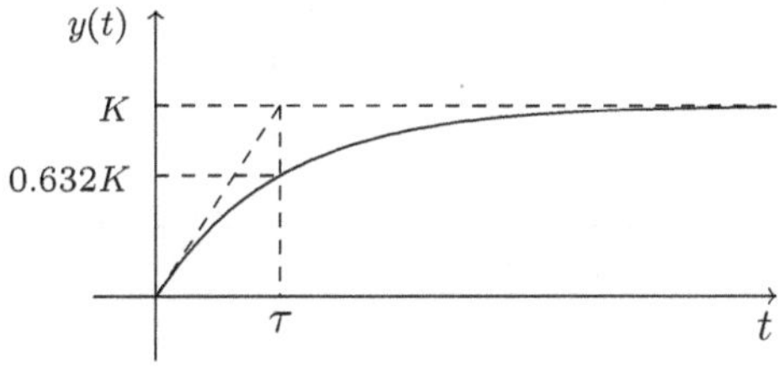

Figure 2.16 Single pole step response.

2.1.8.2 Second-Order System (Quadratic Factor)

The step response of the system under study is completely described by a static gain and a pair of complex conjugate poles. We make this study because this transfer function is, surprisingly, a useful approximation of many closed-loop systems.

Consider the second-order transfer function $G(s)$, where K is the static gain, ω_n is the undamped natural frequency, σ is the damping factor, ζ is the damping ratio, and ω is the frequency, and where ω_n, σ, ζ, and ω are related according to

$$\omega = \omega_n \sqrt{1-\zeta^2} \tag{2.96}$$

and

$$\sigma = \zeta\omega_n \tag{2.97}$$

Then,

$$G(s) = \frac{K\omega_n^2}{(s+\sigma)^2+\omega^2} = \frac{K\omega_n^2}{s^2+2\zeta\omega_n s+\omega_n^2} \Longrightarrow \text{static gain } G(0) = K \tag{2.98}$$

In this case $\zeta < 1$, so that the poles are complex conjugates. The location of these poles in the s-domain is shown in Figure 2.17. Otherwise, the $G(s)$ could be split into two first-order poles and the solution would follow the process for first-order systems. The unit step response can be found as follows:

$$\hat{y}(s) = G(s)\frac{1}{s} = \frac{K\omega_n^2}{s\left((s+\zeta\omega_n)^2+\omega^2\right)} \tag{2.99}$$

$$= \frac{R_1}{s} + \frac{R_2 s + R_3}{(s+\zeta\omega_n)^2+\omega^2} \tag{2.100}$$

(Partial fraction expansion steps not shown here.)

$$y(t) = K\left(1 - e^{-\zeta\omega_n t}\left(\cos(\omega t) + \frac{\zeta}{\sqrt{1-\zeta^2}}\sin(\omega t)\right)\right) \tag{2.101}$$

From Figure 2.17, we can see that there is a geometric relationship between the natural frequency ω_n and the damping factor σ.

$$\cos\theta = \frac{\zeta\omega_n}{\omega_n} \quad \Longrightarrow \quad \zeta = \cos\theta \tag{2.102}$$

Similarly,

$$\sin\theta = \frac{\sqrt{\omega_n^2-\omega^2}}{\omega_n^2} = \frac{\sqrt{\omega_n^2-\omega_n^2(1-\zeta^2)}}{\omega_n^2} \quad \Longrightarrow \quad \sqrt{1-\zeta^2} = \sin\theta \tag{2.103}$$

Then, Eq. (2.101) can be rewritten as:

$$y(t) = K\left(1 - \frac{e^{-\zeta\omega_n t}}{\sin(\theta)}(\cos(\omega t)\sin\theta + \sin(\omega t)\cos\theta)\right) \tag{2.104}$$

which can be further simplified to

$$y(t) = K\left(1 - \frac{e^{-\zeta\omega_n t}}{\sin(\theta)}\sin(\omega t+\theta)\right) \tag{2.105}$$

We characterize this response in Figure 2.18. By that, we mean that figures of merit of the speed of response and the damping are defined in this figure. Both "rise time" and "peak time" are measures of the speed at which response first

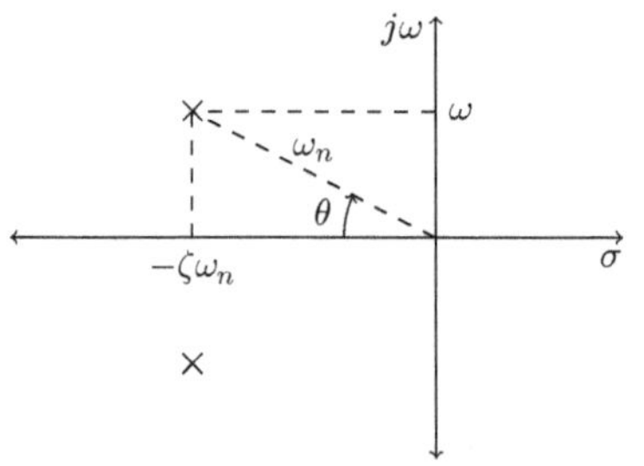

Figure 2.17 Second-order system s-domain.

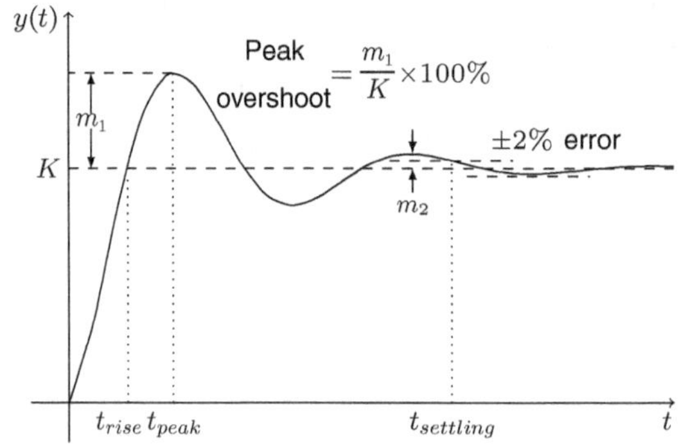

Figure 2.18 Quadratic pole step response.

approaches the steady-state value. The rise time, t_R, is the time at which $y(t) = K$ (that is, the time for $y(t)$ to first reach the steady-state value), occurring at

$$\omega t_R + \theta = \pi$$

so that

$$t_R = \frac{\pi - \theta}{\omega_n\sqrt{1-\zeta^2}} \tag{2.106}$$

A formula for the peak time, t_p, is obtained by setting $dy/dt = 0$:

$$t_p = \frac{\pi}{\omega_n\sqrt{1-\zeta^2}} \tag{2.107}$$

Where the quadratic factor is used to approximate a closed-loop system, we are interested in the degree to which the response overshoots the mark. In many cases, we hope to control this value with our choices of design parameters. The overshoot occurs at $t = t_p$, so it is easy to find that the percent overshoot to be

$$100\frac{m_1}{K} = 100\exp\left[\frac{-\zeta\pi}{\sqrt{1-\zeta^2}}\right] \tag{2.108}$$

The last three relations are plotted in Figure 2.18. Notice a curious fact, *the percent overshoot* depends only on the *damping ratio* (alternatively, upon the pole angle).

A measure of the damping is the ratio of successive maxima, a simple analysis leads to

$$\frac{m_2}{m_1} = \exp\left[\frac{-2\zeta\pi}{\sqrt{1-\zeta^2}}\right] = \frac{m_{k+1}}{m_k} \tag{2.109}$$

The similarity of the right sides of Eqs. (2.108) and (2.109) is obvious, in fact,

$$\frac{m_1}{K} = \sqrt{\frac{m_{k+1}}{m_k}}$$

Once again, a performance measure depends only on the damping ratio.

Finally, another measure of the damping is the rapidity with which the exponential envelopes of the response decay. That decay occurs with the time constant:

$$\begin{aligned}\tau &= 1/\sigma \\ &= 1/(\zeta\omega_n)\end{aligned} \tag{2.110}$$

One final step response metric for the second-order systems is

- **Settling Time**: The time at which the deviation of the response from the final value is bounded by $\pm 2\%$. Alternate definitions of the settling time may restrict or relax the error bound, e.g. a $\pm 5\%$ bound could be used in place of a $\pm 2\%$ bound. For the second-order system, the $\pm 5\%$ settling time is approximately 3τ, and the $\pm 2\%$ settling time is approximately 4τ.

These metrics are illustrated in Figure 2.18.

2.1.9 The Effect of Zeros on Transient Response

The effect of an individual, transfer function pole on the transient response of a system is made clear in the inverse Laplace transformation by Heaviside expansion in Section 1.4: each pole contributes an exponential decay term. The effect of a zero is more subtle. Obviously, the direct effect is on the residues, R_k. However, the indirect effect on the transient response can be profound. We study one example of the effect of a zero in this section.

Incidentally, the entire pattern of pole–zero location affects the value of an individual residue. For this reason, a particular pattern will provide a response unanticipated if individual poles are examined in isolation.

Consider the system described by the transfer function:

$$G(s) = \frac{K\omega_n}{\rho\zeta}\frac{s + \rho\zeta\omega_n}{s^2 + 2\zeta\omega_n s + \omega_n^2} \tag{2.111}$$

Essentially, this is a quadratic factor with a zero.

Consider the step response of this system. In particular, consider the calculation of residues. The numerator contributes a multiplicative numerator term of the form $(-p_1 + \rho\zeta\omega_n)$ to a residue of the transient response. Here, "p_1" is a complex pole of form shown in Figure 2.17. This same factor can be written in the polar form $(Pe^{j\beta})$ where P and β are given the geometric interpretation in Figure 2.19.

Where one goes through the details, it is found that the zero in (2.111) modifies Eq. (2.105) to become

$$\frac{y(t)}{K} = 1 - \frac{e^{-\sigma t}}{\sin(\theta + \beta)}\sin(\omega_d t + \theta + \beta) \tag{2.112}$$

where

$$\beta = \tan^{-1}\left[\frac{\sqrt{1-\zeta^2}}{|\rho - 1|\zeta}\right] \tag{2.113}$$

This step response is plotted in Figure 2.20 for various values of ρ but fixed value of ζ. This corresponds to holding the poles fixed and moving the location of the zero.

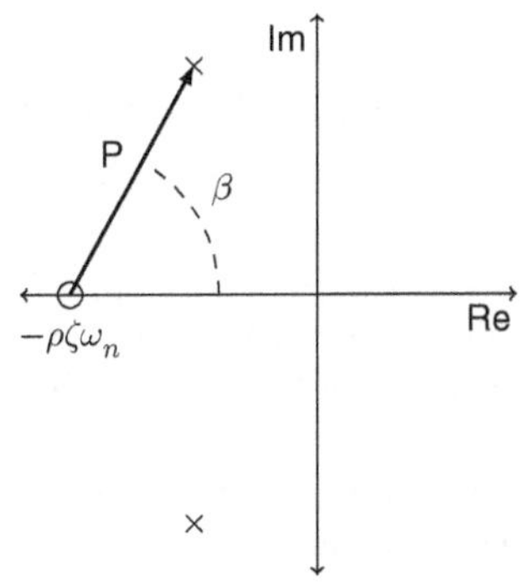

Figure 2.19 Polar form of the factor.

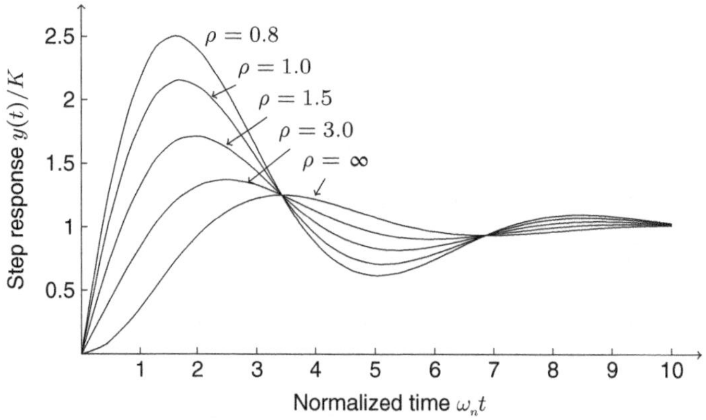

Figure 2.20 Effect of a zero on the second-order response ($\zeta = 0.4$).

Clearly, the response can be sped up by moving the location of the zero. That is, *the rise and peak times* can be favorably influenced in this way. However, some things cannot be influenced. For instance the frequency, time constant (2.110), or the damping per cycle (2.109) are independent of the zero location. It is obvious from this figure that the price one pays for smaller peak or rise time is an increased overshoot. These facts are summarized in the formulas:

$$t_R = \frac{\pi - \theta - \beta}{\omega_n \sqrt{1 - \zeta^2}} \tag{2.114}$$

$$t_p = \frac{\pi - \beta}{\omega_n \sqrt{1 - \zeta^2}} \tag{2.115}$$

$$100\frac{m_1}{K} = 100 \exp\left[\frac{-\zeta(\pi - \beta)}{\sqrt{1 - \zeta^2}}\right] \tag{2.116}$$

Time constant and damping per cycle are not the only things which are independent of the zero location. It is clear from Figure 2.20 that values of the response at a discrete set of times are invariant to the zero location! This fact is established in the next theorem. In Theorem 2.5, the limiting case of $\rho \to \infty$ corresponds to (2.105).

Theorem 2.5 *For the system characterized by a quadratic factor and a single zero, the values of the unit step response at the times:*

$$t_k = \frac{k\pi}{\omega_n \sqrt{1 - \zeta^2}}, \quad k = 0, 1, 2, 3, \ldots \tag{2.117}$$

are invariant, are independent of the zero location as characterized by ρ*. In the limiting case of* $\rho \to \infty$*, these invariants are the relative maxima and minima of the response.*

Proof: Substitute (2.117) into (2.112) to obtain the values

$$y(t_k) = K[1 + (-1)^{k+1} \exp(-\sigma t_k)] \tag{2.118}$$

which are independent of β. ■

Comment: Of course, the response invariants are not peak times for finite ρ.

2.1.10 The Butterworth Pattern

We continue our study of the effect of pattern on response with a study of a famous pole configuration, *the Butterworth pattern*. It is surprising how often this combination recurs in the control theory literature.

Definition 2.6 *The Butterworth* transfer function is a transfer function with unity gain and with poles in the *Butterworth pattern*, which is a set of poles situated in the left half-plane on a circle of radius ω_n and spaced at angles

$$\pm l\frac{\pi}{n}, \quad l = 0, 1, \ldots, \frac{n-1}{2}$$

Table 2.2 The denominator of the kth-order Butterworth transfer function, $n \leq 6$. (The numerator of the nth-order Butterworth transfer function is ω_n^n.)

n	Denominator
$n = 1$	$s + \omega_n$
$n = 2$	$s^2 + \sqrt{2}\omega_n s + \omega_n^2$
$n = 3$	$s^3 + 2\omega_n s^2 + 2\omega_n^2 s + \omega_n^2$
$n = 4$	$s^4 + 2.6\omega_n s^3 + 3.4\omega_n^2 s^2 + 2.6\omega_n^3 s + \omega_n^4$
$n = 5$	$s^5 + 3.24\omega_n s^4 + 5.24\omega_n^2 s^3 + 5.24\omega_n^3 s^2 + 3.24\omega_n^4 s + \omega_n^5$
$n = 6$	$s^6 + 3.86\omega_n s^5 + 7.46\omega_n^2 s^4 + 9.13\omega_n^3 s^3 + 7.46\omega_n^4 s^2 + 3.86\omega_n^5 s + \omega_n^6$

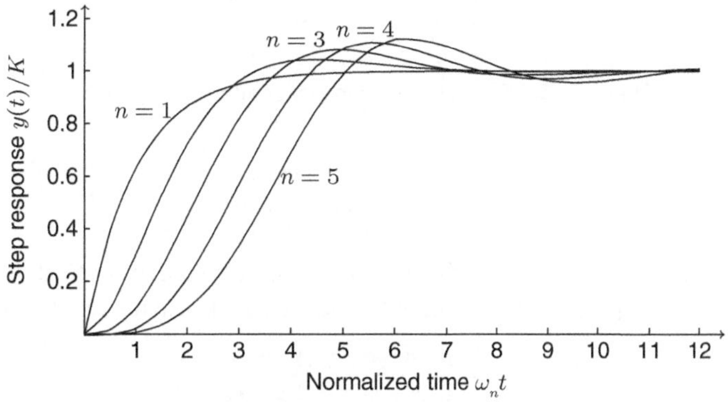

Figure 2.21 Step response of filters in the Butterworth pattern.

if n is odd and

$$\pm\frac{(l+1/2)\pi}{n}, \quad l = 0, 1, \dots, \frac{n}{2} - 1$$

if n is even. These angles are measured from the negative real axis.

Because of the importance of this pattern in control system design, we list the transfer functions for several values of n (Table 2.2).

The step responses associated with *Butterworth* transfer function are illustrated in Figure 2.21 for the first few values of n. These responses are characterized by fair speed of response with good damping per cycle and little overshoot.

2.2 Frequency Response of Dynamic Systems

For the purposes of feedback control, it is crucial to recall the theory and the representations of frequency response. For one thing, this theory provides the generalization of gain, a generalization crucial to feedback design.

2.2.1 Steady-State Frequency Response of LTI systems

Suppose a stable dynamic system is subjected to a sinusoidal input.

$$u(t) = A\cos(\omega t)1(t) \tag{2.119}$$

The system will experience both a transient and steady-state response, as shown in Figure 2.22.

We can determine the transient and steady-state responses analytically by writing the Laplace transform of the input as:

$$\hat{u}(s) = \frac{As}{s^2 + \omega^2} = \frac{As}{(s + j\omega)(s - j\omega)} \tag{2.120}$$

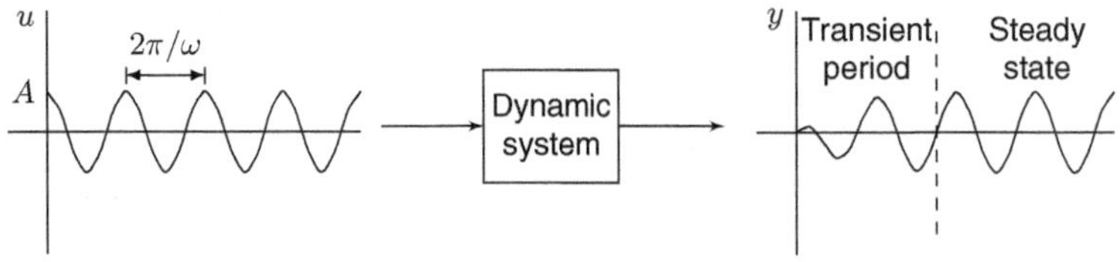

Figure 2.22 Frequency response.

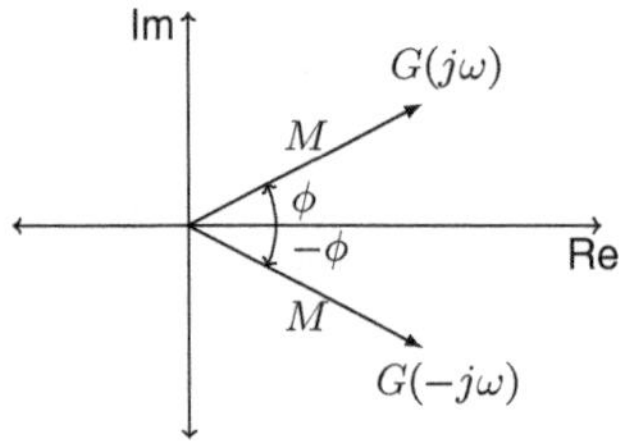

Figure 2.23 Frequency response magnitude and phase.

It follows that the output is

$$\hat{y}(s) = \frac{As}{(s+j\omega)(s-j\omega)}G(s) \tag{2.121}$$

The response is a sum of a transient term and a steady-state response. We study the steady-state response that portion of the total response which is due to the poles of the input, $\hat{u}(s)$. The Heaviside expansion of this steady-state response is

$$\hat{y}_{ss} = \frac{A}{2}\frac{G(-j\omega)}{s+j\omega} + \frac{A}{2}\frac{G(j\omega)}{s-j\omega} \tag{2.122}$$

Next suppose that $G(s)$ is a simple zero such as:

$$G(s) = s+\alpha \Longrightarrow G(j\omega) = \alpha + j\omega, \quad G(-j\omega) = \alpha - j\omega \tag{2.123}$$

Also, if $G(s)$ is a simple pole, we can write

$$G(s) = \frac{1}{s+\alpha} \Longrightarrow G(j\omega) = \frac{1}{\alpha + j\omega}, \quad G(-j\omega) = \frac{1}{\alpha - j\omega} \tag{2.124}$$

In general, $G(-j\omega)$ is equal to the complex conjugate of $G(j\omega)$:

$$G(-j\omega) = \overline{G}(j\omega) \tag{2.125}$$

This result is perfectly general for any rational function since the conjugate of the product is the product of the conjugates. It is left as an exercise to prove that the result also applies where a dead time factor is included in $G(s)$. So, in steady state,

$$\hat{y}_{ss} = \frac{A}{2}\left[\frac{\overline{G}(j\omega)}{s+j\omega} + \frac{G(j\omega)}{s-j\omega}\right] \tag{2.126}$$

We can consider $G(j\omega)$ to change the magnitude of the input by a factor of M and to shift the phase of the input by a phase angle ϕ. $G(j\omega)$ and its conjugate can then be written in polar form:

$$G(j\omega) = Me^{j\phi} \tag{2.127}$$

$$\overline{G}(j\omega) = Me^{-j\phi} \tag{2.128}$$

This is illustrated in Figure 2.23, noting that $M = M(\omega)$ and $\phi = \phi(\omega)$,

It follows that the inverse transform of the steady-state response is

$$y_{ss}(t) = MA\left(\frac{e^{-j\omega t}\cdot e^{-j\phi} + e^{j\omega t}\cdot e^{j\phi}}{2}\right) \tag{2.129}$$

That is,

$$y_{ss}(t) = M(\omega)A\cos(\omega t + \phi(\omega)) \tag{2.130}$$

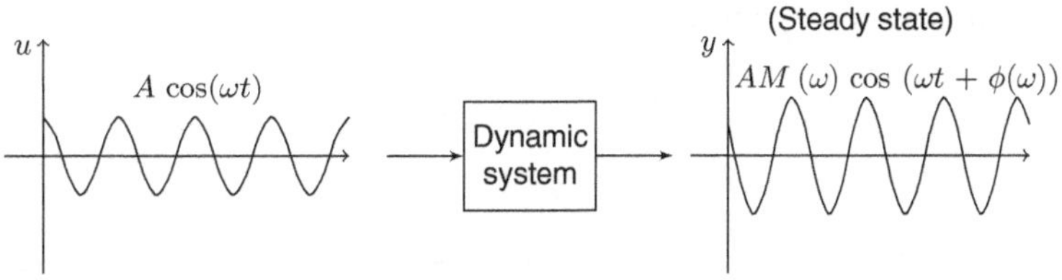

Figure 2.24 Frequency response magnitude and phase in the time domain.

For a frequency input with amplitude A, the steady-state frequency response has an amplitude of MA (where $M(\omega)$ is the gain in the amplitude) and a phase shift $\phi(\omega)$. This is illustrated in Figure 2.24.

Definition 2.7 The modulus of the transfer function evaluated on the imaginary axis, $M(\omega) = |G(j\omega)|$, is the gain and the argument of $G(j\omega)$ is the phase of the steady-state system response. $G(j\omega)$ for $0 \leq \omega \leq \infty$ is called the frequency response of the system.

What we call gain is called "amplitude ratio" in some texts. Our usage is consistent with our definition of "gain" in the previous section because in the limit $\omega \to 0$, the input to Eq. (2.119) is the unit step.

There is no logical way to fix the range of the phase angle, although there are conventions that are associated with certain specific types of transfer functions. In fact, there are an infinite number of equally valid copies of the phase angle variation each starting at a location 2π radians away from the next. We say the copies are *modulo* 2π. It is very useful to remember this fact as one tries to decipher simulation outputs and to interpret stability criteria. There is no reason to assign any more significance to one copy or another but one thing is not allowed: one cannot skip from one copy to another, one must be consistent. Unfortunately, where discontinuities occur in phase (at complex poles or zeros on the imaginary axis), computer programs can switch from one copy to another and that is an error. Simply put *be very careful of the use of software to obtain the phase angle where there are pole or zero pairs on the imaginary axis!* The frequency response of a system is so fundamental to feedback design that all computer aids to control design provide for the calculation and the plotting of $G(j\omega)$. There are a variety of useful graphical representations.

2.2.2 Frequency Response Representation

We consider three ways of graphically representing the frequency response:

1. The graphs of $20\log_{10}(M)$ and ϕ vs. $\log_{10}\omega$ is called the **Bode diagram**.
2. The graph of $20\log_{10}(M)$ vs. $\phi(\omega)$, with ω as parameter, is called the **magnitude-phase plot**. A magnitude-phase plot with two sets of certain contour plots is called a **Nichols Chart.**
3. The graph of Imag$[G(j\omega)]$ vs. Real$[G(j\omega)]$ over the positive frequency range is called the **polar plot**. The same type of plot but vs. all "frequencies" (including negative values of ω) is called a **Nyquist plot**.

Denote

$$\text{Lm } G(j\omega) \triangleq 20\log_{10}[M(\omega)] \tag{2.131}$$

and read this as the *log-magnitude* of the gain. The units of log-magnitude are called *decibels* or, more simply, dB.

The discussion in this section concludes with the detailed study of frequency response of certain canonical transfer functions.

2.2.3 Frequency Response: The Real Pole

Begin with the transfer function:

$$G(s) = \frac{1}{\tau s + 1} \tag{2.132}$$

So that the frequency response is

$$G(j\omega) = \frac{1}{1 + j\tau\omega} \tag{2.133}$$

It is a simple matter to demonstrate that the gain is

$$M(\omega) = \frac{1}{\sqrt{1+\tau^2\omega^2}} \tag{2.134}$$

while the phase is

$$\begin{aligned}\phi(\omega) &= \tan^{-1}(\Im G(j\omega)/\Re G(j\omega)) \\ &= -\tan^{-1}\left(\frac{\tau\omega}{+1}\right) \\ &= -\tan^{-1}(\tau\omega)\end{aligned} \tag{2.135}$$

These relations are useful for estimating the range of gain and phase as ω is varied from 0 to ∞. For the present case, it is clear that, over this frequency range, ϕ varies from 0 to $-90°$, while the gain varies from 1.0 (0 dB) to 0. Detailed results are obtained from standard computer software, while a qualitative understanding is gained from the study of relations (2.134) and (2.135).

It is useful to analyze the asymptotic properties of these equations by looking at the gain response on a logarithmic scale. The log of Eq. (2.134) is

$$\begin{aligned}\text{Lm } M(\omega) &= 20\log_{10}M(\omega) = 20\log_{10}(1) - 20\log_{10}\left(\sqrt{1+\tau^2\omega^2}\right) \\ &= -10\log_{10}(1+\tau^2\omega^2)\end{aligned} \tag{2.136}$$

so that, asymptotically at low frequencies, at $\omega \ll 1/\tau$,

$$\text{Lm } M \approx 0$$

Similarly, where $\omega \gg 1/\tau$,

$$\text{Lm } M \approx -\text{Lm }(\tau) - \text{Lm }(\omega)$$

The low-frequency asymptote is a straight line on the Bode diagram with a slope of 0 dB/decade, while the high-frequency asymptote has a slope of −20 dB/decade. Denote the break frequency

$$\omega_b = 1/\tau \tag{2.137}$$

so-called because the straight-line asymptotes of the log-magnitude curve appear to "break" at this frequency: the asymptotes intersect at this break frequency. Notice that the phase at the break frequency,

$$\phi(\omega_b) = -45° \tag{2.138}$$

Finally, we notice that at the break, the exact value of the gain is $M = 1/\sqrt{2}$ or

$$\text{Lm } M = -3\ \text{dB} \tag{2.139}$$

This −3 dB correction is the maximum deviation of the gain from the straight-line asymptotes.

The Bode diagrams of a single pole are shown in Figures 2.25 and 2.26. An unstable pole will produce the same decrease in slope as a stable pole. However, it will cause a 90° *increase* in phase, rather than decrease.

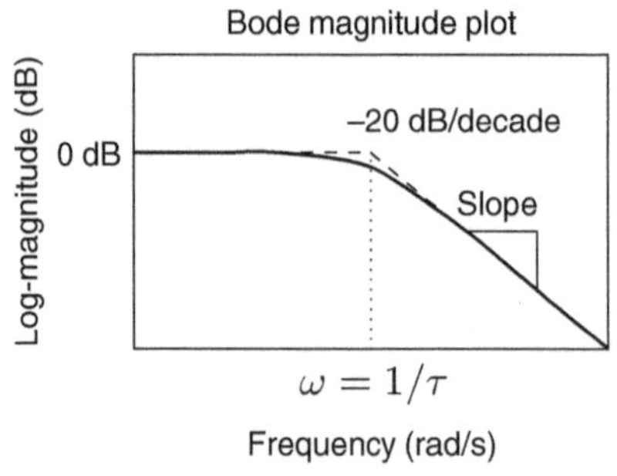

Figure 2.25 Bode magnitude plot, single pole.

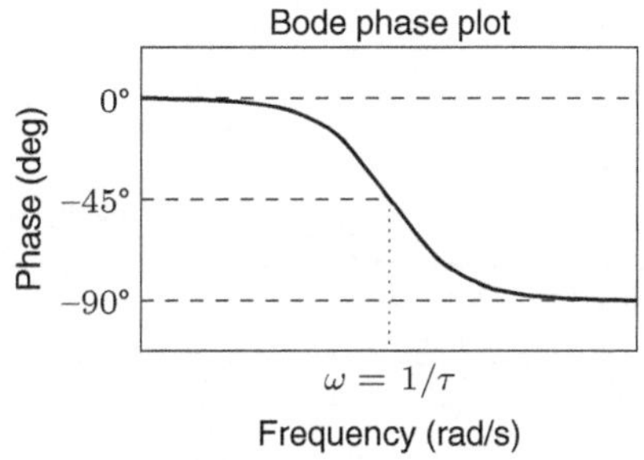

Figure 2.26 Bode phase plot, single pole.

2.2.4 Frequency Response: The Real Zero

The transfer function

$$G(s) = \tau s + 1 \tag{2.140}$$

receives attention at this point, the case of a single real zero. Clearly, the frequency response of the zero is closely related to that of the pole. An important difference is that the zero is a numerator term so an opposite sign is applied to the argument, such that, the phase ranges from 0° to +90°, a *phase advance.*

There is no violation in some principle of causality in this *phase advance* or *phase lead*, as it is called. Remember this is a steady-state frequency response and the transient build-up to the steady-state is such that there is no anticipatory action on the part of the system. Also, remember that for any version of a phase curve, there are equally valid copies at multiples of $\pm 2\pi$, including copies that are interpreted as phase *lag*. It may be that we are misled by terms like "lag" or "lead" because these phase curves are steady-state phase relations and so it is only phase *difference* that can be interpreted in any meaningful way.

Assume, at first, that the zero is in the left half-plane, such that $s = -1/\tau$. Then for a frequency input $s = j\omega$, we have

$$G(j\omega) = 1 + j\tau\omega \tag{2.141}$$

Using the analogy with the pole to quickly outline the main features of the zero frequency response,

$$M(\omega) = \sqrt{1 + \tau^2\omega^2} \tag{2.142}$$

Thus,[1]

$$\phi(\omega) = \tan^{-1}\left(\frac{\tau\omega}{1}\right) \tag{2.143}$$

So one valid copy of the phase curve is in the first quadrant:

$$\phi(\omega) = (0 \leq \phi \leq 90^\circ) \tag{2.144}$$

Additionally, it is conventional to look at the magnitude response on a logarithmic scale. The log of Eq. (2.142) is

$$\text{Lm } M(\omega) = 20\log_{10}\left(\sqrt{1 + \tau^2\omega^2}\right) \tag{2.145}$$

The magnitude formula is graphed in Figure 2.27, and the phase formula is graphed in Figure 2.28. In Figure 2.28, a left half-plane zero is referred to as *minimum phase*, while a zero in the right-half plane is called *nonminimum phase.*

The *non-minimum phase zero*, the zero in the right half-plane has a crucially different frequency response. This is also illustrated in Figure 2.28, while Figure 2.29 shows the zero locations in the *s*-plane. The difference is in the phase curves: first of all, notice that the *positive gain* version of the right half-plane zero transfer function is

$$G(s) = -\tau s + 1 \tag{2.146}$$

Thus, in the case of the non-minimum phase zero

$$\phi(\omega) = \tan^{-1}\left(\frac{-\tau\omega}{+1}\right) \tag{2.147}$$

---

1 A good habit is to write the arctan function as a function with two arguments, a numerator and a denominator. To not do so can lead to error in the composition of a valid copy of the phase curve.

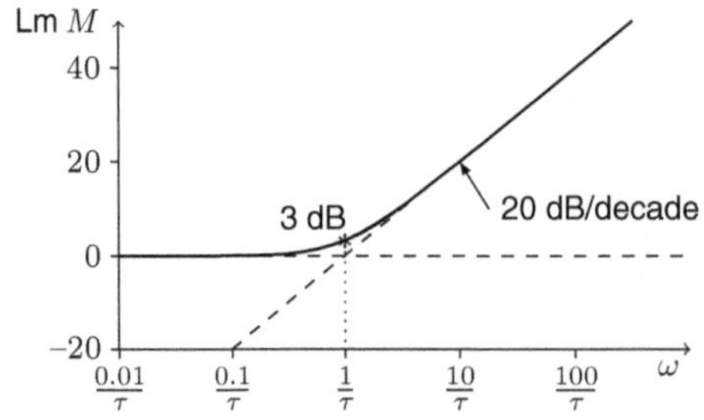

Figure 2.27 Magnitude response for a zero.

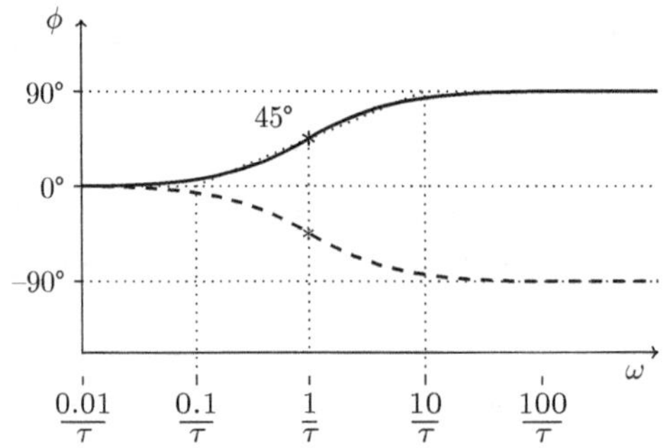

Figure 2.28 Minimum (solid) and non-minimum (dashed) phase response for a zero.

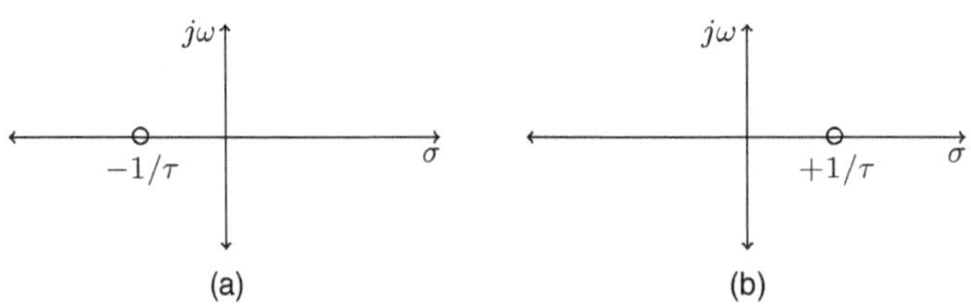

Figure 2.29 (a) Minimum and (b) non-minimum phase zeros.

So that the valid copy of the phase curve is in the *fourth quadrant*:

$$\phi(\omega) = (-90^\circ \leq \phi \leq 0) \tag{2.148}$$

Thus, the phase for the minimum phase zero is less *negative* than that for the non-minimum phase zero. It is for this reason that the terms minimum and non-minimum phase are used. (Clearly, all this a bit of a *misnomer* since a more precise terminology would be "minimum phase-lag zero.")

The gain curves are the same in the two cases, the slope of the log-magnitude curve is 0 dB/decade up to the break frequency, $\omega_b = 1/\tau$, after which the slope increases to +20 dB/decade. The gain correction at this "corner" frequency is +3 dB and the phase is +45° for minimum phase and −45° for non-minimum phase.

Incidentally, where we write a right half-plane zero as:

$$G(j\omega) = -1 + j\tau\omega$$

we have

$$\phi(\omega) = \tan^{-1}\left(\frac{+\tau\omega}{-1}\right) \tag{2.149}$$

So one valid copy of the phase curve is in the third quadrant:

$$\phi(\omega) = (90^\circ \leq \phi \leq 180^\circ) \tag{2.150}$$

However, this is the negative gain version of the transfer function.

2.2.5 Frequency Response: The Quadratic Factor

The quadratic factor of (2.151) is a polynomial factor of many physical transfer functions. The detailed study of the quadratic factor is so important because of its usefulness as an approximation for many closed-loop systems. The frequency response serves as both the explanation and the formulation of this approximation.

Consider a transfer function that is the second-order complex pair of poles:

$$G(s) = \frac{\omega_n^2}{s^2 + 2\zeta\omega_n s + \omega_n^2} \tag{2.151}$$

now substitute $s = j\omega$,

$$G(j\omega) = \frac{\omega_n^2}{\omega_n^2 - \omega^2 + 2\zeta\omega_n j\omega} \tag{2.152}$$

where ω_n is the natural frequency and ζ is the damping ratio as before. Then,

$$M(\omega) = \frac{\omega_n^2}{\sqrt{(\omega_n^2 - \omega^2)^2 + 4\zeta^2\omega_n^2\omega^2}} \tag{2.153}$$

and the phase

$$\phi(\omega) = \tan^{-1}\left(\frac{2\zeta\omega_n\omega}{\omega_n^2 - \omega^2}\right) \tag{2.154}$$

The frequency response of the quadratic factor is illustrated in Figure 2.30 for several different values of pole angle, $\theta = \cos^{-1}\zeta$. Notice the *resonance*, the local maximum of the gain curve. We investigate this effect by *maximizing M* which is the same as minimizing the term under the radical of the denominator of (2.153). Set the derivative of this term, with respect to ω, to zero in order to obtain a formula for the *resonant frequency*, ω_r:

$$-4\omega_r(\omega_n^2 - \omega_r^2) + 8\zeta^2\omega_n^2\omega_r = 0$$

The nonzero solution of this equation is

$$\omega_r = \omega_n\sqrt{1 - 2\zeta^2} \tag{2.155}$$

This frequency is real (resonance actually occurs) only if

$$\zeta \leq \frac{1}{\sqrt{2}} \quad \text{or} \quad \theta \geq 45^\circ \tag{2.156}$$

There is an indication of this fact in Figure 2.30. Notice that resonant frequency is neither the damped natural frequency, ω_d, nor the undamped natural frequency, ω_n. Where the resonant frequency is substituted back into Eq. (2.153), the result is the value of the local gain maximum, the resonant peak:

$$M_p = M(\omega_r) = \frac{1}{2\zeta\sqrt{1 - \zeta^2}} \tag{2.157}$$

This relation is plotted in Figure 2.30 and is fundamental to the approximation of closed-loop systems with a quadratic factor. This idea is developed in Chapter 3. In contrast, we have

$$M_{\omega_n} = \frac{1}{2\zeta} \tag{2.158}$$

It is left as an exercise to show that the break frequency is

$$\omega_b = \omega_n$$

The phase is -90° at this frequency and ranges between the asymptotic values of 0° and -180°. The slope of the low-frequency gain asymptote is 0 dB/decade, while that of the high-frequency asymptote is −40 dB/decade. When these asymptotes are taken together with (2.158), we have one quick procedure for an empirical determination of the damping ratio from the frequency response curve: *Measure the magnification at a phase angle of −90° and take the damping ratio to be one-half the reciprocal of this magnification.*

Definition 2.8 The cross-over frequency, ω_c is that frequency where the modulus is 1 of the return ratio, that is

$$\text{Lm } G(j\omega_c) = 0 \text{ dB} \tag{2.159}$$

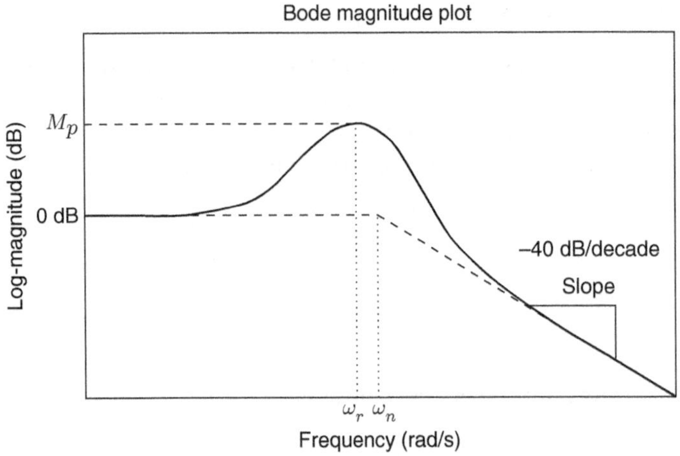

Figure 2.30 Bode magnitude plot, second-order pole.

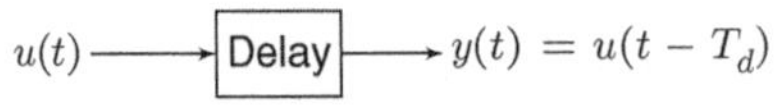

Figure 2.31 Block diagram of time delay.

It is left as exercise to show that, where the static gain is equal to 1.0, the cross-over frequency for the quadratic factor is given by:

$$\omega_c = \sqrt{2}\omega_r \tag{2.160}$$

In general, the cross-over point is varied by simply varying the static gain.

2.2.6 Frequency Response: Pure Time Delay

Consider a dynamic system which represents a pure time delay (also known as dead time), as shown in Figure 2.31.

The impulse response of this system can written as:

$$g(t) = \delta(t - T_d) \tag{2.161}$$

so that

$$y(t) = g(t) * u(t) \tag{2.162}$$

then using Laplace transform,

$$\hat{y}(s) = \mathscr{L}[y(t)] = \underbrace{e^{-sT_d}}_{G(s)} \hat{u}(s) \tag{2.163}$$

Substituting $j\omega$ for s,

$$G(j\omega) = e^{-j\omega T_d} = \cos(\omega T_d) - j\sin(\omega T_d) \tag{2.164}$$

so that the gain and phase are

$$M(\omega) = 1 \tag{2.165}$$

$$\phi = \tan^{-1}\left(\frac{-\sin(\omega T_d)}{\cos(\omega T_d)}\right) = -\tan^{-1}\left(\tan\left(\omega T_d\right)\right) = -T_d\omega \tag{2.166}$$

as shown in Figures 2.32 and 2.33. Pure time delay is an all-pass filter: the transfer function has a gain of 1 (or 0 dB) for all frequencies. The phase angle decreases linearly as a function of frequency (appearing exponential on a semilog plot). The main feature here is the continually decreasing phase (the increasing phase lag).

We demonstrate in the next chapter that excessive phase lag is the cause of instability in feedback control systems. It is for that reason that non-minimum phase zeros and dead time are primary sources of challenge for the control engineer. Think of things this way: poles contribute phase lag but at a decreasing amplitude ratio (tolerable), dead time contributes phase

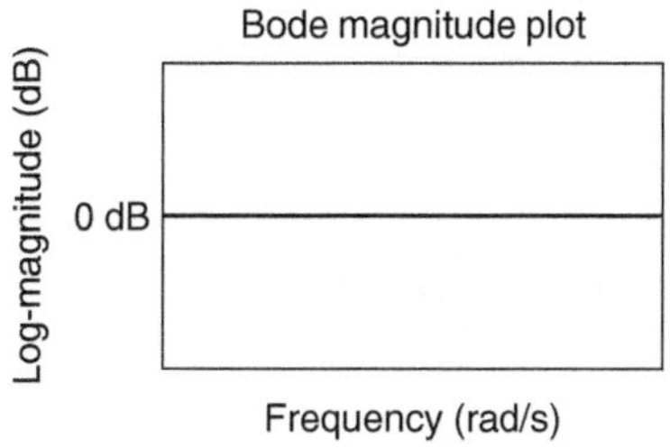

Figure 2.32 Bode magnitude plot, time delay.

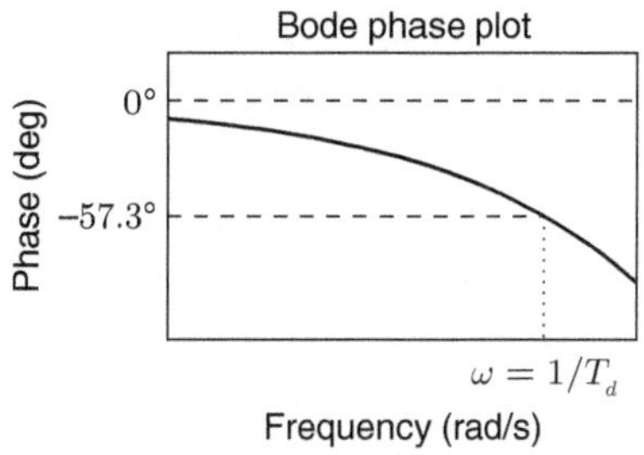

Figure 2.33 Bode phase plot, time delay.

lag at a constant amplitude ratio (bad), while non-minimum phase zeros contribute phase lag at an increasing amplitude ratio (the worst). Minimum phase zeros contribute phase advance (good) at an increasing amplitude ratio.

Considering (2.164), dead time is a nonlinear and irrational function. However, for control system design of LTI dynamic systems, the dead time is required to be linearized and rationalized. The best approximation for this rationalization was developed by Henri Padé in 1890, using the Taylor series expansion. Padé's method approximates functions as a ratio of two polynomial functions, the coefficients of which depend on the approximated function and the order of the approximation:

$$f(x) \approx R_{m,n}(x) = \frac{p_0 + p_1x + p_2x^2 + \cdots + p_mx^m}{q_0 + q_1x + q_2x^2 + \cdots + q_nx^n} \tag{2.167}$$

The full derivation of the Padé approximation is out of the scope of this text. In the case of pure time delay, we must consider the approximation of the function e^x. The approximation uses $n = m$ as:

$$e^x = \frac{e^{x/2}}{e^{-x/2}}$$

so that the degree of the numerator and denominator polynomials are the same. The first-, second-, third-, fourth-, and fifth-order approximations of e^x are shown below [11].

$$e^x \approx R_{1,1}(x) = \frac{1 + \frac{1}{2}x}{1 - \frac{1}{2}x} \tag{2.168}$$

$$e^x \approx R_{2,2}(x) = \frac{1 + \frac{1}{2}x^2 + \frac{1}{12}x}{1 - \frac{1}{2}x^2 + \frac{1}{12}x} \tag{2.169}$$

$$e^x \approx R_{3,3}(x) = \frac{1 + \frac{1}{2}x + \frac{1}{10}x^2 + \frac{1}{120}x^3}{1 - \frac{1}{2}x + \frac{1}{10}x^2 - \frac{1}{120}x^3} \tag{2.170}$$

$$e^x \approx R_{4,4}(x) = \frac{1 + \frac{1}{2}x + \frac{3}{28}x^2 + \frac{1}{84}x^3 + \frac{1}{1680}x^4}{1 - \frac{1}{2}x + \frac{3}{28}x^2 - \frac{1}{84}x^3 + \frac{1}{1680}x^4} \tag{2.171}$$

$$e^x \approx R_{5,5}(x) = \frac{1 + \frac{1}{2}x + \frac{1}{9}x^2 + \frac{1}{72}x^3 + \frac{1}{1008}x^4 + \frac{1}{30\,240}x^5}{1 - \frac{1}{2}x + \frac{1}{9}x^2 - \frac{1}{72}x^3 + \frac{1}{1008}x^4 - \frac{1}{30\,240}x^5} \tag{2.172}$$

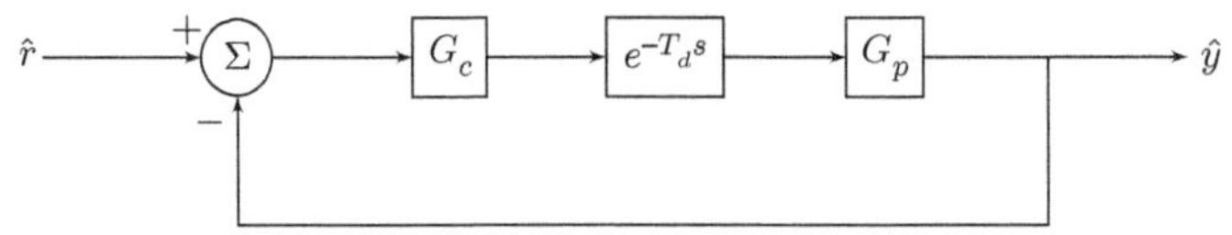

Figure 2.34 Time day in a block diagram.

It is important to note that additional coefficients cannot simply be "tacked on" to obtain a higher-order system. Rather, the values of the coefficients change with the number of coefficients used.

In the case of pure time delay, we have $x = -T_d s$. By making this substitution to an approximation of e^x to obtain $e^{-T_d s}$, time delay is approximated as a linear transfer function. For instance, a first-order approximation of delay is

$$e^{-T_d s} \approx \frac{1 - \frac{T_d}{2}s}{1 + \frac{T_d}{2}s} \tag{2.173}$$

while a second-order approximation is

$$e^{-T_d s} \approx \frac{1 - \frac{T_d}{2}s + \frac{T_d^2}{12}s^2}{1 + \frac{T_d}{2}s + \frac{T_d^2}{12}s^2} \tag{2.174}$$

Higher-order approximations increase the accuracy but add more complexity to the control system design.

Note that these approximations are all-pass filters: they are gain neutral (have no affect on the gain of the feedback loop), regardless of frequency. However, delay introduces an increasing phase lag, as stated earlier, that can drastically degrade the stability margins or even cause instability in a poorly designed feedback system (linear approximations introduce −180° phase shift per order of approximation), so it is very important to account for pure time delay in control system design.

In general, time delay is treated as lag between the controller and the actuator, as shown in Figure 2.34.

Example 2.13 Consider a plant that experiences a pure time delay of $T_d = 0.1$ seconds and has a single pole with time constant $\tau = 0.1$.

$$G_p = e^{-0.1s}\frac{1}{0.1s + 1} \tag{2.175}$$

Then, the first- and second-order approximations of G_p are as follows:

$$G_{p,1} = \frac{20 - s}{s + 20} \cdot \frac{1}{0.1s + 1} \tag{2.176}$$

$$G_{p,2} = \frac{s^2 - 60s + 1200}{s^2 + 60s + 1200} \cdot \frac{1}{0.1s + 1} \tag{2.177}$$

The true system and both approximate systems were simulated, and the results are shown in Figure 2.35.

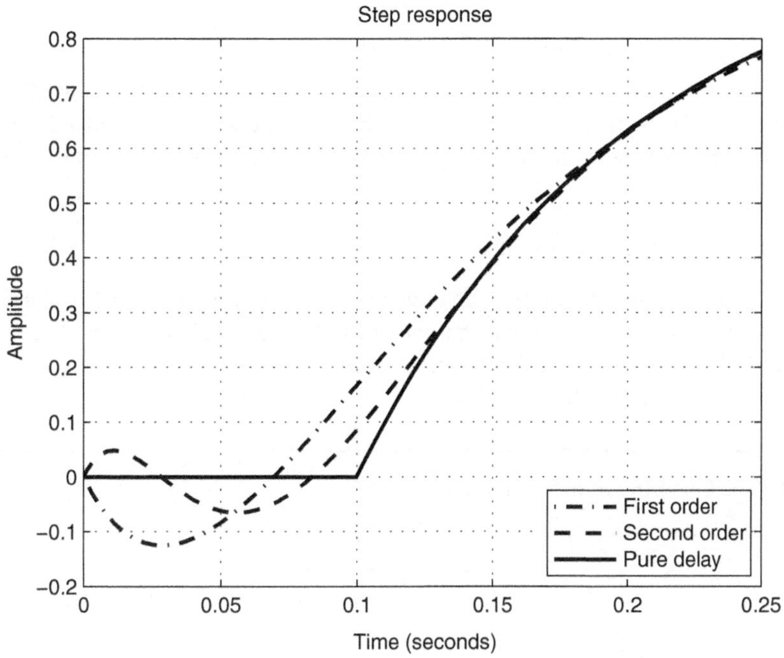

Figure 2.35 Various delay approximations.

2.2.7 Frequency Response: Static Gain

We complete the catalog of frequency response components with the nearly trivial:

$$G(s) = K \tag{2.178}$$

a pure gain. It is obvious that

$$\text{Lm } G(j\omega) = \text{Lm } K, \quad \phi \equiv 0° \tag{2.179}$$

2.2.8 Frequency Response: The Composite Transfer Function

The general transfer function can be written as:

$$G(s) = G_1(s)G_2(s)G_3(s) \dots G_n(s) \tag{2.180}$$

where each $G_k(s)$ is one the items cataloged above. Then, the frequency response

$$G(j\omega) = G_1(j\omega)G_2(j\omega)G_3(j\omega) \dots G_n(j\omega) \tag{2.181}$$

is obtained by superimposing the results obtained to this point. It is no accident that the Bode and magnitude-phase diagrams are structured in such a way that these superpositions can be fully exploited.

Rewrite the basic relation in polar form:

$$\begin{aligned} G(j\omega) &= M_1 e^{(j\phi_1)} M_2 e^{(j\phi_2)} M_3 e^{(j\phi_3)} \dots M_n e^{(j\phi_n)} \\ &= M_1 M_2 M_3 \dots M_n e^{[(j(\phi_1+\phi_2+\phi_3\cdots+\phi_n)]} \end{aligned} \tag{2.182}$$

Therefore, because $\log_{10}(\Pi M_k) = \sum \log_{10} M_k$,

$$\text{Lm } G(j\omega) = \text{Lm } G_1(j\omega) + \text{Lm } G_2(j\omega) + \text{Lm } G_3(j\omega) \cdots + \text{Lm } G_n(j\omega) \tag{2.183}$$

and

$$\phi(\omega) = \phi_1 + \phi_2 + \phi_3 + \cdots + \phi_n \tag{2.184}$$

Clearly, we have a superposition of both the gain and the phase curves on both Bode and Nichols diagrams. Thus, the catalog of curves delineated in the previous sections allows one to sketch the frequency response of a composite transfer function. These sketches are useful as checks to software and, more importantly, as intuitive predictors of system performance and of feedback requirements.

2.2.9 Frequency Response: Asymptote Formulas

It is often sufficient, even useful, to use formulas for the frequency response asymptote rather than a formula for the Bode magnitude itself. It is an easy matter to obtain the asymptote formula and simply apply the following rule to each factor of the form $(\tau s + 1)$ in both the numerator and denominator of the transfer function. Where $1/\tau < \omega$, replace the factor $(\tau s + 1)$ by $\tau\omega$ and where $1/\tau > \omega$, replace the factor by 1.

Example 2.14 Consider the following transfer function:

$$G(s) = \frac{\left(\frac{1}{3}s + 1\right)}{(s+1)\left(\frac{1}{10}s + 1\right)} \tag{2.185}$$

the asymptote formulas are indicated below:

Asymptote formula	Frequency range (rad/s)
1	$\omega < 1$
$1/\omega$	$1 < \omega < 3$
$1/3$	$3 < \omega < 10$
$10/(3\omega)$	$10 < \omega$

A check is that the asymptotes are continuous at the corner frequencies.

2.2.10 Physical Realizability

It is important to obtain results which indicate the range of possibilities for real, physical transfer functions. The most satisfactory results are obtained by imposing a constraint of "cause and effect": the output of a physical system cannot anticipate the input. In Chapter 1, we use the very definition of the limit to demonstrate that instantaneous limit processes, such as differentiation, are physically unrealizable in this sense. (Some limit processes, such as integration, do not require a device that can foresee the future.)

A consideration of frequency response allows one to anticipate the more rigorous result of *the Paley–Wiener Theorem*, which was briefly discussed previously. Intuitively, it seems impossible for a physical system to have increasing gain with increasing frequency: if this sort of thing were possible, the power requirements would be enormous. This leads us to speculate that:

> The degree of the numerator of a transfer function of a physically realizable dynamic system is never greater than the degree of the denominator.

Notice that we satisfy this constraint wherever formula (1.79) is used: state equations (1.75) meet expectations for physical realizability. Something negative can be said about those formulations of the state equations from which improper transfer functions are calculated.

The same sort of reasoning leads some to argue that a physical system must be strictly proper (the *D*-matrix is zero), since it is hard to imagine that a real system can pass a very high frequency without some attenuation. That is fair enough, but this conclusion is not supported by the more rigorous realizability theory based on causality principles and complex analysis.

2.2.11 Non-minimum Phase, All-Pass, and Blaschke Factors

We present a factorization of a rational function which is useful to us in the discussion of feedback control. In particular, this factorization is used to generalize methods developed for *minimum phase*, controlled objects to the *non-minimum phase* case.

Definition 2.9 A system is called *all-pass* if the Bode magnitude of the transfer function is identically a constant over all frequencies.

Example 2.15 It follows from (2.165) that dead time is an all-pass element.

Example 2.16 The non-minimum phase transfer function

$$G(s) = \frac{p - s}{s + p} \tag{2.186}$$

is all-pass as is the unstable transfer function

$$G(s) = \frac{s + p}{p - s} \tag{2.187}$$

Definition 2.10 A rational function either of form (2.186) or of form (2.187) is called a *Blaschke factor*.

Any transfer function can be factored into minimum phase factor and an all-pass factor, where the all-pass factor is a dead time transfer function of a set of Blaschke factors. This obvious result is demonstrated with a simple example:

$$G(s) = \frac{1 - s}{s + 2} = \frac{(s + 1)\,(1 - s)}{(s + 2)\,(s + 1)}$$

We can write this symbolically as:

$$G(s) = G_{mp}(s)G_{ap}(s) \tag{2.188}$$

The minimum phase part, $G_{mp}(s)$, is formed by reflecting all poles and zeros of $G(s)$ in the right half-plane about the imaginary axis into the left half-plane. The Bode magnitude of $G(s)$ is then the Bode magnitude of G_{mp} because G_{ap} is all-pass. In this way, we can imagine that the non-minimum phase factors contribute only to the phase curve.

Where the right half-plane zero is complex, we prefer to write the Blaschke factor as:

$$G(s) = \frac{(z - s)}{(s + \overline{z})} \tag{2.189}$$

and similarly for a right half-plane pole. That is to say, in such cases, the reflection is imagined to be about both the real and the imaginary axes.

2.3 Frequency Response Plotting

The control engineer must be able to estimate the response of the dynamic system from the appearance of the pole–zero diagram of the transfer function. The basic ideas for that sort of estimate are set down in this chapter. These estimates become especially valuable in the following chapters, where it is demonstrated that even high-order, very complicated feedback systems can be made to respond much like the simpler systems described in such detail in this chapter. In fact, it is this eventuality that motivates our detailed investigations here.

The trick to make a complicated feedback system respond much like a simpler system is to properly “shape” the magnitude and phase curves of the feedback system frequency response. As it turns out, this type of consideration is secondary to other more important aspects of control system design, but the primary objectives are obtained from frequency response

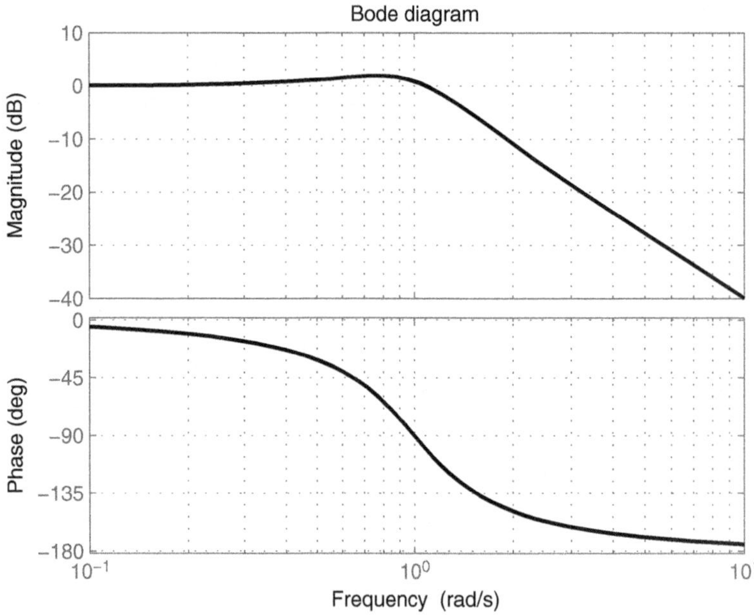

Figure 2.36 Bode Plot.

shaping as well. That is, the steady-state frequency response developed in this chapter is among the most valuable analytic tools of the control engineer.

At this point, we must develop and justify the claims of this summary and, to do that, we turn to Chapter 3, the most interesting and the most important chapter of this volume.

2.3.1 Matlab Codes for Plotting System Frequency Response

2.3.1.1 Bode Plot

These plots can be drawn in Matlab using the bode command, resulting in the plot shown in Figure 2.36.

```
≫ bode(num, den) % (vectors of numerator,
  % denominator coefficients)
OR
≫ bode(A,B,C,D,1) % (state space representation)
OR
≫ bode(G) % (transfer function with s=tf('s') operator)
OR
≫ [MAG,PHASE,w] = bode(num,den);
≫ LM=20*log10(MAG);
≫ semilogx(w,LM)
≫ semilogx(w,PHASE)
```

2.3.1.2 Polar Plot/Nyquist Diagram

The polar plot, shown in Figure 2.38, has the gain of the system as the radial coordinate and the phase as the angular coordinate. A Nyquist diagram, shown in Figure 2.37, plots the real and imaginary parts over the range of positive and negative frequencies. These plots are useful for the stability analysis.

```
≫ nyquist(num,den) % Or other input types
OR
≫ [RE,IM] = nyquist(num,den);
≫ plot(RE,IM)
```

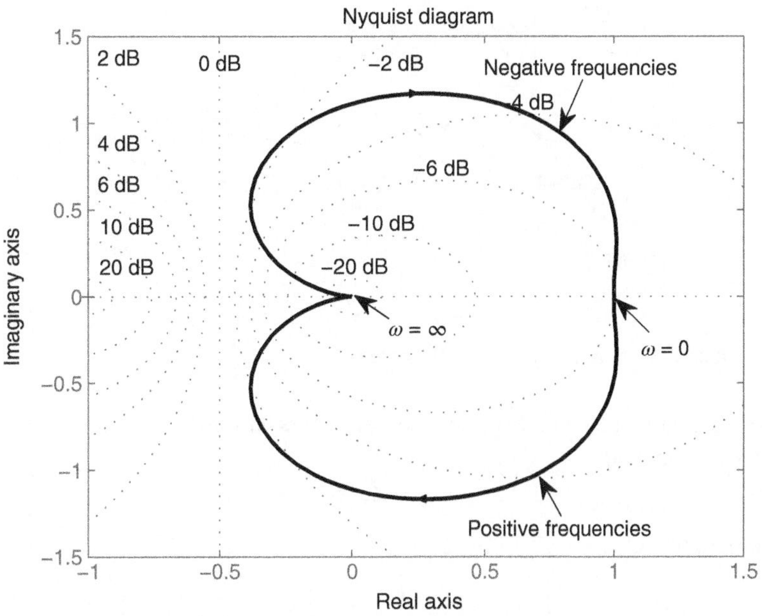

Figure 2.37 Nyquist Plot. The dotted lines show the closed-loop frequency response for the open-loop locus that is drawn.

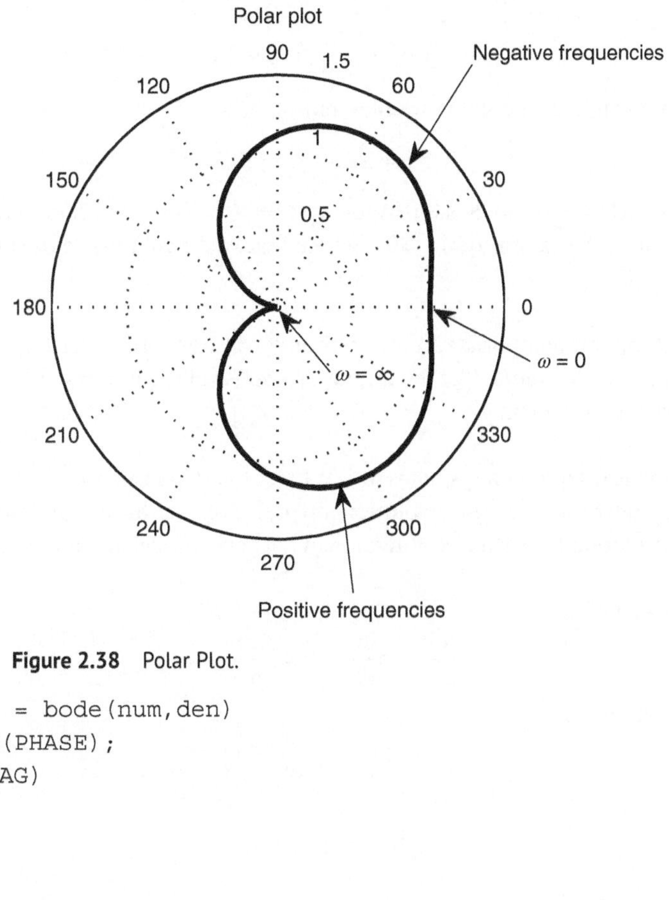

Figure 2.38 Polar Plot.

```
≫ [MAG,PHASE,w] = bode(num,den)
≫ PHASE=deg2rad(PHASE);
≫ polar(PHASE,MAG)
```

2.4 Problems

2.1 Using Heaviside (partial fraction) expansion to find $y(t)$, where

$$\hat{y}(s) = \frac{s^2+1}{s^2-1} \tag{2.190}$$

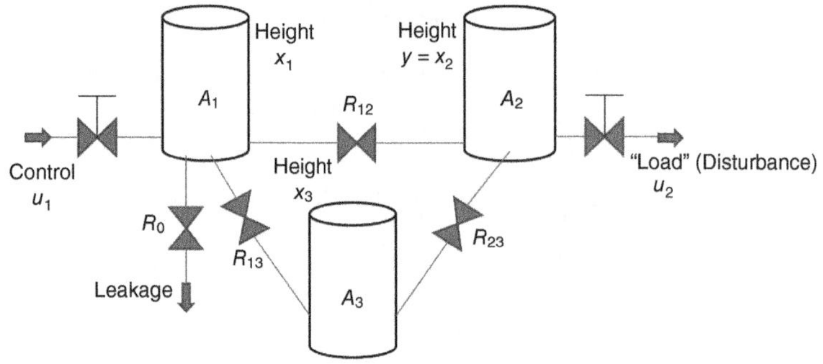

Figure 2.39 Three-tank hydraulic system.

2.2 The hydraulic system of Figure 2.39 will displace pole–zero cancellation whenever the tank areas are equal and the hydraulic resistances are matched:

$$A_k = A,\ R_{i,k} = R \tag{2.191}$$

In this case, the standard model is

$$a = \begin{bmatrix} -\frac{2}{\tau} - \frac{1}{\tau_0} & \frac{1}{\tau} & \frac{1}{\tau} \\ \frac{1}{\tau} & -\frac{2}{\tau} & \frac{1}{\tau} \\ \frac{1}{\tau} & \frac{1}{\tau} & -\frac{2}{\tau} \end{bmatrix}, \quad b = \begin{bmatrix} \frac{1}{A} & 0 \\ 0 & \frac{1}{A} \\ 0 & 0 \end{bmatrix} \tag{2.192}$$

$$a = \begin{bmatrix} 0 & 1 & 0 \end{bmatrix}, \quad b = \begin{bmatrix} 0 & 0 \end{bmatrix} \tag{2.193}$$

Here, the tank heads are taken to be the state variables and

$$\tau = RA, \quad \tau_0 = R_0 A \tag{2.194}$$

Determine the response $y(t) = x_2(t)$ to a step input $u_1(t) = 1(t)$. Take $\tau = 2.0$ seconds, $\tau_0 = 10.0$ seconds, and $A = 10.0\text{ft}^2$. Plot your results. Repeat for $u_2(t) = 1(t)$ (notice that you must rewrite the b-vector here). Plot both step responses on the same graph.

2.3 Find a state-variable description appropriate to the sixth-order Butterworth filter (type ≫ `help TF2SS, help BUTTER` in Matlab). Take $\omega_n = 5/6$ rad/s. Plot the impulse response of the Butterworth filter. Remember to put your system in the continuous time s-domain.

2.4 Suppose the impulse response of a certain system is te^{-t}. What is the step response? Do not use the Laplace transform here! What is the transfer function? Use the convolution integral to obtain the output if the input is e^{-t}. Sketch a graph of both factors in the convolution integrand as a function of t. Check this calculation by using the Laplace transform.

2.5 Consider the differential equation

$$\frac{d^2y}{dt^2} + 4\frac{dy}{dt} + 4y = \frac{d^2u}{dt^2} + 4u \tag{2.195}$$

where the initial conditions are

$$y(0^-) = 0, \quad \frac{dy}{dt}(0^-) = 0 \tag{2.196}$$

and the input is

$$u(t) = \cos(2t)h(t) \tag{2.197}$$

where $h(t)$ is the unit step function.

a) Obtain $\hat{y}(s)$ and use the initial and final value theorems to estimate initial and final values and the initial slope of $y(t)$.
b) Obtain $y(t)$ and verify the results of part (a).

2.6 The goal in this exercise is to give you a chance to become more familiar with Matlab. To start with, consider a dynamic system described by the state-variable equations

$$\frac{dx}{dt} = \begin{bmatrix} -2 & -1 & 0 \\ 1 & 0 & 0 \\ 0 & 1 & 0 \end{bmatrix} x + \begin{bmatrix} 1 \\ 0 \\ 4 \end{bmatrix} u + \begin{bmatrix} 0 \\ 0 \\ 2 \end{bmatrix} d \tag{2.198}$$

$$y = \begin{bmatrix} 0 & 0 & 1 \end{bmatrix} x \tag{2.199}$$

Enter these matrices in Matlab and call them A, B, C, and D. Notice that you need two B-matrices, one for $u(t)$ and one for $d(t)$. Call them B_u and B_d. Use Matlab to determine the transfer functions $\hat{y}/\hat{u}$ and $\hat{y}/\hat{d}$. Ask for the help-screen on SS2TF. Determine the poles, zeros, and gains of these transfer functions. A natural choice is to apply the `ROOTS` command to the numerator polynomial to determine the zeros, but this is often inaccurate: we recommend you use `TZERO` or, even better, use `SS2ZP`.

3

Feedback Principals

We develop a mathematical theory of feedback design in this chapter. The plan is that the principles are laid down in this chapter, based on which specific design methodologies are developed in the following chapters. The theory is made for continuous-time applications since these procedures carry over, with little modification, to the discrete-time control. Of course, the theory of discrete-time control requires additional ideas and techniques.

We start with a dynamic system that we call the "controlled object" or the "object." (Another common name is the "plant," apparently, a contribution of the chemical process control industry.) This dynamic system has an input, $u(t)$, and an output, $y(t)$. The input is always assumed to be limited in some way, say a power limitation or some other type of limitation on its magnitude. The design problem is to place a feedback loop around the controlled object and improve its performance without exceeding the limitations of the input signal, $u(t)$.[1] The *limitation on the input to the object is the essence of the problem* and this should be kept in mind at all times when designing feedback control systems. Think of it this way: Without a limitation of the input, a feedback loop can be affected with a little tinkering and there would be little need for someone trained in engineering science. We call this a *feedback control systems.*

We begin with a review of the reasons for using feedback. A special effort is made to not overemphasize secondary considerations in the design of a control system, say the design for improved transient response. Instead, we emphasize the main points, the design of feedback (i) to attenuate signals input to the feedback loop, and (ii) to provide the capability to perform well in spite of inevitable imperfections in our models of dynamic systems.

By "attenuation of signals," we mean the attenuation of every input signal, at least at some nodes in the feedback loop. It is demonstrated that the attenuation of some signals may be achieved with the use of very low gain, while other signals are attenuated with the use of very *high gain*. Obviously, there are trade-offs in this regard. It is demonstrated that this single notion can be used to achieve all of the primary goals of feedback design. Further, it is shown that high gain is also desirable for "performance robustness," that is, the continued good performance of the closed loop in the face of inadequacy or uncertainty in the mathematical models.

The limitation on $u(t)$ is paramount but there are other fundamental constraints on feedback design. There are stability constraints and the limitation on *bandwidth* (that is, a limitation on the speed of response of a closed-loop relative to that of the open loop). Other feedback design trade-offs are delineated:

1. some signals can be attenuated only at the expense of not attenuating others,
2. assured stability is achieved at the expense of performance robustness,
3. feedback is achieved with the loss of gain and with the introduction of another unwanted signal, the sensor noise.

The primary mathematical objects in our developments are the poles and zeros of the closed-loop transfer functions and, more significantly, the "frequency response" functions, the frequency-dependent magnitude and phase of the transfer functions. The stage is set for succeeding chapters by introducing the notion of *parameterization*. In order to improve the performance of a feedback loop, the designer interjects another dynamic system into the loop, the dynamic component called a "compensator" or "controller." In this chapter, it is demonstrated that all feasible compensators belong to a single family which is characterized by a single representative called a "parameter." This idea has an antiquity but it is only in the last quarter of the twentieth century that its true value was appreciated. Parameterization is the focus or point of departure of all the remaining chapters in our text. For example, it is demonstrated, in the following chapter, that, essentially, all SISO

1 Nowadays, the feedback loop is becoming an inherent part of a dynamic systems, and these dynamic systems simply do not operate without some level an automation such as mechatronics, cyberphysical, and so on.

Robust Control: Youla Parameterization Approach, First Edition. Farhad Assadian and Kevin R. Mallon.

Companion Website: www.wiley.com/go/assadian/robust

design problems are solved with the use of this parameterization. The only case where we must abandon this notion is that case where the compensator is required to be of a special form, say a PID controller. Although, even in this special case, constraints could be imposed to enforce controller structure through the use of optimization.

3.1 The Value of Feedback Control

Why use feedback in the first place? A general feedback control system is shown in Figure 3.1.

It is always possible to use an open-loop system to accomplish many of the same goals often touted for feedback control. If one has achieved a feedback design merely to achieve stability or to speed up the response (such as for an optimized performance index, etc.), someone else could perform a computer simulation and record $u(t)$, the input to the controlled object. An open-loop synthesis can then be achieved in the manner indicated in Figure 3.2.

Incidentally, Figure 3.2 is drawn for a linear, time-invariant system, but the idea applies equally well to any type of control, say a nonlinear, time-varying, digital control.

Note, in passing, that the equivalent open loop has some attractive features, the sometimes-costly sensor is not required and the sensor noise problem is completely avoided. In fact, open-loop control is an important way to realize an "automation" in certain instances. It is interesting to note that any method that is used to design a feedback control for a linear, time-invariant controlled object can also be employed to design a open-loop scheme (refer to Section 3.4 in this chapter).

We shall take as axiomatic that the primary goal of feedback design is to achieve those things that the equivalent open loop cannot do for us. It must be clear that the open-loop controller is blinded to disturbances and will operate properly only if the model used in the simulation is very close to the actual dynamics of the controlled object we eventually synthesize.

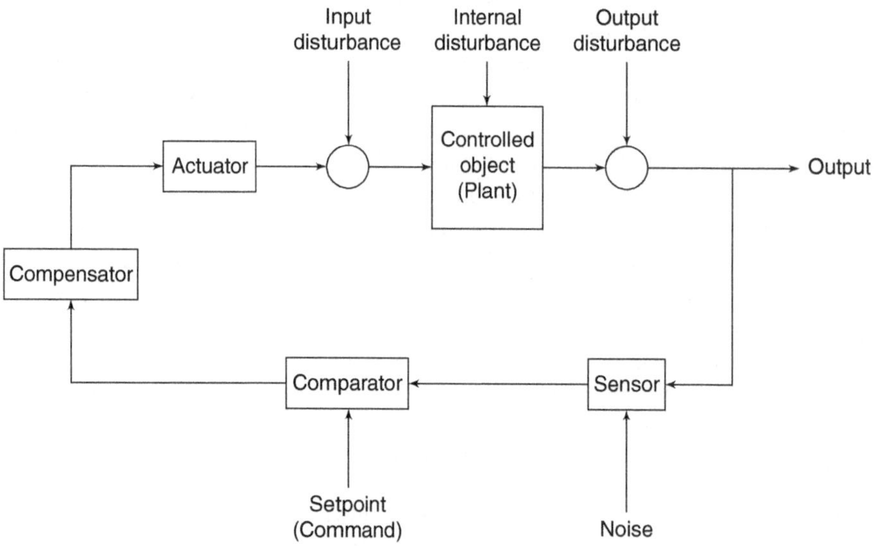

Figure 3.1 Feedback controller block diagram.

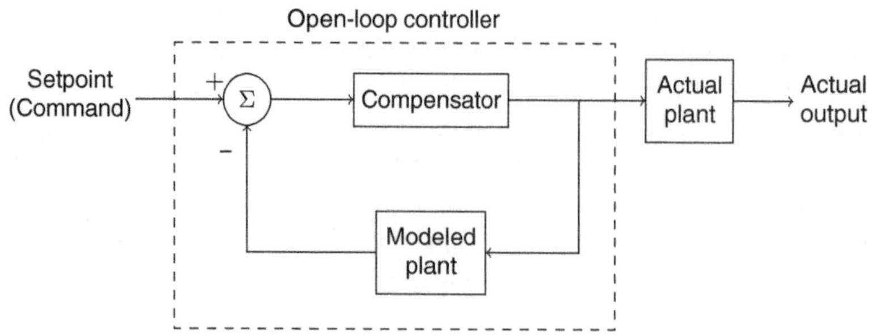

Figure 3.2 Open-loop synthesis.

> The primary goal of feedback design is to attenuate the effects of unwanted inputs and to provide, nearly, the desired response in spite of inadequacies in the model or in spite of changes in physical parameters. Horowitz[2] describes feedback design as the management of uncertainty.

"Feedback," then, is

> the information required to attenuate the unwanted inputs and to adjust for the mismatch of our model with reality.

There are other very good reasons for using feedback but we shall take damping, stability, or optimality, for example, to be secondary goals because such things can be achieved equally well, theoretically, with the recording the input and playing it back into the open loop, as shown in Figure 3.2.

Incidentally, the open loop is a useful way to achieve design goals in certain instances. For example, the equivalent open loop is the way some manufacturing processes (some "automated" machining) are controlled. The open loop works here because the disturbance signals (workpiece applied to the tool) are predictable in all their particulars and the model of the machine tool is accurate enough for most purposes.

An open-loop control (feedforward control) can be combined with a feedback control in an effective overall design; that is, a feedback loop is used to attenuate unwanted inputs and to provide a feedback loop which does vary with parameter change in the object and then an open-loop compensator can be used to improve transient performance. A prescribed speed of response, say, is often achieved with an open-loop compensator after all other objectives are met with a feedback loop.

3.1.1 The Advantages of the Closed Loop

To be more specific, we propose the following list as the main considerations in the use of feedback in control systems.

1. To eliminate the necessity of human intervention. This goal can be due to the incentive to reduce costs through the replacement of labor, or due to a need to control the object in an exotic, remote, or dangerous environment, or to achieve a reliability and accuracy uncommon to human operation. Finally, the need may be a simple desire to relieve humans of burdensome or tedious tasks.
2. To reduce the responses of the controlled object to unwanted external disturbances. One type of disturbance, the "sensor noise," is actually introduced by the feedback designer (the first requirement for achieving feedback control is to provide a measurement device).
3. To desensitize the performance of the system to variations in the dynamics of the controlled object. These variations are unforeseen and unplanned for and are due to the inadequacies of the mathematical model or due to the actual variation of physical parameters (such as friction constants) during the lifetime of the system. A property of a feedback system that is insensitive to parameter variation is said to be *robust*. That is, we speak of *robust stability* or *robust performance*. An interesting aspect to this is that the robustness of one property of a feedback system, say stability, can only be improved at the expense of another, say robust performance.
4. To improve the dynamic response of the controlled object. Particular improvements are the damping and the speed with which the steady state is obtained. It is theoretically possible to speed up the response of a feedback control system but practical limitation, what we call the *bandwidth constraint*, is severe and one is more confident with the improvements in speed achieved with a redesign of the controlled object itself (an increase in power or some other "size" or capability).

This is not to say that stability and damping are to be ignored in feedback design: to be clear on this point, consideration (4) may be crucial for the controlled process itself but, theoretically at least, it can be achieved without the use of feedback and so it is a secondary consideration in the feedback design. Our purpose in presenting this list is to put these things into proper order of priority.

---

2 **Isaac Horowitz** (15 December 1920–2005) was a notable scientist with significant contributions to automatic control theory. He developed and championed the quantitative feedback theory which for the first time introduced a formal combination of the genuine frequency methodology founded by Hendrik Bode with plant uncertainty considerations.

3.2 Closed-Loop Transfer Functions

The simplest and most common feedback arrangement is illustrated in Figure 3.3. We refer to this configuration as the single-loop feedback. There are many other useful configurations; a few others are mentioned here and those are introduced throughout this section.

Take particular notice of the terminology and notation defined in this figure. In particular, notice that the controlled object is that system with input $u(t)$ and output $y(t)$. We intend with this to conform to the state variable notation which is previously introduced. Normally, the controlled object is a "given" in a design and is thought of as the equipment that is fixed by other considerations such as power generation and power transmission. The usual procedure is for a set of design criteria to fix the controlled object using the rules of thermodynamics, or mechanics, and then call in the control engineer to manage the uncertainty and make it work. At least, this is the traditional approach.

The compensator and, perhaps, the prefilter are added to the control system in order to achieve the feedback design. The task of the feedback designer is to obtain the transfer functions of the compensators, $G_c(s)$, and, possibly, $P(s)$ or even $H(s)$. The transfer function of the controlled object, $G_p(s)$, may contain the dynamics of actuators or sensors. At other places, it may be convenient to allocate these effects to other blocks, say the sensor dynamics to the block denoted $H(s)$.

In Figure 3.4, the "disturbance," $\hat{d}_y$, is shown to be injected at the output of the controlled object. This is an appropriate depiction where, say, the disturbance is a "load" such as the variation of tire forces due to different road conditions or the application of a part in a machining operation. We may also indicate a disturbance $\hat{d}_u$. This is appropriate where, say, the disturbance is at the actuator. (Note that we have $P \equiv 1$ in Figure 3.4.)

Another kind of disturbance is the noise in the sensor, $\hat{n}$. A unique feature of this input is that it occurs after the output and should be thought of as the difference between the actual and measured outputs. Also, notice that "noise" is present only after we insert the sensor for purposes of feedback control:

> The addition of sensor noise is a price we pay for feedback control.

The feedback system in which $H(s) \equiv 1$ is called *unity feedback* is shown in Figure 3.5. The reader is admonished to carefully note which developments in this chapter apply to unity feedback and which do not.

Other common terminologies are: The controlled object is commonly called the "plant," especially by those in process control engineering. The command signal, $r(t)$, is sometimes called the "reference target" or, in more limited usage, the "set point." We often refer to the manipulated variable, $u(t)$, as the "controlled input."

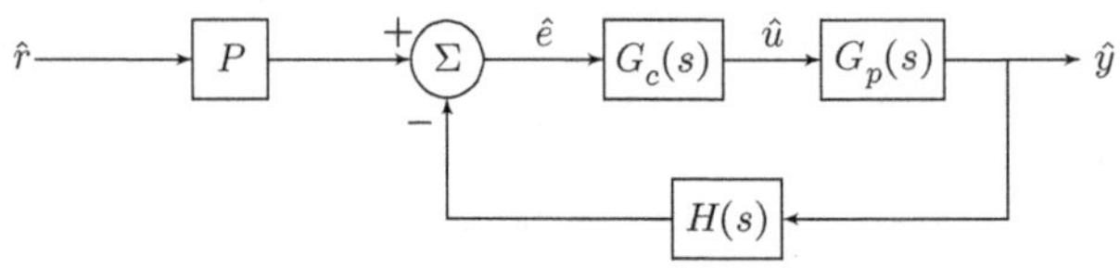

Figure 3.3 Single-loop feedback with prefilter P, controller G_c, plant G_p, and sensor H.

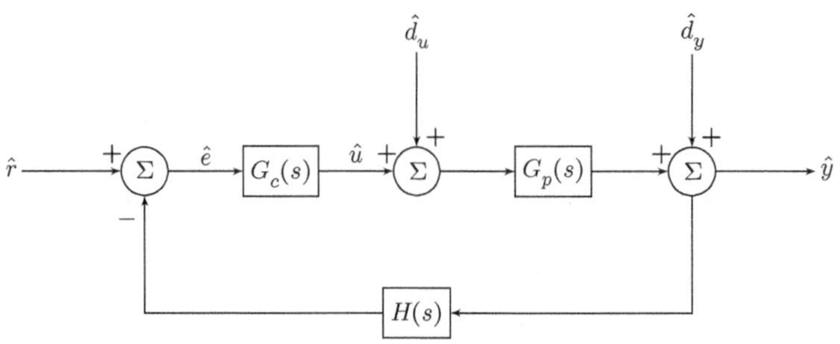

Figure 3.4 Single-loop feedback with disturbances.

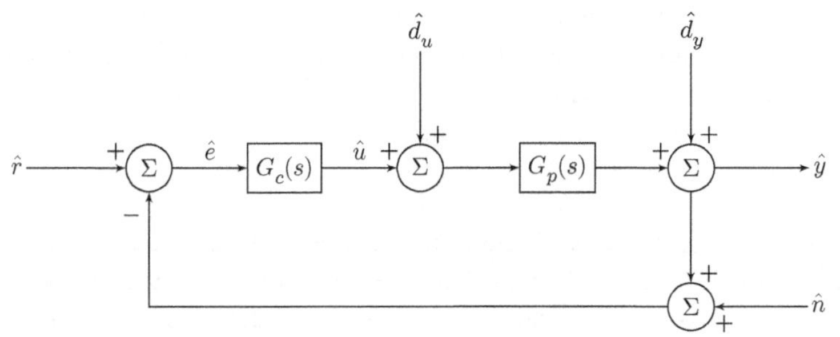

Figure 3.5 Single-loop feedback with unity feedback and with disturbances and noise.

Definition 3.1 The number of compensators and prefilters is called the degree of freedom of the feedback design.

3.2.1 The Return Ratio

The first order of business is to relate closed-loop transfer functions to the open-loop transfer functions. Our device for that purpose is the principle of superposition and the concept of the return ratio.

Definition 3.2 Any individual point in a feedback loop is called a *node*. The return ratio, denoted $L(s)$, is associated with a node and is obtained by imagining a cut made in the loop at the node, the injection of a signal, denote it $\hat{\imath}(s)$, and the measurement of the returned signal. We define the return ratio to be the negative ratio (the transfer function) of the returned signal to the injected signal. Thus, the returned signal $\hat{v}(s)$ may be written as:

$$\hat{v}(s) = -L(s)\hat{\imath}(s) \tag{3.1}$$

The return ratio is also known as the open-loop transfer function because it is the transfer function from input (injected signal) to output (negative returned signal) where the loop is cut or "opened up." Other names are common such as loop transmission or loop transfer function.

We illustrate an imagined experiment in Figure 3.6 for a feedback system. The imagined cut occurs at the manipulated variable. It is obvious that the returned signal:

$$\hat{v}(s) = G_c \cdot (-1) \cdot H(s) \cdot G_p(s) \cdot \hat{\imath}(s)$$

so that the return ratio is

$$L(s) = G_p(s)H(s)G_c(s) \tag{3.2}$$

A moment of reflection is all that is needed to establish that all nodes in a single-loop feedback system have the same return ratio, which is the product of all individual transfer functions in the loop.

3.2.2 Closed-Loop Transfer Functions and the Return Difference

We begin with the **terminology**: By an input to a control system, we mean any autonomous signal injected into the loop. Commands, disturbances, and sensor noise are all inputs.

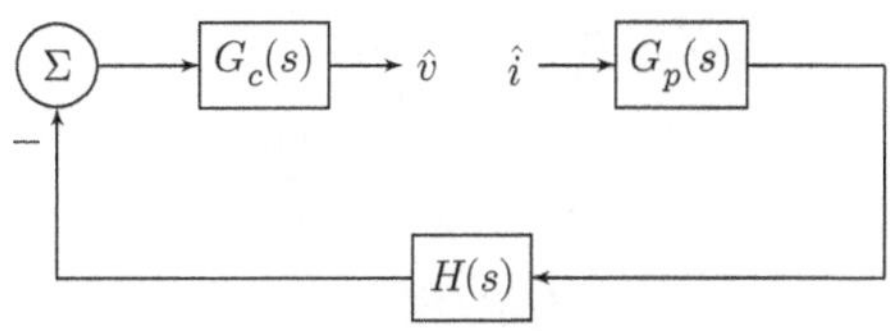

Figure 3.6 Block diagram for determining return ratio.

We wish to characterize the response of any node to any input by the transfer function between any such pair. A transfer function of this kind is given the obvious name, *closed-loop transfer function*. Use $\hat{O}(s)$ to denote some node variable and $\hat{I}(s)$ to denote some input signal. We seek the formula for the closed-loop transfer function

$$\frac{\hat{O}(s)}{\hat{I}(s)}$$

We could imagine that $\hat{I}$ be any input (e.g. $\hat{r}$, $\hat{d}$, or $\hat{n}$) and $\hat{O}(s)$ be any potential output (e.g. $\hat{e}$, $\hat{u}$, or $\hat{y}$).

The guiding principal here is superposition: The node variable is the sum of the signal coming directly from input and the returned signal. By definition, the returned signal is $-L(s)$ times the node variable so that

$$\hat{O}(s) = G_I(s)\hat{I}(s) - L(s)\hat{O}(s) \tag{3.3}$$

where $G_I(s)$ is the product of transfer functions in the *direct path* from the input $\hat{I}$ to the node (output) $\hat{O}$, the path in the direction of signal flow. Hence,

$$\frac{\hat{O}(s)}{\hat{I}(s)} = \frac{G_I(s)}{1 + L(s)} \tag{3.4}$$

This is a remarkably simple and general relationship between open-loop and closed-loop transfer functions. The term $1 + L$ occurs in the denominator of all closed-loop transfer functions and so we give it a name, *the return difference*. The return difference obtains its name from the fact that it is the transfer function $\frac{\hat{i}-\hat{o}}{\hat{i}}$ in the imagined experiment where the loop is cut, Figure 3.6. That is, $1 + L$ is the transfer function from the injected signal to the difference between the injected and returned signal.

3.2.3 Sensitivity, Complementary Sensitivity, and the Youla Parameter

Classical feedback design procedures focused on the return ratio, the open-loop transfer function, in order to achieve closed-loop objectives. Though seemingly indirect, these procedures are quite effective and receive some attention in Chapter 4. However, the more recent (neoclassical) procedures focus directly on the closed-loop transfer functions themselves. These procedures are drawn out throughout this text, but this is a good place to introduce some definitions and propositions fundamental to both design methods.

In addition, we encounter a closed-loop transfer function with a simple appearance but profound significance, the "Youla parameter."

Without the loss of generality, we assume unity feedback, i.e.

$$H(s) \equiv 1 \tag{3.5}$$

Then, the return ratio can be simplified to

$$L(s) = G_c(s)G_p(s) \tag{3.6}$$

Definition 3.3 The sensitivity transfer function is the reciprocal of the return difference:

$$S(s) \triangleq \frac{1}{1 + L(s)} \tag{3.7}$$

Definition 3.4 The complementary sensitivity function is

$$T(s) \triangleq \frac{L(s)}{1 + L(s)} \tag{3.8}$$

Definition 3.5 The *Youla* parameter

$$\begin{aligned} Y(s) &\triangleq \frac{G_c}{1 + L(s)} \\ &= \frac{1}{1 + L(s)} \times \frac{L(s)}{G_p(s)} \\ &= G_c(s)S(s) \\ &= T(s)/G_p(s) \end{aligned} \tag{3.9}$$

All of these names are justified in the ensuing parts of this chapter. In our developments, $Y(s)$ will play an especially fundamental role.

Notice that Figure 3.5 can be used to show that

$$\frac{\hat{y}(s)}{\hat{n}(s)} = -T(s) \tag{3.10}$$

and

$$\frac{\hat{e}(s)}{\hat{r}(s)} = S(s) \tag{3.11}$$

Many other closed-loop relationships are summarized in Table 3.1.

Notice that the complementary sensitivity can be written as:

$$T(s) = \frac{1}{1 + 1/L(s)} \tag{3.12}$$

The denominator, $1 + L^{-1}$, is called the inverse return difference. This may be an unfortunate convention because it would seem, intuitively, that "inverse return difference" should refer to $(1 + L)^{-1}$.

The sensitivity function earns its name in the section on the Bode robustness measures and the complementary sensitivity earns its name there as well. The complementary aspect of these transfer functions is already hinted in the next simple calculation. Consider the obvious equation:

$$\frac{1 + L(s)}{1 + L(s)} = 1$$

which means

$$\frac{1}{1 + L(s)} + \frac{L(s)}{1 + L(s)} = 1$$

We have established a basic constraint:

Proposition 3.1 *In a one-degree-of-freedom feedback system*

$$S(s) + T(s) = 1 \tag{3.13}$$

For example, it is demonstrated in the coming section that the designer will attempt to arrange $T(j\omega) \approx 0$ in order to achieve good sensor noise attenuation at the output. Where that is achieved, a necessary side effect is that $S(j\omega) \approx 1$ and as is demonstrated further on, that means the closed loop will not attenuate the effects of disturbance at the output in this frequency range. Equation (3.13) establishes an unavoidable trade-off in feedback design.

The general shape of S, T, and Y is drawn in Figure 3.7.

Equation (3.13) establishes that there is a constraint on $S(s)$ and $T(s)$. We must be careful, however, as $S(s)$ and $T(s)$ are complex. As shown in Figure 3.8, having $S + T = 1$ does **not** mean that the *magnitudes* of S and T sum to 1. That is, $|T| + |S|$ is not necessarily 1.

Table 3.1 Important rational functions for characterizing closed-loop performance.

Transfer function	From...	To...
$S = 1/(1 + L)$	r	e
	d_y	y
$T = L/(1 + L)$	r	y
	$-n$	y
$Y = T/G_p = G_c S$	r	u
	$-n$	u
	$-d_y$	u

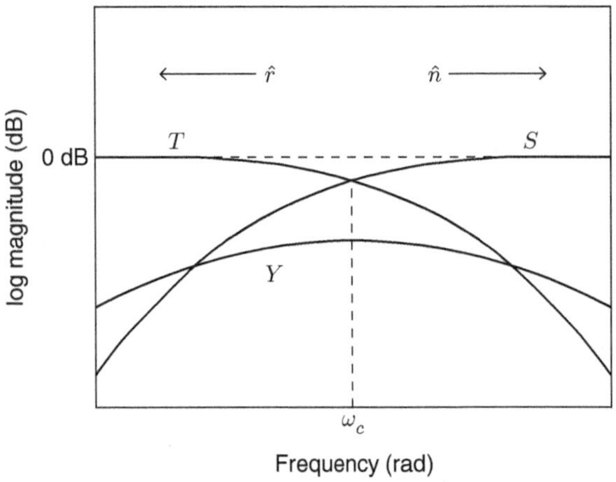

Figure 3.7 Bode magnitude plot for S, T, and Y. Controller commands $\hat{r}$ operate in the low-frequency range, while noise $\hat{n}$ tends to be high frequency. S and T intersect at the crossover frequency, ω_c.

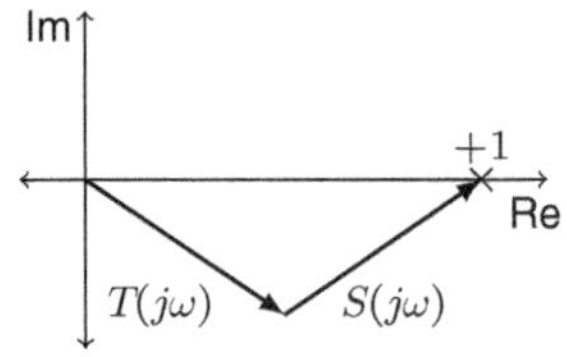

Figure 3.8 S and T relationship.

Additionally, we can see from Eqs. (3.6), (3.8), and (3.9) that

$$T = \frac{L}{1+L} = \frac{G_c G_p}{1+L} = \frac{G_c}{1+L} \cdot G_p \quad \Rightarrow \quad T = Y G_p \tag{3.14}$$

This is another relationship that will prove useful in designing and analyzing control systems.

Example 3.1 Consider a mass with an applied displacement, $u(t)$ and a disturbance force $d(t)$, as shown in Figure 3.9. Suppose that G_c, P, and $H \equiv 1$. We will consider the stability of some feedback transfer functions.

We will first consider the following plant G_p and disturbance G_d dynamics to be as given in Eqs. (3.15) and (3.16) and as shown in Figure 3.10:

$$G_p = \frac{1}{s+1} = \frac{s}{s(s+1)} \tag{3.15}$$

$$G_d = \frac{1}{s(s+1)} \tag{3.16}$$

Then,

$$\frac{\hat{y}}{\hat{r}} = \frac{\frac{1}{s+1}}{1+\frac{1}{s+1}} = \frac{1}{s+2} \tag{3.17}$$

$$\frac{\hat{y}}{\hat{d}} = \frac{\frac{1}{s(s+1)}}{1+\frac{s}{s(s+1)}} = \frac{1}{s(s+2)} \tag{3.18}$$

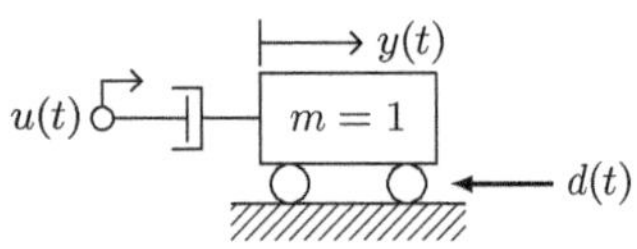

Figure 3.9 A mass with a disturbance and damper force.

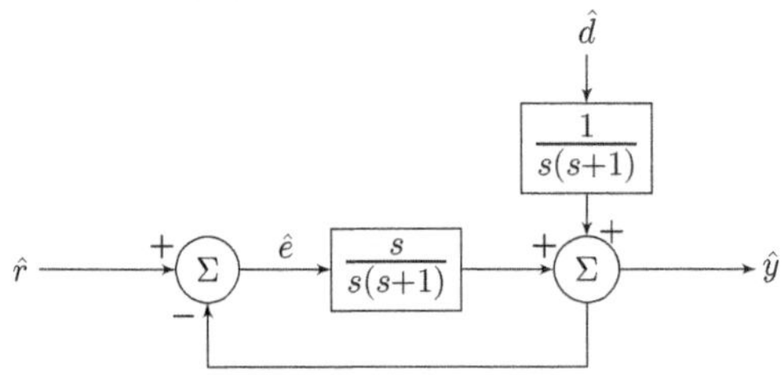

Figure 3.10 Block diagram of the mass and damper system.

The dynamics from $\hat{r}$ to $\hat{y}$ are stable. However, because of the integrator in the dynamics from $\hat{d}$ to $\hat{y}$ (that is, the s^{-1} or the pole at $s = 0$), the response to a disturbance input is BIBO unstable.

Now, we will consider new plant dynamics:

$$G_p = \frac{1}{s} = \frac{s+1}{s(s+1)} \tag{3.19}$$

$$G_d = \frac{1}{s(s+1)} \tag{3.20}$$

Then,

$$\frac{\hat{y}}{\hat{r}} = \frac{\frac{1}{s}}{1+\frac{1}{s}} = \frac{1}{s+1} \tag{3.21}$$

$$\frac{\hat{y}}{\hat{d}} = \frac{\frac{1}{s(s+1)}}{1+\frac{1}{s}} = \frac{1}{(s+1)(s+1)} \tag{3.22}$$

This time, both the response to a commanded input and a disturbance are stable.

What this example shows is that we can use feedback only if the controlled object has cancellations in the left-hand plane.

Internal disturbances can excite components of $G(s)$ that we had assumed were canceled. Suppose we have a plant defined by state matrices A, B, and C (with $D = 0$ so that the system is strictly proper) and that there is a disturbance $d(t)$ with input matrix B_d, so

$$\underline{\dot{x}} = A\underline{x} + Bu + B_d d \tag{3.23}$$

$$y = C^T\underline{x} \tag{3.24}$$

The output y can be written in terms of u and d, by converting the state space representation into transfer function form:

$$\hat{y}(s) = G_p(s)\hat{u}(s) + G_d(s)\hat{d}(s) \tag{3.25}$$

where

$$G_p(s) = C^T(sI - A)^{-1}B = \frac{\eta(s)}{\delta(s)} \tag{3.26}$$

and

$$G_d(s) = C^T(sI - A)^{-1}B_d = \frac{\eta_d(s)}{\delta(s)} \tag{3.27}$$

So, a control system that cancels unwanted dynamics from $G_p(s)$ may have no impact on unwanted dynamics in $G_d(s)$. Various feedback transfer functions can provide insight into how these dynamics are amplified or reduced by feedback systems. These transfer functions will be derived in this section.

First, consider the block diagram in Figure 3.11, which replaces the generic blocks of Figure 3.1 with conventional transfer function and summing junction blocks. Here, G_c represents the controller transfer function, G_p represents the

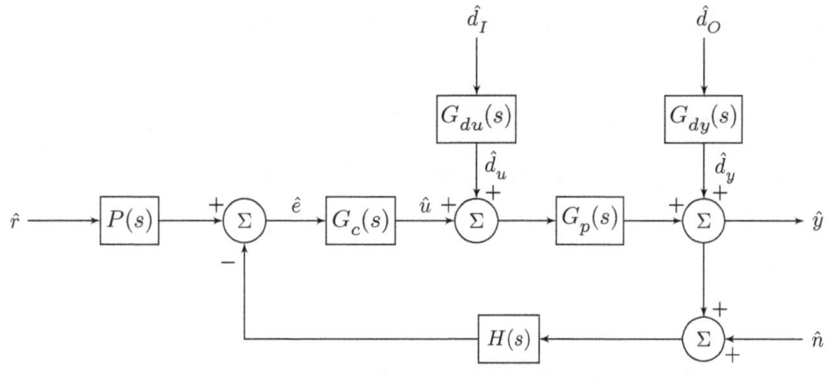

Figure 3.11 SISO block diagram.

plant, G_{du} and G_{dy} represent input and output disturbance dynamics, respectively, and P represents any input signal preconditioning.

3.3 Well-Posedness and Internal Stability

3.3.1 Well-Posedness

Consider the particular feedback system illustrated in Figure 3.12. It is easy to demonstrate that

$$\frac{\hat{y}}{\hat{r}} = \frac{L(s)}{1 + L(s)}$$

where

$$L(s) = \left[1 + \frac{1}{s}\right] \frac{(1 - s)}{(s + 1)} = \frac{(s + 1)(1 - s)}{s(s + 1)} = \frac{1 - s}{s}$$

so

$$\frac{\hat{y}}{\hat{r}} = \frac{(1 - s)/s}{1 + (1 - s)/s} = 1 - s \tag{3.28}$$

an improper, closed-loop transfer function! However, we have demonstrated that improper transfer functions are physically unrealizable. (If nothing else, a dynamic system with an improper transfer function is BIBO unstable.)

Notice that G_c, G_p, and L, in this example, are proper while the closed-loop transfer function, T, is not! We are motivated to establish the following definition:

Definition 3.6 A feedback system is *well-posed* where all the closed-loop transfer functions are proper or strictly proper. Otherwise, the feedback is *ill-posed.*

The example presented above demonstrates that a feedback can indeed be ill-posed and this can happen even where the individual transfer functions, around the loop, are not improper!

Before giving a mathematical condition for well-posedness, consider a description in words as to what well-posedness means and why it would be useful. Referring to Figure 3.12, at first, we must determine under what conditions this interconnection of components makes sense. That is, we need to know when these equations, which result from the feedback loop transfer function, have a solution for an arbitrary input, $\hat{r}$. Then we could say the system of Figure 3.12 is *well-posed* if a unique solution exists for $\hat{y}$ for all input $\hat{r}$; and further, such a unique solution must exist if this realization is perturbed in some neighborhood of the polynomial coefficients of the closed-loop transfer function.

We can demonstrate a mathematical condition for a well-posed feedback. Consider the Heaviside expansion of the return ratio:

$$L(s) = L(\infty) + L'(s) \tag{3.29}$$

where $L'(s)$ is strictly proper and has the same denominator as $L(s)$. It follows from the development in the previous section that the closed-loop transfer function:

$$\frac{\hat{e}(s)}{\hat{r}(s)} = \frac{1}{1 + L(s)} = S(s) \tag{3.30}$$

It is clear that if the denominator, $1 + L(s)$, is strictly proper (if $1 + L(\infty) = 0$) then $1/(1 + L)$ is improper. Thus, $\frac{\hat{e}(s)}{\hat{r}(s)}$ could be improper even where $L(s)$ is not improper. On the other hand, where $1 + L$ is proper, $S(s)$ must be as well. We have proven the following proposition:

Proposition 3.2 *A feedback system is well-posed if and only if* $1 + L(\infty) \neq 0$.

That is, we require that the return difference not have a zero at infinity. Additionally,

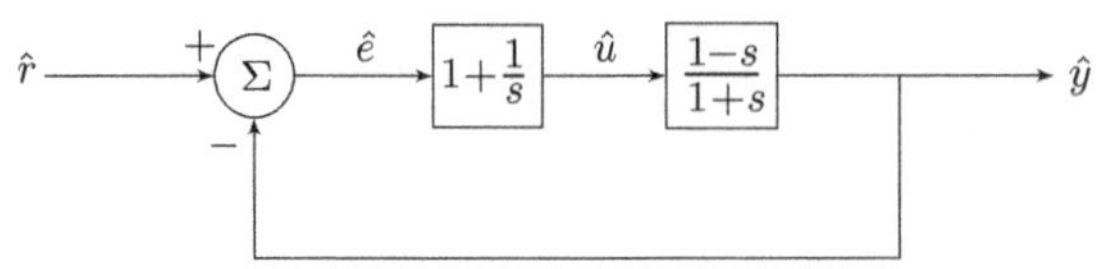

Figure 3.12 Block diagram of system.

Corollary 3.1 *A feedback system is well-posed if $G_c(s)$ or $G_p(s)$ is strictly proper.*

Proof: For example, where there is a single loop,

$$L(\infty) = G_c(\infty)G_p(\infty)$$

where $G_c(s)$ (or $G_p(s)$) is strictly proper, then $G_c(\infty)$ (or $G_p(\infty)$) is zero, so

$$L(\infty) = 0 \neq -1$$

■

Alternative interpretations of an "ill-posed" feedback are available. For example, if we use feedback about a realizable device, we can then ask, is the feedback assured to be realizable or well-posed? The surprising answer is **no**! The possibility for this counterintuitive result will be demonstrated with the next simple example.

Example 3.2 Consider a closed-loop negative feedback system with the simple plant dynamics $G_p = -1$, and let the controller $G_c \equiv 1$, as shown in Figure 3.13.

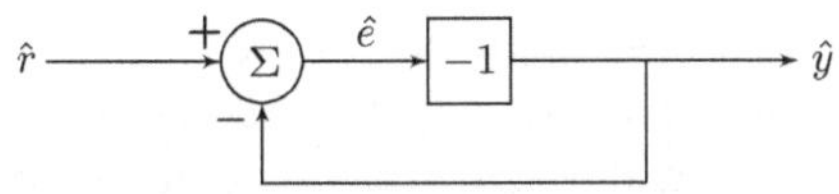

Figure 3.13 Ill-posed feedback.

Then,

$$\hat{e} = \hat{r} - \hat{y} = \hat{r} + \hat{e} \tag{3.31}$$

Or alternatively,

$$\hat{y} = -\hat{e} = \hat{y} - \hat{r} \tag{3.32}$$

Suppose $r = 1$, then what is the error and what is the output? The feedback loop in fact negates the input command, so that the error and output function are independent of the input.

We could resolve this with a memory or the first-order filter element between the error and the output:

$$G_c = \frac{1}{\tau s + 1} \quad \Rightarrow \quad \frac{\hat{y}}{\hat{e}} = \frac{-1}{\tau s + 1} \tag{3.33}$$

$$\hat{e} = \hat{r} - \hat{y} = \hat{r} + \frac{\hat{e}}{\tau s + 1} \Rightarrow \hat{y} = \frac{\tau s + 1}{\tau s}\hat{r} \tag{3.34}$$

It is important to note that "Well-posedness is a necessary but not sufficient condition for internal stability."

3.3.2 The Internal Stability of Feedback Control

We outline one of our two primary mathematical characterizations of feedback systems in this section: The focus is on the poles and zeros of the closed-loop system. The second characterization, which is the focus of much of the remainder of the chapter, is the frequency response, the Bode functions of gain and phase, of the closed-loop transfer functions. Our feedback design theories are based on a combination of these two characterization.

3.3.2.1 The Closed-Loop Characteristic Equation and Closed-Loop Poles

It is clear from the definition of a pole given in Chapter 1 that the following proposition is true.

Proposition 3.3 *The complex number, p_c, is a* closed-loop pole *if and only if*

$$1 + L(p_c) = 0 \tag{3.35}$$

That is, p_c must be a zero of the return difference.

Relation (3.35) is called the rational form of the closed-loop characteristic equation.

$L(s)$ is a rational function,

$$L(s) = \frac{\eta(s)}{\delta(s)} \tag{3.36}$$

where $\eta(s)$ and $\delta(s)$ are polynomials. The return difference $1 + L(s)$ can the be written as a function of these polynomials:

$$1 + L(s) = [\eta(s) + \delta(s)]/\delta(s)$$

Notice that this rational function has the closed-loop characteristic polynomial as numerator and the open-loop characteristic polynomial as denominator. Hence, based on the Proposition 3.3, where $1 + L(p_c) = 0$, we can write

$$\eta(p_c) + \delta(p_c) = 0 \tag{3.37}$$

We call this the *polynomial form of the characteristic equation.*

For one thing, the need for stability requires that any root of this equation must lie in the left half-plane. In the past, research engineers have proposed feedback design procedures based upon the fixing of the roots of this equation. there are the "root-locus" methods and the pole placement, state variable methods. These procedures are of dubious value partly because the focus is on a secondary performance issue, transient response, and partly because the locations of individual closed-loop poles, contrary to the belief of some, are not very the good indicator of transient response. That is to say, even on their own terms, pole placement procedures are of uncertain value. We will discuss more on this in the future.

None of this is to say that we shall disregard the pole-oriented methods. However, their development is made only because these methods can be employed for interesting tasks for which they were not originally intended. In any case, we extract some important information from the location of the poles and zeros of the return ratio and from the location of the closed-loop poles and zeros.

3.3.2.2 Closed-Loop Zeros

We demonstrated in Chapter 2 and will continue to demonstrate that zeros can have a profound but subtle effect on transient response. For that and for other reasons as well, we study the closed-loop zeros.

We begin with the Youla parameter closed-loop transfer function, presented in the previous section, and substitute the ratio of numerator polynomials to denominator polynomials for the various transfer functions:

$$G_k(s) = \eta_k(s)/\delta_k(s) \tag{3.38}$$

The result is

$$Y(s) = \frac{\eta_c(s)\delta_p(s)}{\delta_c(s)\delta_p(s) + \eta_c(s)\eta_p(s)} \tag{3.39}$$

The closed-loop zeros are easily discerned from this equation: The numerator of the closed-loop transfer function is the numerator of the compensator times the denominator of the plant. This proves the following:

Proposition 3.4 *The zeros of a closed-loop transfer function are the zeros in the forward (or "direct") path and the poles in the feedback path and differ from node to node.*

That is, since the "paths" are defined differently for the different nodes, the zeros are not invariant from node to node in the control system. This is in marked contrast to the nodal invariance of the closed-loop poles.

3.3.2.3 Pole–Zero Cancellation and The Internal Stability of Feedback Control

At times, we expect pole–zero cancellations in the transfer functions of some controlled objects, G_p. Pole–zero cancellations can occur in other ways; for example, there is considerable incentive to use the poles or zeros of a compensator to cancel some of the zeros or poles of the controlled object so that a return ratio may actually be designed to be reducible.

Sometimes, we are indifferent to pole–zero cancellations or even find them to be useful things; but, in other instances, they are unfortunate. The feedback theorist must learn the difference. This section is a description of the profound effect that reducibility of an open-loop transfer function or return ratio can have on the *closed-loop response*.[3] (For one thing, it is demonstrated that a feedback system that is actually unstable can pass numerical tests for stability.)

The discussion also focuses on the clever use of pole–zero cancellation and on the not-so-clever use and that occurs near the end of the subsection. The discussion also focuses on the clever use of pole-zero cancellation and on the not so-clever use. This discussion leads to constraints and guidelines on the purposeful use of pole-zero cancellation in feedback design in Sections 3.4 and 3.5.

Example 3.3 Consider the reducible transfer function:

$$\frac{\hat{y}(s)}{\hat{u}(s)} = \frac{s+1}{(s+1)(s+2)} = \frac{s+1}{s^2+3s+2} \tag{3.40}$$

which corresponds to the differential equation:

$$\frac{d^2y}{dt} + 3\frac{dy}{dt} + 2y = \frac{du}{dt} + u \tag{3.41}$$

Retransform this equation but admit the possibility of nonzero initial conditions. The result is

$$\hat{y}(s) = \frac{s+1}{(s+1)(s+2)}\hat{u}(s) + \frac{\dot{y}(0^-) + 3y(0^-) + sy(0^-)}{(s+1)(s+2)} \tag{3.42}$$

a superposition of the force response with the IC response. The canceled pole at $s = -1$ does not play role in the forced response but it does, in general, in the IC response. The point is that a pole–zero cancellation eliminates the effect of the pole in the forced response only.

One might complain of this example that, after all, ICs are the result of past inputs and, for that reason, the particular ICs that would excite the canceled pole cannot be imposed. True enough, but that observation turns out to be more technical than real. The argument only means that the ICs cannot be excited by past values of the particular input, $u(t)$. Recall that the zeros depend on external connections as well as the internal interconnections. This means that just because there is a pole–zero cancellation in one transfer function does not mean that the transfer function from some disturbance to $\hat{y}$ will contain that same cancellation. Now, in any real system, disturbance-like inputs are always acting and one must plan for the worst: *One must plan for the possibility that some unknown or unpredictable disturbance would affect the initial conditions so as to excite the pole we thought had been canceled.*

We turn here to another type of pole–zero cancellation, that cancellation which a designer purposely introduces in the product $L(s) = G_c(s)G_p(s)$ through the selection of a pole or zero of G_c to lie at same location as a corresponding term in G_p.

Example 3.4 Consider an unstable system in which a controller cancels the unstable pole, such as the one shown in Figure 3.14.

The closed-loop transfer function of $\hat{r}$ to $\hat{y}$ is

$$\frac{\hat{y}}{\hat{r}} = \frac{\dfrac{(1-s)}{(1-s)(1+s)}}{1+\dfrac{(1-s)}{(1-s)(1+s)}} = \frac{\dfrac{1}{1+s}}{1+\dfrac{1}{1+s}} = \frac{1}{s+2} \tag{3.43}$$

3 The reducibility of transfer functions is intimately related to the "controllability" and "observability" properties of mathematical models described later in this text.

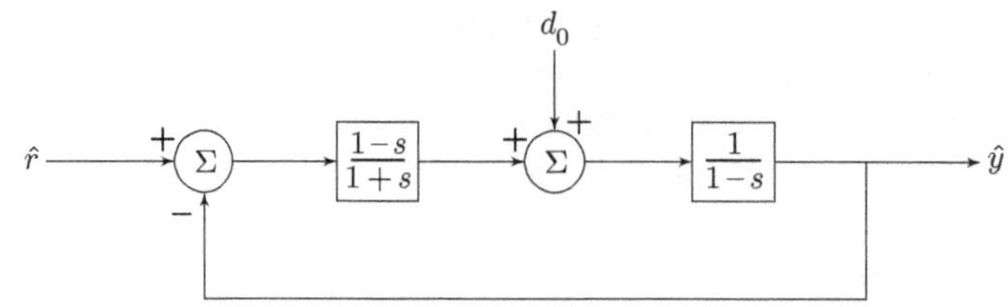

Figure 3.14 Feedback system with an unstable pole/zero cancellation.

which is a stable response due to the cancellation of the unstable pole by the controller's zero. However, now consider the impact of a disturbance at the plant input:

$$\frac{\hat{y}}{\hat{d}_0} = \frac{\frac{1}{1-s}}{1 + \frac{1}{s+1}} = \frac{s+1}{(s+2)(1-s)} \tag{3.44}$$

The unstable pole is no longer canceled, and the result is an unstable response. Although the closed-loop system was nominally stable, a disturbance between the controller and plant still excited the plant's instabilities.

Notice that a pole–zero cancellation in the return ratio (the "open-loop" transfer function) does not necessarily indicate the same cancellation in all closed-loop transfer functions. Finally, notice that

> if a pole of one system component cancels the zero of the component which succeeds it in the feedback loop (say a compensator zero is used to cancel a pole of the controlled object) and if an input signal is injected between these components, the signal will excite the apparently canceled pole at some of the closed-loop nodes. If a pole of one system component cancels the zero of the component which precedes it in the feedback loop (say a compensator pole is used to cancel a zero of the controlled object), any signal will excite the apparently canceled pole at the closed-loop node between the two components.

Example 3.5 Suppose that, in Figure 3.14,

$$G_c = \frac{s+\alpha}{s+\beta} \quad \text{and} \quad G_p = \frac{s+\gamma}{s+\alpha} \tag{3.45}$$

a zero is placed in the compensator to cancel a pole of the controlled object. It is easy to establish

$$\frac{\hat{y}}{\hat{r}} = \frac{s+\gamma}{2s+\beta+\gamma} \quad \text{and} \quad \frac{\hat{y}}{\hat{d}} = \frac{(s+\gamma)(s+\beta)}{(s+\alpha)(2s+\beta+\gamma)} \tag{3.46}$$

So, the cancellation does not occur in the transfer function from the disturbance to the output even though the cancellation is in effect in the transfer function from the command to the output!

Example 3.6 Suppose that, in Figure 3.14,

$$G_c = \frac{s+\delta}{s+\gamma} \quad \text{and} \quad G_p = \frac{s+\gamma}{s+\alpha} \tag{3.47}$$

a zero is placed in the compensator to cancel a pole of the controlled object. It is easy to establish

$$\frac{\hat{y}}{\hat{r}} = \frac{s+\delta}{2s+\alpha+\gamma} \quad \text{and} \quad Y = \frac{\hat{u}}{\hat{r}} = \frac{(s+\delta)(s+\alpha)}{(s+\gamma)(2s+\alpha+\gamma)} \tag{3.48}$$

So, the cancellation does not occur in the transfer function from the reference to the controlled input even though the cancellation is in effect in the transfer function from the command to the output!

Also, note that a computer-aided stability method that is based on the assumption that the closed-loop poles are the zeros of $1 + L(s) = 0$ can give a false indication of stability if $L(s)$ contains a cancellation in the right half-plane. This includes the root locus and Nyquist criterion. This problem is avoided only if common factors in the denominator and numerator of L are not canceled.

One thing is clear: Never effect a pole–zero cancellation in the right half-plane. The design may be adequate with a cancellation in the left half-plane, but only simulation and testing can indicate that possibility.

The concept of internal stability refers to the idea that for a system to be stable, all internal transfer functions must themselves be stable, not just the measurable outputs of the system. For example: consider a room inside of a house that has caught fire. There may be no trace of the fire visible from outside the house, but it is burning nonetheless. Just as the fire will eventually spread to the rest of the house, so to will an internal instability eventually destabilize the system as a whole. To emphasize this point, we give the following definition:

Definition 3.7 A well-posed closed-loop system is *internally stable* if the transfer function from every input to every node has poles strictly in the left half-plane and if there are no right-half-plane pole–zero cancellations in the return ratio.

We constrain ourselves to design only internally stable feedback systems. That is easy enough in SISO, as we constrain ourselves to pole–zero cancellations in the product $L = G_c G_p$ strictly in the left half-plane only. The task is shown to be more interesting in case of neoclassical design where one selects the Youla parameter instead of G_c. We turn to those ideas in Section 3.4.

However, it is clear that there is one situation where no amount of feedback (or anything else) can be used to alleviate an internal instability: if a controlled object has pole–zero cancellation on the imaginary axis or in the right half-plane, any control scheme is internally unstable.

> We require that the controlled object does not have a pole–zero cancellation on the imaginary axis or in the right half-plane in order to even contemplate the possibility of a viable feedback control.

In the older, state variable methods, this requirement was known as the requirement of detectability and stabilizability.

Definition 3.8 A closed-loop system is internally stable if it is well-posed and if for all initial conditions $\underline{x}(0)$ of the plant $G_p(s)$ and $\underline{x}_c(0)$ of the controller G_c, the limits

$$\lim_{t\to\infty} \underline{x}(t) = \underline{0} \tag{3.49}$$

and

$$\lim_{t\to\infty} \underline{x}_c(t) = \underline{0} \tag{3.50}$$

hold when the inputs are zero.

Consider the simplified block diagram shown in Figure 3.5, in which the sensor dynamics $H(s)$, preconditioning $P(s)$, and disturbance dynamics $G_d(s)$ and $G_u(s)$ are all neglected. What are the feedback transfer functions that must be stabilized for internal stability? First, notice that we have four inputs to the system: the commanded input $\hat{r}$, plant input and output disturbances $\hat{d}_u$ and $\hat{d}_y$, and noise $\hat{n}$. Although $\hat{y}$ is the only measured output, conventionally the error signal $\hat{e}$ and the controller output $\hat{u}$ are also treated as outputs. With 4 inputs and 3 outputs, a total of 12 transfer functions can be formed.

These transfer functions can be obtained from Figure 3.5 using block diagram algebra. Of course, some of these have already been derived in Section 3.2, and we know from those earlier derivations to expect that in some cases, different input/output pairings will produce the same transfer function. With that in mind, the 12 transfer functions for internal stability are given in Table 3.2.

Despite considering 12 different input/output pairs, we can characterize the system behavior with only 4 transfer functions: S, T, Y, and G_pS. In fact, additional input/output pairings could be derived from the system in Figure 3.5 – they would still simplify to one of these four transfer functions.

Table 3.2 Transfer functions for internal stability.

		Inputs			
		$\hat{r}$	$\hat{n}$	$\hat{d}_y$	$\hat{d}_u$
Outputs	$\hat{e}$	$\frac{\hat{e}}{\hat{r}} = S$	$\frac{\hat{e}}{\hat{n}} = -S$	$\frac{\hat{e}}{\hat{d}_y} = -S$	$\frac{\hat{e}}{\hat{d}_u} = -G_pS$
	$\hat{y}$	$\frac{\hat{y}}{\hat{r}} = T$	$\frac{\hat{y}}{\hat{n}} = -T$	$\frac{\hat{y}}{\hat{d}_y} = S$	$\frac{\hat{y}}{\hat{d}_u} = G_pS$
	$\hat{u}$	$\frac{\hat{u}}{\hat{r}} = Y$	$\frac{\hat{u}}{\hat{n}} = -Y$	$\frac{\hat{u}}{\hat{d}_y} = -Y$	$\frac{\hat{u}}{\hat{d}_u} = -T$

If we wish to consider the impact of more than one input on an output variable, we can rewrite the contents of Table 3.2 in matrix form:

$$\begin{bmatrix} \hat{e} \\ \hat{y} \\ \hat{u} \end{bmatrix} = \begin{bmatrix} S & -S & -S & -G_pS \\ T & -T & S & G_pS \\ Y & -Y & -Y & -T \end{bmatrix} \begin{bmatrix} \hat{r} \\ \hat{n} \\ \hat{d}_y \\ \hat{d}_u \end{bmatrix} \tag{3.51}$$

If we neglect the inputs $\hat{r}$ and $\hat{d}_y$ and the output $\hat{y}$, all four transfer functions can be fit into a 2×2 matrix expression:

$$\begin{bmatrix} \hat{e} \\ \hat{u} \end{bmatrix} = - \begin{bmatrix} G_pS & S \\ T & Y \end{bmatrix} \begin{bmatrix} \hat{d}_u \\ \hat{n} \end{bmatrix} \tag{3.52}$$

This maps contains all four terms needed to assess internal stability.

3.4 The Youla Parameterization of all Internally Stabilizing Compensators

Given a plant, $G_p(s)$, one typically fixes a preferred closed-loop transfer function, $T(s)$, and then computes a proper and stable Youla parameter, $Y(s)$, knowing that $T(s) = Y(s)G_p(s)$. By shaping the closed-loop transfer function from the injection of the reference signal to the controlled input,

$$\frac{\hat{u}}{\hat{r}} = Y(s) = \frac{G_c(s)}{1 + L(s)} \tag{3.53}$$

where $L(s) = G_c(s)G_p(s)$, we can then solve the above equation for $G_c(s)$:

$$G_c(s) = \frac{Y(s)}{1 - Y(s)G_p(s)} \tag{3.54}$$

Since $G_p(s)$ is fixed, $G_c(s)$ is easily calculated once the value of the rational function $Y(s)$ is given. For this reason, we say Y **parameterizes** G_c **and that (3.54) is the Youla parameterization of** G_c**.**

This would not be a useful observation if it were not for a few remarkable facts that we will now develop. Firstly, we make some incidental observations about the close association of the Youla parameter with the internal model[4] feedback scheme illustrated in Figure 3.15.

Here, $G_p'(s)$ represents the modeled plant while G_p is the true plant. Then, if the plant is modeled well and $G_p' \approx G_p$, a signal equal to $G_p'(s)\hat{u}(s)$ is both added to and subtracted from essentially the same point in a single-loop feedback system. This means that the result is equivalent to single-loop feedback shown in Figure 3.16.

However, we make the simple rearrangement indicated in Figure 3.17 and make it apparent that the single-loop feedback system is the same as the internal model control if "Y" must be selected as in relation (3.54).

We have established this proposition:

4 Notice that a "model" (a simulation) of the controlled object is included in the synthesis of the feedback. It is for this reason that the configuration is called *internal-model control.*

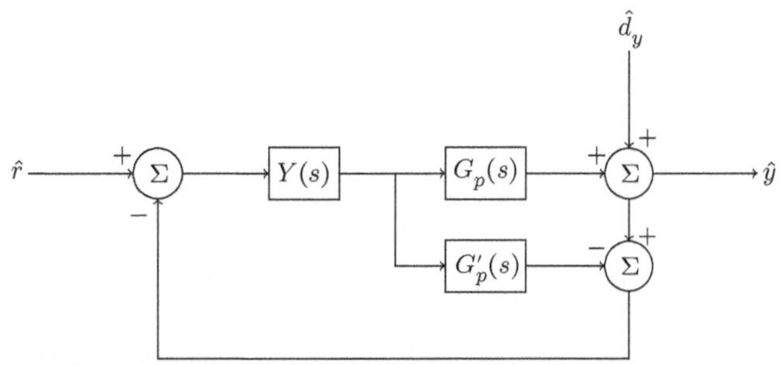

Figure 3.15 The internal model control scheme.

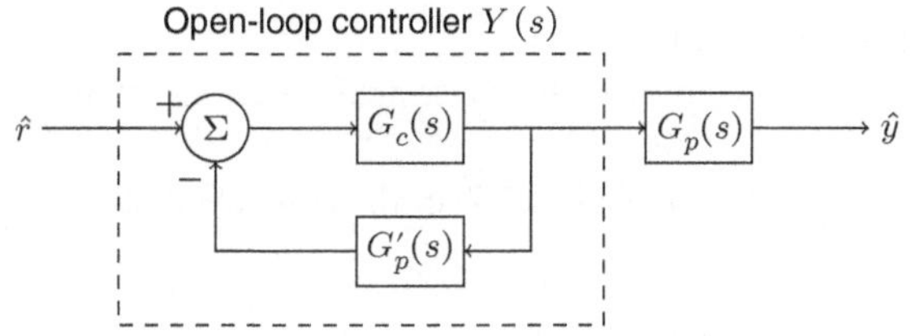

Figure 3.16 Open-loop synthesis for a system with plant $G_p(s)$.

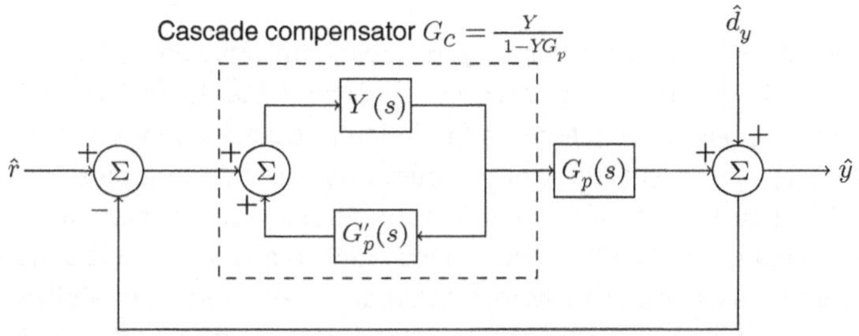

Figure 3.17 Dynamically equivalent single-loop feedback system.

Proposition 3.5 *Single-loop and internal model feedback control, with a model of the controlled object, are equivalent and the connection between the two is the equation pair (3.53) and (3.54).*

Next, as stated above, we can establish a simple closed-loop relation:

$$\frac{\hat{y}}{\hat{r}} = T(s) = G_p(s)Y(s) \tag{3.55}$$

and conclude that the closed loop is stable only if the product YG_p is stable. We want to extend this simple observation to the case of *internal stability* and to the case of an unstable $G_p(s)$.

Remember that in classical control, we select the controller G_c to shape the return ratio L and infer the impact on the closed-loop transfer function T. The advantage of classical control is that we can avoid internal instability by forbidding pole–zero cancellations on the $j\omega$ axis or in the right-hand plane. How might we be able to design internally stabilizing controllers using neoclassical control?

In the previous sections for addressing internal stability, we pointed out that all closed-loop transfer functions can be reduced to only four transfer functions by the following map:

$$\begin{bmatrix} \hat{e} \\ \hat{u} \end{bmatrix} = -\begin{bmatrix} G_pS & S \\ T & Y \end{bmatrix} \begin{bmatrix} \hat{d}_u \\ \hat{n} \end{bmatrix} \tag{3.56}$$

We can rewrite these transfer function in terms of only $Y(s)$ and $G_p(s)$: Since $G_pY = T$, we have that

$$G_pY = 1 - S$$

or

$$S = 1 - G_p Y \tag{3.57}$$

This means that our array of transfer functions or the tableau for internal stability can be rewritten as:

$$\begin{aligned} \frac{\hat{u}(s)}{\hat{n}(s)} &= -Y(s) & \frac{\hat{u}(s)}{\hat{d}_u(s)} &= -G_p(s)Y(s) \\ \frac{\hat{e}(s)}{\hat{n}(s)} &= -[1 - Y(s)G_p(s)] & \frac{\hat{e}(s)}{\hat{d}_u(s)} &= -G_p(s)[1 - Y(s)G_p(s)] \end{aligned} \tag{3.58}$$

We accomplished our task: The array of closed-loop transfer functions has been rewritten entirely in terms of $Y(s)$ and $G_p(s)$.

The closed loop is internally stable if and only if all transfer functions in this tableau are stable. We conclude that a closed loop is internally stable if and only if the following goals are obtained:

1. The Youla parameter, $Y(s)$, itself is BIBO stable.
2. $G_p(s)Y(s)$ is BIBO stable; thus, any unstable pole of the controlled object, $G_p(s)$, must be a zero of the Youla parameter, $Y(s)$.
3. $G_p(s)[1 - G_p(s)Y(s)]$ is BIBO stable; thus, any unstable pole of the controlled object, $G_p(s)$, must be a zero of

$$1 - G_p(s)Y(s) \tag{3.59}$$

It is emphasized in the following that an important feedback design procedure is the direct selection of $Y(s)$ with the requirements (3.59) in mind. It is easy to obtain condition (1), simply by selecting a stable $Y(s)$. Remember that the closed-loop transfer function is

$$T = G_p Y$$

so it is easy to obtain condition (2): Simply fix $Y(s)$ with a zero at each unstable pole of G_p. In fact, an unstable pole of G_p must be selected as a zero of $Y(s)$. On the other hand, since $Y(s)$ must be stable, an unstable pole cannot be selected for $Y(s)$ in order to cancel a non-minimum phase zero in G_p: It is forbidden to cancel a non-minimum phase zero in G_p with the selection of a pole of $Y(s)$. One implication is that the non-minimum phase behavior must be passed on to the closed-loop transfer function, $T(s)$. The tricky part is to specify Y such that condition (3) is also met! This is especially tricky where the unstable pole has a multiplicity greater than 1. (That occurs most often as a repeated pole at the origin.) We say that the set of all $Y(s)$, that satisfy these conditions, is the *Youla parameterization* of all internally stabilizing compensators.

Again, condition (2) means that there is a unstable pole–zero cancellation in $G_p Y$ if G_p has an unstable pole. There might be some confusion generated here because we seem to prohibit right-half plane cancellations in Section 3.3. Careful reading leads to the following precise admonitions: A right half-plane cancellation is forbidden in $L = G_c G_p$ where the classical procedure is to select G_c. However, it is clear from the above development that such a cancellation is required in $T = G_p Y$ where the neoclassical procedure is to select $Y(s)$. Much more on this is provided in Chapter 4.

Once such a parameter is found, then there exists a compensator, $G_c(s)$ that provides an internally stable closed loop and the form of this compensator can be calculated from (3.54). (Incidentally, after G_c is calculated, a check on internal stability calculations is available: merely check that there are no unstable pole cancellations in the product $G_c G_p$.)

We assert that condition (2) is, in fact, an implication of (3). We prove that assertion by contradiction: Suppose that condition (3) holds while (2) does not. This means that $G_p Y$ is unbounded at a unstable pole of G_p, but if that were true, then $G_p(1 - G_p Y)$ would be unbounded there as well, a contradiction. All of this is summarized in the following proposition:

Proposition 3.6 *An internally stabilizing compensator is parameterized as:*

$$G_c(s) = \frac{Y(s)}{1 - G_p(s)Y(s)} \tag{3.60}$$

if and only if $Y(s)$ is any BIBO stable, proper rational function such that $G_p(1 - G_p Y)$ is BIBO stable; That is, such that Y and $1 - GpY$ have a zero at each unstable pole of G_p. ■

Thus, we have:

Corollary 3.2 *Necessary conditions for $Y(s)$ to be internally stabilizing are*

$$Y(p_r) = 0 \tag{3.61}$$

and

$$G_p(p_r)Y(p_r) = 1 \tag{3.62}$$

where "p_r" is any pole of G_p in the right half-plane or on the imaginary axis. These conditions are sufficient only if "p_r" is not repeated. ■

Equations (3.61) and (3.62) may not be sufficient because there is no guarantee that the multiplicity of the zero in $1 - G_pY$ will match the multiplicity of the unstable pole in G_p. Notice that, in this case, $G_c(s)$ may be stable or unstable. That is to say, a stable $T(s)$ is achieved by placing unstable poles in the compensator. Such a thing is always possible, *but note that some object to the use of an unstable compensator because if the feedback fails (a sensor or an actuator fails), then the open-loop system is driven out of bound by the compensator!*

Since $T = G_pY$, relation (3.62) means that internal stability, an unavoidable requirement, is achieved only if

$$T(p_r) = 1 \tag{3.63}$$

Incidentally, condition (3.62) is not inconsistent with (3.61) because of the pole–zero cancellation of the p_r-term on the left side of (3.62). Some may also argue that condition (3.62) is undefined if we directly substitute for $(Y(p_r) = 0)$ and $(G(p_r) = \infty)$. To avoid any confusion, it better to write this condition as:

$$G_pY(p_r) = 1$$

In either case, the Proposition (3.6) is a very remarkable result! Think of this: One is not assured of a stable feedback loop by placing some restriction on, say, the compensator $G_c(s)$. For example, the feedback about a stable compensator in cascade with a stable controlled object can be unstable. This remarkable neoclassical proposition is one of the motivations for focusing on the Youla parameter, $Y(s)$, during design. There are a number of other motivations as will become clear as we continue.

We have a simplification in a special case:

Corollary 3.3 *If the controlled object is stable, the closed loop is internally stable if and only if $Y(s)$ is stable (hence, not improper).* ■

This means that the "Youla parameterization of all stabilizing compensators" is the set of all stable, proper rational functions! That is a way to say that any G_c given by (3.60) will produce an internally stable closed loop provided that the Youla parameter is stable.

Example 3.7 Suppose that the object is stable except for a single pole at the origin. In that case, "Y" is internally stabilizing if and only if "Y" has a zero at the origin and

$$G_p(0)Y(0) = T(0) = 1 \tag{3.64}$$

if and only if the closed-loop static gain is exactly 1.0. For example consider the velocity control of a mass for which

$$G_p(s) = 1/(ms) \tag{3.65}$$

Then, an internally stabilizing Y is

$$Y(s) = ms/(\tau s + 1) \tag{3.66}$$

for arbitrary values of "τ." It follows from (3.60) that the compensator is

$$G_c(s) = \frac{ms/(\tau s + 1)}{1 - 1/(\tau s + 1)} = \frac{m\cancel{s}}{\tau\cancel{s}} = \frac{m}{\tau} \tag{3.67}$$

an arbitrary positive gain!

3.5 Interpolation Conditions

The internal stability condition of the last section, ($T(p_r = 1)$, is very interesting because it is a constraint that must be satisfied by the closed-loop transfer function, a constraint that can never be avoided.

We mentioned that in classical control, the controller G_c is designed to shape the return ration $L = G_c G_p$ so that the performance of the closed-loop transfer function $T(s)$ meet the given requirements. In designing G_c, pole–zero cancellation of unstable (right-hand plane) pole of G_p with an unstable zero (non-minimum phase zero) in G_c is forbidden in order to preserve internal stability. However, we know from Section 3.4 that internally stable feedback is parameterized by the Youla parameter Y in neoclassical control.

In addition, we provided necessary conditions for guaranteeing internal stability and we stated that these conditions become sufficient only when the unstable pole of G_p is not repeated. We pointed out that the tricky part is to specify Y such that the condition $G_p[1 - G_p Y]$ is BIBO stable, especially where the unstable pole of G_p has a multiplicity greater than 1. We will further discuss this below to provide some insights and develop a set of conditions which are relatively easier to apply and that when satisfied guarantee the internal stability for all circumstances including repeated unstable poles of G_p.

Consider the case where G_p has a single unstable pole. We will let that unstable pole be at $s = p_r$ and define $\widetilde{G}_p$ to be the remaining stable part of the transfer function, i.e.

$$G_p = \frac{1}{s - p_r}\widetilde{G}_p \tag{3.68}$$

Then to have a stable $T = YG_p$, we must have

$$Y = (s - p_r)\widetilde{Y} \tag{3.69}$$

In other words, Y must have a non-minimum phase zero at $s = p_r$ to cancel the unstable pole of G_p, while the minimum phase and stable part of Y is denoted $\widetilde{Y}$. Given this factorization for Y, we then have

$$G_p S = G_p(1 - YG_p) = \frac{1}{s - p_r}\widetilde{G}_p(1 - \widetilde{Y}\widetilde{G}_p) \tag{3.70}$$

For internal stability, $G_p S$ must be stable. Therefore, not only must Y cancel the unstable pole in G_p, but so must S. In other words, after including a non-minimum phase zero in Y to cancel the unstable pole of G_p, the unstable pole of G_p must also be a zero of $(1 - YG_p)$ as stated in the last section. This is especially tricky where the unstable pole has a multiplicity greater than 1. To find a set of conditions which when satisfied guarantee internal stability under all conditions, we go back to our very interesting condition of $T(p_r) = 1$.

Suppose the open-loop transfer function L has unstable poles at $s = p_r$ with multiplicity a_p and has unstable zeros at $s = z_r$ with multiplicity a_z. If we let $\widetilde{L}$ denote the stable, minimum phase factor of L then we have

$$L(s) = \frac{(s - z_r)^{a_z}}{(s - p_r)^{a_p}}\widetilde{L}(s) \tag{3.71}$$

Note that at the unstable pole p_r, the denominator of L goes to zero, and so L goes to infinity:

$$L(p_r) = \frac{(p - z_r)^{a_z}}{0^{a_p}}\widetilde{L}(p_r) = \infty \tag{3.72}$$

Additionally, note that at the unstable zero z_r, the numerator of L goes to zero, and so L goes to zero:

$$L(z_r) = \frac{(0)^{a_z}}{(z - p_r)^{a_p}}\widetilde{L}(z_r) = 0 \tag{3.73}$$

The same results apply to minimum phase or stable zero and stable pole of L, such that $L(z_r) = 0$ at the stable zero or minimum phase zero and, $L(p_r) = \infty$ at the stable pole. Therefore, we can write (the proof is a simple exercise),

Proposition 3.7 *If "z" denotes a zero of the return ratio, L, of multiplicity a_z, and "p" denotes a pole of the return ratio of multiplicity a_p, then*

$$T(p) = 1 \qquad S(p) = 0 \tag{3.74}$$

and

$$\frac{d^k}{ds^k}T(p) = 0 \qquad \frac{d^k}{ds^k}S(p) = 0 \qquad 1 \le k \le a_p - 1 \tag{3.75}$$

while

$$T(z) = 0 \qquad S(z) = 1 \tag{3.76}$$

and

$$\frac{d^k}{ds^k}T(z) = 0 \qquad \frac{d^k}{ds^k}S(z) = 0 \qquad 1 \le k \le a_z - 1 \tag{3.77}$$

■

These very important relations are known as *interpolation constraints.* Condition (3.76) subsumes our previous condition $T(p_r) = 1$. Thus, the closed-loop transfer functions, $S(s)$ and $T(s)$, are constrained to take on certain values in the complex plane and these constraints influence the values on the imaginary axis, $S(j\omega)$ and $T(j\omega)$. This proposition imposes irrevocable limits to feedback design, and interestingly, it is these irrevocable limits which guarantee internal stability of the feedback. Hence, *the feedback is internally stable if and only if these constraints are met.*

Example 3.8 In this example, we examine the "Interpolation Constraints" and the consequences of violating these constraints in feedback design.

Consider a force $F(t)$ applied to a mass, as shown in Figure 3.18, to control velocity.

$$\frac{\hat{v}}{\hat{u}} = \frac{1}{ms} \Rightarrow \text{BIBO unstable} \quad \text{where} \quad \hat{u} = F(t) \tag{3.78}$$

We will examine a Youla parameter that cancels the dynamics of the mass but does *not* meet the interpolation condition:

$$Y = \frac{ks}{\tau s + 1} \tag{3.79}$$

Then,

$$T = \frac{k}{m(\tau s + 1)} \tag{3.80}$$

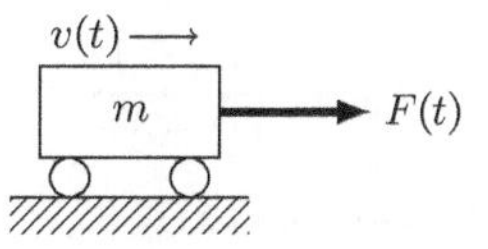

Figure 3.18 Schematic for velocity control.

Suppose that $k = \frac{m}{2}$. Then at $s = 0$, $T = \frac{1}{2}$, so the interpolation condition is *not* met. We will find the corresponding controller.

$$S = 1 - T = \frac{\tau s + \frac{1}{2}}{\tau s + 1} \quad \text{where} \quad S = (1 - G_p Y) \tag{3.81}$$

$$G_c = \frac{Y}{S} = \frac{ks}{\tau s + 1}\,\frac{\tau s + 1}{\tau s + \frac{1}{2}} \tag{3.82}$$

$$\Rightarrow \quad G_c = \frac{ms}{2\tau s + 1} \tag{3.83}$$

This means that the return ratio will have a cancellation of an unstable pole, so the system will not be internally stable.

$$L = \frac{m\not{s}}{2\tau s + 1} \cdot \frac{1}{m\not{s}} \tag{3.84}$$

Specifically, the transfer function $G_p(1 - YG_p)$ will be BIBO unstable.

Example 3.9 Consider a force $F(t)$ applied to a mass again, but this time, to control position rather than velocity, as shown in Figure 3.19.

$$\frac{\hat{y}}{\hat{u}} = \frac{1}{ms^2} \tag{3.85}$$

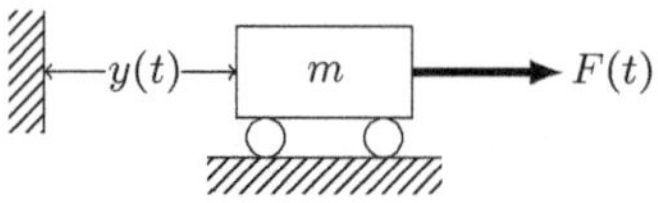

Figure 3.19 Schematic for position control.

We choose a Youla parameter to cancel the dynamics of the mass.

$$Y = \frac{mks^2}{(\tau s + 1)^2} \tag{3.86}$$

$$T = \frac{k}{(\tau s + 1)^2} \tag{3.87}$$

Our interpolation conditions are $T(0) = 1$ and $\frac{dT(0)}{ds} = 0$.

$$T(0) = \frac{k}{(\tau \cdot 0 + 1)^2} = 1 \tag{3.88}$$

Therefore $k = 1$. So,

$$\left.\frac{dT}{ds}\right|_{s=0} = \frac{-2\tau}{(\tau \cdot 0 + 1)^3} = 0 \tag{3.89}$$

Therefore $\tau = 0$. This means we now have

$$Y = ms^2. \tag{3.90}$$

Y is now improper, which is unacceptable.

Next, we consider

$$Y(s) = \frac{ms^2(\tau_1 s + 1)}{(\tau s + 1)^3} \tag{3.91}$$

We added an additional pole and zero to Y with an extra-tuning parameter τ_1. Thus,

$$T(s) = \frac{(\tau_1 s + 1)}{(\tau s + 1)^3} \tag{3.92}$$

The first interpolation condition, $T(0) = 1$, is satisfied and (3.75) is

$$\frac{dT(0)}{ds} = \left[\frac{\tau_1}{(\tau s + 1)^3} - \frac{3\tau(\tau_1 s + 1)}{(\tau s + 1)^4}\right] |_{s=0} \tag{3.93}$$

$$= \tau_1 - 3\tau = 0 \tag{3.94}$$

so that an internally stabilizing value of the parameter is

$$Y(s) = \frac{ms^2(3\tau s + 1)}{(\tau s + 1)^3} \tag{3.95}$$

An internally stabilizing compensator is calculated similar to the Example 3.8

$$G_c(s) = \frac{ms^2(3\tau s + 1)/(\tau s + 1)^3}{1 - (3\tau s + 1)/(\tau s + 1)^3} \tag{3.96}$$

$$= \frac{ms^2(3\tau s + 1)}{\tau^2 s^2(\tau s + 3)} \tag{3.97}$$

$$= \frac{m}{3\tau^2} \frac{(3\tau s + 1)}{(\tau/3)s + 1} \tag{3.98}$$

This could be a lead–lag controller depending of the selection of (τ). As a check, we offer the fact that $G_c G_p$ does not contain the wrong kind of cancellation. Also, $T(s) = (3\tau s + 1)/(\tau s + 1)^3$ so that the closed loop is stable; as a matter of fact, we have placed all three closed-loop poles at $s = -1/\tau$. When we explore this further, we find some unsatisfactory aspects to this solution, but, at least, we have found an internally stable closed loop. We will discuss more satisfactory solutions of this problem in Chapter 4.

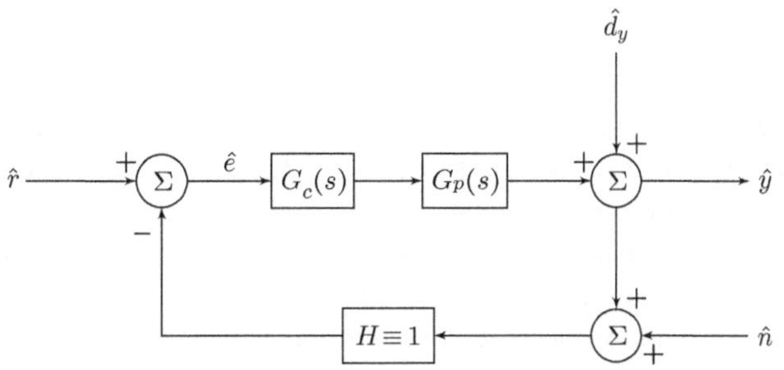

Figure 3.20 Single-loop system with unity feedback.

3.6 Steady-State Error

Before shifting our attention to the application of frequency methods in feedback design, it is important to discuss the time response steady-state error when dealing with the closed-loop transfer functions $S(s)$ and $T(s)$. Consider the block diagram in Figure 3.20.

Steady-state error is found by using the final value theorem.

$$\text{Step input:} \quad r(t) = 1(t) \Rightarrow \hat{r}(s) = \frac{1}{s}$$
$$\text{Ramp input:} \quad r(t) = t(t) \Rightarrow \hat{r}(s) = \frac{1}{s^2}$$

Then we can write

$$\hat{e}(s) = S(s)\hat{r}(s)$$

So, for a step input,

$$e(\infty) = \lim_{s\to 0} s \cdot S(s)\frac{1}{s} = S(0)$$

Hence,

$$e(\infty) = 0 \quad \text{if} \quad S(0) = 0 \Rightarrow T(0) = 1$$

This means, $S(s)$ has at least a single zero at $s = 0$. For a ramp input,

$$e(\infty) = \lim_{s\to 0} s \cdot S(s)\frac{1}{s^2} = \left.\frac{S(s)}{s}\right|_{s=0}$$

This means that $e(\infty)$ is finite if $S(s)$ has a zero at $s = 0$ and is zero if $S(s)$ has at least two zeros at $s = 0$. If so, then

$$\left.\frac{\mathrm{d}S(s)}{\mathrm{d}s}\right|_{s=0} = 0$$

Therefore,

$$T(s) = 1 - S(s) \Rightarrow \left.\frac{\mathrm{d}T(s)}{\mathrm{d}s}\right|_{s=0} = 0$$

3.7 Feedback Design, and Frequency Methods: Input Attenuation and Robustness

In this section, we shift the focus from the poles and zeros of the closed-loop transfer functions to the frequency response. In some design procedures, the pole–zero concepts, such as well-posedness and internal stability, are kept in mind, as constraints, but the main details of the design of the compensators are obtained from the consideration of the frequency response. Those methods are laid down in Chapter 4 and we prepare for them here.

There are two levels of mathematical rigor used in these developments. The more satisfying level is laid down in the later chapters. The less compelling interpretation is developed in Section 3.7.1 and is adequate for the design methods described in the remainder of Chapter 3 and in Chapter 4.

3.7.1 The Frequency Paradigm

One approach to feedback design is the "shaping," or proper alteration, of the frequency response to provide desirable closed-loop performance. A *heuristic* justification for this is that input signals can be thought of, at times, as a superposition of sinusoidal terms of various frequencies, the Fourier series or transform. So that, if one shapes the entire frequency response, the response to all types of inputs are influenced. The point is that the frequency response represents a gain, an amplification or an attenuation, of the various Fourier components of the arbitrary input. This interpretation is an implicit justification for the formulas and techniques designed in the sequel.

The open-loop frequency response gains are easily related to the feedback gains. These feedback gains allow one to infer the ability of the closed loop to attenuate unwanted signals entering at various points in the system.

This paradigm has led to an obviously successful design methodology and one can recognize a limited mathematical justification in some cases. One can also point to important cases where the paradigm breaks down. Some important inputs simply don't have a Fourier transform, the ramp input for example. Even the step input has a Fourier transform in only a limited, least-squares sense.[5] Also, the design methodology is applied equally well to open-loop transfer functions with poles on the imaginary axis or in the right half-plane, and these systems simply don't have a steady-state frequency response.

3.7.2 Input Attenuation and Command Following

How are the design requirements, which are delineated in the previous sections, qualified where frequency methods are applied? That is the question addressed in this subsection.

A recurrent theme is that the answer is framed as the attenuation of some input signal or other at certain nodes of the feedback systems. That is, *we wish to design the feedback system so that all signals are to be attenuated at least at some nodes.* By "attenuation," we mean that the gain (the Bode magnitude) of the closed-loop transfer function is as low as possible at all appropriate frequencies, that frequency range we expect to be present in the input signal. In fact, it is desirable to attenuate most inputs, such as disturbance and noise, at all nodes of the feedback system.

Obviously, we don't wish to attenuate the command signal at the output node because we want $y(t) \approx r(t)$: We call this *good command following* or *good tracking.* But even in this case, the objective can be stated as an attenuation, an attenuation of the *command signal at the error node.* Also, an attenuation of the command at the controlled input node, $u(t)$, is highly desirable in order to reduce the vulnerability of the feedback to a saturation of the controlled input.[6]

There is a difference in interpretation between attenuating unwanted input signals and the attenuation of the command at the error nodes. We wish to attenuate noise in order to prevent degradation of the output or to prevent saturation of the actuators. The attenuation of command at the error node means that the output is rapidly forced to the commanded value. This difference in interpretation is of not of mathematical consequence.

Begin with a very general form of the generic closed-loop transfer function:

$$\frac{\hat{v}}{\hat{r}} = \frac{G_d}{1+L} \tag{3.99}$$

Here, "G_d" is the product of transfer functions in the direct path, and "L" is the return ratio, $\hat{v}$ is an arbitrary node, and $\hat{\imath}$ is an arbitrary input. From this follows:

Proposition 3.8 *An input signal is attenuated at a node where the closed-loop transfer function has a low gain and is obtained where*

1. *the gain of the product of direct path transfer function, G_d, is very small, or*
2. *the gain of the return ratio, L, is very high.* ■

The following facts should be kept in mind:

1. The gain is frequency-dependent.
2. The product of transfer functions in the direct path, "G_d" is dependent on the choice of the node.
3. The return ratio for unity feedback is the product of the compensator and the plant transfer functions, G_cG_p.

---

5 Refer to the "Gibbs phenomenon" in any text on the Fourier transform.
6 Even though, this is a linear control design, this reduction in vulnerability equally applies to the nonlinear system.

There exists some input node pairs that $G_d = L$, such as the output to the sensor noise input. This means the noise is attenuated at the output only where the gain of the return ratio, "L," is made low.

At this point, we refocus on the another formulation of feedback control, the S-T-Y formulation. In this way, we come to a similar conclusion but with slightly different interpretations. We want to attenuate all signals at the controlled input, $u(t)$, but that idea is so important in its ramifications and interpretations that the entirety of Section 3.8 is devoted to the development of those principals. Here, we are content to restrict our attention to the error and output nodes.

Consider the case of *unity feedback*, $H(j\omega) \equiv 1$. Also assume that a disturbance is injected as in Figure 3.20. Under these assumptions, the return ratio is

$$L(j\omega) = G_c(j\omega)G_p(j\omega) \tag{3.100}$$

and simple applications of (3.99) gives

$$\frac{\hat{y}(s)}{\hat{n}(s)} = \frac{-L(s)}{1 + L(s)} = -T(s) \tag{3.101}$$

and

$$\frac{\hat{y}(s)}{\hat{d}(s)} = \frac{1}{1 + L(s)} = S(s) \tag{3.102}$$

To study command following, we form

$$\frac{\hat{e}(s)}{\hat{r}(s)} = \frac{1}{1 + L(s)} = S(s) \tag{3.103}$$

Proposition 3.9 *The feedback loop will attenuate disturbances at the output and follow command signals at those frequencies where $|S(j\omega)|$ is low and it will attenuate noise at the output where $|T(j\omega)|$ is low.* ∎

Knowing that

$$S(j\omega) = 1 - T(j\omega) \tag{3.104}$$

in a single degree of freedom, single feedback loop, the proposition indicates a fundamental trade-off in feedback design.

These equations could just as well be interpreted in terms of the return ratio. For instance, noise is suppressed at the output node at those frequencies where $|L(j\omega)| \ll 1$, but the error is quite large (poor closed loop, command following), and the feedback does not attenuate disturbance at the output at those same frequencies. Similarly, where $|L(j\omega)| \gg 1$, we have good command following and good disturbance attenuation but poor noise attenuation.

Again, a trade-off between good noise attenuation and good disturbance attenuation (with command following) is implied in this statement.

Definition 3.9 The crossover frequency is denoted ω_c and is that frequency at which $|L(j\omega)| = 1$ (0 dB).

We shall discover many significant insights into feedback design about ω_c. It suffices here to point out that this frequency usually separates the range of frequencies where we have good command following and disturbance rejection from the range where noise is suppressed at the output node. Of course, in that range of frequencies about the crossover, we are receiving *none of the benefits of feedback!*

Proposition 3.10 *The crossover frequency is also characterized as that frequency where*

$$|S(j\omega_c)| = |T(j\omega_c)| \tag{3.105}$$

Proof: Immediate consequence of (3.101), (3.102) and Definition 3.9. ∎

3.7.3 Bode Measures of Performance Robustness

In the previous sections, we listed as one of the primary goals of feedback design the assurance (or invariance) of performance in spite of modeling difficulties or parameter variations. This invariance in performance is called *Performance robustness.*[7]

7 In fact, this a relatively new name, a name not used by Bode, for example, or by Perkins and Cruz who generalized this measure to the multivariable case. The older term is "feedback sensitivity."

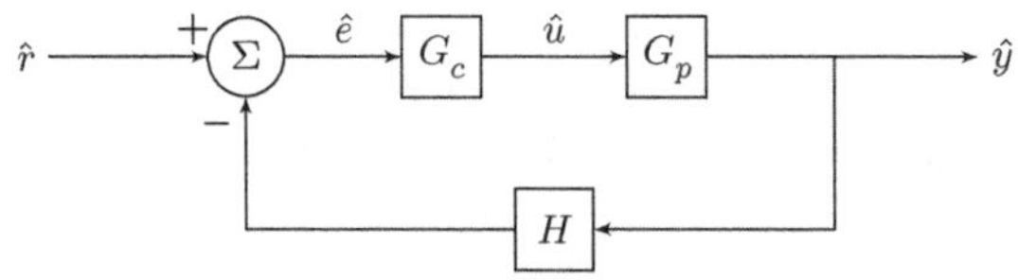

Figure 3.21 Block diagram.

We develop ways to quantify robustness in many places in this text. In the spirit of the axioms developed in this chapter, we develop measures of robustness from the work of Bode.[8] Consider the closed-loop system shown in Figure 3.21.

Two sources of variation in the closed-loop response are variations in the plant dynamics, G_p, and variations in the sensor dynamics, H. We then have a closed-loop response G_{cl} defined by:

$$G_{cl} = \frac{\hat{y}(s)}{\hat{r}(s)} = \frac{G_c(s)G_p(s)}{1 + G_c(s)G_p(s)H(s)} \tag{3.106}$$

Definition 3.10 The Bode differential performance robustness measure (more simply, the Bode measure) is the normalized partial derivative:

$$S_{G_p}^{G_{cl}} = \frac{\%\text{ Change in } G_{cl}}{\%\text{ Change in } G_p} = \frac{\frac{\partial G_{cl}}{G_{cl}}}{\frac{\partial G_p}{G_p}} = \frac{G_p}{G_{cl}} \cdot \frac{\partial G_{cl}}{\partial G_p} \tag{3.107}$$

The Bode measure is read as the sensitivity of G_{cl} with respect to G_p. As given in the equation, this is interpreted as the ratio of a percent change in the closed loop with respect to a percent change that occurs in the controlled object, the change in the system without feedback. The units of this measure are commonly taken to be "percent-per-percent" but it is demonstrated in Proposition 3.11 that we can take the units to be dB per dB. Implicit in this definition is the assumption that the variation occurs in the controlled object: normally, the compensators can be constructed with precisely known parameter values which do not vary in time. Variation in the measurement, $H(s)$ is a possibility and that possibility is accounted for later in this section.

We think of the Bode measure as a gauge of performance robustness as opposed to the stability robustness developed in the next chapter. This is because the Bode measure quantifies the variation in the entire closed-loop frequency response, which determines the response of the system to an arbitrary reference input by our frequency paradigm. The great limitation of this measure is that it only quantifies the effect of differential changes in the controlled object. (There are other derivations of Bode measure in which it seems that large variations are allowed; however, where carefully interpreted, it turns out that these derivations also require small variations.) The method of quantitative feedback theory, not covered in this text, speaks more directly to the situation where large parameter variations are accounted for in design.

Proposition 3.11 *The Bode robustness measure can be rewritten as:*

$$S_{G_p}^{G_{cl}} = \frac{\partial Lm\ G_{cl}(j\omega)}{\partial Lm\ G_p(j\omega)} \tag{3.108}$$

the partial derivative of the log-magnitude of the closed-loop transfer function with respect to the log-magnitude of the open-loop transfer function.

Proof:

$$\begin{aligned} S_{G_p}^{G_{cl}} &= \frac{\partial G_{cl}(j\omega)/G_{cl}(j\omega)}{\partial G_p(j\omega)/G_p(j\omega)} \\ &= \frac{\partial \ln\ G_{cl}(j\omega)}{\partial \ln\ G_p(j\omega)} \end{aligned} \tag{3.109}$$

Now multiply numerator and denominator by $20/\ln(10)$ to convert each to the log-magnitude function. ∎

8 Horowitz complains that Bode himself never used his measure in this way, and that this application of the Bode measure is a more recent invention. Horowitz is a persuasive critic of these popular robustness measures.

How does the Bode measure relate to our closed-loop transfer functions? We start with the definition to the closed-loop transfer function:

$$G_{cl} = \frac{G_c G_p}{1 + G_c G_p H}$$

so that the Bode measure, denoted $S_{G_p}^{G_{cl}}$, is given by:

$$\begin{aligned}
S_{G_p}^{G_{cl}} &= \frac{G_p}{G_{cl}} \cdot \frac{\partial G_{cl}}{\partial G_p} \\
&= \frac{G_p(1 + G_c G_p H)}{G_c G_p} \cdot \frac{\partial}{\partial G_p}\left[\frac{G_c G_p}{1 + G_c G_p H}\right] \\
&= \frac{(1 + G_c G_p H)}{G_c} \cdot \left(\frac{(1 + G_c G_p H)G_c - G_c^2 G_p H}{(1 + G_c G_p H)^2}\right)
\end{aligned}$$

Finally,

$$S_{G_p}^{G_{cl}} = \frac{1}{1 + L} = S \tag{3.110}$$

Notice the correspondence of this Bode robustness measure to the closed-loop transfer function presented in the previous sections. In fact, this correspondence is the reason that the closed-loop transfer function is called the "sensitivity."

From the above, we have established

Proposition 3.12 *The Bode robustness measure depends only on the return difference, $1 + L(j\omega)$. Performance robustness is achieved at those frequencies where the gain of the return ratio is large, at those frequencies where error is suppressed.* ■

We have established that noise is attenuated at the output only where the command is not attenuated at the error node. It follows that noise attenuation must also be traded for performance robustness in the design process.

Next, we consider parameter variation in the feedback path, in the measurement dynamics characterized by $H(s)$. Here, the appropriate sensitivity is

$$\begin{aligned}
S_H^{G_{cl}} &= \frac{H}{G_{cl}} \cdot \frac{\partial G_{cl}}{\partial H} \\
&= \frac{H(1 + G_c G_p H)}{G_c G_p} \cdot \frac{\partial}{\partial H}\left[\frac{G_c G_p}{1 + G_c G_p H}\right] \\
&= \frac{H(1 + G_c G_p H)}{G_c G_p} \cdot \left(\frac{-(G_c G_p)^2}{(1 + G_c G_p H)^2}\right) \\
&= \frac{-G_c G_p H}{1 + G_c G_p H}
\end{aligned} \tag{3.111}$$

Finally, the sensitivity to variations in the sensor is

$$S_H^{G_{cl}} = \frac{-L}{1 + L} = -T \tag{3.112}$$

the negative complementary sensitivity! Once again the way in which $T(s)$ "complements" $S(s)$ is made clear.

Proposition 3.13 *Performance robustness with respect to variation of parameters in the sensor cannot be achieved at the same frequencies as is the performance robustness with respect to parameter variation in the controlled object.* ■

Normally, Lm $T(j\omega)$ is near 0 dB at low frequencies and drops off at higher frequencies. This means that the sensor sensitivity is high at low frequencies (−1 dB/dB) and near the static gain condition and then improves at higher frequencies.

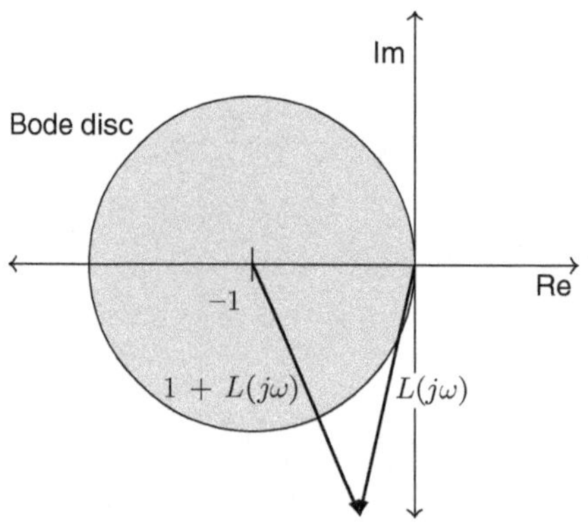

Figure 3.22 Graphical interpretation of the return ratio and return difference.

3.7.4 Graphical Interpretation of Return, Sensitivity, and Complementary Sensitivity

The return difference, $1 + L$, appears in the denominator of every closed-loop transfer function and in both Bode robustness measures, so it deserves a special notation and interpretation. We develop a graphical interpretation here. As side benefits, we develop graphical interpretations of the sensitivity and complementary sensitivity rational functions.

The interpretation is demonstrated in Figure 3.22, which is a picture of the L-plane. The complex number from the origin to any point in the plane is $L(j\omega)$ and the complex number from the -1 point to the polar plot of $L(j\omega)$ is $1 + L(j\omega)$, the return difference. That is because where this complex number is added to -1, the result must be L.

This simple interpretation provides insight into control system design. Those frequencies where the polar plot of $L(j\omega)$ approaches the -1 point are those frequencies where the command signal is not attenuated at the error node and those frequencies for which performance robustness is not achieved. A more precise statement is the sensitivity, $S(j\omega)$, is greater than 1 dB/dB at those frequencies where the polar plot enters the unit circle centered at the -1 point in the L-plane.

Definition 3.11 Call the area inside the unit circle, centered at the -1 point in the complex L-plane, the Bode disc.

Incidentally, a MATLAB script for generating the Bode disc is

```
>> xdisc = 2*[logspace(-4,-1),logspace(-1,0)];
>> ydisc = sqrt(2*xdisc-xdisc.*xdisc);
>> mdisc = 10*log(xdisc.*xdisc+ydisc.*ydisc)/log(10)
>> for k = 1:100;
>> pdisc(k) = 180*(-pi+atan(ydisc(k)/xdisc(k)))/pi;
>> end;
```

The arrays "mdisc" and "pdisc" are the coordinates of the Bode disc on the Lm G chart. These arrays are used in a statement of the form:

```
>> plot(phase,mag,pdisc,mdisc)
```

where arrays "phase" and "mag" are the Lm G chart coordinates of some transfer function. This call only displays the disc from $-90°$ to $-180°$. To display the disc from $-90°$ to $-270°$, generate

```
>> pdiscp = -360*ones(100,1)-pdisc;
```

and use the plot command

```
>> plot(phase,mag,pdisc,mdisc,pdiscp,mdisc)
```

The reader may decide to retain "xdisc" and "ydisc" rather than to clear them. These arrays can be used to generate the Bode disc on the polar plot or Nyquist diagram with the call

```
>> plot(-xdisc,ydisc,-xdisc,-ydisc)
```

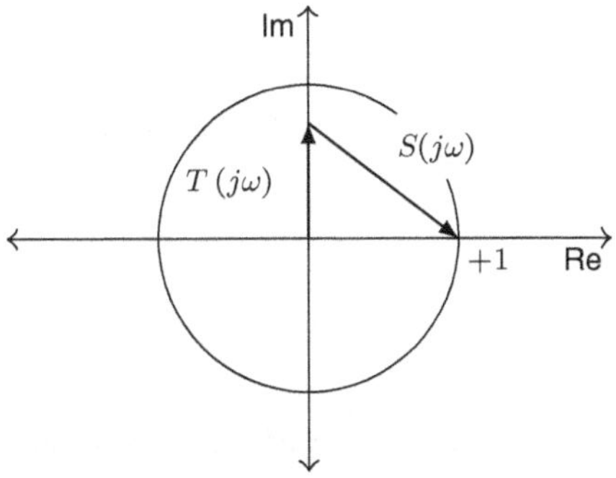

Figure 3.23 Potential S and T yielding $|S| > 1$.

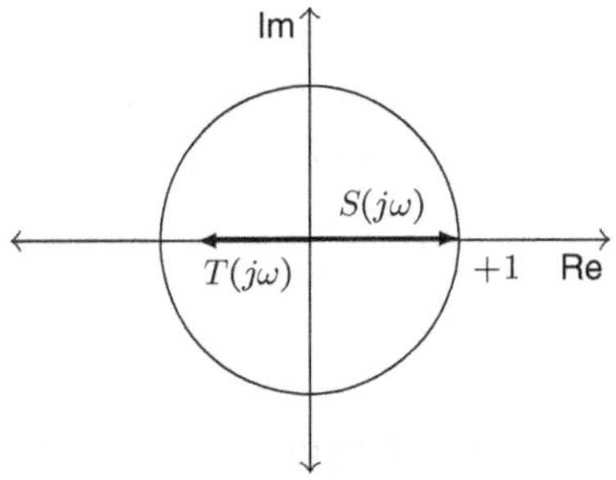

Figure 3.24 Potential S and T where $\angle T = -180^\circ$.

Next, we carefully interpret (3.104)

$$S(j\omega) + T(j\omega) = 1$$

This relation places a constraint on any feedback design. For one thing, it means $|S(j\omega)|$ and $|T(j\omega)|$ cannot both be $\ll 1$(0dB) at the same frequency, although there are situations where we would have preferred that this not be true. There are some tricky aspects to the interpretation of this fundamental constraint because it is a relation between complex numbers. For example, contrary to one's first guess, (3.104) does not mean that $|S(j\omega)|$ and $|T(j\omega)|$ cannot both be $\equiv 1$(0 dB) at the same frequency. Consider the vector diagram in Figure 3.23, which is consistent with (3.104):

Clearly, $|S(j\omega)| = \sqrt{1 + |T(j\omega)|^2} > 1$. This is true even where $|T(j\omega)| > 1$: That is, we can have the highly undesirable situation where both the sensitivity and complementary sensitivity magnitudes are both greater than 1 (0 dB). In fact, we find that this sorry situation will always arise in the case of non-minimum phase objects if we are not careful to limit the closed-loop bandwidth. An extreme case is illustrated by the vector diagram in Figure 3.24, where the argument of the complementary sensitivity is −180°:

Obviously,

$$|S(j\omega)| = 1 + 2|T(j\omega)|$$

in this case. It is these diagrams and our desire to produce small values of $|S(j\omega)|$ and $|T(j\omega)|$ that indicate the importance of the phase of $|S(j\omega)|$ and $|T(j\omega)|$.

3.7.5 Weighting Factors and Performance Robustness

One of the better-explored, neoclassical design methods is to establish upper bounds for the sensitivity and complementary sensitivity. In this procedure, the designer writes

$$|T(j\omega)| \leq |W_{PT}(j\omega)| \tag{3.113}$$

and

$$|S(j\omega)| \leq |W_{PS}(j\omega)| \tag{3.114}$$

The "P" subscript means these constraints are for "performance" criteria. Here, the right sides, called weighting factors, are eventually based on interpretations of the design criteria; for example, $|W_{PT}|$ is chosen to drop off sharply at high frequency and $|W_{PS}|$ is selected to be very low at low frequency. In addition, these weighting factors are chosen to constrain the peak values attained by $T(j\omega)$ or $S(j\omega)$ at any frequency (the so-called ∞-norms).

The ability to make a proper choice of a weighting factor comes with practice. While a weighting factor could be selected to take any arbitrary mathematical form, there are good reasons for taking them to be the Bode magnitudes of rational functions. In addition, it is useful to choose these rational functions to be proper with poles strictly in the left half-plane. That is to say, we select the weighting factors to be the Bode magnitudes of realizable, stable transfer functions. A common choice is the magnitude of the quadratic factor, which we explored in Chapter 2. The reason for this becomes clear in Chapters 5 and 6.

It will also become clear that it is useful to redefine these constraints as

$$|W_{PT}(j\omega)T(j\omega)| \leq 1 \tag{3.115}$$

and

$$|W_{PS}(j\omega)S(j\omega)| \leq 1 \tag{3.116}$$

At this point, we choose to only make some interpretations. First of all, we translate the bound $|S|$ to a bound on $|L|$. Clearly, (3.114) can be rewritten as:

$$\frac{|W_{PS}(j\omega)|}{|1 + L(j\omega)|} \leq 1$$

or

$$|1 + L(j\omega)| \geq |W_{PS}| \tag{3.117}$$

The graphical interpretation of this is: At each frequency ω, the polar plot of the return ratio, $L(j\omega)$, must lie outside the circular disk centered at -1 with radius $|W_{PS}|$. This is remindful of the graphical interpretation of the Nyquist stability criterion (explained in Chapter 4).

Suppose that we extend this procedure by bounding the Youla parameter as well (after all, we reduce the vulnerability of the control system in that way):

$$\frac{|W_{PY}(j\omega)|}{|Y(j\omega)|} \leq 1 \tag{3.118}$$

However, $Y = T/G_p$, so this condition reduces to (3.115) with

$$W_{PT} = W_{PY}/G_p \tag{3.119}$$

The approach here is to take any original weighting on $|T|$, call it W'_{PT}, and compare it with W_{PY}/G_p and then take the new bound on $|T|$ to be the minimum of $|W'_{PT}|$ and $|W_{PY}/G_p|$ at each frequency.

3.8 The Saturation Constraints

It might appear that some design considerations are outside the realm of the feedback designer. For example, it seems, to this point in our development, that power considerations or the sizing of actuators, such as pumps or motors, and plants are left to others. The impression is that these results are merely relayed to the feedback designer, after the fact, in the form of the controlled object transfer function, $G_p(s)$. This is far from the truth; in fact, basic physical constraints are a fundamental part of the feedback design. If that were not so, the academic designs produced on our sketch pads and desktops would likely be of no value. This important matter is the subject of this section.

These other considerations are usually expressed in the feedback design as the consideration of a possible saturation of the controlled input, $u(t)$, in response to an input signal, n, d, or r. (If the saturation were due to the sensor noise, the result would be especially embarrassing because the sensor is added only for purposes of feedback.) In this section, we learn that this consideration tends to limit the bandwidth (hence the speed of response) of the closed-loop system.

3.8.1 Bandwidth and Response Time

A somewhat slippery term in the control literature is that of bandwidth. "Slippery" because the definition is not given in a consistent manner (if it is given at all) and the meaning and use of the term is somewhat vague. Still, the term is useful and an attempt should be made to describe it.

The idea is to convey an estimate of the length of a frequency range which is important on the Bode magnitude curve of some dynamic system. This length is, roughly speaking, proportional to the speed with which the system responds to, say, the unit step input.

Of course, it is possible to give a precise mathematical meaning to bandwidth and we reproduce and develop one such definition in Appendix C. There, ω_c denotes the bandwidth. At that same place, we denote the effective duration of a transient T_r and demonstrate that

$$\omega_c \cdot T_r > 1/4 \tag{3.120}$$

Clearly, the duration of a transient is reduced by increasing the bandwidth.

For present purposes, we choose to give a description instead of a precise definition: The bandwidth is that range of positive frequency bounded by those frequencies where the Bode magnitude falls to 3 dB below the maximum magnitude.

Sometimes, 6 or 12 dB are used instead and so we will refer to the −3 dB- or to the −6 dB bandwidth. Additionally, only positive frequencies are used in this convention so that, as is often the case where the Bode diagram is monotonically decreasing, the bandwidth is taken to lie from the zero frequency to the −3 dB (or the −6 dB) frequency. We denote the bandwidth as "ω_B" (to distinguish it from a break frequency ω_b) and notice that −6 dB corresponds to a gain of 1/2 and −3 dB corresponds to a gain of $1/\sqrt{2}$.

The relationship of bandwidth to speed of response is established in this subsection for two special cases. Begin with the first-order system whose Bode modulus is described by Eq. (2.142). Clearly, the −3 dB bandwidth (the frequency where the magnitude is $1/\sqrt{2}$) is

$$\omega_B = 1/\tau \tag{3.121}$$

Higher bandwidth is related to smaller τ and faster system response to, say, step inputs.

Next, consider the quadratic factor whose Bode modulus is (2.153). A formula for bandwidth can be derived but it is overly complex for most purposes, so we consider the high-frequency asymptote instead. This asymptote is obtained by setting the condition $\omega \gg \omega_n$ to obtain

$$\left[\frac{\omega_n}{\omega}\right]^2$$

This approximation drops to the −6 dB value at

$$\omega_B = \sqrt{2} \cdot \omega_n \tag{3.122}$$

It follows (2.106), (2.107) that, at a fixed value of damping ratio, the rise and settling times diminish as bandwidth is increased.

3.8.2 The Youla Parameter and Saturation

Consider the problem of attenuating disturbance, noise, and command at the node we call the "power-stage" or "controlled-input," $u(t)$. Consider Figure 3.5 and obtain the closed-loop transfer function from the sensor noise to the controlled input,

$$\begin{aligned}\frac{\hat{u}(s)}{\hat{n}(s)} &= \frac{-G_c(s)}{1+L(s)} \\ &= \frac{-1}{1+L(s)}\left[\frac{L(s)}{G_p(s)}\right] \\ &= -Y(s)\end{aligned} \tag{3.123}$$

the "Youla parameter." Similarly, the transfer function form the output disturbance signal to the controlled input is

$$\frac{\hat{u}(s)}{\hat{d}_y(s)} = -Y(s) \tag{3.124}$$

Finally,

$$\begin{aligned}\frac{\hat{u}(s)}{\hat{r}(s)} &= \frac{1}{1+L(s)}\left[\frac{L(s)}{G_p(s)}\right] \\ &= Y(s)\end{aligned} \tag{3.125}$$

These relations can be rewritten in terms of another closed-loop transfer function, the "complementary sensitivity":

$$Y(s) = \frac{T(s)}{G_p(s)} \tag{3.126}$$

The designer needs to know as much as possible about the quantitative properties of the noise signal in order to evaluate the effect of noise on the controlled input. In lieu of that, Eqs. (3.122) and (3.123) can be used to obtain some useful observations. We first make some inferences about the closed-loop transfer functions, $Y(s)$, and then translate these observations into conditions and constraints on the return ratio, $L(s)$.

We take as axiomatic that $|L(j\omega)|$ will be chosen to be much greater than 1 in some range of frequency and much less than 1 in another. The exception is the neighborhood of the crossover frequency where this Bode magnitude is ≈ 1. This approach is certainly justified by the developments of prior sections.

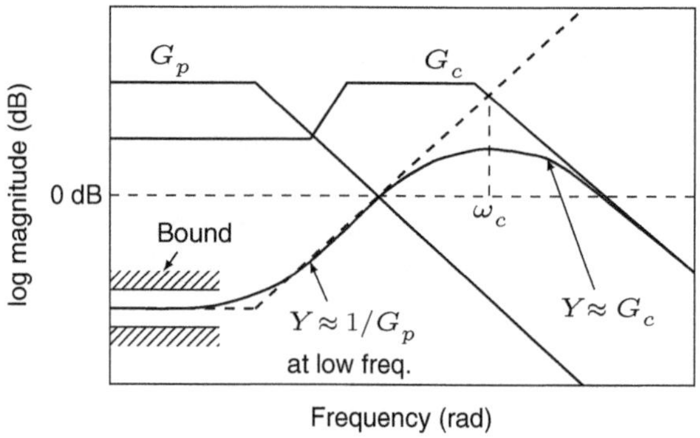

Figure 3.25 Youla design with high-gain plant.

Where $|L(j\omega)| \gg 1$ (at low frequencies), we have that

$$\begin{aligned} Y(j\omega) &\approx L/(LG_p) \\ &= 1/G_p(j\omega) \end{aligned}$$

Thus,

$$\text{Lm } Y \approx -\text{Lm } G_p \tag{3.127}$$

at low frequency (usually at frequencies below crossover).

Where $L(j\omega) \ll 1$ (usually at high frequencies), we have

$$Y(j\omega) \approx L(j\omega)/G_p(j\omega)$$

So,

$$\text{Lm } Y \approx \text{Lm } L - \text{Lm } G_p = \text{Lm } G_c \tag{3.128}$$

Interpret this result to mean that $-\text{Lm } G_p$ is the low-frequency asymptote of $\text{Lm } Y$ and $\text{Lm } L - \text{Lm } G_p$ is the high-frequency asymptote. These asymptotes intersect at that frequency where $-\text{Lm } G_p = \text{Lm } L - \text{Lm } G_p$. That is, the frequency where $\text{Lm } L = 0$ dB, which is the crossover frequency ω_c of the return ratio.

A typical situation is illustrated in Figures 3.25.

We have the obvious implications where the crossover frequency of the return ratio is higher than that of the controlled object:

Proposition 3.14 *Below the crossover frequency of the controlled object, the tendency of the controlled input, $u(t)$, to saturate is dictated by the transfer function of the controlled object and is not affected by the feedback.* ■

and

Proposition 3.15 *Above the crossover frequency of the return ratio, the tendency of the controlled input to saturate is dictated by Bode modulus of the return ratio less the Bode modulus of the controlled object (Bode modulus of the compensator) and, so, is affected by the feedback.* ■

and

Proposition 3.16 *Between the crossover frequency of the controlled object and the return ratio, the tendency of the controlled input to saturate makes the transition from being independent of feedback to being dependent.* ■

The conclusions are modified where the crossover frequency of the return ratio is set lower than the crossover of the controlled object (the closed loop responds slower than the open loop).

We expect the reference and the disturbance to be low frequency and we conclude that there is nothing the feedback designer can do about saturation to these inputs! This is a task for the designer of the controlled object: that person may be asked to increase the open-loop gain. It is intuitively satisfying that a saturation of this nature is due to the fact that the controlled object is not "powerful" enough.

This begs the question: is there a benefit to simultaneous design of the controlled object and the compensator? After all, a controlled object is a mechatronics system, where mechatronics systems are synergistic systems that include mechanical, electronics, computer, telecommunications, and control systems. We don't pursue answering this question in this book, but we should say that there are recent research studies [12] that showed the simultaneous design of a controlled object and a compensator could decrease the demand on requiring a more powerful controlled object.

On the other hand, the sensor noise is usually high frequency and something can be done about saturation in this frequency range. That "something" is to design the feedback loop so that $L(j\omega) \approx G_p(j\omega)$ after crossover. This has direct consequences when designing feedback loops using the neoclassical technique with Youla parameterization, as will be discussed in Chapter 4.

If the crossover frequency of the return ratio is much greater than that of the controlled object, Lm L will be large at the latter frequency and the feedback system is vulnerable to saturation in response to any of the three inputs with content in this frequency range.

The last two conclusions dictate that the return ratio must be shaped close to the controlled object after the crossover frequency of the controlled object.

This means that:

> The closed-loop system cannot respond much faster than the controlled object! Major improvements in the speed of response can be obtained only with a redesign of the controlled object, as improvements in the crossover frequency of G_p. This usually involves a boost in power or some other property of the actuators.

Think of it this way: feedback is an information process which insures the automatic system will adjust for disturbance or model uncertainty: Feedback can never be used to circumvent, say, the conservation of mass or energy. Another way to think of this is that a premium is placed on high-bandwidth controlled objects because a faster-responding feedback loop can be designed for such objects. Also, the value of a highly accurate and stable sensor is apparent because, in that case, the designer can allow greater deviations of Lm (L) above Lm (G_p), greater values of Lm (Y) at the higher frequencies.

3.9 Problems

3.1 Consider a closed-loop transfer function:

$$T = \frac{100K(2.5s - 1)}{(10s + 1)(5s + 1)(2.5s + 1)} \tag{3.129}$$

that corresponds to a plant:

$$G_p = \frac{100(2.5s + 1)}{(10s + 1)(5s + 1)(1.25s + 1)} \tag{3.130}$$

(a) What K will meet the interpolation condition of $T(0) = 1$ for command tracking?
(b) Verify the interpolation conditions for $S = 1 - T$.

3.2 For each controller/plant pair, indicate whether the closed-loop system is internally stable and whether it is well-posed.

(a) $G_p = \dfrac{10(1 - s/2)}{(1 - s/2)}$, $G_c = \dfrac{1}{(1 - s/2)}$

(b) $G_p = \dfrac{1}{(0.25s + 1)^2}$, $G_c = \dfrac{\frac{3}{4}(s/4 + 1)^2}{s(s/6 + 1)}$

(c) $G_p = \dfrac{10(1 - s)}{(1 - s/2)}$, $G_c = (1 - s/2)$

3.3 Consider the unstable and non-minimum phase plant,

$$G_p = \frac{10(1+s)}{(1-s/2)} \tag{3.131}$$

Consider a Youla

$$Y = \frac{K(\tau_z s+1)}{(\tau_1 s+1)(\tau_2 s+1)} \tag{3.132}$$

Check the interpolation conditions for this plant and Youla. Is it possible to achieve internal stability with the given Youla parameter? What ranges of τ_z, τ_1, and τ_2 are required?

3.4 Consider the transfer function

$$G_p = \frac{(10+s)}{s^3} \tag{3.133}$$

with proposed Youla

$$Y = \frac{s^3(\tau_z s+1)}{(\tau + ps+1)^4} \tag{3.134}$$

Will this Youla be able to meet the interpolation conditions on T?

4

Feedback Design For SISO: Shaping and Parameterization

This chapter introduces the neoclassical method of controller design. It will discuss measures of stability and robustness followed by algebraic and analytical constraints in feedback design, revealing how Youla parameter could be utilized in designing a feedback control by considering the constraints and trade-offs in SISO control design.

4.1 Closed-Loop Stability Under Uncertain Conditions

To this point, the concept of gain seemed the only important idea for feedback control because it is this parameter which determines the extent to which an input signal is attenuated at the nodes of the feedback loop. Also, it is the gains of the sensitivity functions that characterize the robustness. We demonstrate in this section that it is the phase curve *in combination* with the gain curve that determines the stability of the closed loop.

The use of high gain in the return ratio must be abandoned after some frequency. For one reason, the gain must be lowered in order to attenuate high-frequency sensor noise; that is, in order to effect the basic trade-off. For another reason, the gain must be lowered in order to establish closed-loop stability. It is the latter fact that is developed in this section. In particular, we study the extent to which the gain must be diminished at high frequency in order to obtain a robust stability.

In the final part of this section, we develop the implications of all this for $T(j\omega)$ and $S(j\omega)$ and in that way, the significance of the H_∞ norm is introduced.

The first task is to obtain a stability limit condition, a condition written in terms of frequency response characteristics. The simplest way to do that is to recognize that, at the stability limit, $s = j\omega$, is a closed-loop pole. It then follows from (3.35) (the closed-loop characteristic equation), that

$$1 + L(j\omega) = 0 \tag{4.1}$$

This stability limit condition is the basis of our study of stability in the frequency formalism.

4.1.1 Harmonic Consistency

There is another way to obtain the condition in Eq. (4.1), a way that has an intuitive appeal.

Imagine the situation illustrated in Figure 4.1: a closed-loop system is at a stability limit so that, in steady state, the output has the oscillatory form:

$$y(t) = A\cos(\omega t) \tag{4.2}$$

for some amplitude "A" and some frequency "ω".

Each component in the feedback loop modifies its input with amplitude and phase changes dictated by the frequency response of that component's transfer function. This sort of modification is shown after each block in Figure 4.2.

There is one other important shift in phase: the negative feedback means that a minus sign, a $\pm 180°$ phase shift occurs at the summing junction. The *principal of harmonic consistency* is that the amplitude and phase changes around the loop must be such that the signal leaving the controlled object must match up with the form (4.2) assumed for the output, that is,

$$M_p M_c M_H A = A$$

and

$$\phi_H + \phi_c + \phi_p \pm 180° = 0 \tag{4.3}$$

Robust Control: Youla Parameterization Approach, First Edition. Farhad Assadian and Kevin R. Mallon.

Companion Website: www.wiley.com/go/assadian/robust

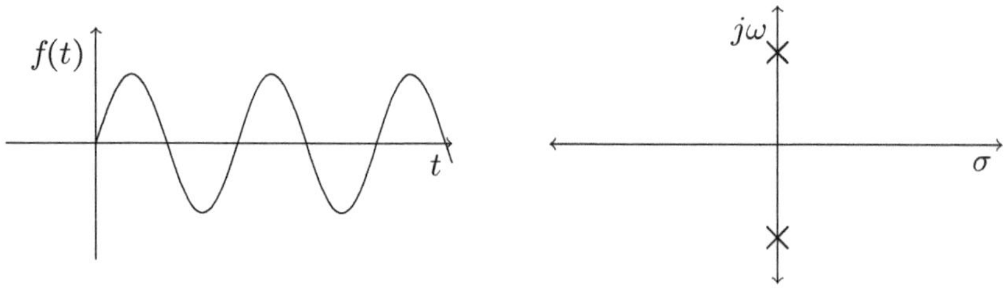

Figure 4.1 Marginally stable system.

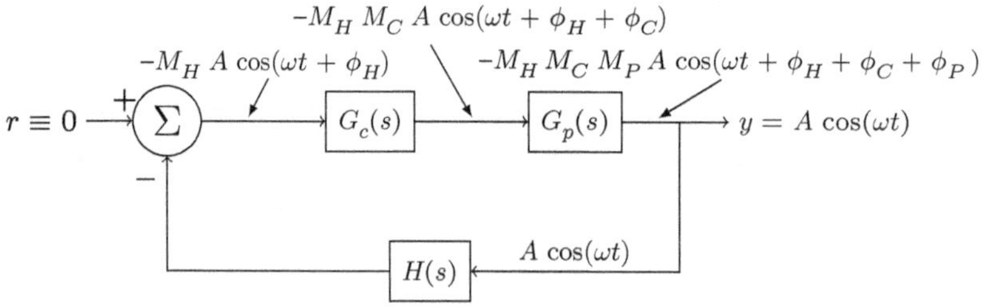

Figure 4.2 Block diagram for harmonic consistency.

These conditions can easily be rewritten in terms of the frequency response of the return ratio.

Theorem 4.1 *The stability limit conditions for the closed loop are*

$$|L(j\omega)| = 1 \tag{4.4}$$

or

$$Lm\ L(j\omega) = 0\ dB \tag{4.5}$$

and the phase condition:

$$\phi_L = \pm k180^\circ, \quad where\ k = \pm 1, \pm 3, \pm 5, \dots \tag{4.6}$$

where "k" is any odd integer. ∎

The conditions of Eqs. (4.4) and (4.6) are equivalent to (4.1). Normally, but not always, the system is stabilized by reducing the gain from this condition. Mathematically, $\pm 180^\circ$ are the same angle and so, depending on the software package, one must be alert where any one of these phases occur near a Lm $L = 0$ dB condition. Normally, the system is stabilized by keeping the phase less than $+180^\circ$ or greater (less negative) than -180°. Finally, one can be fooled by an uncritical use of this simple criterion: the wrong indication is given if a right half-plane cancellation occurs in the return ratio (more on this further on).

A more precise and a more complete statement of closed-loop stability, the *Nyquist Stability Criterion*, is discussed in Section 4.1.2.

4.1.2 Nyquist Stability Criterion: Heuristic Justification

There is a more satisfying way to study closed-loop stability, the Nyquist stability criterion. The rigorous proof of this criterion is not provided in this chapter. However, the utility of the criterion is required at this point so we provide a more intuitive justification here, at a point where the required intricacies of complex analysis are not given.

Rewrite the rational form of the characteristic equation:

$$1 + L(s) = 0 \tag{4.7}$$

by placing both terms over a common denominator and factoring both the numerator and denominator polynomials:

$$1 + \frac{\eta(s)}{\delta(s)} = \frac{k(s+p_1)(s+p_2)\dots(s+p_n)}{(s+\delta_1)(s+\delta_2)\dots(s+\delta_n)} = 0 \tag{4.8}$$

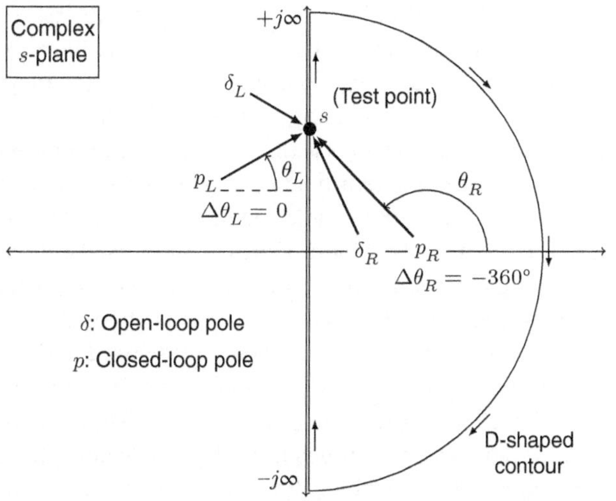

Figure 4.3 An intuitive justification of the Nyquist stability criterion. Imagine that the test point moves about the D-shaped contour in the clockwise direction.

The "δ_k" are the open-loop poles and the "p_k" are closed-loop poles (the zeros of the characteristic equation).

Consider the situation illustrated in Figure 4.3 and imagine carrying the "test point" through one complete clockwise tour around the D-shaped contour which encloses the right half-plane. Clearly, the vector $s - p_k$ will not contribute a net rotation to $1 + L$ if p_k lies in the left half-plane – starting at the origin, for instance, θ_L increases from 0° to +90°, then decreases to −90°, and returns to 0°. So, the net change in θ_L is zero. On the other hand, there is a net clockwise rotation if p_k lies in the right half-plane if the closed-loop pole, p_k, is in the unstable region – the net change in θ_R will be −360°.

The same thing can be said about the vector $s - \delta_k$ except that, since this is a denominator term, $s - \delta_k$ contributes a counterclockwise rotation to $1 + L$ should the open-loop pole, δ_k, lie in the right half-plane.

Our summary is that the number of clockwise rotations of $1 + L$, the test point makes one complete clockwise tour (that is, in the direction of increasing frequency), is $N_c - N_o$, where N_c is the number of unstable closed-loop poles and N_o is the number of unstable open-loop poles:

$$N = N_c - N_o \tag{4.9}$$

Here, a negative integer value of N corresponds to a counterclockwise rotation of $1 + L$.

The natural thing is to imagine these rotations as the motion of a vector drawn from the origin of the $1 + L$ plane. However, we prefer to imagine the motion of a vector in the L-plane (the polar or Nyquist plot). A rotation of $1 + L$ about its origin (the −1 point) is equivalent to the encirclement by L of the −1 point in the L-plane so we have the following stability criterion:

Theorem 4.2 (Nyquist stability criterion) *The number of unstable close-loop poles is*

$$N_c = N_o + N \tag{4.10}$$

where

$$N_o \triangleq \text{the number of unstable open-loop poles} \tag{4.11}$$

and

$$N \triangleq \text{the number of times the Nyquist locus encircles the } -1 \text{ point in a clockwise manner:} \tag{4.12}$$

"N" is determined by moving along $L(j\omega)$ in the direction of increasing ω. ■

We assume here that the curved portion of the D-shaped contour maps to the origin of the L-plane; that is, we assume that L is strictly proper. Notice that if there are any unstable closed-loop poles, then we have closed-loop stability if and only if $N = -N_o$; that is, if and only if the Nyquist locus encircles the −1 point in a counterclockwise manner a number of times exactly equal to the number of unstable poles in the open loop.

We must modify the D-shaped contour where a pole or zero is on the imaginary axis (in either case, there is a discontinuity in phase at that frequency). The usual procedure is to make a small semi-circular indentation in the contour at the complex pole or zero. The indentation is made into the left half-plane. Let "ϵ" denote the radius of the semicircle and imagine constructing the Nyquist locus of L on this closed contour. The parameter, ϵ, will have a geometric interpretation on this mapping. Finally, allow $\epsilon \to 0$ and draw the usual conclusions about stability from the asymptotic Nyquist locus.

It is now clear that the closed loop is stabilized in the manner indicated in Theorem 4.1 only if there are no *open-loop poles in the right half-plane.* If the open loop is unstable (as is the case with high-performance aircraft), the closed loop is stabilized by modifying the frequency response of the return ratio, $L(j\omega)$, in quite another way.

How great should we make the adjustment for stability indicated in Theorem 4.1? What are the recommendations that have evolved in this regard? These questions are among those addressed in this subsection.

In the Bode diagrams of minimum phase systems, an asymptotic phase of $-90°$ is associated with each multiple of slope of -20 dB/decade in the gain curve and, asymptotically, $+90°$ is associated with each multiple of slope of $+20$ dB/decade. This follows from the superposition of the frequency response curves, as discussed in Chapter 2.

Combine Eqs. (4.4) and (4.5) with this principle to draw some conclusions about the closed-loop instability. The goal is to be able to predict instability by examining only the gain curve on the Bode diagram. This can only be done for minimum phase return ratios because it is for these ratios that this principle can be used to draw the required inferences about phase.

A minimum phase feedback system is close to the stability limit condition if the gain crosses the 0 dB line with a slope of -40 dB/decade. Normally, a slope more negative that this indicated instability while a slope of more positive than -40 dB/decade indicates stability. If the asymptotic phase has been realized at this crossover frequency, then the phase margin is 90°.

4.1.3 Stability Margins and Stability Robustness

An important practical observation must be injected here. It is impossible to have perfect confidence in the predictions of stability made in this way because mathematical models are always an approximation of the truth. Our next explorations of feedback design are based on that observation. In particular, we wish to provide a closed-loop response that is stable even in face of the inevitable uncertainties.

Mathematical models are always an approximation of the truth. This is due to necessity, convenience, or ignorance. This will always be true. Also, inevitable shortcuts are made in the development of a model; for example, one such shortcut is to neglect bilateral coupling effects. This means that even the true order of the controlled object (the number of poles) is underestimated and uncertain. Usually, the effect can be characterized as an uncertainty in the high-frequency portion of the Bode (or Nyquist) diagram. Since we know next to nothing about this uncertainty, except for its inevitable existence, we call it *unstructured uncertainty*. Another name is parasitic dynamics because it is "parasitic" to the stability of high-gain feedback.

Even where we assume the model is of the correct order, we are unsure of the true values of parameters. This is due to uncertainty in measurement or due to the actual change of parameter values due to wear, to warm-up, or due to changes in operating conditions. This kind of uncertainty is different than unstructured uncertainty because, in this instance, we can bound the uncertainties in the magnitude and phase of $G_p(j\omega)$ given estimates of the parameter variations. Also, the bound on phase uncertainty are functions of the bounds on gain uncertainty (with frequency as the intermediate variable). For this reason, we refer to this as *structured uncertainty*.

A special concern is the unstructured uncertainty, for which we must install some kind of factor of safety for stability. We begin that development with the following:

Definition 4.1 The phase margin, ϕ_m, is related to a specific value of the phase, ϕ, of $L(j\omega)$; namely, the phase of the return ratio at the crossover frequency, ω_c:

$$-180° + \phi_m = \phi(\omega_c) \tag{4.13}$$

The gain margin is denoted K_m and is defined by:

$$K_m = 1/|L(j\tilde{\omega})| \tag{4.14}$$

where $\tilde{\omega}$ is that frequency where the phase of L is exactly $\pm 180°$. The two margins are referred to, collectively, as *stability margins*.

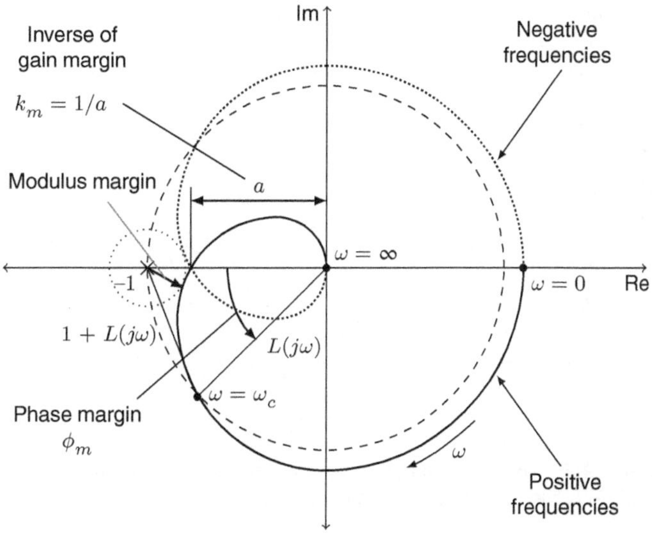

Figure 4.4 Nyquist plot with stability margins.

In some cases, there may be several crossover frequencies and several values for the stability margins. Also, a particular software package may return +180° where the convention is to use −180°. In that case, the phase margin is

$$\phi_m = 180^\circ - \phi(\omega_c) \tag{4.15}$$

Gain and phase margins are easily visualized and read from the common frequency response diagrams. For example, the definitions are illustrated in Figure 4.4 for the polar plot. The margins, first of all, indicate how close our design is to satisfy one of the stability limit conditions (4.4) and (4.5) and, as such, can be looked on as factors of safety against instability. Much more is indicated by these stability margins and that is drawn out later.

An important part of any feedback design is to assure minimum values for the gain and phase margins. For various reasons, this value of ϕ_m is the lower bound usually recommended for phase margin: It is recommended that

$$\phi_m \geq 60^\circ \tag{4.16}$$

We have demonstrated that high gain is desirable at any frequency in order to provide input attenuation and robustness. This means that the steepest possible slopes in the gain curve near crossover, the −40 dB/decade slopes, are desirable and that should be kept in mind when selecting the compensator.

4.1.4 Margins, $T(j\omega)$ and $S(j\omega)$, and H_∞ Norms (Relationships Between Classical and Neoclassical Approaches)

There is an interesting relationships between the stability margins and the magnitude of the closed-loop transfer functions, $T(j\omega)$ and $S(j\omega)$. Also, there are interesting relationships at the crossover frequency of the return ratio, $L(j\omega)$.

The crossover frequency – the frequency at which $|L(j\omega)| = 1$ – is denoted ω_c and the phase margin, ϕ_m, is defined such that $-180^\circ + \phi_m$ is the argument of the return ratio at this frequency. That is,

$$\begin{aligned} L(j\omega_c) &= 1 \cdot e^{-j(\pi-\phi_m)} \\ &= \cos(\pi - \phi_m) - j\sin(\pi - \phi_m) \\ &= -\cos(\phi_m) - j\sin(\phi_m) \end{aligned} \tag{4.17}$$

Next we observe that, since $T = LS$ and $|L(j\omega_c)| \triangleq 1$,

$$|T(j\omega_c)| = |S(j\omega_c)| \tag{4.18}$$

That is, the intersection of the sensitivity and complementary sensitivity magnitude curves occurs at the crossover frequency of the return ratio.

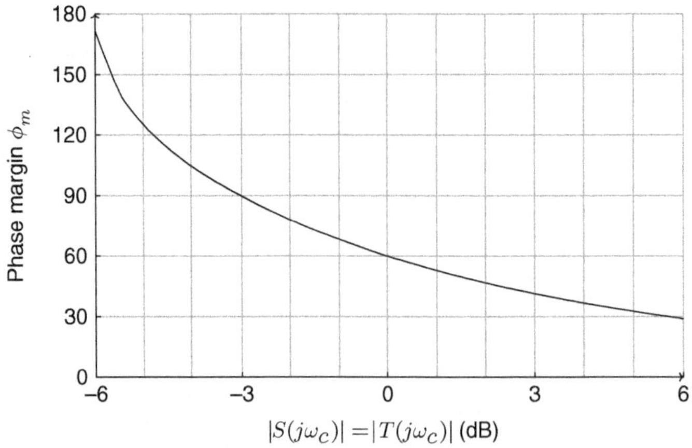

Figure 4.5 The phase margin from the intersection of $|S(j\omega_c)|$ and $|T(j\omega_c)|$ (valid only if G_p is minimum phase).

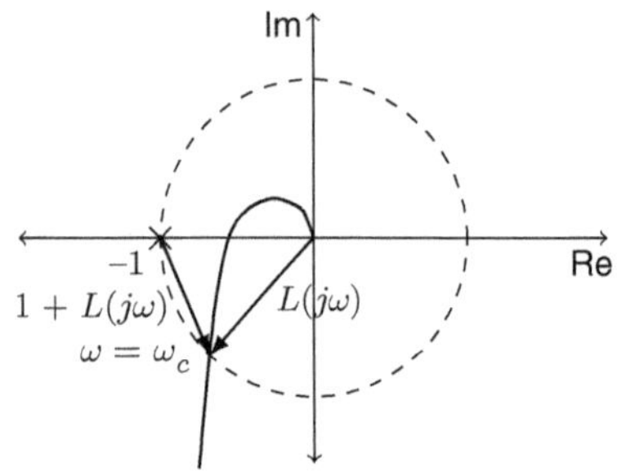

Figure 4.6 Crossover frequency on Nyquist plot.

Equation (4.17) means that the magnitude of the sensitivity is

$$\begin{aligned}|S(j\omega_c)| &= \frac{1}{|1 + L(j\omega_c)|}\\ &= \frac{1}{|1 - \cos(\phi_m) - j\sin(\phi_m)|}\\ &= \frac{1}{\sqrt{[1 - \cos(\phi_m)]^2 + \sin^2(\phi_m)}}\\ &= \frac{1}{\sqrt{2 - 2\cos(\phi_m)}}\end{aligned}$$

At this point, we turn to a table of trigonometric identities, ($\cos(2\theta) = 1 - 2\sin^2(\theta)$), to establish that

$$|S(j\omega_c)| = \frac{1}{2\sin(\phi_m/2)} = |T(j\omega_c)| \tag{4.19}$$

That a relationship between ϕ_m and T should exist is probably obvious from an examination of Figure 4.4, but the simplicity of the relationship is a surprise. This relationship is plotted in Figure 4.5.

Consider Figure 4.6 Clearly, the crossover frequency is where $|L| = 1$; that is, where $L(j\omega)$ crosses the unit circle. The value of $|S(j\omega_c)| = |T(j\omega_c)|$ is the ratio of "1" to the length $|1 + L|$.

This observation allows us to conclude with the following:

Proposition 4.1 *The phase margin is* $> 60°$ *if* $T(j\omega)$ *intersects* $S(j\omega)$ *at a magnitude below* 1 *(0 dB) and is* $< 60°$ *if* $T(j\omega)$ *intersects* $S(j\omega)$ *at a magnitude above* 1 *(0 dB).* ■

Next, let $\tilde{\omega}$ denote that frequency at which the argument of the return ratio is $\pm 180°$ (where the Nyquist locus crosses the negative real axis). By definition, the Bode magnitude at this frequency is $1/K_m$ (Figure 4.4), the reciprocal gain-margin.

That is to say,

$$L(j\tilde{\omega}) \triangleq -1/K_m \tag{4.20}$$

This leads to the conclusion that

$$T(j\tilde{\omega}) = \frac{1}{1-K_m} \tag{4.21}$$

so

$$|T(j\tilde{\omega})| = \frac{1}{|K_m - 1|} \tag{4.22}$$

Similarly,

$$|S(j\tilde{\omega})| = \frac{K_m}{|K_m - 1|} \tag{4.23}$$

We usually assume that the closed loop is stable, that $K_m > 1$. The significance of $|T(j\tilde{\omega})|$ for stability robustness is among the features illustrated in Figure 4.6. Clearly,

$$\text{Lm } S(j\tilde{\omega}) = \text{Lm } T(j\tilde{\omega}) + \text{Lm } K_m \tag{4.24}$$

The question arises, how do we find $\tilde{\omega}$ on say, a $T - S$ diagram? (Clearly, $\tilde{\omega}$ is the frequency where $\arg(L(j\omega)) = -180°$ on an L-diagram.) The answer is supplied by an examination of (4.21): If we assume that $K_m > 1$, then $\arg(T(j\tilde{\omega})) = -180°$. That is we seek the frequency where this argument obtains and the gain margin, in dB, is the distance of the sensitivity above the complementary sensitivity (in log-magnitude).

Incidentally, there is an alternative to the use of these classical stability margins, an alternative in which only a single number is needed to account for unstructured uncertainty.

4.1.4.1 Neoclassical Approach

Although gain margins and phase margins are useful measures of stability, they do no not paint a complete picture regarding stability. These margins measure the amount of gain *or* phase change a system can handle before going unstable, but they do not say anything about how *simultaneous* gain and phase changes can destabilize a system. For instance, consider the Nyquist locus shown in Figure 4.7. This system has an infinite gain margin and has a phase margin near 90°, yet a small amount and gain and phase change at the same time could easily cause the locus to encircle −1 and make the system unstable.

For this reason, neoclassical modeling of uncertainties uses a single number to account for unstructured uncertainty.

Definition 4.2 The stability margin, M_2, is the minimum distance from the plot to the −1 point in the $L(j\omega)$-plane, as illustrated in Figure 4.4.

It is easy to prove the following:

Proposition 4.2 *The M_2-margin is related to the return difference by:*

$$M_2 = \min_{\omega} |1 + L(j\omega)| \tag{4.25}$$

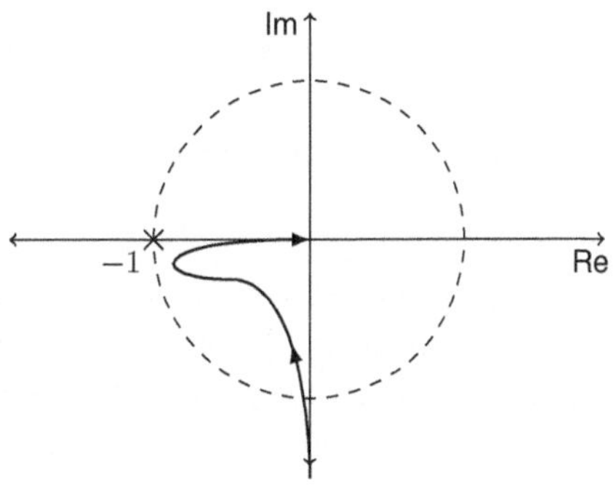

Figure 4.7 Nyquist plot.

and to the sensitivity by:

$$1/M_2 = \max_{\omega} |S(j\omega)| \tag{4.26}$$

■

Currently, this measure is gaining favor over the classical phase and gain margins. For one thing, this number is easy to pick off the $S(j\omega)$ curve and, for another, it is a reliable measure of just how close the $L(j\omega)$ locus comes to the -1 point. As discussed earlier, in contrast, it is conceivable that the classical measures could be adequate and the locus still comes dangerously close to the critical point, as shown in Figure 4.7. And besides, by forcing the locus to lie outside some circle about -1, de facto values of phase and gain margin are imposed on the design.

Example 4.1 Now, consider a feedback loop for which that maximum magnitude of the sensitivity function, max $|S(j\omega)|$ is less than 2 (where a magnitude of 2 is roughly 6 dB). Then, show the classical gain margin will be at least 2.0 (6 dB) and the phase margin will be at least 29°:

$$\max_{\omega} |S(j\omega)| < 2$$

Then, from Eq. (4.26),

$$M_2 > \frac{1}{2}$$

Then, we can write

$$\begin{aligned} |S(j\tilde{\omega})| &\leq \max_{\omega} |S(j\omega)| \\ \frac{k_m}{k_m - 1} &\leq \frac{1}{M_2} < 2 \\ k_m &< 2(k_m - 1) \\ k_m &> 2 \approx 6\text{dB} \end{aligned}$$

Therefore, if $M_2 > 1/2$, then the gain margin has a lower bound of 2 (about 6 dB). The derivation of the phase margin is left as an exercise.

We can follow this same procedure to find other bounds on the gain margin. For instance, if max $|S(j\omega)| < 4$ dB, then we can find that $k_m > 2.71$ or approximately 8.7 dB and the phase margin will be at least 36.8°. And where max $|S(j\omega)| < 3$ dB, the gain margin will be greater than 3.42 (10.7 dB) and the phase margin will be at least 41.5°.

The denominator of every closed-loop transfer function is the return difference, so we have

Proposition 4.3 *The frequency that corresponds to M_2 is a closed-loop resonant frequency.* ■

A new norm is introduced in Chapter 5, the H_∞-norm for rational functions of a complex variable (or in short ∞-norm):

$$||G(s)||_\infty \triangleq \max_{\omega} |G(j\omega)| \tag{4.27}$$

The significance of this norm begins to become apparent by the following statement:

> The H_∞-norm of the sensitivity, $S(s)$, is the M_2 stability margin and the frequency of this maximum is a closed-loop resonant frequency.

It should be noted that while this "norm" plays a pivotal role in the neoclassical theory, some of the significance of this number was already appreciated in the classical theory where the H_∞-norm of $T(s)$ is called the "peak magnitude." This number plays various purposes in feedback design: clearly, the H_∞-norm of $T(s)$ occurs at a closed-loop resonant frequency and can be related, in an approximate way, to the closed-loop damping ratio. The H_∞-norms of T and S are illustrated in Figure 4.8.

As a result of these observations, it becomes clear that

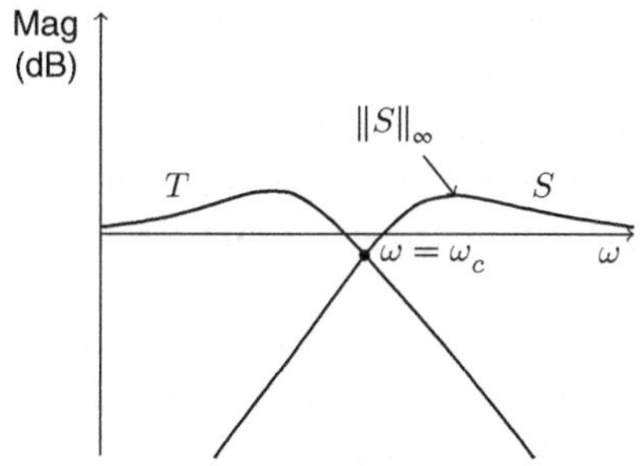

Figure 4.8 Crossover frequency on magnitude plot of S and T.

we always seek to reduce the H_∞-norms of both $T(s)$ and $S(s)$ in order to obtain sufficient damping and stability margin and to reduce closed-loop resonances.

We will finish this chapter by discussing and bringing to the attention of all the mathematical design constraints. These constraints heavily influence neoclassical feedback design covered in Chapter 4.

4.2 Mathematical Design Constraints

There are certain constraints with more of a mathematical nature than a technical. This is not to say that they are any less real; in fact, these mathematical constraints are equally important considerations in design. Some can be developed in this section but others can only be derived in Chapter 5 because some mathematical background is required for that task. However, it makes sense to at least catalog them here.

4.2.1 Sensitivity/Complementary Sensitivity Point-wise Constraints

Classical feedback design procedures focus on the return ratio, the open-loop transfer function, in order to achieve closed-loop objectives. Though seemingly indirect, these procedures are quite effective and receive some attention in Chapter 4. In addition, there are the neoclassical procedures which focus directly on the closed-loop transfer functions, $T(s)$, $S(s)$, and $Y(s)$ themselves. These procedures are drawn out in the next chapters in this sequel, but this is a good point to introduce some definitions and propositions fundamental to those design methods. Knowing the result (3.104)

$$S(j\omega) = 1 - T(j\omega)$$

and combine this equation with $|S(j\omega)|^2 + |T(j\omega)|^2 = 1$ to establish the Bode magnitude relations of the

Proposition 4.4

$$|1 - |T(j\omega)|| \leq |S(j\omega)| \leq 1 + |T(j\omega)| \tag{4.28}$$

Similarly,

$$|1 - |S(j\omega)|| \leq |T(j\omega)| \leq 1 + |S(j\omega)| \tag{4.29}$$

■

For example, where we choose $S(0) = 0$, for good command-following and robustness to parameter changes in the controlled object, inequalities (4.28), (4.29) mean that $T(0) = 1$. This means that we will be susceptible to sensor parameter variation in this low-frequency range.

It follows from this proposition that the gain of either $S(s)$ or $T(s)$ will constrain the value of the other. Roughly speaking, one gain must be near unity where the other is small; for example, those frequencies where error is suppressed are those frequencies of noise which are not suppressed. Often, one can choose to suppress noise at the lower frequencies and attenuate the noise signal at higher frequencies, since one normally expects sensor noise to be at relatively higher frequencies. Obviously, some designs can run into difficulty over this point. This elementary consideration is a good example of the trade-offs that constitute feedback control design.

Other important constraints are the "interpolation conditions" expressed in (3.74)–(3.77). Clearly, the $S(s)$ and $T(s)$ transfer functions are constrained to take on certain values in the complex plane and these constraints influence the values on the imaginary axis. It has been demonstrated that these constraints cannot be removed by a pole–zero cancellation where "z" or "p" lie in the right half-plane, where a factor of the return ratio is non-minimum phase. These positions place constraints on the values that the sensitivity and the complementary sensitivity can take on at specific frequencies and, as such, they impose limits to feedback design.[1]

One final point on this matter: Where the gain of $L \approx 1$ (this occurs at the crossover frequency), the gains of both $S(s) = 1/(1+L)$ and $T(s) = L/(1+L)$ are greater than 1/2 (−6 dB). Neither gain is very small in this range; that is to say **feedback is not very effective near the crossover frequency.**

4.2.2 Sensitivity, Complementary Sensitivity, and Analytic Constraints

4.2.2.1 Non-minimum Phase Constraints on Design

One can often improve a feedback design with the use of a pole–zero cancellation, a cancellation of a pole on the controlled object with a zero in a compensator. It is clear from the principle of internal stability that this is not an option where a pole or a zero lies in the right half-plane, where the controlled object is characterized by a non-minimum phase zero or an unstable pole. Of course, even if the pole–zero cancellation is effected in the left half-plane, the bandwidth constraints, described in Chapter 3, place a practical constraint on what can be done. As a practical matter, the cancellation should not be effected in the case of stable poles where the canceled pole lies very near the imaginary axis since it may be excited by some disturbance signal. Where cancellation is impossible, improvement must be sought in some other, severely constrained way.

In this subsection, that constraint is described in more detail. This discussion is aided where we employ factorization for the return ratio:

$$L(s) = L_{mp}(s)L_{ap}(s) \tag{4.30}$$

$L_{mp}(s)$ is a minimum phase factor and $L_{ap}(s)$ is all-pass and this factorization will provide us with some tools to shape $L(s)$ in case of right half-plane pole or zero or pure delay element. The effect of the all-pass factor on $L(j\omega)$ is merely to contribute frequency-dependent phase lag while not modifying the modulus in any way. Clearly, this effect is to reduce the stability margins so that a lower crossover frequency must be selected in order to meet design constraints on stability robustness.

The case that is often described is that where the Lm L curve is monotonically decreasing. That argument is repeated here with admonition that things are a bit different where the curve is allowed a more general shape.

4.3 The Neoclassical Approach to Internal Stability

Classical and neoclassical approaches to feedback design are described in this chapter. The emphasis is on the neoclassical. It is likely that, in the twentieth century, a major portion of actual designs were performed using classical techniques. The neoclassical principles, much like described here, hold out much promise.

The choice of the "shape" of a frequency response curve in the G and or Lm G plane is what we call a shaping design method. The curve being "shaped" is usually Bode magnitude of the Youla parameter, $|Y(j\omega)|$, or of the return ratio, $|L(j\omega)|$. The frequency, ω, is the independent variable in this process. In other words, "shaping" procedures are methods in which a frequency response is manipulated in order to satisfy the design criteria described in Chapter 3. The criteria are the usual: robustness, attenuated effects of unwanted inputs, stability margins, etc. Let's also not forget improved speed of response and damping to the limited extent that is possible.

An important constraint is that of *internal stability* described in great detail in Chapter 3. That has to do with pole–zero cancellation; in particular, the cancellation of a pole in the right half-plane or on the imaginary axis with a zero in the product:

$$L(s) = G_c(s)G_p(s) \tag{4.31}$$

Incidentally, if such a pole–zero cancellation occurs in G_p itself, then nothing can be done; that is,

---

1 In contrast, we have constraints that relate the sensitivity in one frequency range to the sensitivity in another range. Freudenberg and Looze have named these constraints analytical constraints and they are developed in Chapter 6.

A necessary condition for successful design of a feedback control is that all pole–zero cancellation in G_p occur strictly in the left half-plane.

This is a restatement of an old ("modern") control concept: The requirement that an object be *detectable and stabilizable.*

Where "L" is being shaped (the classical method), internal instability is easy to avoid: simply select G_c so that there is no cancellation of an unstable pole. In this classical approach, one draw inferences about the all-important closed-loop transfer functions, the *sensitivity*:

$$S(s) = 1/(1 + L) \tag{4.32}$$

the *complementary sensitivity*

$$T(s) = L/(1 + L) \tag{4.33}$$

and the *Youla parameter*

$$Y(s) = G_c/(1 + L) \tag{4.34}$$

$$= G_c S \tag{4.35}$$

$$= T/G_p \tag{4.36}$$

In that sense, the classical method is an *indirect method.*

The simple truth is that the superior methods are those associated with the Youla parameter. For one thing, there is much less cutting and fitting with the neoclassical methods. This procedure is a direct method in which the focus is on

$$T(s) = Y(s)G_p(s) \tag{4.37}$$

the neoclassical counter part of (4.31). We call this a "direct" method because both "Y" and "T" are closed-loop transfer functions. Notice that *one can easily place closed-loop poles and zeros with the proper choice of $Y(s)$ for a given G_p.* This sort of thing can also be accomplished with the "modern" methods but in an unnecessarily complicated way: the neoclassical procedures are highly transparent and require nothing more than an understanding of some basic mathematics.

Alternatively, one can choose "T" and infer "Y" with this equation. That is fine but remember that in order that $Y(s)$ not be improper, *you must choose the relative degree (number of poles less the number of zeros) of $T(s)$ to be no less than that of $G_p(s)$.* Also, an absolute requirement is that any non-minimum phase zero of G_p must be included in "T."

Another appeal of the neoclassical approach is that we directly reduce the vulnerability of the controlled variable to saturation by selecting the Bode magnitude of the parameter, $|Y(j\omega)|$, to have low values at all frequencies if possible. In this way, we eliminate Horowitz' criticism of most feedback design procedures – the failure of most feedback design procedures to deal directly with bandwidth constraints. Another appealing aspect of this approach is that many other considerations can be automatically fixed during this selection of $Y(s)$. These principles were highlighted in Chapter 3 and are illustrated with examples in this chapter.

In either the classical or neoclassical procedures, the goal is to provide low values of $|S(j\omega)|$ at "low frequencies" and low values of $|T(j\omega)|$ at "high frequencies." "High" and "low" with respect to the crossover frequency, ω_c, the frequency where $|S(j\omega_c)| = |T(j\omega_c)|$. That frequency, ω_c, is fixed as the reciprocal of the characteristic response time of the closed loop. "High" and "low" values of Bode magnitude mean larger or smaller than 1 (0 dB). Of course, all this must be accomplished while keeping $|Y(j\omega)|$ low at all frequencies. Only in this way will the actuators be protected from unreasonable demands in response to the signals input at various points in the feedback loop. Another convenient way to think of ω_c is that frequency where $|Y(j\omega)|$ begins a significant deviation from $1/|G_p(j\omega)|$.

Other details of $|S(j\omega)|$ and $|T(j\omega)|$ are important; in particular, we are interested in obtaining low values of the maxima of these functions (the so-called H_∞ norm). The H_∞ norm of $|T(j\omega)|$ is related, in an approximate way, to the damping ratio of the closed loop and for that purpose we use Eq. (2.157). The H_∞ norm of $|S(j\omega)|$ is related to the M_2 stability margin (4.26). The use of this margin is gaining favor over the classical *phase and gain* margins. However, the classical margins can be obtained from the Bode diagrams of $|S(j\omega)|$ and $|T(j\omega)|$ as well as M_2: the phase margin is obtained from the common value of $|S(j\omega_c)|$ and $|T(j\omega_c)|$ at crossover. If G_p is minimum phase, the phase margin is obtained from Eq. (4.19). (Incidentally, since $S + T = 1$, it is impossible for $|T|$ and $|S|$ to be *simultaneously* less than 1/2, that is less than −6 dB.) If G_p is minimum phase, the gain margin is $|S(j\omega_c)| - |T(j\omega_c)|$ at that frequency, ω, where the phase of T is −180°.

Perhaps the most intriguing thing about the neoclassical approach is the idea of parameterization: as demonstrated in Chapter 3, *an internally stabilizing compensator is parameterized as:*

$$G_c(s) = \frac{Y(s)}{1 - Y(s)G_p(s)} \tag{4.38}$$

if and only if $Y(s)$ is any BIBO stable, proper rational function such that $G_p(1 - YG_p)$ is BIBO stable; that is, such that Y and $1 - YG_p$ have a zero at each unstable pole of G_p.

This is a very remarkable result! Think of this: One is not assured of a stable feedback loop by placing some restriction on, say the compensator, $G_c(s)$. For example, feedback about a stable compensator in cascade with a stable controlled object can be unstable. This remarkable proposition is one of the motivations for focusing on the Youla parameter, $Y(s)$, during design. There are a number of other motivations.

We demonstrated in Chapter 3 that necessary conditions for $Y(s)$ to be internally stabilizing are

$$Y(p_r) = 0 \tag{4.39}$$

and

$$Y(p_r)G_p(p_r) = 1 \tag{4.40}$$

where "p_r" is any pole of G_p in the right half-plane or on the imaginary axis. These conditions are sufficient only if "p_r" is not repeated. Equations (4.39) and (4.40) may not be sufficient because there is no guarantee that the multiplicity of the zero in $1 - YG_p$ will match the multiplicity of the unstable pole in G_p. Notice that $G_c(s)$ may be stable or unstable. That is to say, a stable $T(s)$ is achieved by placing unstable poles in the compensator (such a thing is always possible). Use of a unstable compensator has double negative effects. First, given a bounded input, the Youla parameter in the feedback loop will be unbounded, this in turn results in actuator saturation. Next, if the feedback loop fails (a sensor or an actuator fails), then the opened-loop system is driven out of bounds by the controller!

If the controlled object is stable, the closed loop is internally stable if and only if $Y(s)$ is stable and proper. This means that, in this special case, the Youla parameterization of all stabilizing compensators is simply the set of all stable, proper rational functions! In other words, any G_c given by (4.38) will produce an internally stable closed loop provided that the parameter is stable.

The *interpolation conditions* were derived in Chapter 3: If "z" denotes an uncanceled multiplicity a_z of the return ratio, L, and "p" denotes an uncanceled pole of multiplicity a_p of the return ratio, then

$$T(p) = 1 \qquad S(p) = 0 \tag{4.41}$$

and

$$\frac{d^k}{ds^k}T(p) = 0 \qquad \frac{d^k}{ds^k}S(p) = 0 \qquad 1 \le k \le a_p - 1 \tag{4.42}$$

while

$$T(z) = 0 \qquad S(z) = 1 \tag{4.43}$$

and

$$\frac{d^k}{ds^k}T(z) = 0 \qquad \frac{d^k}{ds^k}S(z) = 0 \qquad 1 \le k \le a_z - 1 \tag{4.44}$$

Since unstable poles of L cannot be canceled, these conditions must always be obtained for poles or zeros on the imaginary axis or in the right half-plane. Fixing these conditions with selection of $Y(s)$ is one way to obtain internal stability.

Also, these conditions mean that the closed-loop transfer functions, $S(s)$ and $T(s)$, are constrained to take on certain values in the complex plane. Clearly, these constraints influence the values on the imaginary axis, $S(j\omega)$ and $T(j\omega)$. These constraints cannot be removed by a pole–zero cancellation where "z" or "p" lies in the right half-plane, where a factor of the return ratio is non-minimum phase. These interpolation conditions impose irrevocable limits to feedback design.

4.4 Feedback Design And Parameterization: Stable Objects

The description of the neoclassical "shaping" of closed-loop transfer functions begins in this section. The treatment is complete and the procedures are quite satisfying in that the vulnerability of the actuators is one of very first considerations in the shaping process. A transfer-function-based procedure for closed-loop pole–zero assignment is also featured. The main emphasis is on obtaining a satisfactory $S(s)$, $T(s)$, and $Y(s)$.

The shaping of closed-loop transfer functions seems the most direct procedure for finding an internally stabilizing proper compensator which meets all design criteria. The most obvious form of the closed-loop shaping is described as follows: it is easy to demonstrate that

$$G_c = \frac{1}{G_p}\frac{T}{1-T} \tag{4.45}$$

The idea is that we can choose the closed-loop transfer function, $T(s)$, we want and then use G_p to calculate the required compensator transfer function. The problem is that we do not assure the feasibility of the procedure because there many important aspects, such as the vulnerability of the actuators, that are concealed and obscured in this formulation. Also, it is likely that G_c would not even be proper (realizable) if calculated in this way. Finally, if G_p is non-minimum phase or unstable, then this form of compensation would produce an internally unstable closed loop. These criticisms are removed in the alternative procedure we describe in this section.

Incidentally, notice the complete cancellation of the plant in this procedure (4.45); that is, the poles and zeros of G_p are absent in the return ratio, G_cG_p. A cancellation of a portion of G_p will also appear in our procedure, but that cancellation does not appear in G_c. (Actually, most feedback design methods involve some cancellation, even in the state variable methods where the cancellations are obscured.)

The focus in the method proposed in this section is the Youla parameter:

$$Y(s) = u(s)/r(s) = -u(s)/d(s) = -u(s)/n(s) \tag{4.46}$$

This is the closed-loop transfer function that quantifies the vulnerability of the actuators to saturation. We expect $r(t)$ and $d(t)$ to contain low and intermediate frequencies, while $n(t)$ will have a relatively high-frequency content. That is, the actuators are vulnerable at nearly all frequencies so that $|Y(j\omega)|$ should be kept as low as possible at nearly all frequencies.

It follows from (4.37) that *if G_p is stable, then the closed-loop is internally stable if and only if the parameter, $Y(s)$, is stable.* There is no parallel result in the classical approach; that is, there is no way to ensure even the stability of the closed loop – let alone *the internal stability* – by restricting the choice of $G_c(s)$ in some way! Recall that everywhere that we say "stable" we mean bounded input–bound output (BIBO) stability. This means that a transfer function might be stable only if it is not improper.

We write the parameter as:

$$Y(s) = \frac{1}{1+L(s)}\frac{L(s)}{G_p(s)} \tag{4.47}$$

and conclude that

$$Y(j\omega) \approx 1/G_p(j\omega) \qquad \text{where} \qquad |L(j\omega)| \gg 1 \tag{4.48}$$

and

$$Y(j\omega) \approx L(j\omega)/G_p(j\omega) = G_c(j\omega) \qquad \text{where} \qquad |L(j\omega)| \gg 1 \tag{4.49}$$

Recall that we design $|L|$ to have very large values at low frequency in order to attenuate disturbances, to improve command following, and to achieve robustness to variations in the controlled object. We design $|L|$ to have very small values at high frequency in order to attenuate sensor noise and achieve stability robustness. It is clear that:

> Our goals are also attained by shaping $|Y(j\omega)|$ to $1/|G_p(j\omega)|$ at low frequencies and by keeping $|Y(j\omega)|$ low at high frequencies.

This is the essence of the shaping procedure. We will examine this procedure by providing several examples in the following sections.

Notice that where we accomplish this shaping, $1/|G_p|$ is the low-frequency asymptote for $|Y(j\omega)|$ and $|L(j\omega)|/|G_p(j\omega)| = |G_c(j\omega)|$ is the high-frequency asymptote. (We set the latter before we even calculate $L(s)$.) This means that the high- and low-frequency asymptotes for $|Y(j\omega)|$ meet at

$$\omega_{Yb} = \omega_c \tag{4.50}$$

the crossover frequency of the return ratio! *This means by setting the break frequency of the parameter, we essentially fix the closed-loop response time.*

We begin with an important observation.

4.4.1 Renormalization of Gains

We try to use a high static gain in G_c because that will reduce the vulnerability of the actuators and will help to attenuate the response at the output to the disturbance.

In a sense, any disturbance rejection criterion can be satisfied by magically increasing the gain in any controlled object by redefining the units of the controlled input. For example, suppose you take the units of the controlled input to be millimeters and the gain of G_p is 1.0 but you need that gain to be 10^3. You can merely redefine the input to be measured in meters and, voilà, the gain is a 1000.

The reader senses that there is something very wrong in this conclusion and in that you are correct. The explanation is that, if before you waved your wand, $u(t)$ saturated at a magnitude of 2, it now saturates at a magnitude of 2×10^{-3}. That is, you now have a more severe restriction on $|Y|$. Think of it this way, you compensated for an apparent attenuation of a disturbance response by a corresponding tightening of the restrictions $Y(s)$. From this point of view, it is the nonlinearity (the restriction on u) that gives an absolute meaning to "gain."

4.4.2 Shaping of the Closed-Loop: Stable SISO

The basic result is (4.37) and we have a simplification of that in a special case.

Corollary 4.1 *If the controlled object is stable, the closed loop is internally stable if and only if $Y(s)$ is stable (hence, not improper).* ■

This means, as stated earlier, that the Youla parameterization:

$$G_c(s) = \frac{Y(s)}{1 - Y(s)G_p(s)}$$

of all internally stabilizing compensators is simply the set of all stable (hence, not improper) rational functions, $Y(s)$! In other words, any G_c given by (4.38) will produce an internally stable closed loop provided that the parameter is stable. Equally important, any internally stabilizing compensator can be expressed as in (4.38) for some $Y(s)$. Think of it this way, if you restrict yourself to form (4.38), then no other designer can accomplish a better design using some other form. Of course, this is assuming you will search over all possible $Y(s)$, which is not likely.

Where G_p is stable (possibly non-minimum phase), the formulation is as follows for single-degree-of-freedom design.

(a) **Placement of Closed-loop Poles, Zeros, and Gain**: Begin with the relation:

$$T(s) = Y(s)G_p(s)$$

and "shape" $Y(s)$. You begin that by canceling unwanted poles and zeros of G_p and then including desired location for the closed-loop poles and zeros (we will discuss the impact of the cancellation of poles and zero in the following section). Next make an adjustment in the gain of $Y(s)$ in order to fix a desired value of the closed-loop gain $|T|$. For example, take $Y = 1/G_p$ at low frequency so that $T = 1$ in that range. This gives good low-frequency command following and, since $S = 1 - T$ is thereby forced near zero, there will be good disturbance rejection there as well. (Of course, the mirror image of a non-minimum phase zero of G_p must be used as a pole of $Y(s)$ in that range because $Y(s)$ must be stable.) Always remember that $Y(s)$ is a closed-loop transfer function, so the necessary and sufficient condition for internal stability (BIBO) is that "Y" is stable, hence proper.

(b) **Adjust the Frequency Response of** $Y(s)$: Observe the Bode magnitude of $|Y(j\omega)|$ and recognize that this diagram indicates the vulnerability of your actuators. For example, you pay a price where you cancel a LHP pole of G_p because the corresponding zero in $Y(s)$ will make you more vulnerable to saturation (the slope of the Bode magnitude $|Y(j\omega)|$ will increase in that way, more on this in the following sections). It may be that you were too ambitious in your selection of the closed-loop poles and so the diagram indicates values for Lm $|Y(j\omega)|$ which are too large. You may have to re-evaluate the choices made in step (a). Generally, it is far better if you can design by canceling only a limited number of LHP poles of G_p near the crossover frequency.

(c) **Adjust the Stability Margins**: The phase margin is obtained where $|T(j\omega_c)| = |S(j\omega_c)|$ and is related to this common value (see previous sections). Normally, you want this intersection to occur below 0 dB line since that assure a phase margin greater than 60°. Also, this crossover frequency, ω_c, provides an estimate of the closed-loop response time

$t_r \approx 1/(4\omega_c)$. Contrary to expectations, this estimate is often more accurate than an estimate based on the closed-loop poles themselves. The gain margin is the difference:

$$\text{Lm } |S(j\omega)| - |\text{Lm } |T(j\omega)| \tag{4.51}$$

at that frequency, ω, where the phase of $|T(j\omega)|$ is $-180°$. Another, perhaps better, measure is the M_2 stability margin which is equal to max $|S(j\omega)|$. Alternatively, recall that unstructured uncertainty can be modeled such that stability margins (stability robustness) are characterized as constraints on the weighted version of $|T(j\omega)|$. That means these uncertainties will help fix the shape of $|Y(j\omega)|$ at high frequencies.

(d) **Shape $Y(s)$ at High Frequencies to Offset Sensor noise Problems: Add High-frequency Closed-loop Poles**: The constraint $Y(s)$ must be proper or strictly proper will help in this regard, but you may have to take additional steps by adding some high-frequency poles to $Y(s)$ (hence, to $T(s)$).

(e) **Evaluate the Design in the Linear Domain**: Examine the Bode magnitudes $|T(j\omega_c)|$ and $|S(j\omega_c)|$ and evaluate the capability of your design for robustness, command following, and unwanted signal attenuation utilizing the same linear plant model G_p used for control design.

(f) **Obtain the Transfer Function of the Compensator**: From

$$G_c(s) = \frac{Y(s)}{S(s)} \tag{4.52}$$

This must be synthesized in some way in order to realize the feedback design.

(g) **Evaluate the Design with the More Complete Plant Model, in the Nonlinear Domain**: Integrate G_c in the complete simulation model and again evaluate the capability of your design for robustness, command following, and unwanted signal attenuation. Are your actuators capable of satisfying the requirements you have placed upon them?

With regard to pole–zero placement in step (a): it is good to cancel a zero in G_p with a pole in Y since that causes a decrease in the slope of $Lm|Y(j\omega)|$, a lessening of the demand on the actuators! Of course, since $Y(s)$ must be stable, that can occur only in the left half-plane. This point illustrates the simple fact that minimum phase zeros in the controlled object can be put to good use in feedback design. In a similar vein, one pays a price for canceling a pole in G_p with a zero in Y because that causes an *increase* in the slope of Lm $|Y(j\omega)|$; an increase in the demand on the actuators. That makes sense because where we cancel an object pole, we are attempting to produce a closed loop that responds faster than the open loop and the cost is immediately made apparent by an increase in the demand on the actuators. Incidentally, since $T = YG_p$ must be stable, any unstable pole of G_p must be canceled in this way. That fact makes very clear the limitation that an unstable pole in the object places on feedback design. *It is inherent that actuators are called upon to do more in the case of unstable objects.* More will be in the following sections.

With regard to the inclusion of **high-frequency poles** in step (d). "High frequency" means poles with break frequencies above crossover, and that means that these poles will not influence closed-loop transient response. It is probably for that reason that this concept is neglected in some developments of the feedback theory. Notice the way that this procedure keeps

$$|T(j\omega_c)| \approx |G_p(j\omega_c)| \tag{4.53}$$

near crossover since that is the way you prevent excessive growth of $|Y|$. There is another, very good reason to add high-frequency poles. Recall that $u(s) = Y(s)r(s)$ and suppose that $r(s)$ is a step change in command. It follows from the initial value theorem that if $Y(s)$ has "k" more poles than zeros, then $u(0^-)$ and the initial values of $k - 1$ derivatives of $u(t)$ will be zero. Thus, the larger the value of "k," the less usage demanded in the initial range of the actuator output. That is to say, we increase "k" to diminish the kind of excessive actuator usage.

We will summarize the procedure to select $Y(j\omega)$ and the impact of pole–zero cancellation in feedback design, including several examples in the following section. Before that, however, we lay out the competing objectives and constraints in shaping the response of a system.

4.4.3 Neoclassical Design Principles

Recall the various feedback transfer functions for the system shown in Figure 4.9.

$$S = \frac{\hat{e}}{\hat{r}} = -\frac{\hat{e}}{\hat{n}} = \frac{\hat{y}}{\hat{d}_y} \tag{4.54}$$

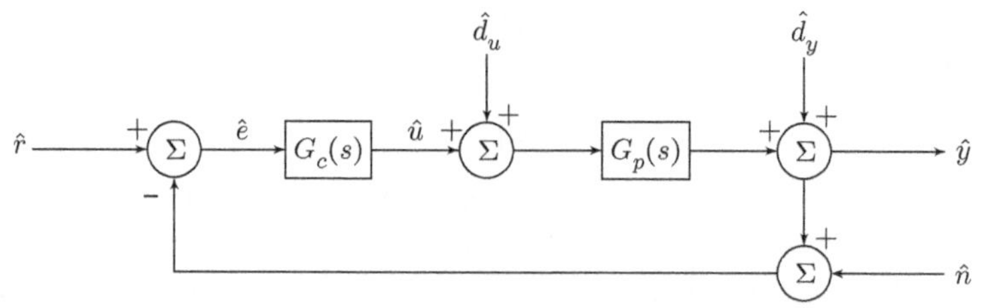

Figure 4.9 Block diagram of a controlled system, $H(s)$, $P(s) \equiv 1$.

$$T = \frac{\hat{y}}{\hat{r}} = -\frac{\hat{y}}{\hat{n}} = -\frac{\hat{u}}{\hat{d}_u} \tag{4.55}$$

$$Y = \frac{\hat{u}}{\hat{r}} = -\frac{\hat{u}}{\hat{n}} = -\frac{\hat{u}}{\hat{d}_y} \tag{4.56}$$

The shape of these feedback transfer functions impacts the performance and robustness of the system; however, there are generally competing objectives:

1. For disturbance rejection, we must make S small.
2. For noise attenuation, we must make T small.
3. For reference tracking, we must make $T \approx 1$.
4. For reducing control effort, we must make Y small.

Let us consider these objectives in terms of the return ratio L.

1. $S = 1/(1 + L)$, so for small S we need $L \gg 1$.
2. $T = L/(1 + L)^{-1}$, so for small T we need $L \ll 1$.
3. $T = L/(1 + L)^{-1}$, so for $T \approx 1$ we need $L \gg 1$.
4. $Y = G_c/(1 + L)$, so for small Y we need $G_c \ll 1$ at high frequencies.

In meeting these objectives, we are subjected to constraints:

- Algebraic:
  - $S + T = 1$.
  - Internal stability: Interpolation conditions.
- Analytic:
  - $\max_\omega |T(j\omega)| \geq \left|\frac{p+z}{p-z}\right|$, $\max_\omega |S(j\omega)| \geq \left|\frac{z+p}{z-p}\right|$. This is an lower bound on both $\|T\|_\infty$ and $\|S\|_\infty$ when there are unstable poles and zeros. For instance, a system with an unstable zero at $s = 2$ and an unstable pole at $s = 1$ will have

$$\max_\omega |S(j\omega)| \geq \frac{2+1}{2-1} = 3 \approx 10 \text{ dB} \tag{4.57}$$

To design a controller with the Youla parameterization method, we select a Y considering the above objectives and constraints, calculate $T = YG_p$, then calculate $G_c = Y/S = Y/(1 - T)$. The selection of Y will determine the performance and robustness of the system. A robust system will have S small in the low-frequency regions and T small in high-frequency regions. Additionally, the infinity norm should be small relative to 0 dB.

If we are aiming for $Y = 1/G_p$, how do we select $Y(j\omega)$? Do we cancel poles and zeros in G_p? What do we do with unstable poles and zeros? This will be discussed in the following section.

Another important analytic constraint is related to the **Bode Integral Theorem**. These analytic constraints will be further discussed in Chapter 6.

In Sections 4.5 and 4.6, the some of the important aspects of the neoclassical feedback control concepts are discussed. In Section 4.7, we provide several control design examples.

4.5 Loop Shaping Using Youla Parameterization

We discussed some of the impact of pole–zero cancellation and shaping the Youla parameter including the consequences of this shaping in feedback control design. This section will detail the basic steps of shaping the Youla parameter to build the desired closed-loop response.

When we speak of pole–zero cancellation to shape the Youla parameter, there are four principal cases that need to be considered. These are cancellation of ...

a) Stable zeros (LHP zeros) of G_p,
b) Stable poles (LHP poles) of G_p,
c) Non-minimum phase zeros (RHP zeros) of G_p,
d) Unstable poles (RHP poles) of G_p.

4.5.1 LHP Zeros of G_p

Minimum phase zeros of G_p can be canceled by placing a stable (LHP) pole in the Youla parameter, knowing that

$$T(s) = Y(s) \cdot G_p(s) \tag{4.58}$$

But what is the impact of this cancellation and when should we decide to cancel?

If a LHP zero of a plant is located at a lower frequency than its slowest pole (let's take a simple case of one LHP zero with two poles), then this zero will result in an overshoot due to a step input, and hence the zero dynamics decreases the M_2 stability margin. In addition, as we will show in the example below, not canceling this type of zero results in an increase in actuator demand at high frequencies, and sensor noise amplification in this frequency range. If this is the case, then this zero must be canceled by adding a pole to the Youla parameter. Addition of a pole to the Youla parameter has a secondary benefit of lowering the magnitude frequency response and, hence, lowering the actuator demand.

The impact of this cancellation is demonstrated in (Figures 4.10–4.13) on the closed-loop complementary sensitivity, sensitivity, and Youla transfer functions for a simple case of one pole and one zero G_p. It should be noted that if the T and S vectors are greater than 1, then the M_2 stability margin decreases. It should also be clear that the vector addition of T and S is equal to 1 .

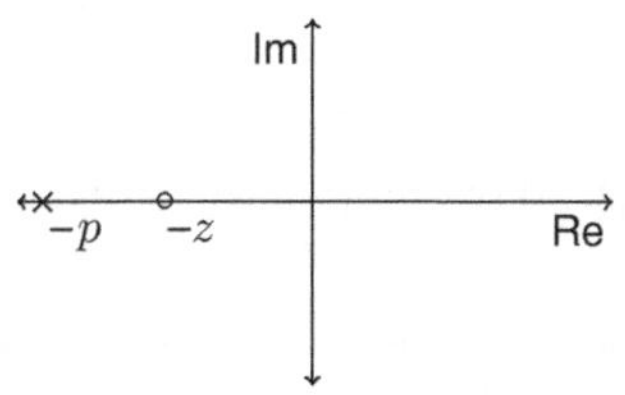

Figure 4.10 s-Plane location of LHP zero.

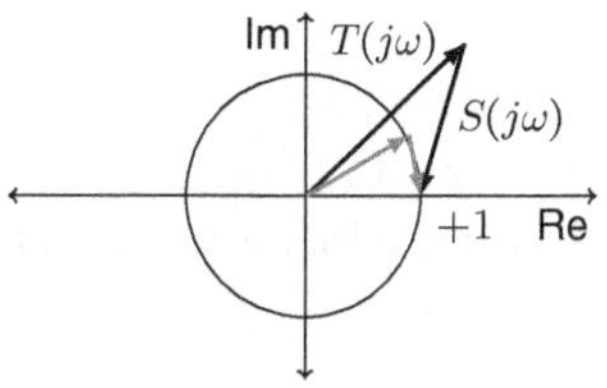

Figure 4.11 S and T for a LHP without cancellation in black and with cancellation in gray.

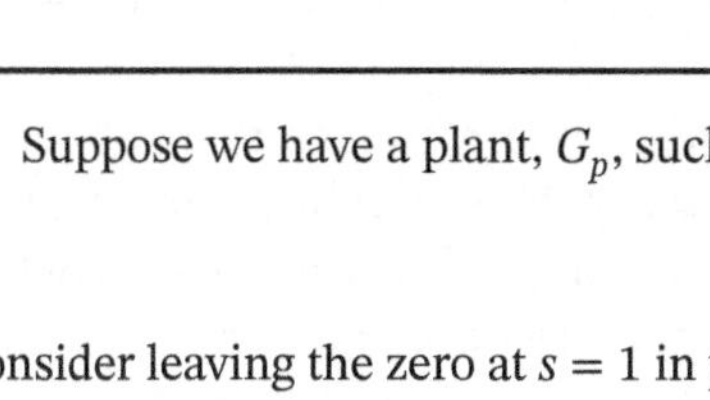

Figure 4.12 $T = YG_p$ when canceling a zero.

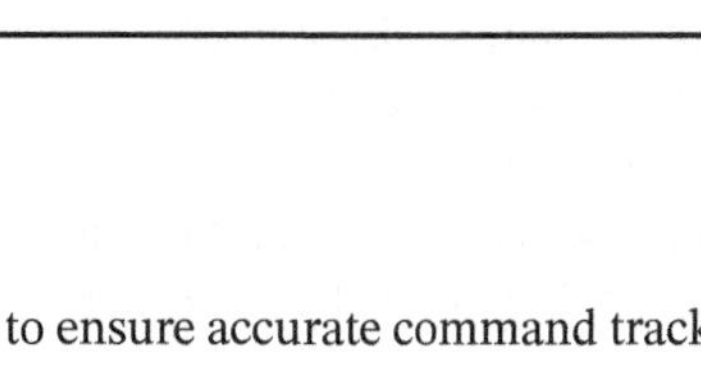

Figure 4.13 Y when canceling a zero.

Example 4.2 Suppose we have a plant, G_p, such that

$$G_p = \frac{s+1}{s+2} \tag{4.59}$$

We will first consider leaving the zero at $s = 1$ in place. We set $Y_1 = 2$ to ensure accurate command tracking at steady state ($T = 1$ at low frequencies). This gives a closed-loop response of

$$T_1 = \frac{s+1}{\frac{1}{2}s+1} \tag{4.60}$$

Y_1 and T_1 have the magnitude frequency responses shown in Figures 4.14 and 4.15

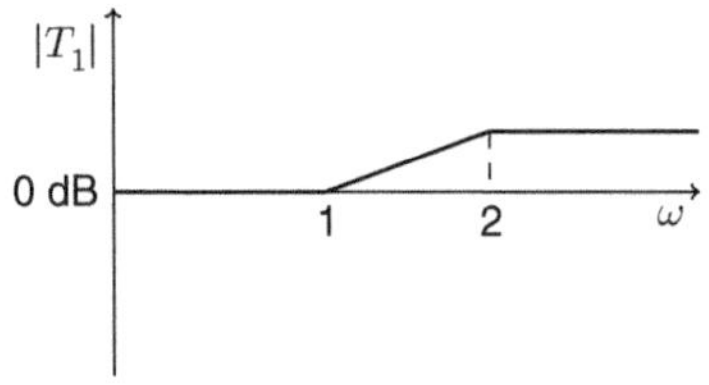

Figure 4.14 Magnitude plot of T_1 for Example 4.2.

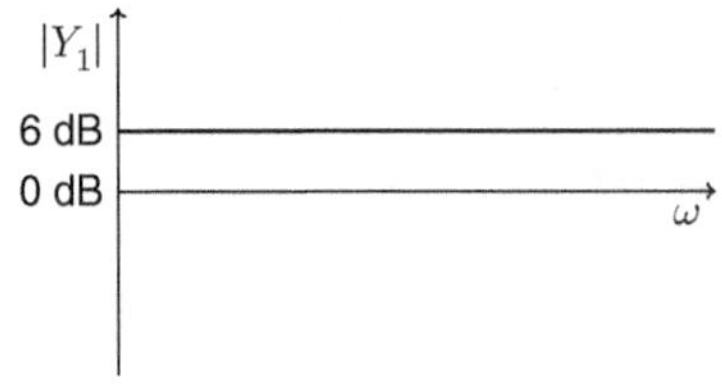

Figure 4.15 Magnitude plot of Y_1 for Example 4.2.

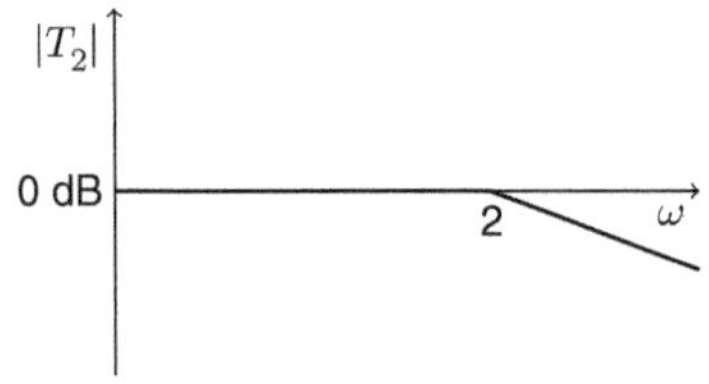

Figure 4.16 Magnitude plot of T_2 for Example 4.2.

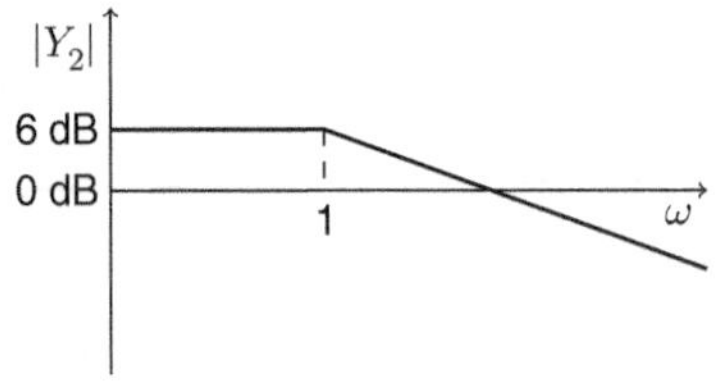

Figure 4.17 Magnitude plot of Y_2 for Example 4.2.

If we don't cancel the zero, the closed-loop response will amplify noise. In addition, the M_2 stability margin reduces and actuator demand is constant throughout the entire frequency range. *This is not a good design.* Instead, we shape Y to cancel the zero of the plant, G_p:

$$Y_2 = \frac{2}{s+1} \tag{4.61}$$

Then,

$$T_2 = \frac{1}{\frac{1}{2}s+1} \tag{4.62}$$

As shown in Figures 4.16 and 4.17, the actuator and closed-loop response both now roll off at high frequencies, limiting the bandwidth of the system so that it will reject the high-frequency sensor noise and reduces actuator demand at high frequency.

4.5.2 Non-minimum Phase Zeros

Canceling a zero in the right-hand plane (that is, an unstable or non-minimum phase zero) would require introducing a pole in the right-hand plane (an unstable pole) to the Youla transfer function. This would violate internal stability, so cancellations of non-minimum phase zeros is forbidden. Therefore, the existence of a non-minimum phase zero in the plant is bad news and we have to live with this.

In the earlier chapters, we illustrated that a non-minimum phase zero results in a reverse action. For a high-gain controller, this reverse action could have a destabilizing effect, as at higher gains, poles of a transfer function migrate to its zeros, as can be shown by its root locus plot. So, if there exists one non-minimum phase zero, at high gain, one pole will cross the $j\omega$-axis and migrates toward this non-minimum phase zero. This will result in destabilizing the system. Therefore, one can observe that the non-minimum phase zero will have a band-limiting effect on the closed-loop feedback control. Is there anything could be done to alleviate this problem?

The answer to this question is yes. Instead of canceling the non-minimum phase zero, we minimize its effects by placing a stable pole, at the same or lower than the non-minimum phase zero frequency, in the Youla transfer function. As mentioned, the non-minimum phase zero in the plant restricts or constrains the speed of the response or the system bandwidth. The effective lower bound of the time constant τ_p of poles in a non-minimum phase system is given by:

$$\tau_p \geq \frac{1}{|z_k|} \tag{4.63}$$

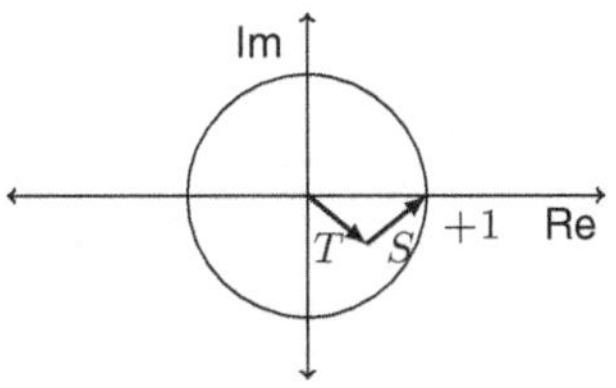

Figure 4.18 S and T with pole placement.

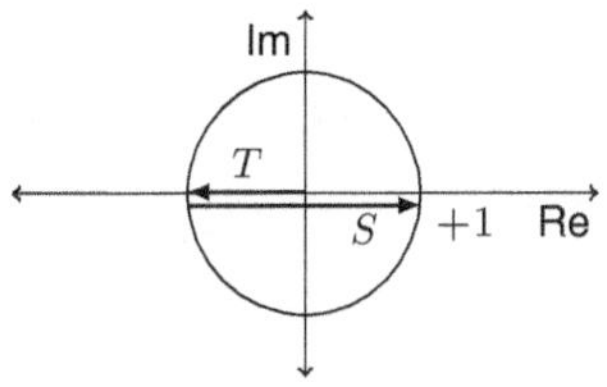

Figure 4.19 S and T without pole placement.

where z_k is the closest non-minimum phase zero to the $j\omega$-axis. An alternative interpretation of this limit is that a stable closed-loop pole must be placed closer to the $j\omega$-axis than a non-minimum phase zero. The reason for this is briefly illustrated in Figures 4.18 and 4.19. At first glance, Figure 4.19 could be confusing as the sensitivity transfer function, S is equal to 2, and we clearly stated that the algebraic constraint, $T + S = 1$. How could this be?

We should remind ourselves that T and S are vectors with magnitudes and phases. So, the algebraic constraint is a vector addition. With the presence of a non-minimum phase zero, the phase of T could drop to $-180°$ while its magnitude at very low frequency, when the magnitude of S is very small, can be equal to 1. Therefore, $-1 + S = 1$ where, S is now equal to 2 or 6 dB. This large magnitude in S drastically reduces the M_2 stability margin. We will discuss this effect by presenting several examples later in this chapter. This is precisely why we need to include a low-frequency pole in Y to cancel the negative effect of the plant non-minimum phase zero.

Example 4.3 Suppose we have a plant

$$G_p = \frac{1-s}{s+2} \tag{4.64}$$

We want to minimize the effects of the non-minimum phase zero. However, canceling it with a $1 - s$ in the denominator would introduce an RHP pole, destabilizing the system. We first set $Y_1 = 2$ and observe the response.

$$T_1 = \frac{1-s}{\frac{1}{2}s+1} \tag{4.65}$$

These magnitude responses are shown in (Figures 4.20 and 4.21).

We can't cancel the zero, but we need to eliminate the amplification of high-frequency noise. Observe that the time constant of the close-loop pole is $\tau = \frac{1}{2}$. This is too fast – faster than the time constant of the non-minimum phase zero. We can replace the pole with a slower one, for instance $\tau = 2$.

$$Y_2 = \frac{\frac{1}{2}(s+2)}{s+\frac{1}{2}} \tag{4.66}$$

The new magnitude responses are shown in (Figures 4.22 and 4.23).

$$T_2 = \frac{1-s}{2s+1} \tag{4.67}$$

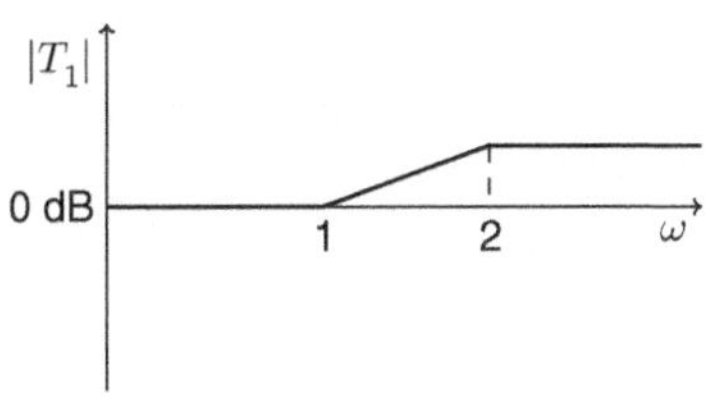

Figure 4.20 Magnitude plot of T_1 for Example 4.3.

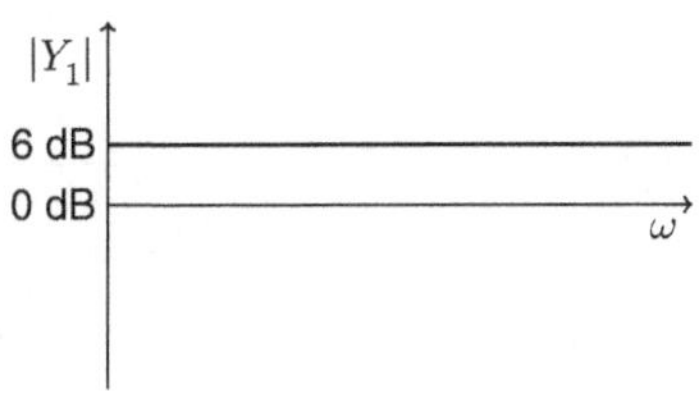

Figure 4.21 Magnitude plot of Y_1 for Example 4.3.

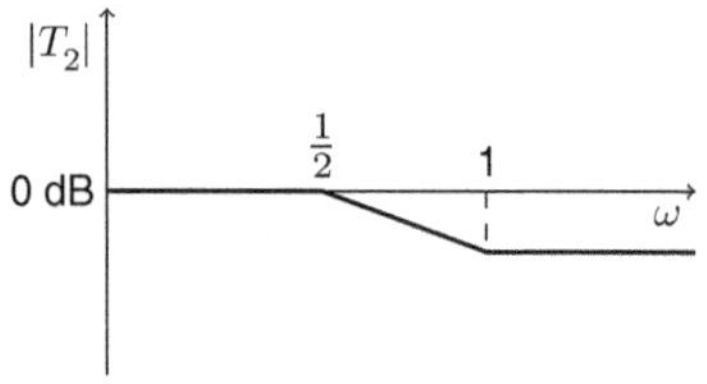

Figure 4.22 Magnitude plot of T_2 for Example 4.3.

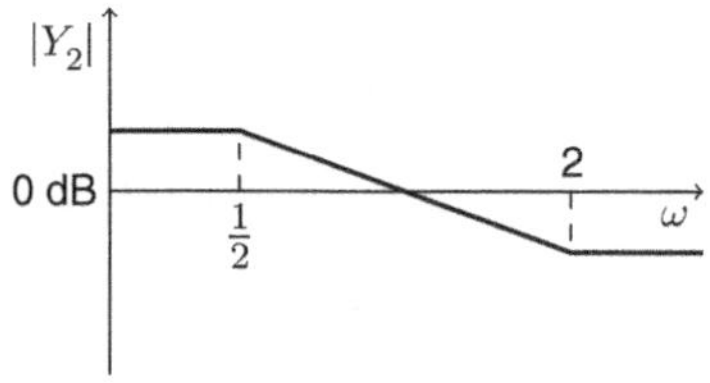

Figure 4.23 Magnitude plot of Y_2 for Example 4.3.

4.5.3 LHP Poles of G_p

Stable poles of G_p are canceled with minimum phase zeros in Y. In this case, We have the option of canceling or keeping these stable poles. Therefore, our design requirement including the closed-loop desired response, in this case, will determine this cancellation. For instance, we may choose to cancel poles to speed up the response or to improve actuator demand and stability margins as illustrated in Figures 4.24 and 4.25. Faster poles increase the actuator demand and speed up the response, while slower poles have an opposite effect. Y must be proper or strictly proper, so canceled poles of G_p must always be replaced with the same number or more new poles in Y.

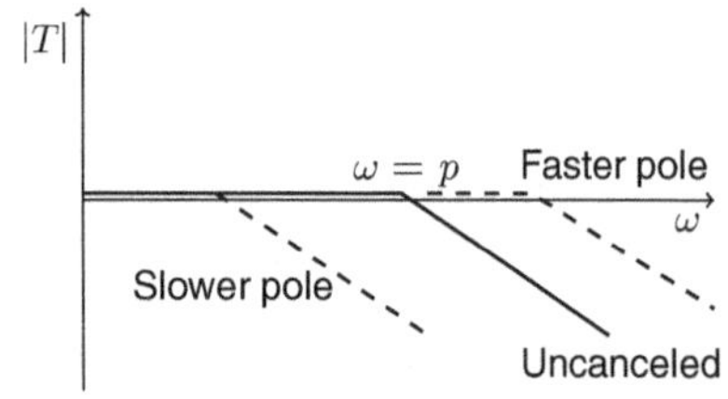

Figure 4.24 $T = YG_p$ when canceling a pole.

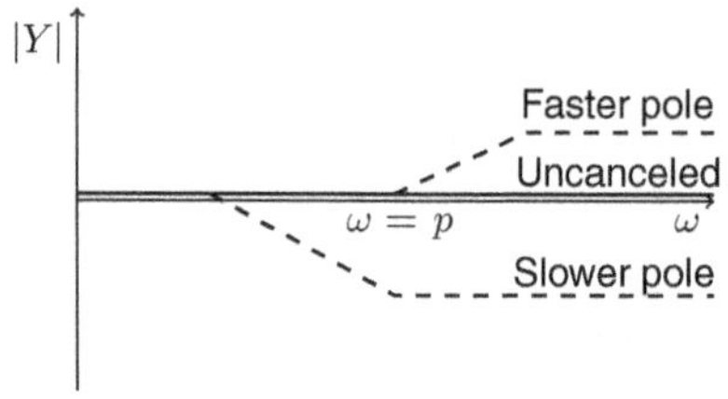

Figure 4.25 Y when canceling a pole.

Example 4.4 Suppose $G_p = \frac{1}{s+2}$.

First, set $Y = 2$, yielding no cancellations and zero steady-state error.

$$T_1 = \frac{2}{s+2} \tag{4.68}$$

These magnitude responses are shown in (Figures 4.26 and 4.27).

Perhaps we want to increase the bandwidth of this plant. We will cancel the pole and replace it with a pole that has a faster time constant. The new magnitude responses are shown in (Figures 4.28 and 4.29). As can be seen in Figure 4.29, actuator demand has increased:

$$Y_2 = \frac{5(s+2)}{s+5} \Rightarrow T_2 = \frac{5}{s+5} \tag{4.69}$$

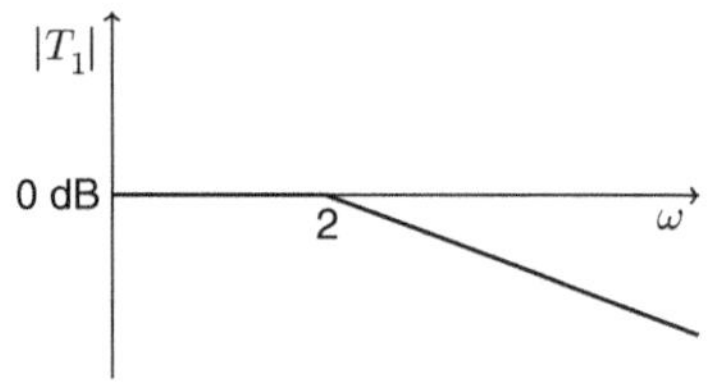

Figure 4.26 Magnitude plot of T_1 for Example 4.4.

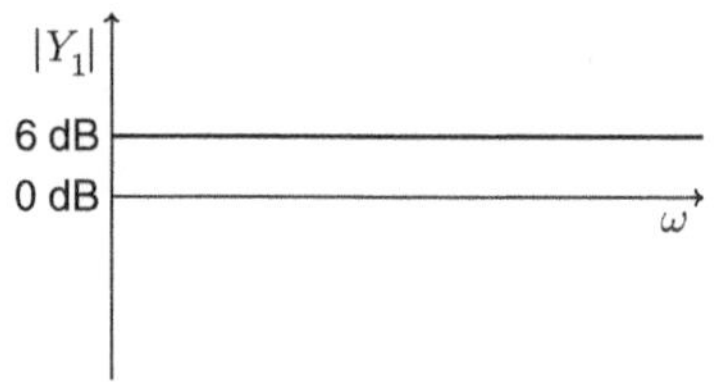

Figure 4.27 Magnitude plot of Y_1 for Example 4.4.

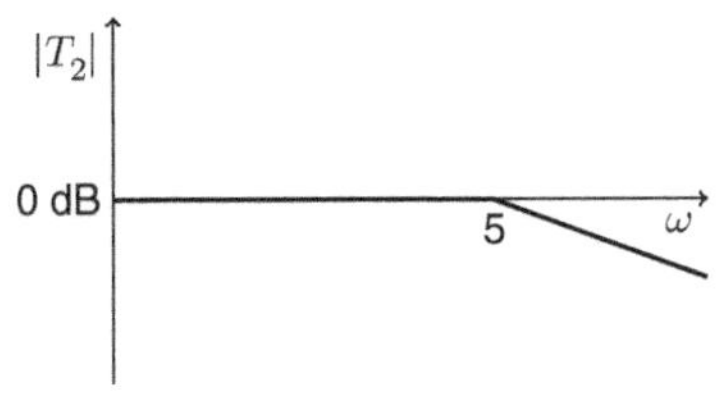

Figure 4.28 Magnitude plot of T_2 for Example 4.4.

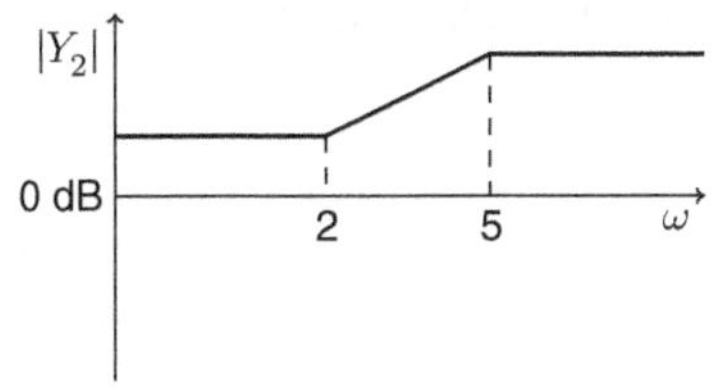

Figure 4.29 Magnitude plot of Y_2 for Example 4.4.

4.5.4 Unstable Poles

In order to stabilize a plant, we have to cancel its unstable poles. This can be done by introducing zeros in Y, and as these poles are in the RHP, we have no other option but to introduce a non-minimum phase zero in Y for each unstable pole of G_p. Furthermore, we must make sure to respect the interpolation conditions, discussed in the previous chapters, so that the final controller results in an internally stable feedback system. We discuss the case of an unstable plant in Example 4.5.

Example 4.5 Suppose we have

$$G_p = \frac{1}{s-1} \tag{4.70}$$

We first select a Y that will cancel the unstable pole and that has tuning parameters that we can use to meet the interpolation conditions. In this example, we will use a gain K and a time constant τ of the new pole as our tuning parameters. Then, the Youla parameter can be written as:

$$Y_1 = \frac{K(s-1)}{\tau s+1} \tag{4.71}$$

Therefore,

$$T_1 = \frac{K}{\tau s+1} \tag{4.72}$$

Next, we attempt to verify the interpolation conditions and expose any constraints on K and τ for meeting the interpolation conditions. We have an unstable pole at $s = 1$, so we know that we have to meet the following interpolation condition to guarantee internal stability:

$$T(1) = 1 = \frac{K}{\tau+1} \tag{4.73}$$

Therefore, we have the following constraint:

$$K = \tau + 1 \tag{4.74}$$

We can select a K, τ pair and examine the result. For instance, let's consider $\tau = 2$ and, hence, $K = 3$.

$$T_1 = \frac{3}{2s+1} \tag{4.75}$$

This closed-loop transfer function results in a steady-state error due to a step input. To eliminate this steady-state error, we can introduce the constraint $T(0) = 1$ to ensure that L has an integrator:

$$T(0) = 1 = \frac{K}{1} \tag{4.76}$$

This means $K = 1$, which in turn means that $\tau = 0$. This results in ill-posed feedback and is unacceptable. Alternatively, decreasing steady-state error without eliminating it will increase the magnitude of Y, meaning greater and greater control effort is required.

This means that the original Y has to be modified to meet all the given constraints. It should not be too difficult to observe that we need to add an additional zero to Y to meet the given constraints, but we will discuss this later. By adding an additional zero to Y, we also need to add an extra pole to enforce properness. We propose the following Y:

$$Y_2 = \frac{K(s-1)(s+z)}{(\tau s+1)^2} \tag{4.77}$$

Therefore,

$$T_2 = \frac{K(s+z)}{(\tau s+1)^2} \tag{4.78}$$

From the first interpolation condition

$$T(0) = 1 = \frac{K(z+0)}{(\tau \cdot 0+1)^2} \tag{4.79}$$

we have the requirement that $Kz = 1$. The next interpolation condition gives

$$T(p) = 1 = \frac{K(z+1)}{(\tau+1)^2} \tag{4.80}$$

From this, we can obtain a relationship between K and z in terms of the closed-loop time constant τ:

$$K = (\tau+1)^2 - 1 \tag{4.81}$$

$$z = \frac{1}{(\tau+1)^2 - 1} \tag{4.82}$$

The selected Youla can meet the interpolation conditions and eliminate steady-state error, while affording us some leeway in the design. The interpolation conditions in fact expose an interesting characteristic of the design: if we slow the response time by increasing τ, we move the non-minimum phase zero closer to the origin, which in turn increases the overshoot and can induce oscillations. On the other hand, speeding up the response time by lowering τ, the non-minimum phase zero moves away from the origin, which results in less overshoot or oscillations.

For example, selecting $\tau = 2$ gives us $z = 1/8$ and $K = 8$:

$$Y_2 = \frac{(s-1)(8s+1)}{(2s+1)^2} \tag{4.83}$$

$$T_2 = \frac{8s+1}{(2s+1)^2} \tag{4.84}$$

Having obtained a desirable shape for T, we can then compute the corresponding controller, G_c:

$$G_c = \frac{Y}{1-T} = 2 + \frac{1}{4}s^{-1} \tag{4.85}$$

Interestingly enough, the result is PI control.

4.6 Design Guidelines

In this section, we discuss the impact of our design on the shapes of $T(s)$, $S(s)$, and $Y(s)$. This means that by observing these shapes, we learn to decide if our feedback control design is acceptable.

Recall the typical shape of S and T, as shown in Figure 4.30.

Good design keeps the magnitude of S and T at the crossover frequency less than 0 dB. This ensures phase margins $\geq 60°$. Good design also keeps the overshoot in S small for a good M_2 margin, therefore, gain margin.

Recall from Section 3.8 that generally the return ratio is large at low frequencies and small at high frequencies, as shown in Figure 4.31.

Does it matter, then, where high gain comes from? At low frequencies,

$$T(j\omega) \approx 1 \Rightarrow Y = \frac{1}{G_p} \tag{4.86}$$

Whereas at high frequencies,

$$L \to 0 \Rightarrow Y = \frac{G_c}{1+L} \approx G_c \tag{4.87}$$

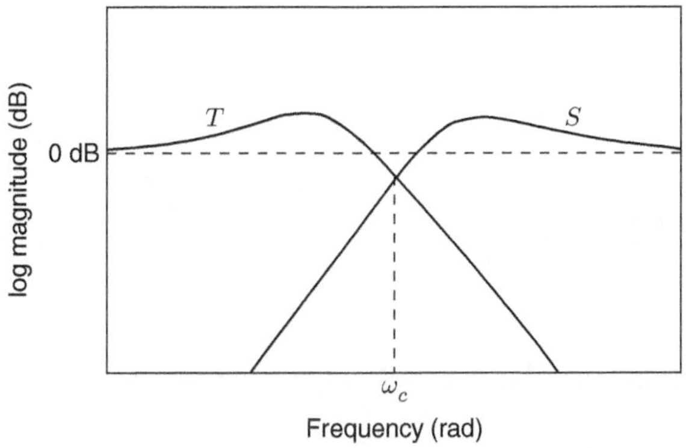

Figure 4.30 *S* and *T* for a typical system.

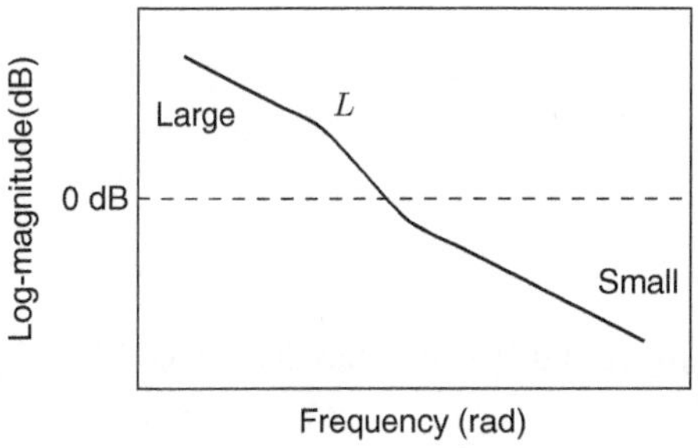

Figure 4.31 Return ratio *L* for a typical system.

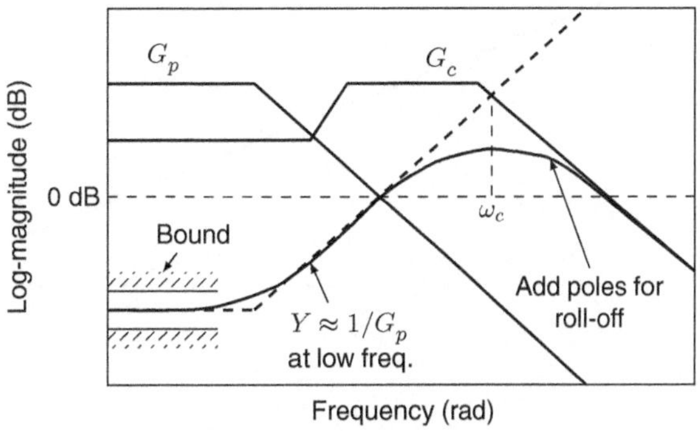

Figure 4.32 Youla design with high-gain plant.

Suppose that the magnitude of Y must be bounded at low frequencies. If so, then the plant must have high gain. Then, to control the gain of Y at high frequency, we must make the crossover frequency smaller (move it to the left) to slow the response time and we must include a pole (or poles) at high frequency to ensure roll-off, as shown in Figure 4.32.

In general, at low frequencies $L(j\omega) \gg 1$, so $Y \approxeq 1/G_p$. Therefore, at low frequencies, performance is heavily dependent on the plant dynamics (or perhaps, the engineers who designed the plant), rather than your controller. On the other hand, at high frequencies $L(j\omega) \ll 1$, so $Y \approxeq G_c$. Performance at high frequencies is most heavily dependent on the controller you design.

4.7 Design Examples

This section consists of several example problems of neoclassical design.

Example 4.6 Consider a basic plant model:

$$G_p = \frac{10}{s+1}, \quad T = Y\frac{10}{s+1} \tag{4.88}$$

The simplest choice for Y will leave the plant pole in place while ensuring that there is no steady-state error:

$$Y = \frac{1}{10} \tag{4.89}$$

Then,

$$T = \frac{1}{s+1} \tag{4.90}$$

$$S = 1 - T = \frac{s}{s+1} \tag{4.91}$$

The resulting compensator is simply PI control:

$$G_c = \frac{Y}{S} = \frac{\frac{1}{10}}{\frac{s}{s+1}} = \frac{s+10}{10s} = \frac{1}{10} + \frac{1}{10}\frac{1}{s} \tag{4.92}$$

We then check for internal stability:

$$L = G_c G_p = \frac{1}{10}\frac{s+1}{s} \cdot \frac{10}{s+1} = \frac{1}{s} \tag{4.93}$$

$$1 + L = 1 + \frac{1}{s} = \frac{s+1}{s} \tag{4.94}$$

So, Y is proper and internal stability is met. The magnitude responses of the controlled system are given in (Figures 4.33 and 4.34).

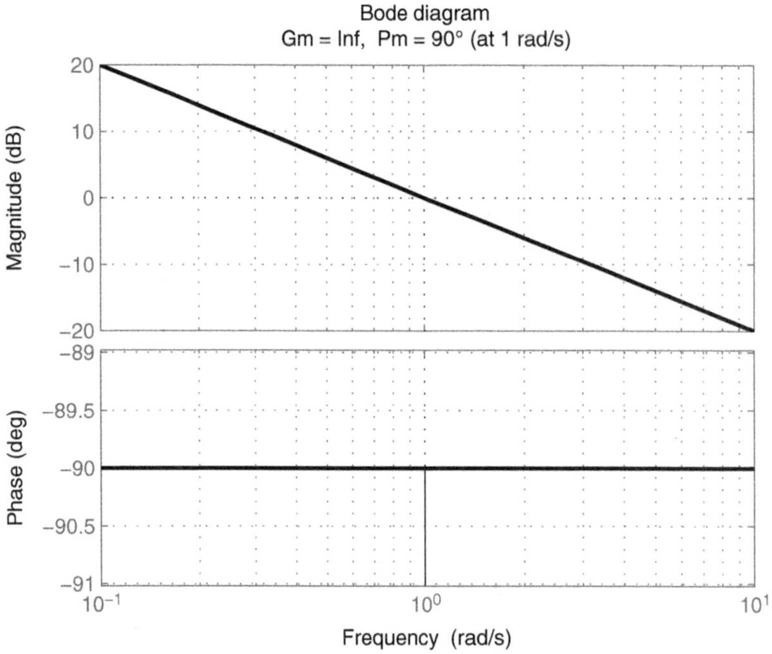

Figure 4.33 Stability margins of the controlled system.

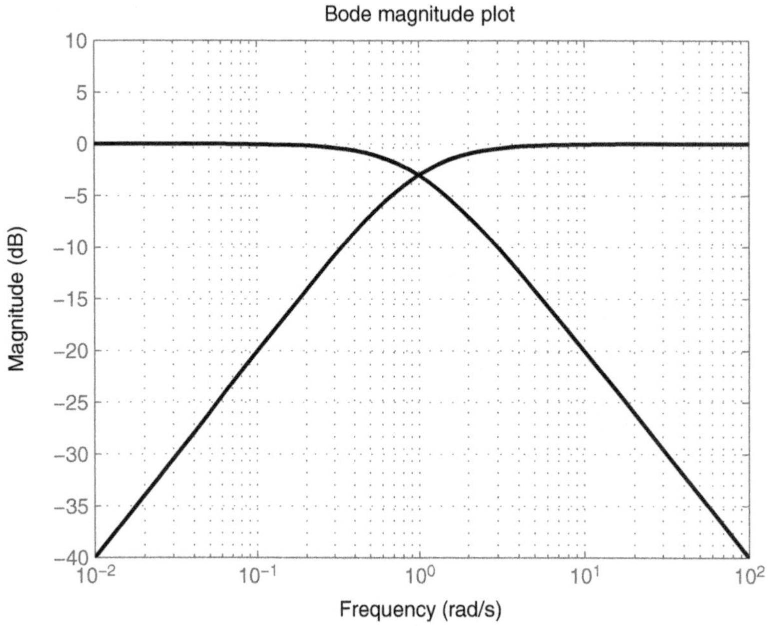

Figure 4.34 Magnitude plot of S and T.

Example 4.7 We will consider the same plant, but now choose to speed up the system. We will place the closed-loop pole at $s = -2\,(\tau = \frac{1}{2})$:

$$Y = \frac{1}{10}\frac{s+1}{\frac{s}{2}+1} \tag{4.95}$$

$$T = \frac{1}{\frac{s}{2}+1}, \quad S = 1 - T = \frac{s}{s+2} \tag{4.96}$$

$$G_c = \frac{Y}{S} = \frac{\frac{1}{5}\frac{s+1}{s+2}}{\frac{s}{s+2}} = \frac{s+1}{5s} = \frac{1}{5} + \frac{1}{5}\frac{1}{s} \tag{4.97}$$

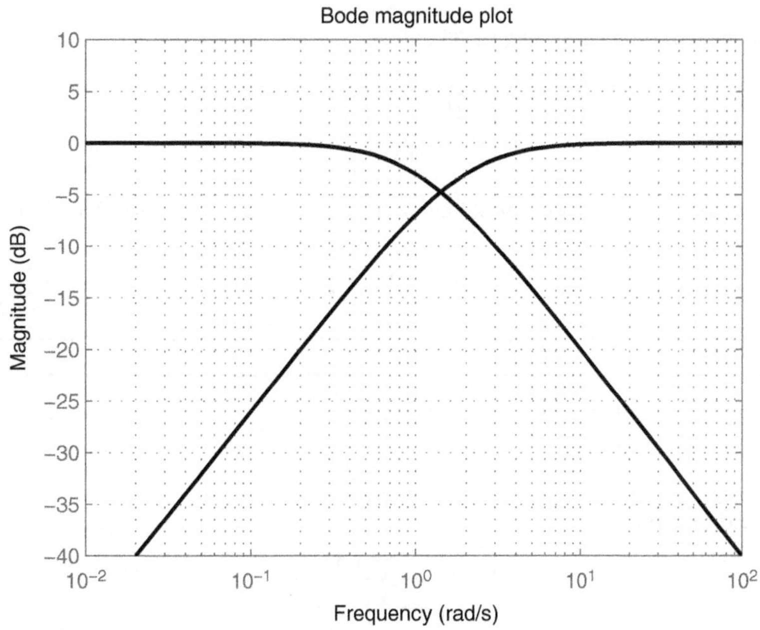

Figure 4.35 Magnitude plot of S and T. ω_c is now at 2 rad/s.

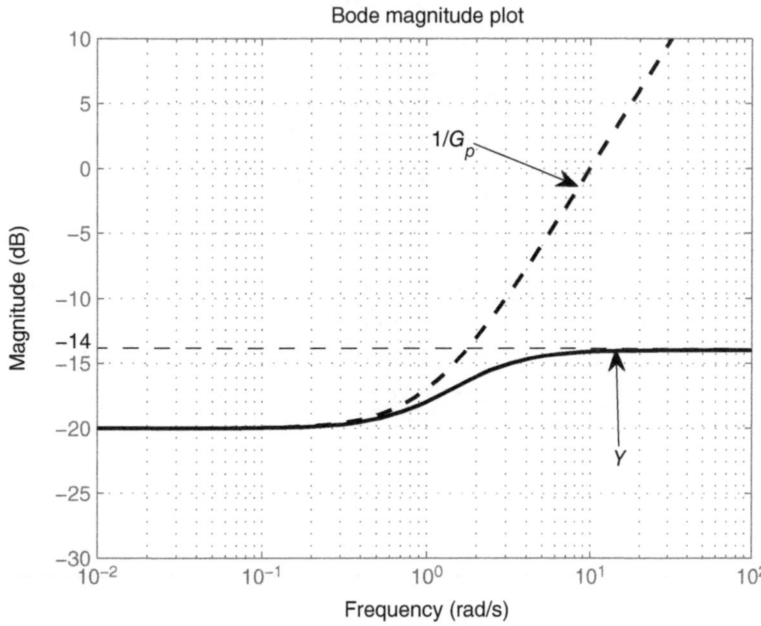

Figure 4.36 Magnitude plot of Y.

Again, we check for internal stability:

$$L = G_c G_p = \frac{1}{5}\frac{s+1}{s} \cdot \frac{10}{s+1} = \frac{2}{s} \tag{4.98}$$

$$1 + L = 1 + \frac{1}{s} = \frac{s+2}{s} \tag{4.99}$$

The magnitude responses of the controlled system are given in (Figures 4.35 and 4.36)

An additional pole could be placed at $s = -5$ to attenuate the effect of high-frequency disturbances.

Example 4.8 What if our plant does not have high gain, and our system requires small Y at low frequencies?

$$G_p = \frac{1}{s+1} \tag{4.100}$$

For instance, let's say that we require $Y(0) = -20$ dB. Then,

$$Y = \frac{1}{10} \Rightarrow T = \frac{1}{10(s+1)} \tag{4.101}$$

$$T(0) = \frac{1}{10} \tag{4.102}$$

This result is not good: the closed-loop response will have significant steady-state error. The problem arises from low gain in G_p:

$$T(0) = Y(0)G_p(0) \tag{4.103}$$

Example 4.9 Consider a plant with two poles, one at the origin:

$$G_p = \frac{1}{s(s+1)} \tag{4.104}$$

We want to place two closed-loop poles at $s = -2$:

$$Y = \frac{s(s+1)}{\left(\frac{s}{2}+1\right)\left(\frac{s}{2}+1\right)} \tag{4.105}$$

Then,

$$T(s) = \frac{1}{\left(\frac{s}{2}+1\right)\left(\frac{s}{2}+1\right)} \tag{4.106}$$

$$S(s) = \frac{s\left(\frac{s}{4}+1\right)}{\left(\frac{s}{2}+2\right)\left(\frac{s}{2}+1\right)} \tag{4.107}$$

$$G_c = \frac{Y}{S} = \frac{4(s+1)}{s+4} \tag{4.108}$$

Verify the interpolation conditions:

$$T(0) = \frac{1}{\left(\frac{0}{2}+1\right)\left(\frac{0}{2}+1\right)} = 1 \tag{4.109}$$

Verify internal stability:

$$L = G_c G_p = \frac{4(s+1)}{s+4} \cdot \frac{1}{s(s+1)} = \frac{4}{s(s+4)} \tag{4.110}$$

$$1 + L = 1 + \frac{4}{s(s+4)} = \frac{s^2+4s+4}{s^2+4s} \tag{4.111}$$

Y is proper, internal stability is met.

Example 4.10 Consider the dynamics of a satellite positioning system:

$$G_{p,1} = G_{p,0} \cdot G_{p,Ap} \tag{4.112}$$

$$G_{p,0} = 0.1 \cdot \frac{1}{s(100s+1)(0.1s+1)} \tag{4.113}$$

$$G_{p,Ap} = \frac{1-100s}{1+100s} \tag{4.114}$$

Classical controller design would require hand-tuning poles, zeros, and integrators for phase correction, high-frequency roll-off, boosting low-frequency gain, and so on. With neoclassical control, the process is much simpler.

The actuator delay (the non-minimum phase zero) requires that we slow the system down ($\tau > 100$). So, we can cancel out each of the poles and replace them with slower ones. While it is not necessary to slow down *every* pole, we can obtain better stability margins by doing so:

$$Y = \frac{10(1+100s)^2(0.1s+1)s}{(200s+1)^4} \tag{4.115}$$

$$T = \frac{1-100s}{(200s+1)^4} \tag{4.116}$$

Example 4.11 Consider the following non-minimum phase plant:

$$G_p(s) = \frac{20\left(1-\frac{s}{2}\right)}{(s+1)\left(\frac{s}{10}+1\right)} \tag{4.117}$$

The current closed-loop time constants are $\tau = 1$ and $\tau = 1/10$ seconds. We want to design a closed-loop response with time constants $\tau = 1/2$ and $\tau = 1/100$ seconds:

$$T = YG_p = Y\frac{20\left(1-\frac{s}{2}\right)}{(s+1)\left(\frac{s}{10}+1\right)} \tag{4.118}$$

$$Y = \frac{(s+1)\left(\frac{s}{10}+1\right)}{20\left(\frac{s}{2}+1\right)\left(\frac{s}{100}+1\right)} \tag{4.119}$$

Therefore,

$$T = \frac{1-\frac{s}{2}}{\left(\frac{s}{2}+1\right)\left(\frac{s}{100}+1\right)} \tag{4.120}$$

$$S = \frac{1.01s\left(\frac{s}{202}+1\right)}{\left(\frac{s}{2}+1\right)\left(\frac{s}{100}+1\right)} \tag{4.121}$$

$$G_c = \frac{Y}{S} = \frac{1}{20.2}\cdot\frac{(s+1)\left(\frac{s}{10}+1\right)}{s\left(\frac{s}{202}+1\right)} \tag{4.122}$$

Verify the well-posedness and the interpolation conditions:

$$L = G_cG_p = \frac{1}{20.2}\frac{(s+1)\left(\frac{s}{10}+1\right)}{s\left(\frac{s}{202}+1\right)}\cdot\frac{20\left(1-\frac{s}{2}\right)}{(s+1)\left(\frac{s}{10}+1\right)} = \frac{\frac{20}{20.2}\left(1-\frac{s}{2}\right)}{s\left(\frac{s}{202}+1\right)} \tag{4.123}$$

$$1+L = 1+\frac{\frac{20}{20.2}\left(1-\frac{s}{2}\right)}{s\left(\frac{s}{202}+1\right)} = \frac{\frac{20}{20.2}\left(\frac{s}{2}+1\right)\left(\frac{s}{100}+1\right)}{s\left(\frac{s}{202}+1\right)} \tag{4.124}$$

$$S(2) = \frac{1.01s\left(\frac{s}{202}+1\right)}{\left(\frac{2}{2}+1\right)\left(\frac{2}{100}+1\right)} = 1 \tag{4.125}$$

Y is proper, internal stability and the interpolation conditions are met. However, there are still problems with this design. Consider the frequency and step responses shown in Figures 4.37 and 4.38.

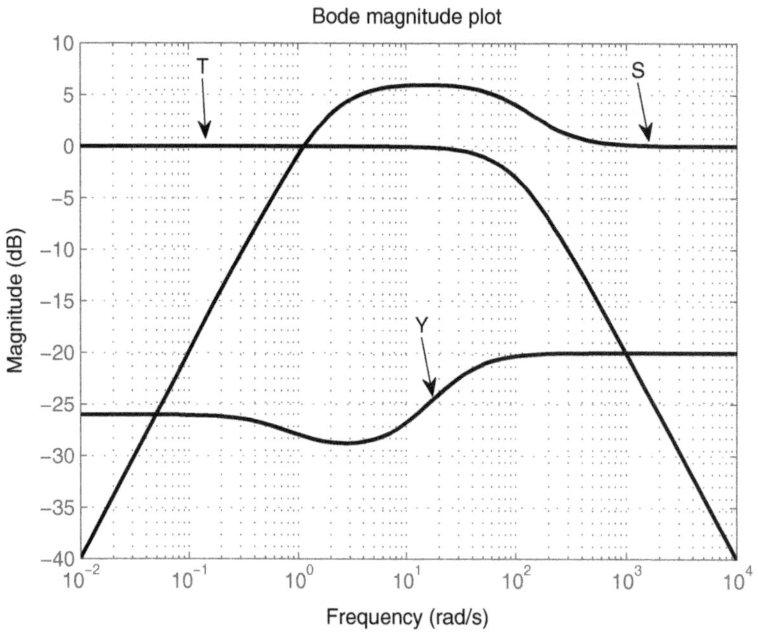

Figure 4.37 Magnitude plot of S, T, and Y.

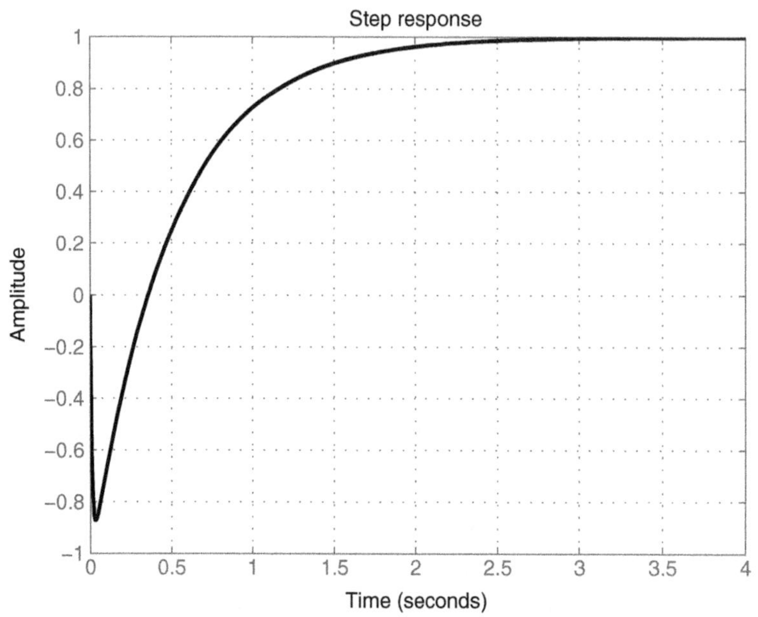

Figure 4.38 Closed-loop step response. The controller produces unacceptably large undershoot.

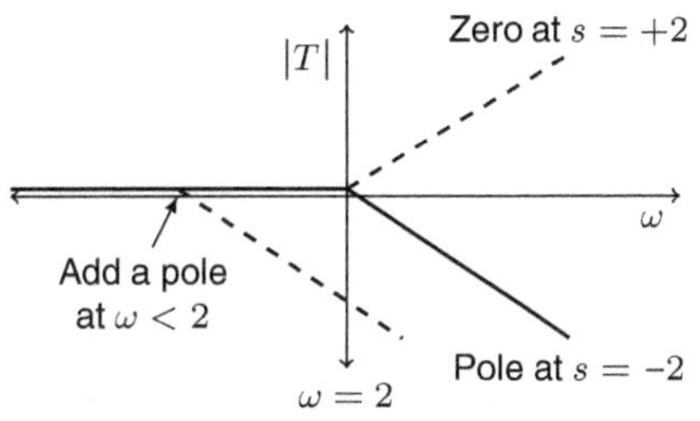

Figure 4.39 $T = YG_p$ when adding a pole.

$\|S\|_\infty \approx 2$, which means the M_2 margin is too low. Output disturbances will be amplified and the system will have poor command following (T alone cannot tell you about good command following – one must look at S as well). Overall, it is a poor design.

The pole at $s = -2$ and zero at $s = +2$ both occur at the frequency $\omega = 2$. The response could be corrected (for instance) by adding a pole at $\omega < 2$ to correct the non-minimum phase zero and reduce the M_2 margin, as shown in (Figure 4.39).

Adding this pole and zero results in a new Y:

$$Y = \frac{1 + \frac{s}{2}}{1 + \frac{s}{10}} \times \frac{(s+1)\left(\frac{s}{10}+1\right)}{20\left(\frac{s}{2}+1\right)\left(\frac{s}{100}+1\right)} \tag{4.126}$$

Example 4.12 Consider the plant,

$$G_p = \frac{s^2}{s^2 + 2\zeta\omega_n s + \omega_n^2} \tag{4.127}$$

Typically, we want new poles to have $\tau > 1/|z|$ for non-minimum phase zeros, but here the zeros are at the origin – an appropriate time constant would be infinitely large. That is not an option, so instead we must settle for a time constant of new poles that is very large ($\tau \gg 1$). True command tracking will not be possible, with large τ, we will effectively make T and bandpass filter that covers

$$Y = \frac{\omega_n^2 \tau^2}{(1+\tau s)^2} \tag{4.128}$$

Then,

$$T = \frac{\omega_n^2 \tau^2 s^2}{(1+\tau s)^2 (s^2 + 2\zeta\omega_n s + \omega_n^2)} \tag{4.129}$$

We can check our interpolation conditions and clearly see that they are met:

$$T(0) = \frac{\omega_n^2 \tau^2 0^2}{(1+\tau 0)^2 (0^2 + 2\zeta\omega_n 0 + \omega_n^2)} = 0 \tag{4.130}$$

$$\left.\frac{dT}{ds}\right|_{s=0} = 2s(\cdots) + s^2(\cdots)\Big|_{s=0} = 0 \tag{4.131}$$

(The complete definition of the derivative of T is not given, but it should be clear that all terms will have an s in their numerator.):

$$T(0) \neq 1 \tag{4.132}$$

So we know immediately that this control system will have steady-state error.

Then,

$$S = 1 - T = \frac{(1+\tau s)^2 (s^2 + 2\zeta\omega_n s + \omega_n^2) - \omega_n^2 \tau^2 s^2}{(1+\tau s)^2 (s^2 + 2\zeta\omega_n s + \omega_n^2)} \tag{4.133}$$

$$G_c = \frac{Y}{S} = \frac{(s^2 + 2\zeta\omega_n s + \omega_n^2)(\omega_n^2 \tau^2)}{(1+\tau s)^2 (s^2 + 2\zeta\omega_n s + \omega_n^2) - \omega_n^2 \tau^2 s^2} \tag{4.134}$$

$$= \frac{s^2 + 2\zeta\omega_n s + \omega_n^2}{(\tau^2 s^2 + 2\tau s + 1)(s^2 + 2\zeta\omega_n s + \omega_n^2) - \omega_n^2 \tau^2 s^2} \tag{4.135}$$

$$\Rightarrow \quad G_c = \frac{(s^2 + 2\zeta\omega_n s + \omega_n^2)(\omega_n^2 \tau^2)}{\tau^2 s^4 + (2\zeta\omega_n \tau^2 + 2\tau)s^3 + (1 + 4\tau\zeta\omega_n)s^2 + 2\tau\omega_n^2 s + \omega_n^2} \tag{4.136}$$

Example 4.13 Consider the plant

$$G_p = \frac{4(0.4 - s)}{(s + 0.1)(s + 0.2)(s + 0.8)} \tag{4.137}$$

that is subject to actuator saturation design constraints $|u(t)| \leq 0.10$ for noise $|n(t)| \leq 0.01$ at $\omega > 0.1$ rad/s and input $|r(t)| \leq 1$ at $\omega < 0.1$ rad/s. First consider the constraint on saturation due to noise:

$$\frac{|\hat{u}|}{|\hat{n}|} = \frac{0.10}{0.01} = 10 = 20 \text{ dB} \tag{4.138}$$

So, $|Y(\omega)|$ must be less than 20 dB at $\omega > 0.1$ rad/s. For the constraint on saturation due to commanded inputs,

$$\frac{|\hat{u}|}{|\hat{r}|} = \frac{0.10}{1} = 0.1 = -20 \text{ dB} \tag{4.139}$$

So, $|Y(\omega)|$ must be less than -20 dB at $\omega < 0.1$ rad/s. We start by choosing a shape for Y. We cannot cancel the non-minimum phase zero, but will cancel all three poles and replace them with one first-order pole and one second-order pair.

$$Y = \frac{K(s + 0.1)(s + 0.2)(s + 0.8)}{4(\tau s + 1)(s^2 + 2\zeta\omega_n + \omega_n^2)} \tag{4.140}$$

Then,

$$T = \frac{K(0.4 - s)}{(\tau s + 1)(s^2 + 2\zeta\omega_n + \omega_n^2)} \tag{4.141}$$

Our interpolation conditions are $T(0) = 1$ (for command tracking) and $T(0.4) = 0$. $T(0.4) = 0$ is clearly met, while $T(0) = 1$ requires constraints on the tuning parameters K and ω_n:

$$T(0) = 1 = \frac{0.4K}{\omega_n^2} \tag{4.142}$$

Therefore, $K = \omega_n^2/0.4$.

We can assume that the peak actuator effort will occur at the start of a step response. Then, use the initial value theorem to obtain an additional constraint.

$$\begin{aligned} Y(0^+) = 0.1 &= \lim_{s\to\infty} sY(s)\frac{1}{s} \\ &= \lim_{s\to\infty} \frac{K(s + 0.1)(s + 0.2)(s + 0.8)}{4(\tau s + 1)(s^2 + 2\zeta\omega_n + \omega_n^2)} = \frac{K}{4\tau} \end{aligned} \tag{4.143}$$

Therefore, $\tau = K/0.4 = (\omega_n/0.4)^2$. At this point, we can tune the system with ζ and ω_n until the design requirements are met. (Note that would could tune with τ or ζ instead of ω_n, but we have already defined K and τ in terms of ω_n.)

Let's select $\omega_n = \sqrt{0.4}$ (resulting in $K = 1$) and $\zeta = 0.707$. The resulting Youla magnitude plot is shown in Figure 4.40.

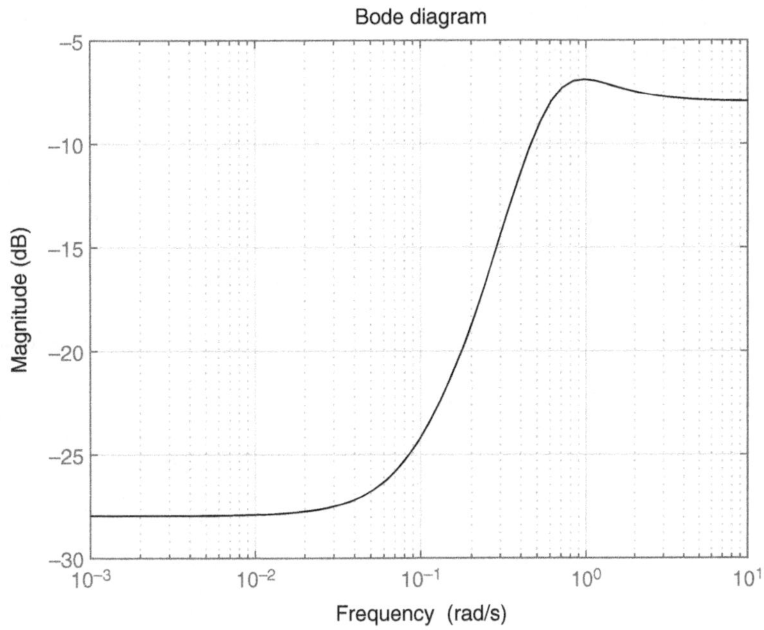

Figure 4.40 Magnitude response of Y.

4.8 Problems

4.1 Consider the unstable and non-minimum phase plant,

$$G_p = \frac{10(1-s)}{(1-s/2)}$$

(a) Design a compensator, G_c, that stabilizes the plant by selecting an appropriate $Y(s)$ to ensure closed-loop internal stability using interpolation conditions. Compute T, S, and G_c. Provide Bode magnitude plots of T and Y. Is your design acceptable? Briefly comment on your design.

(b) Propose a new $Y(s)$ that not just ensures internal stability but also provides better performance and command tracking. For this case, compute only T and briefly comment on your design.

4.2 Consider the dynamics of a satellite positioning system,

$$G_p = G_{p,0} \cdot G_{p,Ap}$$

$$G_{p,0} = \frac{0.1}{s(100s+1)(0.1s+1)}$$

$$G_{p,Ap} = \frac{1-100s}{1+100s}$$

Classical controller design would require hand-tuning poles, zeros, and integrators for phase correction, high-frequency roll-off, boosted low-frequency gain, and so on. With neoclassical control, the process is much simpler.

Use Youla parameterization to design a feedback controller that results in a relatively fast response time, by selecting an appropriate closed-loop time constant, with zero percent overshoot, and meeting the internal stability constraint. You need to only provide your choice of $Y(s)$ transfer function and compute $T(s)$. There is no need to provide any Bode plots or compute the controller transfer function.

4.3 This problem considers a system with M_2 stability measure of $M_2 = 1$. Your task is to compute max $|S(j\omega)|$ in dB and to compute the classical gain margin associated with $M2 = 1$.
Hint: For computing the classical gain margin, you can do this geometrically by plotting an arbitrary $L(j\omega)$ on a polar plot and then superimposing the $M_2 = 1$ circle on this plot to illustrate the classical gain margin. Or, use the equation that relates $|S(j\omega)|$ to the classical gain margin.

4.4 The transfer function $G_p(s)$ describes the simplified dynamics of a transmission clutch:

$$G_p = \frac{-(150)(s^2 - 0.12s + 144)}{(s+20)(s^2 + 30s + 10\ 000)}$$

Use Youla parameterization to design a feedback controller that meets the specifications in Table 4.1.

Table 4.1 Requirements.

Metric	Specification
Settling time	2.0 s
Overshoot	< 10%
Steady-state error	0
Gain margin	> 10 dB
Phase margin	> 60°
Crossover frequency	> 5 rad/s

Justify your design by providing all the relevant Bode magnitude plots and step responses.

4.5 Consider the block diagram shown in Figure 4.41. Design $G_c(s)$ so that the closed-loop system behaves approximately like the second-order system, which, for a unit step input, has
- an overshoot that does not exceed 16% ($\zeta \geq 0.5$),
- a settling time less than or equal to 5.3 seconds ($\omega_n \geq 1.5$ rad/s), and
- a steady state error below 5%.

Design the controller using the Youla method. You need to fully justify your choice of compensator and illustrate that the given requirements have been met.

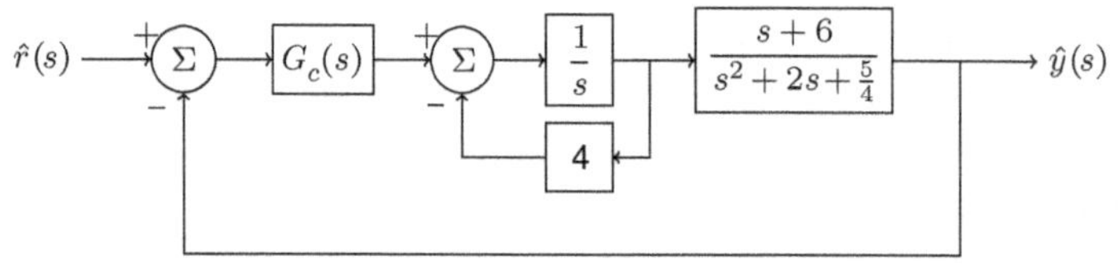

Figure 4.41 Block diagram for Problem 4.5.

4.6 The open-loop block diagram for the active suspension problem in Figure 4.42.
In this problem, your task is to design a Youla controller by hand to minimize the effect of v_d on $\dot{x}_s$. Provide the following Bode plots.
(a) One plot of T, S, Y, G_p, and G_c.
(b) One plot of T and $G_{p,d}S$.
(c) One magnitude plot of $\frac{\dot{x}_s}{v_d}$ (both controlled and uncontrolled).
(d) One magnitude plot of $\frac{\dot{x}_u}{v_d}$ (both controlled and uncontrolled).
(e) One magnitude plot of $\frac{v_{rel}}{v_d}$ (both controlled and uncontrolled).
(f) One magnitude plot of $\frac{q_s}{v_d}$ (both controlled and uncontrolled).

Figure 4.42 Block diagram of the active suspension with sprung mass velocity as an output.

4.7 Consider the unstable and non-minimum phase plant,

$$G_p = \frac{e^{-\tau s}}{(s-1)}$$

Design a compensator, G_c, by selecting an appropriate $Y(s)$, where $0 < \tau \leq 0.1$, to ensure closed-loop internal stability using interpolation conditions and good target following (small steady-state errors are acceptable, such as ≤ 0.2), and compute T, S, and G_c. Provide Bode plots of T, S, and Y, including your design phase and gain margin, and the step response of the closed-loop transfer function T. Is your design acceptable? Briefly comment on your design.

4.8 Apply the method of Youla parameterization to the object characterized by:

$$G_p = \frac{(10+s)}{s^3}$$

Evaluate the effectiveness of your design by examining the Bode magnitude of $Y(j\omega)$, $S(j\omega)$, $T(j\omega)$, and by plotting the closed-loop step response.
Hint: Instead of interpolation conditions, try to make the four closed-loop transfer functions, $Y, YG_p, 1-YG_p, G_p(1-YG_p)$, BIBO stable to meet internally stability. This can be accomplished by enforcing that S transfer function includes s^3 in its numerator.

4.9 Consider the plant

$$G_p = \frac{s^2}{s^2 + 2\zeta\omega_n s + \omega_n^2} \tag{4.144}$$

Do your best using Youla parameterization to obtain a controller that can give command tracking.
Note: You cannot cancel the zeros at $s = 0$, so you will not be able to enforce an interpolation condition on $T(0) = 1$.

5

Norms of Feedback Systems

Much of this material is a prerequisite for the next chapter, the application of *optimization* in control theory. This application is not new and can be regarded as a branch of mathematics whose goal is to improve a variable in order to maximize a benefit or minimize a cost.

We tried to avoid the necessary mathematical background, especially in complex analysis, for some of developments in this chapter and the chapters to follow. However, there is sufficient background to allow us to retrace certain of our steps and develop rigorous relationships alluded to only in an heuristic manner in previous chapters. In addition, the development of rigorous relationships shed more light into some of the concepts of this chapter and the following chapters. Therefore, we included the background material, "Complex Analysis" and "Real and Complex Analysis in Signal, Dynamic system and Feedback Theory," in the first two appendices of this book.

In the rest of this chapter, the concentration is on the signal and system norms for SISO system. But, before discussing this topic, we will cover some required concepts in Laplace and Fourier transform, state-space representations, and controllability and observability.

5.1 The Laplace and Fourier Transform

This section will establish some basic properties of the Laplace and Fourier transforms. These properties will be used throughout the chapter to describe various norms and their uses.

5.1.1 The Inverse Laplace Transform

The Cauchy integral formula, as discussed in Appendix A, is

$$f(\lambda_0) = \frac{1}{2\pi j}\int_{\partial\Omega} \frac{f(\lambda)}{\lambda - \lambda_0} \mathrm{d}\lambda \tag{5.1}$$

if $f(\lambda)$ is analytic everywhere on and within the contour "$\partial\Omega$" and if λ_0 is interior to the contour. We say that the value of a function that is analytic in some region is given entirely by the values it attains on the boundary of that region. The counter-clockwise direction is the positive direction of integration in this formula. We wish to apply this formula to the Laplace transform

$$\hat{y}(s) = \int_{0^-}^{\infty} e^{-st} y(t)\mathrm{d}t \tag{5.2}$$

This integral definition converges for "s" in the "region of convergence", which is the half-plane to the right of all poles of $\hat{y}(s)$. That is to say, $\hat{y}(s)$ is analytic everywhere in the region of convergence. This means that where the Cauchy integral formula is applied to (5.2), we have

$$\hat{y}(s) = \frac{1}{2\pi j}\int_{D} \frac{\hat{y}(\lambda)}{\lambda - s} \mathrm{d}\lambda \tag{5.3}$$

Here,"D" is the D-shaped contour (Nyquist contour) illustrated in Figure 4.3 and as given in more detail in Appendix A. Imagine taking the inverse transform of both sides of Eq. (5.3). The only function of "s" on the right side is $\frac{1}{(\lambda - s)}$, and the inverse transform of this term is $-e^{(\lambda t)}$.

$$y(t) = -\frac{1}{2\pi j}\oint_{D} e^{\lambda t}\hat{y}(\lambda)\mathrm{d}\lambda \tag{5.4}$$

Robust Control: Youla Parameterization Approach, First Edition. Farhad Assadian and Kevin R. Mallon.

Companion Website: www.wiley.com/go/assadian/robust

We can now replace λ with s and reverse the direction of the path integral.

$$y(t) = \frac{1}{2\pi j}\oint_D e^{st}\hat{y}(s)\mathrm{d}s \tag{5.5}$$

Notice that the direction of integration is indicated on the integration symbol itself to serve as a reminder.

The summary of this is given in the following theorem.

Theorem 5.1 The Laplace Transformation

$$\hat{y}(s) = \int_{0^-}^{\infty} e^{-st}y(t)\mathrm{d}t \tag{5.6}$$

is associated with the inverse Laplace transformation

$$y(t) = \frac{1}{2\pi j}\oint_D e^{st}\hat{y}(s)\mathrm{d}s \tag{5.7}$$

where the contour "D" encloses the entire half-plane lying to the right of all poles of $\hat{y}(s)$ and the integration is taken in the clockwise direction. ■

The integral

$$\oint_D e^{st}\hat{y}(s)\mathrm{d}s \tag{5.8}$$

is evaluated using Cauchy's residue theorem. This theorem states that

$$2\pi j\sum_K \text{(Residues of } \hat{y}(s) \text{ at } s = s_K) = \oint_D e^{st}\hat{y}(s)\mathrm{d}s = \lim_{\omega\to\infty}\int_{\sigma-j\omega}^{\sigma+j\omega} \hat{y}(s)e^{st}\mathrm{d}s \tag{5.9}$$

σ is a real number so that the contour path of integration is in the region of convergence of $\hat{y}(s)$.

Example 5.1 Use the Nyquist contour to find the value of

$$\mathcal{L}^{-1}\left[\frac{1}{(s+1)(s-2)^2}\right]$$

It is easy to find this particular inverse Laplace Transform using Heaviside expansion as covered in Chapter 2; however, this example serves as an illustration of the use of the contour integral method. It should be noted that there are examples that there are no alternative except using the contour integral method.

Now,

$$\oint_D \frac{e^{st}}{(s+1)(s-2)^2}\mathrm{d}s = 2\pi j\sum_K \text{(Residues of } \hat{y}(s) \text{ at } s = s_K)$$

The residue at $s = -1$ is given by

$$\lim_{s\to-1}(s+1)\frac{e^{st}}{(s+1)(s-2)^2} = \frac{1}{9}e^{-t}$$

The residue at $s = 2$ is given by

$$\lim_{s\to2}\frac{\mathrm{d}}{\mathrm{d}s}\frac{e^{st}}{(s+1)} = \left[\frac{(s+1)te^{st} - e^{st}}{(s+1)^2}\right]_{s=2} = \frac{1}{9}(3te^{2t} - e^{2t})$$

Thus,

$$\oint_D \frac{e^{st}}{(s+1)(s-2)^2}\mathrm{d}s = 2\pi j\left[\frac{1}{9}(e^{-t} + 3te^{2t} - e^{2t})\right]$$

Evaluate the integral as the radius of the Nyquist contour goes to ∞. Thus,

$$\mathcal{L}^{-1}\left[\frac{1}{(s+1)(s-2)^2}\right] = \frac{1}{2\pi j}\int_{\sigma-j\infty}^{\sigma+j\infty}\frac{e^{st}}{(s+1)(s-2)^2}\mathrm{d}s$$
$$= \frac{1}{9}(e^{-t} + 3te^{2t} - e^{2t})$$

A result easily verified by the use of Heaviside expansion.

5.1.2 Parseval's Theorem

There are several extremely important implications of the formulas developed in Section 5.1.1. These implications are generally called the Parseval theorems.

Begin with the following lemma.

Lemma 5.1 *If $\hat{y}(s)$ is the Laplace transform of a real function, then it is symmetric about the real axis so that*

$$\overline{\hat{y}(j\omega)} = \hat{y}(-j\omega) \tag{5.10}$$

Proof: It follows from (5.6) that

$$\hat{y}(j\omega) = \int_{0^-}^{\infty} e^{-j\omega t} y(t)\mathrm{d}t = \int_{0^-}^{\infty} \cos(\omega t)y(t)\mathrm{d}t - j\int_{0^-}^{\infty} \sin(\omega t)y(t)\mathrm{d}t \tag{5.11}$$

We conclude the lemma. ∎

In particular, this result must hold for a transfer function because a transfer function is the Laplace transform of the impulse response.

Often we are interested in an integral called the "energy" of a signal that takes the form

$$E = \int_{0^-}^{\infty} y^2(t)\mathrm{d}t \tag{5.12}$$

If $\hat{y}(s)$ is not strictly proper, $y(t)$ contains an impulse. Then, the Eq. (5.12) for $y(t)$ would be unbounded or undefined:

$$\int_{0}^{\infty} \delta^2(t)\mathrm{d}t = \delta(0) \tag{5.13}$$

The integral for E is finite only if $\hat{y}(s)$ (i) is strictly proper, (ii) is characterized by poles that lie strictly in the left half-plane. In this case, use (5.7) to write

$$E = \frac{1}{2\pi j}\int_{0^-}^{\infty} y(t) \oint_D e^{st}\hat{y}(s)\mathrm{d}s\ \mathrm{d}t \tag{5.14}$$

Reverse the order of integration to find

$$E = \frac{1}{2\pi j}\oint_D \hat{y}(s)\int_{0^-}^{\infty} e^{st}y(t)\mathrm{d}t\ \mathrm{d}s \tag{5.15}$$

We have established that the energy of a signal

$$E = \int_{0^-}^{\infty} y^2(t)\mathrm{d}t = \frac{1}{2\pi j}\oint_D \hat{y}(s)\hat{y}(-s)\mathrm{d}s \tag{5.16}$$

This integral exists where the poles of $\hat{y}(s)$ lie strictly in the left half-plane. The D-contour encircles the entire right half-plane and the integration is taken clockwise. The other restriction is that $\hat{y}(s)$ be strictly proper.

However, if $\hat{y}(s)$ is strictly proper, the contribution of the curved portion of the contour to the integral is exactly zero; that is, contour integral (5.16) reduces to an simple integration along imaginary axis. On this axis,

$$\mathrm{d}s = j\ \mathrm{d}\omega \tag{5.17}$$

and (5.10) means

$$\hat{y}(j\omega)\hat{y}(-j\omega) = |\hat{y}(j\omega)|^2 \tag{5.18}$$

Apply all this to (5.16) to prove:

Theorem 5.2 **Parseval's Theorem** *Where it exists, the energy of a signal is*

$$E = \int_{0^-}^{\infty} y^2(t)\mathrm{d}t = \frac{1}{2\pi}\int_{-\infty}^{\infty} |\hat{y}(j\omega)|^2\mathrm{d}\omega \tag{5.19}$$

where $\hat{y}(s) \triangleq \mathscr{L}[y(t)]$. ∎

5.1.3 The Fourier Transform

Much of the following requires an understanding of another integral transform, a transform that can be applied to signals not already encountered in our developments but which characterize those that commonly impinge on our control systems.

Definition 5.1 The Fourier transform of a function $v(t)$ is denoted $\mathscr{F}[v(t)]$ or $\hat{v}(j\omega)$ and is given by the improper integral[1]

$$\mathscr{F}[v(t)] = \int_{-\infty}^{\infty} e^{-j\omega t} v(t) dt \tag{5.20}$$

We don't restrict our signals to be zero for negative times so the lower limit is taken to be $-\infty$. That is, we expand our horizons beyond that of command signals and begin to study the properties of noise and disturbance signals that have no clear beginning.

Of course, a physical signal at $t = -\infty$ is an idealization, much as it is to say, the idealization of a signal with a discontinuity: The philosophy is the same, it is easier to deal with the idealization that it is to attempt an analysis of the real thing. Of course, the price paid is the manipulation of an improper integral.

An immediate implication is obtained by comparing the definitions of Laplace and Fourier transforms.

Proposition 5.1 *Suppose that $v(t) \equiv 0, t \leq 0$, and if $\sigma = 0$ is in the domain of convergence of the Laplace transform, $\mathscr{L}[v(t)]$, then*

$$\mathscr{F}[v(t)] = \hat{v}(s)|_{s=j\omega} \tag{5.21}$$

The point is that we wish to include in our study those signals that do not satisfy the hypothesis of this proposition. This means, for example, that any function with a Laplace transform with all poles strictly in the left half-plane has a Fourier transform, such as in the example below.

Example 5.2 Consider the impulse response, $g(t)$, of an asymptotically stable, linear system has a Fourier transform, as shown in (Figure 5.1).

$$\mathscr{F}[g(t)] = G(s)|_{s=j\omega} \tag{5.22}$$

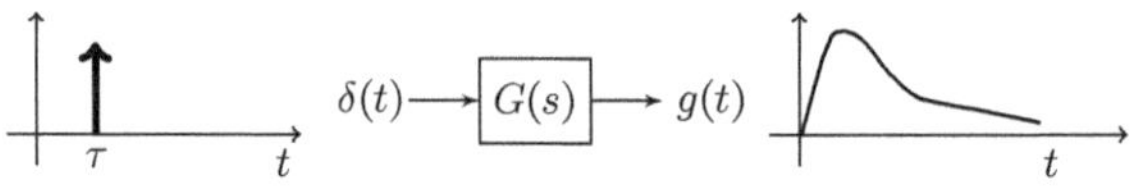

Figure 5.1 Impulse response for Example 5.1.

We will demonstrate that we can associate with the Fourier transform, the inverse Fourier transformation

$$v(t) \triangleq \mathscr{F}^{-1}[\hat{v}(j\omega)] = \frac{1}{2\pi} \int_{-\infty}^{\infty} e^{j\omega t} \hat{v}(j\omega) d\omega \tag{5.23}$$

It is usually more convenient to rewrite this as

$$v(t) \triangleq \mathscr{F}^{-1}[\hat{v}(j\omega)] = \frac{1}{2\pi j} \int_{-\infty}^{\infty} e^{j\omega t} \hat{v}(j\omega) d(j\omega) \tag{5.24}$$

So that we can imagine the inverse-transform integral as an integral along the imaginary axis in the s-plane.

This trick will allow us to take advantage of the Cauchy residue theorem. More importantly, we are now able to imagine the Fourier transform as a mapping of sets of time functions onto the imaginary axis of the s-plane. The "domain" (the set of functions to be transformed) is not coincident with the domain of the Laplace transform and neither set is a subset of the other.

---

1 One encounters variations in the definition of the Fourier transform. The differences are in the constant-multiplier placed in front of the integral. Obviously, we have chosen 1.0, but, in another text, the choice might be $\frac{1}{2\pi}$.

5.1.3.1 Properties of the Fourier Transform

As in the case of the Laplace and Fourier transforms, there are a number of useful facts that are immediate consequences of the definition.

- **Homogeneity**:

$$\mathcal{F}[v(t/\alpha)] = \alpha\hat{v}(\alpha j\omega) \quad \text{if } \alpha > 0 \tag{5.25}$$

- **Time shifting**:

$$\mathcal{F}[v(t-\tau)] = e^{-j\omega\tau}\,\hat{v}(j\omega) \tag{5.26}$$

- **Frequency shifting**:

$$\mathcal{F}[e^{-j\omega_0\tau}v(t)] = \hat{v}(j\omega + \omega_0) \tag{5.27}$$

- **Symmetry**:

$$\hat{v}(-j\omega) = \overline{\hat{v}}(j\omega) \tag{5.28}$$

- **Differentiation**:

$$\mathcal{F}[\mathrm{d}^k v(t)/\mathrm{d}t^k] = (j\omega)^k\hat{v}(j\omega) \tag{5.29}$$

- **Convolution**:

$$\mathcal{F}\left[\int_{-\infty}^{\infty} v_1(t-\tau)v_2(\tau)\mathrm{d}t\right] = \hat{v}_1(j\omega)\hat{v}_2(j\omega) \tag{5.30}$$

Where we idealize the signal as extending back to $-\infty$, we have the generalization of Theorem 2.3.

5.1.3.2 Inverse Fourier Transformation By Heaviside Expansion

It is clear that the imaginary axis must be in the region of convergence of the Fourier transformation. A point that requires some explanation is how the poles of the transform in the two half-planes are related to the time-domain signal.

First notice that we can rewrite any signal as

$$v(t) = v_n(t) + v_p(t) \tag{5.31}$$

where

$$v_n(t) = 0, \quad t \geq 0 \tag{5.32}$$

and

$$v_p(t) = 0, \quad t < 0 \tag{5.33}$$

That is, $v_n(t)$ is the signal at the negative time interval, and $v_p(t)$ is the signal at the positive time interval, indicated by our choice of subscripts.

Think of this decomposition as converting the Fourier transform into the sum of two Laplace-like transforms:

$$\mathcal{F}[v(t)] = \int_{-\infty}^{0} e^{-j\omega t}v_n(t)\mathrm{d}t + \int_{0}^{\infty} e^{-j\omega t}v_p(t)\mathrm{d}t \tag{5.34}$$

The interesting turn of events is that the first integral is over negative values of time, t. If this integral is to exist, it must be true, at the very least, that $v_n(t) \to 0$ as $t \to -\infty$.

Replace "t" with "$-t$" in the first integral to conclude that

$$\mathcal{F}[v(t)] = \int_{0}^{\infty} e^{-j\omega t}v_n(-t)\mathrm{d}t + \int_{0}^{\infty} e^{-j\omega t}v_p(t)\mathrm{d}t \tag{5.35}$$

This decomposition has the appearance of two Laplace transforms and so we can bring our understanding of the latter transform to develop further our Fourier transform.

Notice that we must have the imaginary axis in the region of convergence of both components of (5.35) and "t" is replaced by "$-t$" in the first integral.

This leads the conclusion that

> The poles of the Fourier transform that lie in the left half-plane are associated with only that portion of $v(t)$ for $t \geq 0$, while those poles in the right half-plane are associated with only that portion of $v(t)$ where $t < 0$.

This observation leads immediately to a Heaviside-expansion scheme for the inverse Fourier transformation.

5.2 Norms of Signals and Systems

Control system analysis and design using optimization techniques require translation of the system properties and its associated signals into the language of mathematics. One fundamental property of a system and its signal in the analysis and design of control systems is their size. The way we compare the size of different objects mathematically is through real numbers. For example, it is interesting to note that if two real numbers are given, such as 4 and 3, we immediately know 4 is greater than 3. However, if we are given two complex numbers such as $5i$ and $3 + 4i$, an association (a "mapping") to real numbers, such as the absolute value of a complex number, is required to differentiate their sizes. The idea of "minimum" and "maximum" is a reference to a property of a set which can be well ordered such as a set of real numbers. There are many ways of interpreting the size of a signal or a system. Norms are the geometrical measures of the size of signals and systems that are found to be useful in the analysis and design of control systems. These measures of size are the mapping of scalar functions, vector functions, or matrices of rational functions onto the set of real positive numbers.

In this chapter, we discuss the properties of scalar signals and SISO systems. We will use the information on Laplace and Fourier transforms presented in Section 5.1 to provide some insight into these norms. We define the related norms that are used for the development of performance/robustness measures for optimal robust controllers.

We will extend and generalize these results to vector signals and MIMO systems in the second volume of this book including several concepts related to systems such as linearity, controllability and observability and the state-space approach for computation of the related norms.

5.2.1 Signal Norms

Consider the two disturbances shown in (Figure 5.2). Which is "worse"?

The answer depends on the context. A norm is the measure of the "size" of the signal. To determine how big (or "bad") a signal is, we need to define what we mean by size. Let $V = \{V_1(\cdot)\}$ be a signal and let $\|V_1(\cdot)\|$ be a mapping to real number.

Definition 5.2 $\|V\|$ is a semi-norm if it is nonnegative, homogeneous, and does not violate the triangle inequality.

$$\textbf{Non-negativity:}\ \|V\| \geq 0 \tag{5.36}$$

$$\textbf{Homogeneity:}\ \|\alpha V\| = |\alpha| \cdot \|V\| \quad \text{for all } \|V\| < \infty \tag{5.37}$$

$$\textbf{Triangle Inequality:}\ \|V + U\| \leq \|V\| + \|U\| \tag{5.38}$$

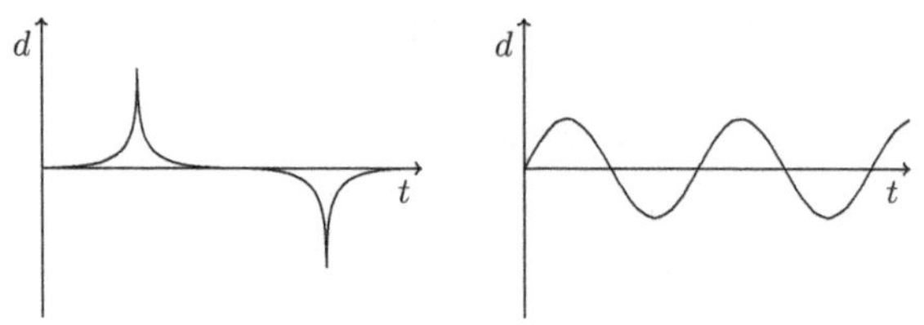

Figure 5.2 Two disturbances.

Definition 5.3 $||V||$ is a norm if it is a semi-norm and

$$||V|| = 0 \text{ only if } V \equiv 0 \tag{5.39}$$

5.2.1.1 Particular Norms

Definition 5.4 The 1-norm (known as L_1 norm [3]) measures the integrated absolute error (IAE) and is also a measure of resource consumption.

$$||y||_1 = \int_{-\infty}^{\infty} |y(t)|\mathrm{d}t \tag{5.40}$$

As explained in Section 5.1, the lower limit is taken here to be $-\infty$.

Definition 5.5 The 2-norm (L_2 norm) measures the integrated square error (ISE) and is also a measure of signal energy.

$$||y||_2 = \sqrt{\int_{-\infty}^{\infty} y^2(t)\mathrm{d}t} \tag{5.41}$$

Definition 5.6 The ∞-norm (L_∞ norm) measures peak magnitude.

$$||y||_\infty = \sup_t |y(t)| \tag{5.42}$$

where "sup" stands for supremum. It should be noted that if the maximum of a set exists, then it is the supremum. However, if the maximum does not exist, then the supremum is used.

Definition 5.7 The root mean squared-norm (RMS-norm) is a semi-norm (it can be zero for nonzero signals) that is a measure of power in a repeated signal (where τ is the period). It is sometimes written as pow $(y(t)) = ||y||_{RMS}^2$.

$$||y||_{RMS} = \sqrt{\lim_{\tau\to\infty} \frac{1}{\tau} \int_0^{\tau} y^2(t)\mathrm{d}t} \tag{5.43}$$

Using Parseval's theorem, we can relate this to a norm in the frequency domain.

$$P_y = ||y||_{RMS}^2 = \frac{1}{2\pi} \int_{-\infty}^{\infty} S_y(\omega)\mathrm{d}\omega \tag{5.44}$$

where $S_y(\omega)$ is the power spectral density (PSD) of $y(t)$ such that

$$S_y(\omega) = \lim_{\tau\to\infty} \frac{|y_\tau(\omega)|^2}{\tau} \tag{5.45}$$

and $\int_{-\infty}^{\infty} y_\tau^2(t)\mathrm{d}t = \int_0^{\tau} y^2(t)\mathrm{d}t$, this is the truncated version of the everlasting signal $y(t)$.

Example 5.3 Consider the function

$$y(t) = \begin{cases} 0, & t \le 0 \\ t, & 0 < t \le 1 \\ 0, & t > 1 \end{cases} \tag{5.46}$$

Then,

$$||y||_1 = \int_0^1 t\ \mathrm{d}t = \left.\frac{t^2}{2}\right|_0^1 = \frac{1}{2} \tag{5.47}$$

$$||y||_2 = \sqrt{\int_0^1 t^2\ \mathrm{d}t} = \sqrt{\left.\frac{t^3}{3}\right|_0^1} = \frac{1}{\sqrt{3}} \tag{5.48}$$

$$||y||_\infty = \sup_t t = 1 \tag{5.49}$$

(This is not a repeated signal, so we will not consider the RMS-norm.)

Example 5.4 Consider the function

$$y(t) = A \sin t \tag{5.50}$$

This is a repeated signal with period 2π. Therefore,

$$\|y\|_{RMS} = \sqrt{\lim_{\tau\to\infty} \frac{1}{\tau} \int_0^{\tau} A^2 \sin^2 t \, dt} \tag{5.51}$$

$$\sin^2 t = \frac{1}{2} - \frac{1}{2}\cos 2t \tag{5.52}$$

$$\frac{1}{\tau}\int_0^{\tau} A^2 \sin^2 t \, dt = \frac{1}{\tau}\int_0^{\tau} \frac{A^2}{2} dt - \underbrace{\frac{1}{\tau}\int_0^{\tau} A^2 \cos 2t \, dt}_{=0} \tag{5.53}$$

$$\|y\|_{RMS} = \sqrt{\frac{A^2}{2}} = \frac{A}{\sqrt{2}} \tag{5.54}$$

Signals can typically be classified as *energy* or *power* signals. For an energy signal, the 1-norm, 2-norm, and ∞-norm are finite and nonzero, while the RMS-norm is zero. On the other hand, a power signal has an infinite 1-norm and 2-norm and a finite, nonzero ∞-norm and RMS-norm. Energy and power signals are illustrated in (Figures 5.3 and 5.4).

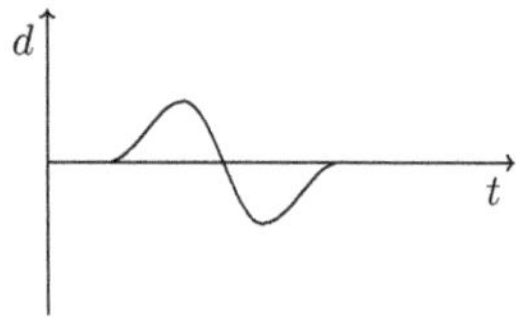

Figure 5.3 An energy signal.

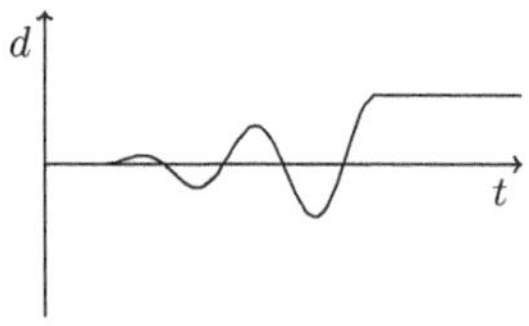

Figure 5.4 A power signal.

5.2.1.2 Properties of Norms

Proposition 5.2 *The 2-norm is bounded by the product of 1 and ∞-norms according to*

$$\|y\|_2^2 \le \|y\|_\infty \cdot \|y\|_1 \tag{5.55}$$

Proof:

$$\|y\|_2^2 = \int_{-\infty}^{\infty} y^2 dt$$
$$= \int_{-\infty}^{\infty} |y| \cdot |y| dt$$
$$\int_{-\infty}^{\infty} |y| \cdot |y| dt \le \|y\|_\infty \int_{-\infty}^{\infty} |y| dt$$
$$\|y\|_\infty \int_{-\infty}^{\infty} |y| dt = \|y\|_\infty \|y\|_1$$
$$\|y\|_2^2 \le \|y\|_\infty \|y\|_1$$

■

Proposition 5.3 *If a signal has a finite 2-norm, then the RMS-norm is equal to zero.*

Proof:

$$\|y\|_{RMS}^2 = \lim_{\tau\to\infty} \frac{1}{\tau} \int_0^{\tau} y^2(t) dt$$
$$= \left(\lim_{\tau\to\infty} \frac{1}{\tau}\right) \cdot \left(\lim_{\tau\to\infty} \int_0^{\tau} y^2(t) dt\right)$$
$$= \|y\|_2^2 \cdot \lim_{\tau\to\infty} \frac{1}{\tau}$$

■

Proposition 5.4 *The power is bounded by* $||y||_\infty$.

$$||y||^2_{RMS} \leq ||y||^2_\infty \tag{5.56}$$

Proof:

$$\lim_{t\to\infty}\frac{1}{\tau}\int_0^\tau y^2(t)\mathrm{d}t \leq ||y||^2_\infty \lim_{t\to\infty}\frac{1}{\tau}\int_0^\tau \mathrm{d}t$$
$$||y||^2_{RMS} \leq ||y||^2_\infty$$

■

5.2.2 Norms of Dynamic Systems

We will now consider the norms of a dynamic system, $G(s)$. Specifically, we are concerned with the 2-norm (H_2) and the ∞-norm (H_∞), as these are the norms that will be used when optimization is discussed in Chapter 6. Note that "H_p" denotes Hardy spaces – the set of all well behaved functions (analytic functions).[2] The H_2 and H_∞ norms are defined as follows:

$$H_2 : \ ||G||_2 = \sqrt{\frac{1}{2\pi}\int_{-\infty}^{\infty} |G(j\omega)|^2 \mathrm{d}\omega} \tag{5.57}$$

$$H_\infty : \ ||G||_\infty = \sup_\omega |G(j\omega)| \tag{5.58}$$

The 2-norm of a dynamic system is calculated as follows:

$$||G||^2_2 = \frac{1}{2\pi}\int_{-\infty}^{\infty} |G(j\omega)|^2 \mathrm{d}\omega \tag{5.59}$$

Recalling that $G(-j\omega) = \overline{G(j\omega)}$, and $|G(j\omega)|^2 = G(j\omega)\overline{G(j\omega)}$, then

$$||G||^2_2 = \frac{1}{2\pi j}\int_{-\infty}^{\infty} G(j\omega)G(-j\omega)\mathrm{d}(j\omega) \tag{5.60}$$

Then by replacing the limits of integration with the Nyquist D-contour (this can be done as $G(j\omega)$ is strictly proper, as per Section 5.1), we can write

$$||G||^2_2 = \frac{1}{2\pi j}\int_D G(s)G(-s)\mathrm{d}s \tag{5.61}$$

By the residue theorem,

$$||G||^2_2 = 2\pi j\left(\frac{1}{2\pi j}\right)\sum_k R_k \tag{5.62}$$

where

$$\int_D G(s)G(-s)\mathrm{d}s = 2\pi j\sum_k R_k \tag{5.63}$$

at its poles in the LHP. This leads to the following theorem:

Theorem 5.3 *The 2-norm of a system is*

$$||G||^2_2 = \sum_k R_k \tag{5.64}$$

where R_k are the residues of $G(s)G(-s)$ in the LHP. ■

Example 5.5 Consider the dynamic system $G(s)$.

$$G(s) = \frac{K}{\tau s + 1} \tag{5.65}$$

2 We have used L_p and H_p for signals and systems; however, one should note that these norm spaces are equivalent for well-behaved functions.

We can find the 2-norm as follows:

$$G(s)G(-s) = \frac{K^2}{(\tau s+1)(-\tau s+1)} \tag{5.66}$$

$$G(s)G(-s) = \frac{K^2}{\tau^2} \cdot \frac{1}{\left(s+\frac{1}{\tau}\right)\left(-s+\frac{1}{\tau}\right)} \tag{5.67}$$

$$\|G\|_2^2 = R_1 = \left(s+\frac{1}{\tau}\right) G(s)G(-s)|_{s=-\frac{1}{\tau}} \tag{5.68}$$

$$R_1 = \frac{K^2}{2\tau} \tag{5.69}$$

$$\|G\|_2 = \frac{K}{\sqrt{2\tau}} \tag{5.70}$$

5.2.3 Input–Output Norms

5.2.3.1 Transient Inputs (Energy Bounded)

What system property ("gain") relates an output to an input norm?

Suppose that some dynamic system is given an input $u(t) = \delta(t)$ so that $y(t) = g(t)$, where $g(t)$ is the system impulse response, as shown in Figure 5.5.

Then,

$$\int_0^\infty y^2(t)\mathrm{d}t = \int_0^\infty g^2(t)\mathrm{d}t \tag{5.71}$$

So,

$$\|y\|_2^2 = \|g\|_2^2 \tag{5.72}$$

Per Parseval's theorem,

$$\|y\|_2^2 = \|G\|_2^2 \tag{5.73}$$

In the case of in impulse function input $u(t) = \delta(t)$, the 2-norm of the output norm is equal to the 2-norm of the dynamic system.

Now let us consider the more general case, where $u(t)$ is any arbitrary input. Then,

$$\|y\|_2^2 = \frac{1}{2\pi}\int_{-\infty}^{\infty} |G(j\omega)|^2 |u(j\omega)|^2 \mathrm{d}\omega \tag{5.74}$$

$$\leq \frac{\|G\|_\infty^2}{2\pi}\int_{-\infty}^{\infty} |u(j\omega)|^2 \mathrm{d}\omega \tag{5.75}$$

$$\|y\|_2^2 \leq \|G\|_\infty^2 \, \|u\|_2^2 \tag{5.76}$$

So, for any input $u(t)$, the "gain" $\|y\|_2/\|u\|_2$ is bounded by the infinity norm of the dynamic system $G(s)$.

$$\frac{\|y\|_2}{\|u\|_2} \leq \|G\|_\infty \tag{5.77}$$

Then there exists a some input, u, such that we can define the norm of the dynamic system in terms of the input and output norms, per Eq. (5.77),

$$\|G\|_\infty = \sup_{\|u\|_2 \neq 0} \frac{\|y\|_2}{\|u\|_2} \tag{5.78}$$

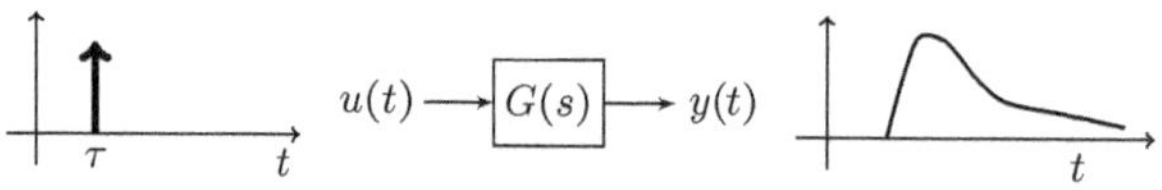

Figure 5.5 Impulse response for an arbitrary system.

We can place a bound on the ∞-norm of $y(t)$ in a similar manner. We begin with

$$y(t) = \int_0^\infty g(t-\tau)u(\tau)\mathrm{d}\tau \tag{5.79}$$

using convolution to express the behavior of the dynamic system. Then,

$$|y(t)| = \left|\int_0^\infty g(t-\tau)u(\tau)\mathrm{d}\tau\right| \tag{5.80}$$

We can then use the Cauchy–Schwarz [3] inequality to obtain

$$|y(t)| \le \sqrt{\int_0^\infty g^2(t)\mathrm{d}\tau} \cdot \sqrt{\int_0^\infty u^2(t)\mathrm{d}\tau} \tag{5.81}$$

$$\le \|g\|_2 \|u\|_2 \tag{5.82}$$

$$\le \|G\|_2 \|u\|_2 \tag{5.83}$$

Therefore, the maximum value of $y(t)$ is bounded the product of the input $u(t)$ 2-norm and the dynamic system $G(s)$ 2-norm.

$$\frac{\|y\|_\infty}{\|u\|_2} \le \|G\|_2 \tag{5.84}$$

5.2.3.2 Persistent Inputs (Energy Unbounded)

Let $S_u(\omega)$ and $S_y(\omega)$ indicate the PSD of the input and output signals of a dynamic system with a transfer function $G(j\omega)$, respectively. Then,

$$S_y(\omega) = |G(j\omega)|^2 S_u(\omega) \tag{5.85}$$

Let us assume white noise input with $S_u(\omega) = 1$. Then,

$$S_y(\omega) = |G(j\omega)|^2 \Leftrightarrow \frac{1}{2\pi}\int_{-\infty}^{\infty} S_y(\omega)\mathrm{d}\omega = \frac{1}{2\pi}\int_{-\infty}^{\infty} |G(j\omega)|^2\mathrm{d}\omega \tag{5.86}$$

Then, the power of the output, $y(t)$ is given by

$$P_y = \|G\|_2^2 \tag{5.87}$$

Therefore, the H_2 norm of a transfer function measures the RMS response of its output when it is driven by a white noise excitation input.

For these persistent signals, we can also write,

$$\|y\|_{RMS}^2 = \frac{1}{2\pi}\int_{-\infty}^{\infty} S_y(\omega)\mathrm{d}\omega \tag{5.88}$$

$$= \frac{1}{2\pi}\int_{-\infty}^{\infty} |G(j\omega)|^2 S_u(\omega)\mathrm{d}\omega \tag{5.89}$$

$$\le \frac{\|G\|_\infty^2}{2\pi}\int_{-\infty}^{\infty} S_u(\omega)\mathrm{d}\omega \tag{5.90}$$

$$\|y\|_{RMS}^2 \le \|G\|_\infty^2 \, \|u\|_{RMS}^2 \tag{5.91}$$

Therefore,

$$\frac{\|y\|_{RMS}}{\|u\|_{RMS}} \le \|G\|_\infty \tag{5.92}$$

By concentrating the PSD of the input u near frequencies at which $|G(j\omega)| \approx \|G_\infty\|$, then the following is obtained,

$$\|G\|_\infty = \sup_{\|u\|_{RMS}\neq 0} \frac{\|y\|_{RMS}}{\|u\|_{RMS}} \tag{5.93}$$

The norms described in this section will be utilized in Chapter 6 for the controller design using optimization techniques.

5.3 Quantifying Uncertainty

Consider some arbitrary complex system model that could be nonlinear or infinite-dimensional or both, and which is some representation of a real world system. Let w be an input to the system, and let z be an output of the system, as shown in Figure 5.6.

Next, consider G to be a simplified model of this complex system, so that we have the dynamics shown in Figure 5.7. By "simplified model" we mean that the original complex model is reduced so that it could be represented by a transfer function for the purpose of control design using the techniques that are laid out in this book.

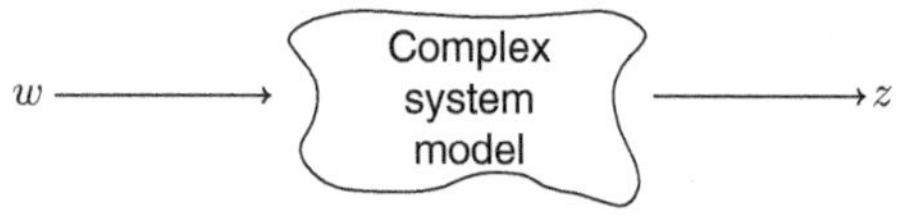

Figure 5.6 An arbitrary complex dynamic system.

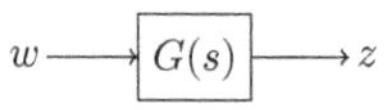

Figure 5.7 A simplified dynamic system.

There are two fundamental questions that we should ask before attempting to design a control system:

1. Is the complex model an exact picture of reality?
2. Is the simplified model a good representation of this complex model?

The short answer to both of these questions is **no**. The following quote by George Box[3] is particularly fitting:

> All models are wrong, but some are useful.

In other words, it is impossible to develop an exact model of reality, no matter how sophisticated and complex our models are. Furthermore, we have to proceed with care when to simplify a model so that we do not discard the very thing that we are after or trying to control in this case. In any act of simplification (or developing a simple model), there is a trade-off between complexity and uncertainty.[4] As we simplify a model of a system, the complexity is reduced but inherently, the uncertainty is increased. However, this is not necessarily a bad thing. If we could quantify and model the uncertainty, we can design or analyze a control system for its performance and robustness to these uncertainties. For instance, we could use the system shown in Figure 5.8, which uses a simplified model G and some model of uncertainty Δ, where v and η are internal signals of G. By utilizing these signals, the Δ model acts as a perturbation to the nominal model G.

How do these uncertainties relate to model simplification and how do we characterize and quantify these uncertainties? These are the questions we try to address in this section.

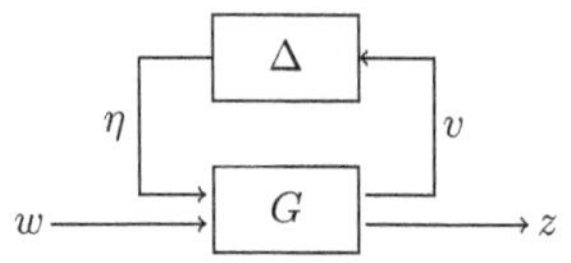

Figure 5.8 A dynamic system with modeled uncertainty Δ.

5.3.1 The Characterization of Uncertainty in Models

In Chapter 4 and the Section 5.2, we discussed M_2-margin and the H_∞-norm. We stated that "the H_∞-norm of the sensitivity, $S(s)$, is the M_2 stability margin and the frequency of this maximum determines a closed-loop resonant frequency." In addition, we pointed out that "we always seek to reduce the H_∞-norms of $T(s)$ and $S(s)$ to obtain sufficient stability margins and to reduce closed-loop resonances."

It has been pointed out that the specification of the M_2 margin is conservative because it is independent of frequency. That is, it may be that higher values of max $|S(j\omega)|$ can be tolerated in some frequency ranges. In this section, we begin a development that will end with a more satisfactory treatment of these matters.

3 18 October 1919–28 March 2013, was a British statistician, who worked in the areas of quality control, time-series analysis, design of experiments, and Bayesian inference.

4 Here, what we mean by uncertainty is the level of confidence in our model.

First of all, we attempt to depict our uncertainty in a more precise way. We do that by finding some statement that characterizes all possible values of $G(s)$. That is to say, we cannot hope to design without laying down something (say, some bounds) on what might be the range of possibilities for the input–output relation. We are "uncertain" about all aspects of the model of a dynamic system, and we choose to characterize that as an uncertainty about the frequency response, the magnitude, and phase of $G(j\omega)$.

In characterizing uncertainty about the frequency response, the uncertainties are categorized as follows:

- **Unstructured Uncertainty**: This means a disc is specified in the G-plane (Nyquist plot of G) for each frequency ω. The disc, which is specified in terms of the location of its center and its radius, is such that all variations of $G(j\omega)$ are expected to lie within the disc.
- **Structure Uncertainty**: This means that we specify the specific set of points that represent all possible values of $G(j\omega)$.

Think of it this way: the unstructured uncertainty disc is a conservative covering of the structured uncertainty. There are many ways to obtain an uncertainty statement. A pragmatic way of defining uncertainty is to roughly divide the frequency response region into:

- **Low-frequency Uncertainty**: An example of this is parameter uncertainty, which is a statement of the bounds on the parameters of the state variables or transfer-function model.

  "Parameter uncertainty is an example of a structured uncertainty statement," because one obtains the unique values of the magnitude and phase give the model. In fact, this is the most highly structured form of an uncertainty statement and is considered in the low-frequency region.
- **High-frequency Uncertainty**: An example of this is unmodeled dynamics, unmodeled either because high frequency dynamics would have no significant consequences or because the dynamics were to difficult to identify and/or model.

  "The greatest kind of uncertainty occurs at high frequency and that is considered an unstructured uncertainty." We are confident about a structured form of uncertainty only at lower frequencies. Implicit in a statement of parameter uncertainty is the assumption that the model is quite good; so good, in fact, that it is only the precise values of parameters are in the doubt. The unstructured uncertainty at high frequencies, by its very nature, must be based upon the experience of successful designers and may have very little purely scientific foundations.

Here is one point-of-view, the attitude of which adopted for this text:

- Any design method that does not account for high frequency uncertainty (a form of unstructured uncertainty) is not viable.
- Any design method that is based on unstructured uncertainty at all frequencies is overly conservative.
- A design method based on low frequency uncertainty (a form of structured uncertainty) is viable only if there is confidence in the model of the controlled object at high frequencies.

Anyone that accepts this view will recognize that the classical method is viable because the use of stability margins is, in fact, accounting for the high frequency uncertainty. One can argue that the use of margins may not be an adequate accounting, but there is no denial of the intent of the use of these numbers in design.

5.3.2 Weighting Factors and Stability Robustness

The classical stability margins only bound the model uncertainty at two distinct frequencies. Thus, it is possible to imagine situations wherein the margins are accurate measures of the unstructured uncertainty and the margin bounds are met by the designer, but the variations at other frequencies are sufficient to cause instability.

One of the features of the neoclassical design is that a more satisfactory representation is made of the "high frequency/unstructured" uncertainty. We shall explore the case where the statement of uncertainty is unstructured at all frequencies in spite of our observation that this is a very conservative approach. The results are too interesting to ignore.

There are number of ways to represent unstructured uncertainty and it is rewarding to examine all of them. The first formulation begins as follows: Denote the nominal value of the transfer function by $G_p(s)$ and any other possible input–output relation by $\tilde{G}_p(s)$, then

$$\frac{\tilde{G}_p}{G_p} - 1$$

is a measure of the variation of the model. Imagine factoring this rational function as

$$\frac{\tilde{G}_p}{G_p} - 1 = \Delta(s)W_s(s) \tag{5.94}$$

where the factor $\Delta(s)$ is our model of unstructured uncertainty at all frequencies and the factor $W_s(s)$ is a weighting factor that emphasizes the importance of this uncertainty in different frequency ranges. The factors are selected such that the poles of Δ lie strictly in the left half-plane (strictly stable unstructured uncertainty)[5] and

$$0 \le |\Delta(j\omega)| \le 1 \quad \text{or} \quad ||\Delta||_\infty \le 1 \tag{5.95}$$

Note that he subscript "s" is used to denote the fact that all this is done to obtain information about the stability robustness of feedback control. Rewrite the defining equation for Δ and W_s:

$$\tilde{G}_p = (1 + \Delta W_s)G_p \tag{5.96}$$

It is in this way that this formulation earns its name, *multiplicative* variation. The impact of the variation on feedback control system is investigated in Section 5.3.3.

5.3.3 Robust Stability (Complementary Sensitivity) and Uncertainty

We considered some unstructured uncertainty Δ, constrained to $||\Delta||_\infty \le 1$, paired with frequency weighting W_s. If we have a plant model G_p and the uncertainty as shown in Figure 5.9, then we can say that the "true" plant is

$$\tilde{G}_p = (1 + W_s\Delta)G_p \tag{5.97}$$

derived as *multiplicative* uncertainty in Section 5.3.2. Similarly, if the modeled return ratio is

$$L_0 = G_cG_p \tag{5.98}$$

then the "true" return ratio is

$$L = G_c\tilde{G}_p = L_0(1 + W_s\Delta) \tag{5.99}$$

Now, let v be the input to the uncertain dynamics $W_s\Delta$ and let i be the output. We then consider the closed-loop transfer function from i to v.

$$v = (-G_cG_p)(i + v) \tag{5.100}$$

So,

$$\frac{v}{i} = -\frac{G_cG_p}{1 + G_cG_p} = -T_0 \tag{5.101}$$

Here, T_0 indicates the closed-loop transfer function when G_p is not subject to any uncertainties. We can then condense the system to the block diagram, as shown in Figure 5.10.

So, for some bounded perturbation to the closed-loop system, we will capture its magnitude behavior through W_s.

This brings us to the concept of robust feedback: A compensator is robust with respect to a feedback property (e.g. stability) if that property obtains for all G_p in some set of plants called P – that is, all plants within a defined range of uncertainty.

The maximum loop gain in Figure 5.10 equals $\|-\Delta W_sT_0\|_\infty$ which is < 1 for all allowable Δs (e.g. $||\Delta||_\infty \le 1$) if and only if the small-gain condition $\|W_sT_0\|_\infty < 1$ holds. The foregoing discussion is related to the *small-gain theorem* such as $\|W_sT_0\|_\infty < 1$ assures robust stability. A more rigorous proof for robust stability is provided below.

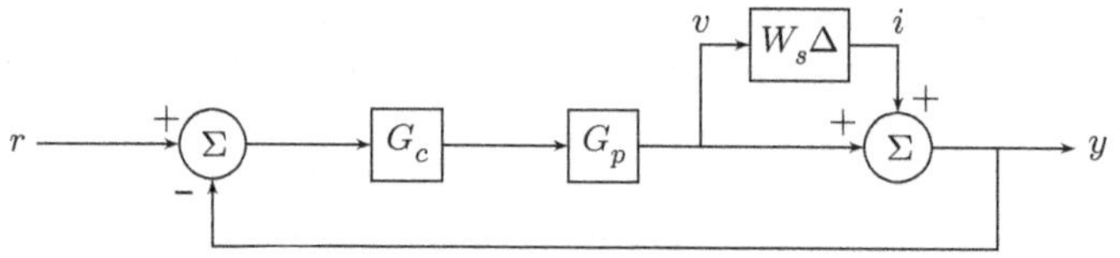

Figure 5.9 True feedback.

5 Constraining the Δ to be stable simplifies the mathematics, however, equivalently, one can relax this constraint.

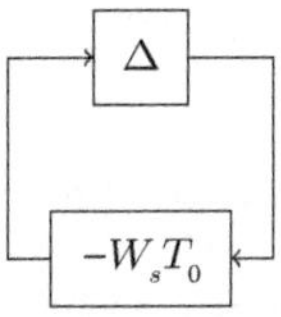

Figure 5.10 Impact of uncertainty on closed-loop.

Consider for some unknown perturbation Δ, where

$$\|\Delta\|_\infty \leq 1 \tag{5.102}$$

From Eq. (5.97),

$$\frac{\tilde{G}_p(j\omega)}{G_p(j\omega)} - 1 = W_s(j\omega)\Delta \tag{5.103}$$

Accounting for Eq. (5.102), we can then say

$$\left|\frac{\tilde{G}_p(j\omega)}{G_p(j\omega)} - 1\right| \leq |W_s(j\omega)| \tag{5.104}$$

Using the property that norms are homogeneous and multiplying the ratio $\frac{\tilde{G}_p}{G_p}$ by $\frac{G_c}{G_c}$, we can then write

$$\frac{|G_c\tilde{G}_p - G_cG_p|}{|G_cG_p|} = \frac{|L - L_0|}{|L_0|} \leq |W_s(j\omega)| \tag{5.105}$$

Theorem 5.4 Multiplicative Uncertainty *The controller G_c provides robust stability for stable $W_s\Delta$ if and only if*

$$\|W_s\Delta T_0\|_\infty < 1$$

and where $\|\Delta\|_\infty \leq 1$, then

$$\|W_sT_0\|_\infty < 1$$

Proof: Let L be the true return ratio and let L_0 be the nominal return ratio. Then,

$$|L(j\omega) - L_0(j\omega)| < |1 + L_0(j\omega)| \text{for all } \omega \in \mathbb{R}$$

Since L and L_0 have the same right half-plane poles and zeros. This is equivalent to

$$\frac{|L(j\omega) - L_0(j\omega)|}{|L_0(j\omega)|} \cdot \frac{|L_0(j\omega)|}{|L_0(j\omega) + 1|} < 1$$

As we defined earlier,

$$\frac{|L - L_0|}{|L_0|} \leq |W_s|$$

Additionally,

$$|T_0| = \frac{|L_0|}{|L_0 + 1|}$$

So,

$$|W_s\Delta| \cdot |T_0| < 1.$$

Assuming (without proof) that the Cauchy–Schwarz inequality applies, then $\|W_sT_0\|_\infty < 1$ and knowing $\|\Delta\|_\infty \leq 1$, the closed-loop system is robustly stable. To prove the converse, assume

$$\|W_sT_0\|_\infty \geq 1$$

and we prove the result if we can find an allowable "Δ" such that the perturbed loop is unstable. Since "T_0" is strictly proper, there must exist some frequency, ω_1, such that

$$\|W_s(j\omega_1)T_0(j\omega_1)\|_\infty = 1$$

suppose $\omega_1 = 0$ (the more general case is left for the exercises). In that case, $W_s(0)T_0(0)$ is a real number, either +1 or −1. If we take $\Delta = -W_s(0)T_0(0)$, then Δ is allowable and

$$1 + W_s(0)T_0(0) = 0$$

the Nyquist locus passes through the critical point so the perturbed loop is not stable; hence, the closed loop is not internally stable. ■

The above characterizes robust stability for a multiplicative uncertainty, that is

$$\tilde{G}_p = (1 + W_s\Delta)G_p \tag{5.106}$$

What about other types of uncertainty? Consider the different ways uncertainty can act on the plot, as shown in Figure 5.11.

The conditions for robust stability for each of the cases in Figure 5.11 can be derived in a similar manner to the condition for multiplicative uncertainty. These conditions are summarized in Table 5.1.

To briefly recap, we modeled the unknown high frequency or unstructured uncertainty as Δ. The only assumption that we made regarding this unknown that it is upper bounded (stable) and normalized such that $\|\Delta\|_\infty \le 1$. This unknown acts as a perturbation to a nominal plant, G_p. The perturbation can generate a family of plants, $\tilde{G}_p$ but because the perturbation is stable, the right half plane poles and zeros of this family of plant are exactly the same as our nominal plant, G_p. For the multiplicative uncertainty model as given in Eq. (5.104) and using the *small gain theorem*, we derived a condition for robust stability of the closed loop system, $\|W_sT_0\|_\infty < 1$, where T_0 is the closed loop transfer function and W_s is our filter or weighting function. This weighting function is selected based on our unstructured uncertainty and gives an upper bound to $\frac{\tilde{G}_p(j\omega)}{G_p(j\omega)} - 1$. More on the design of this filter is given in the example later and in the following chapters.

We found if the designed feedback controller G_c satisfies the condition above, then not only the feedback loop for the nominal plant would be stable but the stability of the feedback loop for the family of plants due to the perturbation by the unknown uncertainty would be guaranteed. This is a remarkable outcome. Of course if the norm of the unknown is not normalized such that $\|\Delta\|_\infty \le \beta$, then the robust stability condition will be equivalently scaled, $\|W_sT_0\|_\infty < \frac{1}{\beta}$.

Assume that we design a controller that mathematically meets the robust stability condition. A good question to ask at this stage is, what physical changes occur in the feedback loop so that this controller guarantees robust stability? We will investigate this in Chapter 8 of this text, when we discuss the application-oriented problems and solutions.

We will now consider the following example for designing the robust stability filter, W_s.

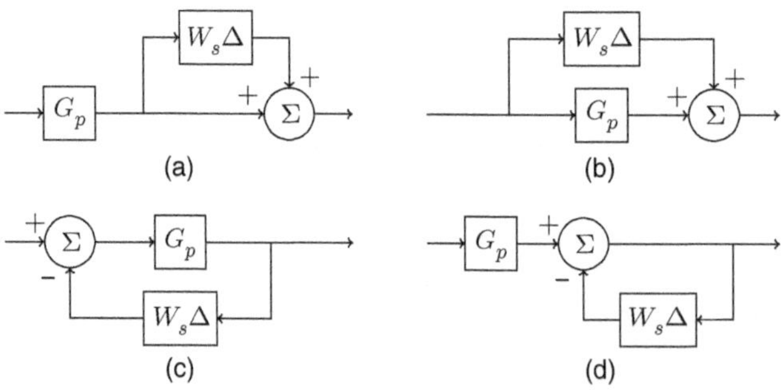

Figure 5.11 Types of uncertainty.

Table 5.1 Robust stability conditions.

	Characterization	Graphical interpretation	Robust stability condition
(a)	$\tilde{G}_p = (1 + W_s\Delta)G_p$	Figure 5.11a	$\|W_sT\|_\infty < 1$
(b)	$\tilde{G}_p = G_p + W_s\Delta$	Figure 5.11b	$\|W_sY\|_\infty < 1$
(c)	$\tilde{G}_p = \frac{G_p}{1+W_s\Delta G_p}$	Figure 5.11c	$\|W_sSG_p\|_\infty < 1$
(d)	$\tilde{G}_p = \frac{G_p}{1+W_s\Delta}$	Figure 5.11d	$\|W_sS\|_\infty < 1$

Example 5.6 Consider a plant subject to uncertain delay.

$$G_p = \frac{1}{s^2} \tag{5.107}$$

$$\tilde{G}_p = e^{-\tau s} G_p, \quad 0 \le \tau \le 0.1 \tag{5.108}$$

With $\|\Delta\|_\infty \le 1$, we want to select W_s so that

$$\left| \frac{\tilde{G}_p}{G_p} - 1 \right| = |e^{-\tau s} - 1| \le |W_s| \quad \forall\, \omega, \tau \tag{5.109}$$

Although the work is not shown here, it can be found that

$$|e^{-\tau s} - 1| = \sqrt{2(1 - \cos(\tau\omega))} \tag{5.110}$$

The corresponding magnitude plot of $e^{-\tau s} - 1$ shown in Figure 5.12. Clearly, a high pass filter should be chosen for W_s, such as

$$W_s = \frac{\alpha s}{\tau_f s + 1} \tag{5.111}$$

where α and τ_f are tuning parameters. The magnitude peaks at a value of 2 (or about 6 dB), so we need $\alpha > 2\tau_f$ to ensure $|W_s|$ goes to 2 as ω approaches infinity. Additionally, $|W_s|$ must bound $|e^{-\tau s} - 1|$ for the range of possible τ. We can see from Eq. (5.110) that as τ increases, the peak magnitude of $|e^{-\tau s} - 1|$ occurs at lower frequencies. Therefore, we shape $|W_s|$ for the worst case delay, where $\tau = 0.1$, the upper limit of τ.

As an example, Figure 5.12 shows $|W_s|$ for $\alpha = 0.25$ and $\tau_f = \tau = 0.1$.

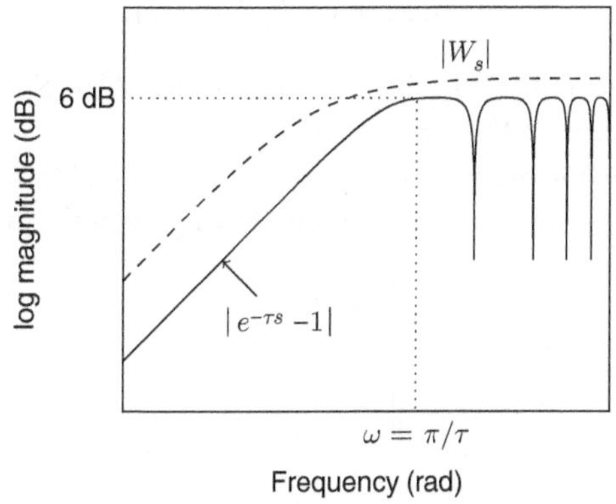

Figure 5.12 Bode plot of $e^{-\tau s} - 1$ and appropriate filter.

5.3.4 Sensitivity and Performance

Consider an error signal

$$e = w - Pu \tag{5.112}$$

Recall that P is a set of plants within a defined range of uncertainty of the nominal plant G_p. As this error signal is a controlled signal that we want to minimize, we will define an output z to be

$$z = W_p e = W_p w - W_p P u \tag{5.113}$$

where W_p is some weighting filter applied to the error signal. We use the subscript p to indicate that this is a *performance* weighting factor. We can write these equations in matrix form.

$$\begin{bmatrix} z \\ e \end{bmatrix} = \begin{bmatrix} W_p & -W_p P \\ 1 & -P \end{bmatrix} \begin{bmatrix} w \\ u \end{bmatrix} \tag{5.114}$$

In this case, the transfer function from w to z is

$$T_{zw} = W_p - W_p PK(1 + PK)^{-1} = W_p \left(1 - PK(1 + PK)^{-1}\right) \tag{5.115}$$

letting $u = Ke$ define the controller dynamics. That means we have

$$T_{zw} = W_p S \tag{5.116}$$

To minimize the filtered error z, we must minimize $T_{zw} = W_p S$. In particular, we will choose to minimize the infinity norm.

$$\min_K \|T_{zw}\|_\infty = \min_K \left\|W_p S\right\|_\infty \tag{5.117}$$

In other words, we want to find the controller K that minimizes the peak magnitude of $W_p S$. The weighting filter W_p can be shaped so that the resulting controller K and sensitivity function S meet the competing design objectives outlined in Section 4.4.3. Conventionally,

$$|W_p S| < 1 \tag{5.118}$$

Therefore,

$$|S| < |W_p^{-1}| \tag{5.119}$$

In other words, the weighting filter W_p should be chosen so that its inverse bounds the desired shape of S.

5.3.5 Performance and Stability

Generally, robust feedback refers to robust stability and nominal performance. In this subsection, we develop a conservative criteria that, if met, will ensure robust performance. The H_∞ control problem seeks to find a controller that achieves robust stability while guaranteeing some level of performance. The simplest H_∞ control problem was posed in Eq. (5.117) and is restated below.

$$\underset{K}{\text{minimize}} \max_\omega |W_p(j\omega)S(j\omega)| \tag{5.120}$$

In other words, find the controller K that minimizes the maximum value of $|W_p S|$ across all frequencies. S was chosen as the transfer function to minimize to minimize the error signal.

However, designing around S itself is not enough for good design, and in addition, the problem posed in Eq. (5.120) is not easily solved even for linear systems. Therefore, conventionally, as previously shown, that for robust stability, we want

$$\|W_s T\|_\infty < 1 \tag{5.121}$$

Equivalently for S, we have

$$\left\|W_p S\right\|_\infty < 1 \tag{5.122}$$

for nominal performance.

Before continuing on, note that we have weighting filter W_p paired with S to affect performance and weighting filter W_s paired with T to affect stability. From this point onward, we will define W_s as W_Δ to avoid confusion.

To address robust performance, we must simultaneously shape the weighting filters W_p and W_Δ (sometimes called W_1 and W_3 in other works, respectively). We start the perturbed sensitivity function

$$\tilde{S} = \frac{1}{1+\tilde{L}} = \frac{1}{1 + L_0(1 + W_\Delta \Delta)} \tag{5.123}$$

so then,

$$\tilde{S} = \frac{\frac{1}{1+L_0}}{\frac{1}{1+L_0} + (1 + W_\Delta \Delta)\frac{L_0}{1+L_0}} = \frac{S_0}{1 + W_\Delta \Delta T_0} \tag{5.124}$$

where T_0 was defined in Eq. (5.101) and S_0 is similarly defined as

$$S_0 = \frac{1}{1 + L_0} \tag{5.125}$$

Given that $\|\Delta\|_\infty \le 1$, for robust performance we must therefore have

$$\left\|W_p \tilde{S}\right\|_\infty = \left\|\frac{W_p S_0}{1 + W_\Delta T_0}\right\|_\infty \tag{5.126}$$

Theorem 5.5 *The controller G_c provides robust performance if and only if*

$$\left\| |W_p S_0| + |W_\Delta T_0| \right\|_\infty < 1 \tag{5.127}$$

■

Theorem 5.5 is often simplified to

$$\left\| W_p S_0 \right\|_\infty < \frac{1}{2} \tag{5.128}$$

and

$$\| W_\Delta T_0 \|_\infty < \frac{1}{2} \tag{5.129}$$

to enforce robust performance [13].

In addition, we must penalize the control input u through a third weighting filter W_u or W_2.

$$\| W_u G_c S_0 \|_\infty = \| W_u Y_0 \|_\infty < 1 \tag{5.130}$$

where S_0 and Y_0 denote the nominal sensitivity and Youla transfer functions, respectively. These requirements may be combined into the stacked H_∞ problem:

$$T_{zw} = \begin{bmatrix} W_p S \\ W_u Y \\ W_\Delta T \end{bmatrix} \tag{5.131}$$

5.4 Problems

5.1 On one hand, we know that signal power can be written as the square of the RMS of a signal such that

$$||y||_{RMS} = \sqrt{\lim_{\tau \to \infty} \frac{1}{\tau} \int_{\frac{-\tau}{2}}^{\frac{\tau}{2}} y^2 dt}$$

so

$$P_y = ||y||_{RMS}^2$$

On the other hand, for a random signal (where signal energy is not finite), signal power can be written as

$$P_y = 2 \int_0^\infty S_y(2\pi f) df$$

This equation, in discrete case, can be written as

$$P_{y_i} = (2\Delta f) S_y(2\pi f_i)$$

Given the information above and knowing that a random profile of a road can be approximated by a superposition of $(N \to \infty)$ sine waves

$$z_R(t) = \sum_{i=1}^{N} A_i \sin(\omega_i t - \theta_i)$$

you are asked to find a relation between the discrete road amplitude (A_i) and the road PSD (Power Spectral Density), $S_{z_R}(2\pi f_i)$ over the narrowband, Δf.

5.2 A system with the transfer function $\frac{K}{\tau s+1}$ has a disturbance input, $d(t)$, known to have the energy bound $||d||_2 \le \alpha$. Suppose that we want to find the best estimate of the ∞-norm of the output $y(t)$. Knowing that $\frac{\infty\text{-norm}}{2\text{-norm}}$ = 2-norm of the transfer function. Find the bound for $||y||_\infty$.

5.3 Apply the method of Youla parameterization to the object characterized by

$$G_p = \frac{4(0.4 - s)}{(s + 0.1)(s + 0.2)(s + 0.8)}$$

Your closed-loop system should behave like a Butterworth filter. Assume that you must keep $|u(t)| \leq 0.10$ while the sensor noise peaks at $|n(t)| = 0.01$ above 0.25 rad/s. Evaluate the effectiveness of your design by examining the Bode magnitude of $Y(j\omega)$, $S(j\omega)$, $T(j\omega)$. What are the stability margins (phase, gain, and M_2 margins)? Approximate the closed-loop rise time T_R using the equation

$$\omega_c \times T_R > \frac{1}{4} \tag{5.132}$$

where and ω_c is the cross-over frequency.

6

Feedback Design By the Optimization of Closed-Loop Norms

In this chapter, we discuss the H_∞ method for SISO robust control design. We provide insights to the frequency domain solution of H_∞ control and selection of the weighting filters in shaping complementary sensitivity, sensitivity, and Youla transfer functions.

6.1 Introduction

The foundations of neoclassical feedback design are developed in the previous chapters. These basic notions are referred to throughout the remainder of this text.

Feedback design objectives can be stated in terms of three primary rational functions: the sensitivity function, $S = 1/(1 + L)$, the complementary sensitivity, $T = L/(1 + L)$, and the Youla parameter, $Y = G_c S = T/G_p$. These rational functions are interpreted as closed-loop transfer functions. In particular, disturbance attenuation can be characterized by the sensitivity and the complementary sensitivity.

To achieve the objective of good feedback design (robustness and attenuation of the effects of unwanted inputs), we arrange things so that the Bode magnitudes of S and Y are as small as possible at all frequencies; however, there are basic trade-offs. We also attempt to arrange things so that the Bode magnitude Lm T is small ($\ll 0$ dB) for some purposes and near 0 dB for others. Complicating all this is the fact that Bode magnitudes of these functions cannot be selected independently since we have a series of relationships such as

$$T(s) + S(s) = 1 \tag{6.1}$$

$$T(s) = Y(s)G_p(s) \tag{6.2}$$

$$S(s) = Y(s)/G_c(s) \tag{6.3}$$

The point is that $S(j\omega)$ and $T(j\omega)$, for example, cannot be varied completely independently so that many of the trade-offs of feedback design are clearly delineated by the Bode diagrams of these rational functions.

It is important to observe that the frequency content is different for the different disturbance and noise signals. This observation indicates an important opportunity, the opportunity to attenuate the effects of some input signals in one frequency range and the effects of other signals in other frequency ranges: A compromise is not required at all frequencies. In the usual case, we expect reference and disturbance signals to be dominated by low frequencies and the sensor noise by relatively high frequency.

There are other aspects of $S(j\omega)$ and $T(j\omega)$ that must be considered in design. First, there is the *cross-over frequency*, the frequency ω_c where $|S(j\omega_c)| = |T(j\omega_c)|$. To any extent that we are capable of selecting this frequency, we wish it to separate the sensor-noise frequencies from those frequencies typical of command signals. Also, the reciprocal of the cross-over frequency, $1/\omega_c$, is a measure of the speed of response of the closed loop and that makes things more interesting. Incidentally, there are two other ways to think of the cross-over frequencies:

- ω_c is that frequency where the return ratio $|L| = 1$ (0 dB).
- An approximation of ω_c is that frequency where there is a separation of $|Y(j\omega)|$ from $|1/G_p(j\omega)|$.

Robust Control: Youla Parameterization Approach, First Edition. Farhad Assadian and Kevin R. Mallon.

Companion Website: www.wiley.com/go/assadian/robust

The common value of sensitivity and complementary sensitivity $|S(j\omega_c)| = |T(j\omega_c)|$ (where these two curves intersect) is directly related to the phase margin (see Figure 4.5). Another stability margin is the reciprocal of the peak magnitude (or "H_∞ norm") of $S(s)$.

The H_∞ norm of $T(s)$ is a measure of the approximate damping ratio of closed-loop response and for that reason we wish to reduce this number where possible as well.

6.1.1 Frequency Domain Control Design Approaches

The two basic design methods are characterized by a fundamental equation: First, there is the scheme characterized by the equation for the return ratio,

$$L(s) = G_c(s)G_p(s) \tag{6.4}$$

The compensator $G_c(s)$ is selected to "shape" the open-loop ("return ratio") transfer function. This is an indirect method because one is continually drawing inference about T, S, and Y for each modification made to L with a change in G_c. The classical procedure is described in many control text books and is based upon fixing or shaping the return ratio, $L(j\omega)$. This is the only alternative where the specific form of the compensator is fixed, say fixed as a PID controller or as a lead-lag compensator.

The designer must also provide for the stability of the feedback loop. If the return ratio is minimum phase, that means the slope of the gain curve, $|L(j\omega)|$, should be −20 dB/decade at the cross-over frequency and −40 dB/decade on either side of the cross-over frequency. The phase margin will likely be between 45° and 90° in this case. The larger value of the phase margin will occur where the cross-over lies near the middle of the −20 dB asymptote.

A larger concept is that of internal stability: It is possible for some closed-loop transfer functions to be stable while the internal stability condition is violated. We say the feedback is internally stable where *all* closed-loop transfer functions are stable. In the classical procedure, internal stability is assured by selecting $G_c(s)$ so that there are no pole–zero cancellations in $L(s)$ in the right half-plane or on the imaginary axis.

It is best not to be too critical of the highly successful classical method; however, it is possible to notice some things that are a bit undesirable. The classical procedure is indirect and that always poses some problems for the beginning student. Also, it is not exactly clear how to select $G_c(s)$ so that *all* the closed-loop transfer functions are stable. Of course, one can use trial and error to eventually stumble upon something that works and there are guides, rules, and tests to aid the designer in this way. The more experienced designer will accomplish this task in an admirably rapid way. However, the procedure is hardly straightforward and this means that after one is finished, there is no guarantee that there is not some easier and better way to achieve the compensation. Also, after stability is assured, the other design criteria must be checked and the procedure started all over where some other kind of problem is indicated. Think of it this way: there is no immediate indication of the trade-offs between stability, and say, demand on the actuator where the classical approach is used. *The neoclassical procedure suffers from none of these defects.*

The focus of the neoclassical procedure is the equation

$$T(s) = Y(s)G_p(s) \tag{6.5}$$

All feedback design methods involve some form of pole–zero cancellation (this is certainly true in the "modern" control theory – especially where asymptotic observers are employed – but that simple fact is completely obscured in that formulation of things). Sometimes cancellation is an excellent idea, sometimes a bad idea, and sometimes it is absolutely forbidden. The constraints are internal stability and the demand placed on the actuators (as characterized by $|Y(j\omega)|$). It is *forbidden* to have a pole–zero cancellation in the right half plane or on the axis in the right side of (6.4). On the other hand, it is *absolutely* required that all closed-loop poles of $G_p(s)$ not strictly in the left half-plane be canceled by the zeros selected for $Y(s)$. However, it is *forbidden* to cancel a zero of $G_p(s)$ with a pole selected for $Y(s)$ if that zero is on the axis or in the right half-plane.

So far, we have tried to address all the feedback control design objectives and constraints (some of which are discussed above) through hand-computing an appropriate feedback compensator. In this chapter, we will accomplish feedback control design through various *norm optimization* techniques.

Control design using an optimization approach is discussed in Sections 6.4 and 6.5. Before that, we briefly discuss and summarize feedback control design objectives and constraints. In Section 6.3, we introduce the linear fractional transformation (LFT), a platform utilized for the design and norm optimization of optimal feedback controllers.

6.2 Optimization Design Objectives and Constraints

In this section, we will discuss control design objectives and constraints. In Section 4.2, we briefly discussed control objectives and constraints for hand computation of the controller using the Youla parameter. Similar objectives and constraints are discussed below for the case of control design by norm optimization. The fundamental question is, how do these objectives and constraints translate to the language of mathematics to be used by optimization for proper design of a feedback controller? This is addressed later.

As discussed in Chapter 4, the design objectives of neoclassical control are fundamentally are accomplished by shaping three closed-loop transfer functions, S, T, and Y. However, these design objectives are generally conflicting,

1. Acceptable disturbance rejection and robustness to structured uncertainty (plant parameter variations) require that S be small.
2. Acceptable sensor noise attenuation and robustness to unstructured uncertainty (high frequency unmodeled dynamics) require that T be small.
3. Acceptable reference tracking (good performance) requires that $T \approx 1$.
4. Acceptable control effort requires that Y be small.

It is important to note that objectives 2 and 3 are in conflict and that these two objectives are both in conflict with (1). In addition, as $Y \approx \frac{1}{G_p}$ in the low frequency region, there is nothing that can be done in this region to reduce control effort!

Let us investigate why such a conflict exists at the first place. Both complementary sensitivity T and sensitivity S transfer functions cannot be kept small simultaneously. This is due to one of the algebraic constraints which will be discussed later. In addition, T cannot simultaneously be kept both small and ≈ 1, as suggested by objectives 2 and 3. However, because sensor noise attenuation and reference tracking are accomplished in two distinct frequency regions (sensor noise attenuation in high frequency while reference tracking in low frequency regions), it is therefore possible to avoid this conflict between the objective 2 and 3. We will also see that it is precisely due to this frequency division that the algebraic constraint of S and T could be addressed.

We are in a sense lucky that feedback control design is possible because we can find a suitable compromise between these conflicting design objectives due to this frequency domain division!

In neoclassical feedback control design, there exist two types of constraints, algebraic constraints and analytical constraints. We will address both in this section.

6.2.1 Algebraic Constraints

The following constraints are said to be algebraic constraints.

- The sensitivity and complementary sensitivity transfer functions are two complex vectors in the s-plane when computed at each complex value of s with the following relationship,

$$T(s) + S(s) = 1 \tag{6.6}$$

- The interpolation conditions discussed in Chapter 4 are considered algebraic constraints,

$$T(p) = 1 \qquad S(p) = 0 \tag{6.7}$$

and

$$\frac{d^k}{ds^k}T(p) = 0 \qquad \frac{d^k}{ds^k}S(p) = 0 \qquad 1 \le k \le a_p - 1 \tag{6.8}$$

while

$$T(z) = 0 \qquad S(z) = 1 \tag{6.9}$$

and

$$\frac{d^k}{ds^k}T(z) = 0 \qquad \frac{d^k}{ds^k}S(z) = 0 \qquad 1 \le k \le a_z - 1 \tag{6.10}$$

where "z" denotes an uncanceled zero of the return ratio, L, of multiplicity a_z, and "p" denotes an uncanceled pole of the return ratio of multiplicity a_p.

6.2.2 Analytic Constraints

The greatest limitations on stability and disturbance rejection are related to the presence of unstable poles and nonminimum-phase zeros (poles and zeros in the Right Half Plane) and time delays. It should be pointed out, of course, that the rational function approximation of time delay using the Padé approximation results in a transfer functions with zeros in RHP.

6.2.2.1 Nonminimum Phase Effect

The task that is begun here is to explore the limitations and constraints on the design of feedback systems associated with a nonminimum phase controlled object.

The first observation is that it is always possible to factor a nonminimum phase controlled object into the product of a minimum phase and an "all-pass" transfer function. By "all-pass," we mean a transfer function for which the Bode magnitude is identically 0 dB for all frequencies. For example,

$$\frac{(1-s)(s+2)}{(s+3)(s+4)} = \overbrace{\frac{(s+1)(s+2)}{(s+3)(s+4)}}^{\text{minimum phase}} \cdot \overbrace{\frac{(1-s)}{(s+1)}}^{\text{all-pass}} \tag{6.11}$$

In general, this means that for any nonminimum phase $T(s)$ and $S(s)$,

$$T(s) = T_{ap}(s)T_{mp}(s) \qquad S(s) = S_{ap}(s)S_{mp}(s) \tag{6.12}$$

where the subscript "*ap*" denotes an "all-pass" or *Blaschke factor* and "*mp*" denotes a *minimum-phase factor.*

Next we call upon the maximum modulus theorem (Appendix A) and apply it to the D-shaped contour that effectively borders the entire right half plane. We assume that the design for internal stability is successful so that there are no poles of either $S(s)$ and $T(s)$ in the right half plane. The maximum modulus theorem means that $|S(s)|$ and $|T(s)|$ will achieve their respective maxima on the contour itself. Specifically speaking, the maxima of $S(s)$ and $T(s)$ occurs on the imaginary axis.

However, the modulus of an all-pass factor is unity on the imaginary axis, so

$$\max_{\omega} |T(j\omega)| = \max_{\omega} |T_{mp}(j\omega)|$$

Similarly,

$$\max_{\omega} |S(j\omega)| = \max_{\omega} |S_{mp}(j\omega)| \tag{6.13}$$

Now suppose that $L(s)$ (the return ratio) has a right half-plane zero at $s = z$ and a right half-plane pole at $s = p$. Then the all-pass factors have the form

$$T_{ap}(s) = \frac{s-z}{s+z} \qquad S_{ap}(s) = \frac{s-p}{s+p} \tag{6.14}$$

On the other hand, interpolation conditions are that $S(z) = 1$ and $T(p) = 1$. So, we have

$$T(p) = 1 = T_{mp}(1)T_{ap}(1) \tag{6.15}$$

$$S(z) = 1 = S_{mp}(1)S_{ap}(1) \tag{6.16}$$

Combine (6.12) and (6.14) to conclude that

$$T_{mp}(p) = T_{ap}^{-1}(p) = \frac{p+z}{p-z} \tag{6.17}$$

$$S_{mp}(z) = S_{ap}^{-1}(z) = \frac{z+p}{z-p} \tag{6.18}$$

It then follows from (6.17) and (6.18) that the peak magnitudes (H_∞ norms) have bounds

$$\max_{\omega} |T(j\omega)| \geq \left|\frac{p+z}{p-z}\right| \tag{6.19}$$

and

$$\max_{\omega} |S(j\omega)| \geq \left|\frac{z+p}{z-p}\right| \tag{6.20}$$

If we use frequency weighting functions as we discussed in Chapter 5 to shape both $T(s)$ and $S(s)$ for performance and robustness objectives, then we can write the first **Analytic Constraints** as (6.21) and (6.22) (for more information on bounds on the weights W_p and W_Δ see [14]),

$$\max_{\omega} |W_\Delta(j\omega)T(j\omega)| \geq \left| W_\Delta(p)\frac{p+z}{p-z} \right| \tag{6.21}$$

and

$$\max_{\omega} |W_P(j\omega)S(j\omega)| \geq \left| W_P(z)\frac{z+p}{z-p} \right| \tag{6.22}$$

Robustness weight (W_Δ) and performance weight (W_p), which are evaluated at the unstable pole and zero, enter these constraints. As we saw in Chapter 4, this means if an unstable pole of a plant is close to an unstable zero, then a lower bound is established for the H_∞ norms. So if a performance criteria is set such as $\max_{\omega}|W_P(j\omega)S(j\omega)| < 1$, and the plant includes an unstable zero and pole, then to achieve this criteria, the performance weight must satisfy $W_p(z)\frac{z+p}{z-p} < 1$. If z and p are close together, then this makes $W_p(z)$ very small, which could result in reducing the bandwidth for avoiding a large overshoot in $S(s)$. This is a horrendous constraint, but it must be respected!

6.2.2.2 Bode Sensitivity Integral Theorem

The Bode Sensitivity Integral Theorem is a conservation theorem formula that quantifies some of the limitations of feedback control systems. The consequences of this theorem are the "Area Formula" or "Waterbed Effect" which apply to nonminimum phase controlled objects. These consequences are elegantly summarized by Bode's Integral Theorem provided later.

Recall the definition of sensitivity,

$$S \triangleq (1+L)^{-1} \tag{6.23}$$

Among other things, this closed-loop transfer function characterizes the tracking ability of the closed loop. Of course, robustness (disturbance rejection and low frequency parameter variations) is another important feedback criterion quantified by this function.

Theorem 6.1 *Suppose that the denominator of the return ratio is of degree at least two larger than the numerator and let p_k denote the set of poles of the return ratio which lie strictly in the right half-plane, then the sensitivity is constrained by*

$$\int_0^\infty Lm\ |S(j\omega)| d\omega = 20\pi \sum_k p_k \tag{6.24}$$

For the proof, see [14].

If the return ratio is stable, however, but possess a zero in the RHP (nonminimum phase zero) then,

$$\int_0^\infty \text{Lm}\ |S(j\omega)| d\omega = 0 \tag{6.25}$$

This effect is illustrated in Figure 6.1.

If $L(s)$ is unstable, the area b exceeds the area a by $\pi \sum_k p_k$. Any reduction in $|S|$ at low frequencies is purchased at the price of an increase in $|S|$ at high frequencies; overshoot in $|S(j\omega)|$ (and $|T(j\omega)|$) is inevitable.

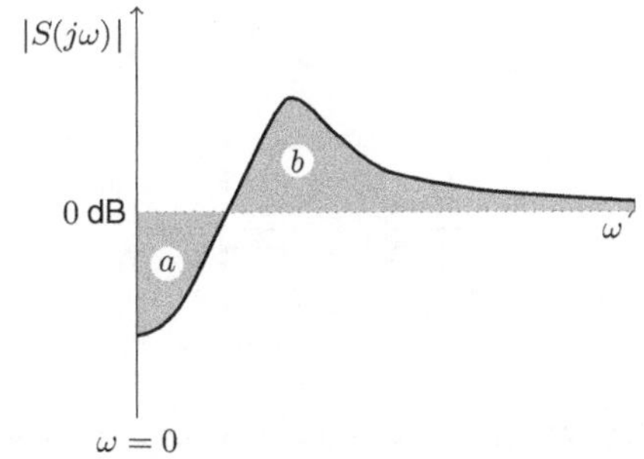

Figure 6.1 Frequency response constraint.

These algebraic and analytic constraints, at times, put very stringent limitations on feedback control design, and they have to be dealt with when designing a feedback controller using either norm optimization or hand computation by Youla parameterization. It is worth mentioning that if the controlled object is stable and minimum phase, the only constraint that we have to deal with is selecting a stable Youla parameter and, of course, $T(s) + S(s) = 1$.

6.3 The Linear Fractional Transformation

LFT, as is given by its name, is simply a transformation given by

$$x \to \frac{ax+b}{cx+d} \tag{6.26}$$

We will discuss this transformation in more detail in the MIMO part of this book. However, it suffices here to say that this transformation interestingly results in a platform which has been utilized for robust optimal control development such as $||H||_2$ and $||H||_\infty$ norm control optimization methods. In Section 6.4, we will introduce the LFT and its implication for SISO robust control development.

The LFT framework for modeling and designing controllers with uncertainty is illustrated in Figure 6.2 and uses the following notation:

- P: Plant or system model.
- K: Controller.
- Δ: The set of all possible uncertainties.
- w: The vector signal of all exogenous inputs (i.e. noises, disturbances, and commands).
- z: The vector signal including all controlled signals and tracking errors.
- u: The controller output (actuator input) signals.
- y: The measurement signals.
- η: The vector signal of all fictitious outputs.
- v: The vector signal of all fictitious inputs.

This platform is generalized for both MIMO and SISO control development. However, in this section, we will only consider LFT for SISO (where SISO is referred to one actuator input, u, and a single sensor measurement, y) control analysis and design.

Another way of structuring the LFT is to use a matrix form for P, such that

$$\begin{bmatrix} z \\ y \end{bmatrix} = \begin{bmatrix} P_{11} & P_{12} \\ P_{21} & P_{22} \end{bmatrix} \begin{bmatrix} w \\ u \end{bmatrix} \tag{6.27}$$

as shown in Figure 6.3.

Then, the response of z to w and u is

$$z = P_{11}w + P_{12}u \tag{6.28}$$

while the response of y to w and u is

$$y = P_{21}w + P_{22}u \tag{6.29}$$

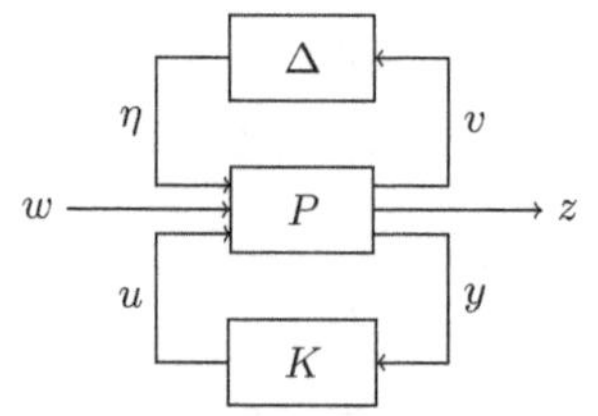

Figure 6.2 A dynamic system with upper and lower LFTs.

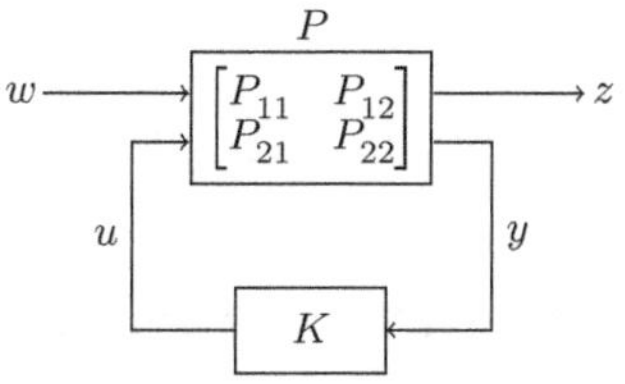

Figure 6.3 LFT with matrix representation of P.

With $u = Ky$, we can then form the closed-loop definition for u.

$$y = P_{21}w + P_{22}Ky \tag{6.30}$$

$$\left(I - P_{22}K\right) y = P_{21}w \tag{6.31}$$

$$y = \left(I - P_{22}K\right)^{-1} P_{21}w \tag{6.32}$$

$$u = K\left(I - P_{22}K\right)^{-1} P_{21}w \tag{6.33}$$

Using this, we can form the closed-loop response of z to the input w, denoted by the transfer function matrix T_{zw}.

$$T_{zw} = P_{11} + P_{12}K\left(I - P_{22}K\right)^{-1} P_{21} \tag{6.34}$$

This is known as the lower LFT, where $T_{zw} = F_{\ell}(P, K)$, therefore

$$z = F_l(P, K)w \tag{6.35}$$

We can relate this framework to a conventional block diagram representation as follows. First, consider the system shown in Figure 6.4.

Without the loss of generality, we have replaced the controller K in Figure 6.2 with G_c in Figure 6.4. In Figure 6.4, we have four inputs to this system: the reference signal input, r, the input disturbance, d_u, the output disturbance, d_y, and senor noise n. We can then assign $n = w_1$, $d_y = w_2$, $d_u = w_3$, and $r = w_4$ to form the input vector w. The vector of controlled signals z is simply any set of tracked variables, often the same as the measured output y, but in this case, it includes the error signal "e" and the control output "u." We can then redraw Figure 6.4 as Figures 6.5 and 6.6.

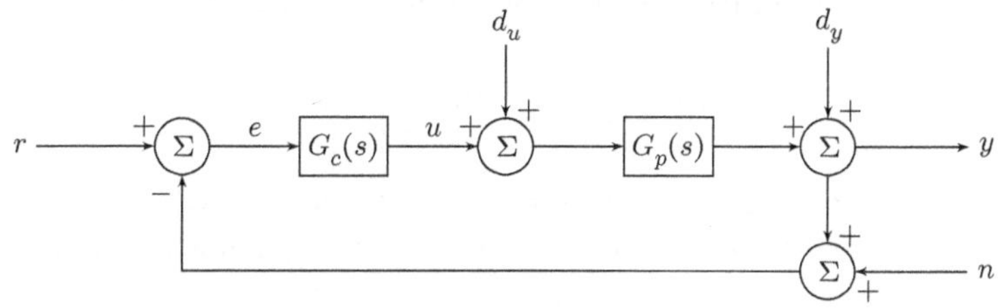

Figure 6.4 Block diagram of a controlled system.

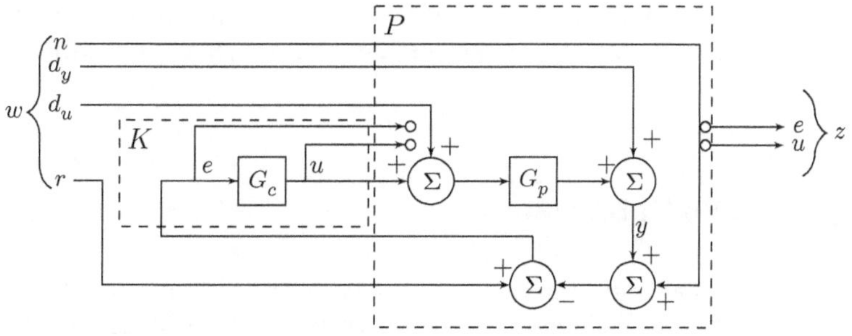

Figure 6.5 Block diagram organized in the LFT framework.

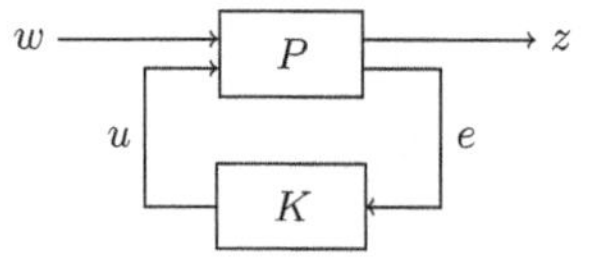

Figure 6.6 The dynamic system manipulated to match the LFT form.

The plant P in Figure 6.5 is the augmented version of G_p and the augment controller K in this case is equal to the original controller G_c. The input–output dynamic equations for the augmented plant P can be written as

$$z\left\{\begin{bmatrix} e \\ u \\ \hdashline e \end{bmatrix}\right. = \left[\begin{array}{cccc:c} -1 & -1 & -G_p & 1 & -G_p \\ 0 & 0 & 0 & 0 & 1 \\ \hdashline -1 & -1 & -G_p & 1 & -G_p \end{array}\right] \left.\begin{bmatrix} n \\ d_y \\ d_u \\ r \\ \hdashline u \end{bmatrix}\right\} w \tag{6.36}$$

This dynamic system can then be represented by the following LFT, as illustrated in Figure 6.6.
Where in Figure 6.6,

$$P = \left[\begin{array}{cccc:c} -1 & -1 & -G_p & 1 & -G_p \\ 0 & 0 & 0 & 0 & 1 \\ \hdashline -1 & -1 & -G_p & 1 & -G_p \end{array}\right], \quad G_c = K \tag{6.37}$$

$$z = \begin{bmatrix} e \\ u \end{bmatrix}, \quad w = \begin{bmatrix} n \\ d_y \\ d_u \\ r \end{bmatrix}, \quad u = Ke$$

Then it is left as an exercise to show that the closed-loop map T_{zw} can be written as

$$z = \begin{bmatrix} -S & -S & -G_pS & S \\ -Y & -Y & -T & Y \end{bmatrix} w \tag{6.38}$$

The transfer function matrix given above includes repetitive transfer functions. Therefore, we define a unique and non-repetitive map from the reference and disturbance inputs r and d_u to the outputs $z = \begin{bmatrix} e \\ u \end{bmatrix}$ as

$$\begin{bmatrix} e \\ u \end{bmatrix} = \begin{bmatrix} S & -G_pS \\ Y & -T \end{bmatrix} \begin{bmatrix} r \\ d_u \end{bmatrix} \tag{6.39}$$

As defined in the previous chapters, T is the complementary transfer function, S is the sensitivity transfer function, and Y is the Youla transfer function. These four transfer functions can be written as a function of the Youla transfer function,

$$\frac{u}{r} = Y, \quad \frac{e}{r} = 1 - G_pY, \quad \frac{u}{d_u} = -G_PY, \quad \frac{e}{d_u} = -G_p(1 - GpY)$$

It is important to note that we accomplished both internal stability and SISO control design performance and robustness objectives through hand computation of the Youla transfer function in the previous chapters. Now, we are ready to accomplish the same objectives by shaping these four transfer functions through weighting filters using optimization of the system norms as discussed next.

6.4 Setup for Loop-Shaping Optimization

There are several norm optimization methods developed by robust control researchers in the past several decades. One of these methods is called the "Loop Shaping Technique," and in this subsection we discuss how LFT is utilized to setup this optimization approach.

In the loop shaping technique, we are interested in minimizing a norm of the closed-loop map T_{zw} (as given in Section 6.3) for designing a controller to accomplish all the required feedback control objectives. Before we pose this minimization

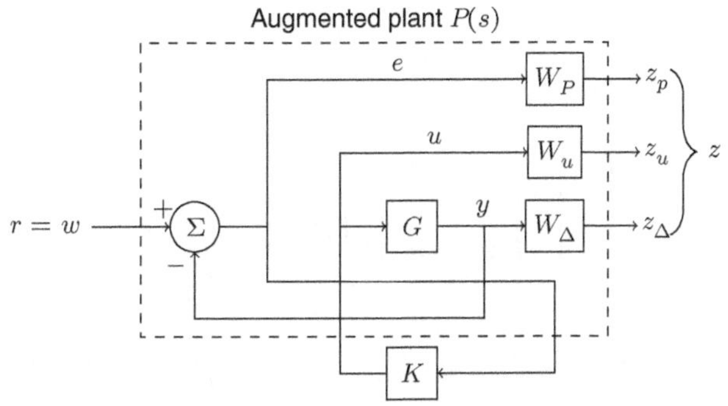

Figure 6.7 Controlled system in form for H_∞ synthesis.

problem, we need to show how to augment the general feedback loop, as discussed in Section 6.3, to obtain the closed-loop map T_{zw}.

In Section 6.3, in Eq. (6.39), we showed that there are only four closed-loop transfer functions that need to be shaped for accomplishing the feedback control loop objectives. In the loop shaping technique, three of the four transfer functions (S, T, and Y) are shaped through appropriate selection of shaping or tuning filters (W_P, W_Δ, and W_u) as illustrated in Figure 6.7.

A couple of points are worth mentioning here. First, as discussed in Section 6.2, the performance, the performance and robustness objectives are divided in such a way in the frequency domain that makes using these weighing filters feasible. The second point is about meeting the internal stability, the most important objective of the feedback control design. In the previous chapters, we discussed that when hand computing the Youla transfer function, it is not difficult to guarantee the stability of three out of four transfer functions presented in Eq. (6.39). However, to guarantee the stability of the fourth transfer function, G_pS, is not an easy task. This was the main reason that we introduced the "interpolation conditions" for assuring internal stability. However, we stated in the paragraph above that this transfer function, G_pS, is not even being directly considered in the loop shaping approach. So, how does the loop shaping method guarantees internal stability? We will provide the answer to this question in in Section 6.5 after discussing the solution to the loop shaping approach.

The synthesis of an H_∞ or H_2 controller requires the system to be in a standard form with augmented plant $P(s)$ with the use of LFT, as shown in Figure 6.7. Here, the only input is the reference input r. We will show that by having this input and augmenting the plant with three weighting filters (W_P, W_Δ, and W_u), the norm optimization problem is correctly set up for obtaining the desired shapes of the three transfer functions S, T, and Y.

This augmented plant can be written in matrix form as

$$\begin{bmatrix} z_p \\ z_u \\ z_\Delta \\ \hline e \end{bmatrix} = \underbrace{\left[\begin{array}{c|c} W_P & -W_PG \\ 0 & W_u \\ 0 & W_\Delta G \\ \hline 1 & -G \end{array}\right]}_{P(s)} \begin{bmatrix} w \\ u \end{bmatrix} \tag{6.40}$$

It is shown (through tedious algebra) [3] that this plant has the following state-space realization,

$$\begin{bmatrix} \dot{x} \\ \hline z \\ e \end{bmatrix} = \left[\begin{array}{c|cc} A & B_1 & B_2 \\ \hline C_1 & D_{11} & D_{12} \\ C_2 & D_{21} & D_{22} \end{array}\right] \begin{bmatrix} x \\ \hline w \\ u \end{bmatrix} \tag{6.41}$$

Additionally, because u is dependent on the error e, these representations can be further condensed, by knowing that

$$u = Ke$$

and

$$e = w - Gu$$

therefore

$$\begin{aligned} e &= w - GKe \\ &= (1 + GK)^{-1}w \end{aligned}$$

and

$$u = K(1 + GK)^{-1}w \tag{6.42}$$

So, the augmented plant then becomes

$$\begin{bmatrix} z_p \\ z_u \\ z_\Delta \end{bmatrix} = \left[\begin{array}{c:c} W_P & -W_PG \\ 0 & W_u \\ 0 & W_\Delta G \end{array}\right] \begin{bmatrix} w \\ K(1+GK)^{-1}w \end{bmatrix} \tag{6.43}$$

and therefore,

$$z_p = \left(W_P - W_PGK(1+GK)^{-1}\right) w = W_P\left(1 - \frac{GK}{1+GK}\right) w = W_PSw \tag{6.44}$$

$$z_u = \left(W_uK(1+GK)^{-1}\right) w = W_u\left(\frac{K}{1+GK}\right) w = W_uYw \tag{6.45}$$

$$z_\Delta = \left(W_\Delta GK(1+GK)^{-1}\right) w = W_P\left(\frac{GK}{1+GK}\right) w = W_\Delta Tw \tag{6.46}$$

Finally,

$$\begin{bmatrix} z_p \\ z_u \\ z_\Delta \end{bmatrix} = \begin{bmatrix} W_PS \\ W_uY \\ W_\Delta T \end{bmatrix} w. \tag{6.47}$$

Then, the closed-loop map T_{zw} is the stacked form, $\begin{bmatrix} W_PS & W_uY & W_\Delta T \end{bmatrix}^T$, as is presented in Eq. (6.47). This form is the frequency weighting of the three closed-loop transfer functions of interest (S, Y, and T).

With this setup, we are able to simultaneously consider bounds on S (for performance), T (for robustness), and Y (to penalize large actuator usage), by posing the following optimization objective (the H_∞ control problem),

$$\underset{K}{\text{minimize}}\ \|T_{zw}\|_\infty \tag{6.48}$$

In other words, find the controller K that minimizes the infinity-norm of T_{zw}. Alternatively, we could choose to minimize the 2-norm. Then, the control problem would instead be

$$\underset{K}{\text{minimize}}\ \|T_{zw}\|_2 \tag{6.49}$$

We will discuss the difficulties, benefits, and the methods for obtaining solutions to these norm optimization problems in the second volume of this book.

6.4.1 Setup for Youla Parameter Loop Shaping

Our approach in designing controllers proposed in earlier chapters included computing the Youla parameter and deriving a feedback controller utilizing the Youla parameter. However, the proposed loop shaping technique directly searches for an optimal controller that minimizes the 2- or ∞-norms given in Eqs. (6.48) and (6.49), without utilizing the Youla parameter.

In this subsection, we propose an optimization approach where instead of searching directly for an optimal controller, we search for an optimal Youla parameter by minimizing the 2- or ∞-norms as given below,

$$\underset{Y}{\text{minimize}}\ \|T_{zw}\|_{2,\infty} \tag{6.50}$$

Youla parameterization is discussed in many articles and textbooks. The solutions proposed in these references include techniques such as co-prime factorization for computing the Youla parameter. However, here we propose a different approach. We are simply searching for an optimization method that would be equivalent to the hand computation technique for Youla parameter presented in the earlier chapters. Then the optimal Youla parameter could be used in computing a feedback controller as proposed earlier. In this context, internal model control (IMC) is an appropriate platform, as illustrated in Figure 6.8, for this task.

IMC presents an elegant approach to the principal reason for utilizing a feedback control in the first place. In Figure 6.8, G_{p_a} represents the actual plant, while, the transfer function, G_{p_m}, is the nominal plant model. The output disturbance is represented by d, the measurement by y, the reference input by r, and the actual plant and the nominal plant model input by u. It can be shown that

$$y = \frac{G_{p_a}Y}{1 + Y(G_{p_a} - G_{p_m})} r + \frac{1 - G_{p_m}Y}{1 + Y(G_{p_a} - G_{p_m})} d \tag{6.51}$$

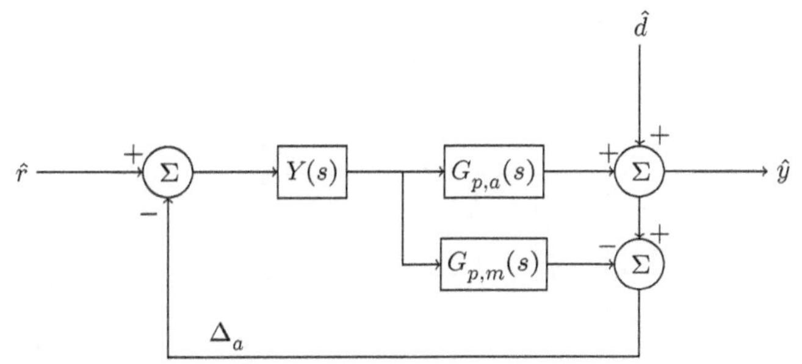

Figure 6.8 Simplified block diagram for internal model control.

where the complementary sensitivity, sensitivity, and Youla parameter transfer functions are

$$T_{IMC} = \frac{G_{p_a} Y}{1 + Y(G_{p_a} - G_{p_m})} \tag{6.52}$$

$$S_{IMC} = \frac{1 - G_{p_a} Y}{1 + Y(G_{p_a} - G_{p_m})} \tag{6.53}$$

$$Y_{IMC} = \frac{Y}{1 + Y(G_{p_a} - G_{p_m})} \tag{6.54}$$

respectively.

The difference between the actual plant and nominal plant model outputs includes some important uncertainties. These uncertainties are both at the plant level due to lack of having a perfect nominal plant model and environmental uncertainties, which are modeled by the output disturbance, d. Feedback signal Δ_a represents these uncertainties. Here lies the fundamental benefit of the feedback: the feedback signal Δ_a is exactly equal to zero if there are no uncertainties (no output disturbances and a perfect nominal plant model). This states that the only reason to use a feedback loop in control design is simply the existence of uncertainties. Without uncertainties, the IMC is transformed to a feedforward control loop design with Youla parameter serving as a feedforward controller.

In this section, we propose two different approaches to optimize the 2- and ∞-norms of the closed loop as given in Eq. (6.50) through directly computing the optimal Youla parameter,

- Loop shaping using IMC structure as illustrated in Figure 6.9.
- Model matching using IMC structure as shown in Figure 6.10.

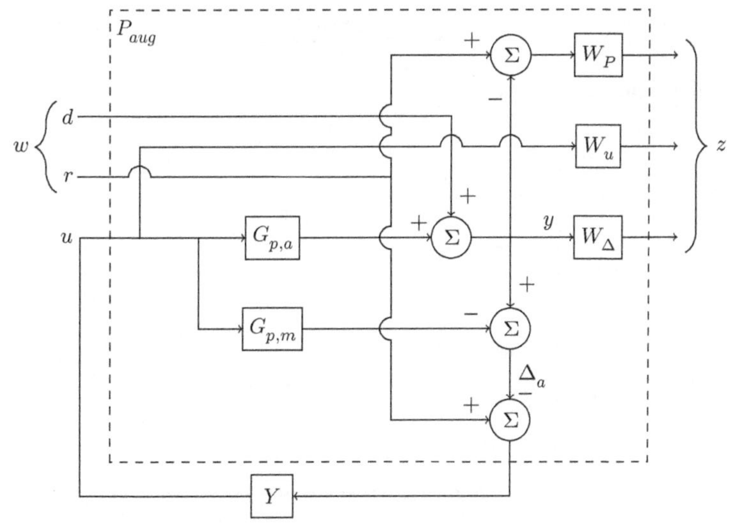

Figure 6.9 Direct loop shaping with Youla parameter.

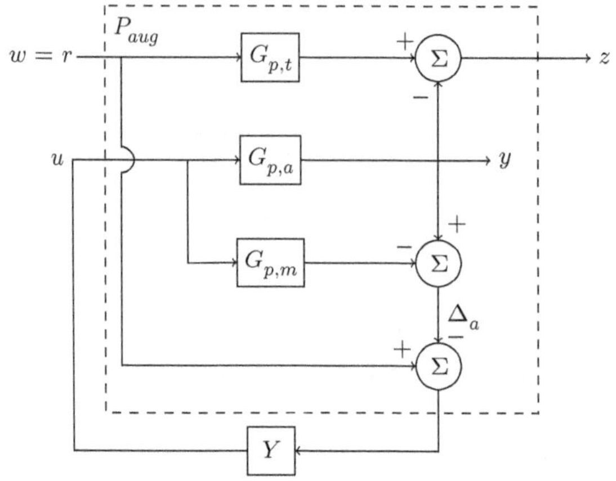

Figure 6.10 Model matching with Youla parameter.

In the model matching problem given in Figure 6.10, defining filters W_P, W_u, and W_Δ is optional and not necessary. The computation of the Youla parameter, Y, is performed by minimizing a norm of the closed-loop transfer function, T_{zw}, which forces our actual plant $G_{p,a}$ to follow our virtual model $G_{p,t}$ in the presence or absence of the uncertainties Δ_a.

In Section 6.5, we briefly discuss the solutions to SISO norm optimization. A more complete solution to the norm optimization will be discussed in the second volume of this book when discussing MIMO control design.

6.5 H_∞-norm Optimization Problem

We will discuss the solutions to both H_∞ and H_2 solutions in detail in the next volume of this book. However, in this section, our intent is to provide some insight into the frequency domain solution of H_∞ problem before presenting its equivalent and more detailed state-space solution in the next volume.

We start this section by first discussing a simplified version of H_∞ problem in the frequency domain:

$$\text{minimize} \left[\max_{\omega} |W(j\omega)S(j\omega)|\right] \tag{6.55}$$

where $S(s)$ is the sensitivity and $W(s)$ is a weighting function. We choose the weighting function $W(s)$ to be proper, stable, and with zeros strictly in LHP. The weighting function is chosen so that weighting is large at low frequency and small at high frequency so that the solution to Eq. (6.55) results in an all pass filter as will be discussed later.

Example 6.1 Let

$$W(s) = \frac{\alpha\tau s + 1}{\tau s + 1}$$

with magnitude plot shown in (Figure 6.11)

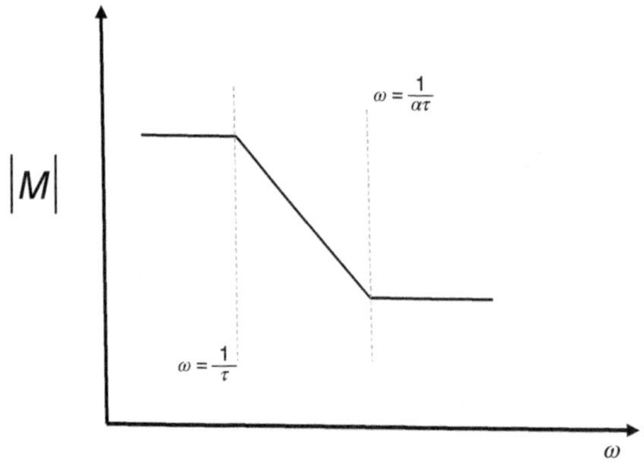

Figure 6.11 Sensitivity transfer function weighting function.

In this simple problem, we need to consider the following three constraints:

1. Well-posedness
2. Internal stability
3. Feedback controller must be proper or strictly proper so it is realizable.

In the first instance, it is not clear how to ensure the third constraint. In fact, we can ignore this constraint and instead use an ad hoc approximation to fix things after the fact .

6.5.1 Solution to a Simple Optimization Problem

Minimizing Eq. (6.55) is similar to equating,

$$W(s)S(s) = KS_{ap}(s)$$

where S_{ap} (an all pass filter,) and K are selected to satisfy the interpolation conditions on $S(s)$. Here we must assume that the plant transfer function, $G_p(s)$, has no poles on $j\omega$-axis (however, we will relax this constraint soon). Therefore,

$$S(s) = \frac{KS_{ap}(s)}{W(s)} \tag{6.56}$$

Interpolation conditions state that,

$$S(z_i) = 1, \quad i = 1,2, \dots, n$$
$$S(p_k) = 0, \quad k = 1,2, \dots, m$$

where z_i and p_k are zeros and poles in the RHP, respectively, and where n is the number of zeros in the RHP and m is the number of poles in the RHP. Assuming that no repeated poles and zeros in RHP, we therefore have

$$S_{ap} = \frac{(-s + p_1)(-s + p_2) \dots (-s + p_m)}{(s + p_1)(s + p_2) \dots (s + p_m)} B(s) \tag{6.57}$$

If $n = 1$ (where n is the number of zeros in RHP of G_p), then we select $B(s)$ to be also equal to 1, and we select K so that $S(z_i) = 1$. If $n > 1$, then we select,

$$B(s) = \frac{(-s + a_1)(-s + a_2) \dots (-s + a_{n-1})}{(s + a_1)(s + a_2) \dots (s + a_{n-1})} \tag{6.58}$$

Then we select a_k so that

$$S(z_i) = 1$$

Example 6.2 Let

$$G_p = \frac{(-s + 1)}{(-s + 2)(s + 3)}$$

and we select $W(s) = \frac{\alpha\tau s+1}{\tau s+1}$ then,

$$S(s) = K\frac{S_{ap}(s)}{W(s)}$$

Here, $B(s) = 1$ as $n = 1$. Therefore,

$$S_{ap} = \frac{(-s + 2)}{(s + 2)} B(s) = \frac{(-s + 2)}{(s + 2)}$$

We can write the sensitivity transfer function, $S(s)$ as

$$S(s) = K\frac{(-s + 2)}{(s + 2)} \frac{(\tau s + 1)}{(\alpha\tau s + 1)}$$

and due to interpolation condition, $S(1) = 1$ then $K = 3\frac{\alpha\tau+1}{\tau+1}$. But, we know $T(s) = 1 - S(s)$, so

$$\begin{aligned} T(s) &= 1 - K\frac{(-s + 2)}{(s + 2)} \frac{(\tau s + 1)}{(\alpha\tau s + 1)} \\ &= \frac{(s + 2)(\alpha\tau s + 1) - K(-s + 2)(\tau s + 1)}{(s + 2)(\alpha\tau s + 1)} \end{aligned}$$

The relative degree of T is 0. However, it can be shown that for controller G_c to be realizable (satisfying the third constraint), the relative degree of T must be greater than or equal to 1. Modifying T by adding a fast pole $(\tau_0 s + 1)$ and scaling by K_T to satisfy interpolation condition gives us

$$T(s) = K_T \frac{(s+2)(\alpha\tau s + 1) - K(-s+2)(\tau s + 1)}{(s+2)(\alpha\tau s + 1)(\tau_0 s + 1)}$$

We select τ_0 to be small such that,

$$\frac{1}{\tau_0} \gg \frac{1}{\alpha\tau}$$

Next, we compute K_T to satisfy interpolation condition, $T(2) = 1$:

$$K_T \frac{1}{2\tau_0 + 1} = 1$$

Therefore, $K_T = 2\tau_0 + 1$.

With this T, it can be shown that we have satisfied all three constraints as discussed above.

What if $G_p(s)$ has poles on $j\omega$-axis? Then we can write,

$$G_p(s) = \frac{\widetilde{G_p}(s)}{P(s)} \tag{6.59}$$

where $P(s)$ is the polynomial of poles on $j\omega$-axis. Thus, $\widetilde{G_p}(s)$ has no poles on $j\omega$-axis. We can write,

$$S(s) = P(s)\widetilde{S}(s) \tag{6.60}$$

Then we select $W(s) = \frac{\widetilde{W}(s)}{P(s)}$. In this case,

$$W(s)S(s) = \frac{\widetilde{W}(s)}{P(s)} P(s)\widetilde{S}(s) \tag{6.61}$$

Therefore, the solution for $\widetilde{W}(s)\widetilde{S}(s)$ is the same as the solution for $W(s)S(s)$.

Example 6.3 Let

$$G_p = \frac{(-s+1)}{s(s+1)}$$

We select

$$W(s) = \frac{\alpha\tau s + 1}{s(\tau s + 1)}$$

Then,

$$\widetilde{W}(s) = \frac{\alpha\tau s + 1}{\tau s + 1}$$

Knowing that $G_p(s) = \widetilde{G_p}(s)/P(s)$, we have $P(s) = s$. We can then write,

$$\widetilde{W}(s)\widetilde{S}(s) = KS_{ap}(s) \tag{6.62}$$

The interpolation conditions are

$$S(1) = 1 \rightarrow P(1)\widetilde{S}(1) = 1 \cdot \widetilde{S}(1)$$

and

$$S(0) = 0 \rightarrow P(0)\widetilde{S}(0) = 0 \cdot \widetilde{S}(0)$$

Taking Eq. (6.62) and knowing $S_{ap}(s) = 1$, we have

$$\widetilde{W}(s)\widetilde{S}(s) = K$$

Therefore,

$$\widetilde{S}(s) = \frac{K(\tau s + 1)}{(\alpha\tau s + 1}$$

But interpolation condition enforces that $\tilde{S}(1) = 1$. So, we can compute K to be

$$K = \frac{\alpha\tau + 1}{\tau + 1}$$

We can now compute $S(s)$ and $T(s)$.

$$S(s) = \frac{Ks(\tau s + 1)}{(\alpha\tau s + 1)}$$

It is evident that this transfer function is improper. However, we continue by computing $T(s)$,

$$T(s) = 1 - S(s) = \frac{(\alpha\tau s + 1) - Ks(\tau s + 1)}{(\alpha\tau s + 1)}$$

Given an improper $S(s)$, it is expected that $T(s)$ would also be improper as the relative degree of $T(s) = -1$.

Similar to our previous approach, we modify $T(s)$ by adding two fast poles,

$$T(s) = \frac{K_T}{(\tau_0 s + 1)^2}\left\{\frac{(\alpha\tau s + 1) - Ks(\tau s + 1)}{(\alpha\tau s + 1)}\right\}$$

The interpolation conditions for $T(s)$ are

$$T(0) = 1 = K_T\{1\} \rightarrow K_T = 1$$

and

$$T(1) = 0 = \frac{K_T}{(\tau_0 + 1)^2}\{0\} \rightarrow T(1) = 0$$

Therefore,

$$T(s) = \frac{(\alpha\tau s + 1) - Ks(\tau s + 1)}{(\alpha\tau s + 1)(\tau_0 s + 1)^2} \quad \frac{1}{\tau_0} \gg \frac{1}{\alpha\tau}$$

Then we can recalculate $S(s) = 1 - T(s)$. Knowing that $Y(s) = T(s)/G_p(s)$ and $G_c(s) = Y(s)/S(s)$, the final controller can be computed.

$$G_c(s) = \frac{Y(s)}{S(s)}$$

6.6 H_∞ Design

Imagine that we devise upper bounds for $|T(j\omega)|$ and $|S(j\omega)|$ as illustrated in Figure 6.12.

The idea is to obtain,

$$|T(j\omega)| < |W_{DT}(j\omega)| \tag{6.63}$$

$$|S(j\omega)| < |W_{DS}(j\omega)| \tag{6.64}$$

We prefer to write the above equations as

$$|W_{PT}T(j\omega)| < 1, \quad W_{PT} = \frac{1}{W_{DT}} \tag{6.65}$$

$$|W_{PS}S(j\omega)| < 1, \quad W_{PS} = \frac{1}{W_{DS}} \tag{6.66}$$

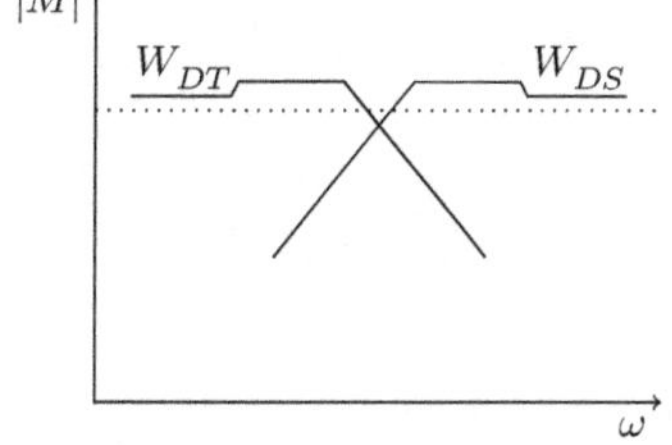

Figure 6.12 Sensitivity and complementary sensitivity transfer functions upper bounds.

Satisfaction of both these constraints is equivalent to satisfaction of

$$\max_{\omega} \{|W_{PS}S(j\omega)|, |W_{PT}T(j\omega)|\} < 1 \tag{6.67}$$

However, the theory cannot easily deal with this type of constraint. Instead, we find a new constraint that approximates the above, but which the theory (i.e. loop shaping technique) can handle. This approximate constraint is

$$\left| \begin{matrix} W_{PS}S(j\omega) \\ W_{PT}T(j\omega) \end{matrix} \right| < 1 \tag{6.68}$$

It can be shown that

$$\frac{1}{\sqrt{2}} \left| \begin{matrix} W_{PS}S(j\omega) \\ W_{PT}T(j\omega) \end{matrix} \right| (j\omega) \leq \max_{\omega} \{|W_{PS}S(j\omega)|, |W_{PT}T(j\omega)|\} \leq \left| \begin{matrix} W_{PS}S(j\omega) \\ W_{PT}T(j\omega) \end{matrix} \right| (j\omega) \tag{6.69}$$

Meaning, the approximation is reasonable.

To address the actuator constraint, one utilizes a similar weighting function for the Youla transfer function, for instance $W_{PY} = \frac{1}{W_{DY}}$. However, we know that $Y = \frac{T}{G_p}$, so

$$|W_{PY}Y| = \left| \frac{W_{PT}}{G_p} T \right| \tag{6.70}$$

It is usual to incorporate the weighting transfer functions in the generalized plant (as discussed in the loop shaping section of this chapter), so that the controlled variable z (the output of the generalized plant) now satisfies:

$$z = \begin{bmatrix} W_{PS}S \\ W_{PT}T \\ W_{DY}Y \end{bmatrix} w \tag{6.71}$$

Then, the nominal performance constraint becomes

$$\|T_{zw}\|_{\infty} < 1 \tag{6.72}$$

It should be evident that there is a difference between this objective (nominal performance constraint) and the one given in the loop shaping in Section 6.3,

$$\underset{K}{\text{minimize}} \|T_{zw}\|_{\infty} \tag{6.73}$$

We will discuss this in the second volume of this book when presenting the MIMO H_{∞} solution. Furthermore, we will discuss how the nominal performance constraint has to change for addressing robust performance and robust stability and performance objectives.

6.7 H_{∞} Solutions Using Matlab Robust Control Toolbox for SISO Systems

In this section, we investigate the use of Matlab Robust Control Toolbox in solving the H_{∞} optimization problem. The problem that we interested to solve was posed in Sections 6.4 though 6.6 as a mixed sensitivity shaping loop shaping problem,

$$z = \begin{bmatrix} W_{P}S \\ W_{\Delta}T \\ W_{Y}Y \end{bmatrix} w \tag{6.74}$$

Then the optimization objective becomes,

$$\|T_{zw}\|_{\infty} < 1 \tag{6.75}$$

6.7.1 Defining Frequency Weights

In this section, we learn to define the weights W_P and W_{Δ} using simple first-order transfer functions. These weights will then utilized to compute a controller using commands in robust control toolbox in Matlab. The intent of these frequency domain weights is to penalize the closed-loop transfer functions S, T, and Y, so that by solving the optimization problem given above, the desired shapes of these closed-loop transfer functions are obtained. The shaping of these closed-loop transfer functions is actually what we have accomplished using an indirect hand-computation approach in the previous chapters.

Before defining these weighting functions, it is again beneficial to discuss the desired shapes of S and T closed-loop transfer functions.

In Figure 6.13, control design requirements are transformed in each frequency region to prescribe a specific shape for the sensitivity transfer function. In the low frequency region, when the magnitude of this transfer function is small relative to 0 dB, then this enforces good command following and low frequency disturbance rejection capability of the closed-loop system. On the other hand, for good stability measures, e.g. gain and M_2 margins, the overshoot in S must be minimized. To reject high frequency disturbances, e.g. sensor noise rejection and unmodeled dynamics, for meeting robust stability, S should approached unity, as the complementary sensitivity transfer function, T, becomes very small.

In addition to the previous requirements, the peak resonance in T, as shown in Figure 6.14, results in a an overshoot in time domain transient response, e.g. due to a step input. Therefore, to minimize this time domain overshoot, one has to minimize the peak resonance of T in frequency domain.

To obtain the desired shapes of S, the following inverse weighting transfer function, illustrated in Figure 6.15, could be utilized.

In this case, the optimization meets its objective if,

$$|S| < |W_P|^{-1} \tag{6.76}$$

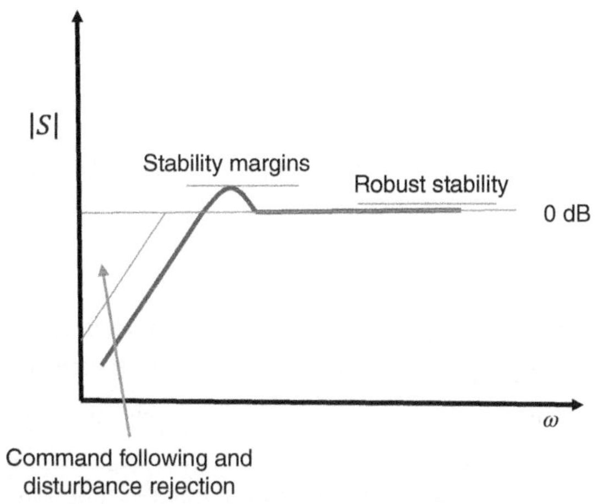

Figure 6.13 Frequency division of the sensitivity transfer function for meeting control design requirements.

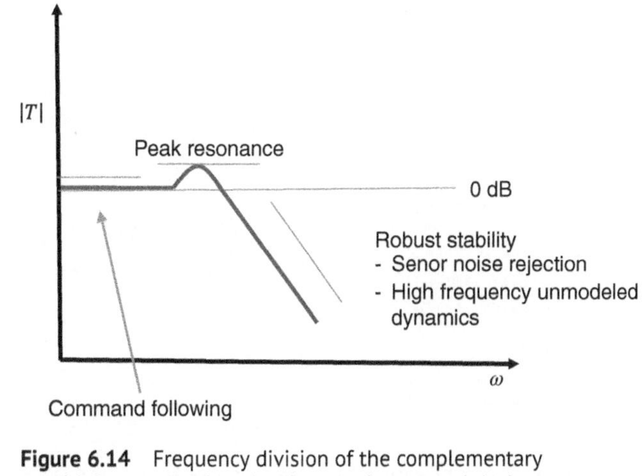

Figure 6.14 Frequency division of the complementary sensitivity transfer function for meeting control design requirements.

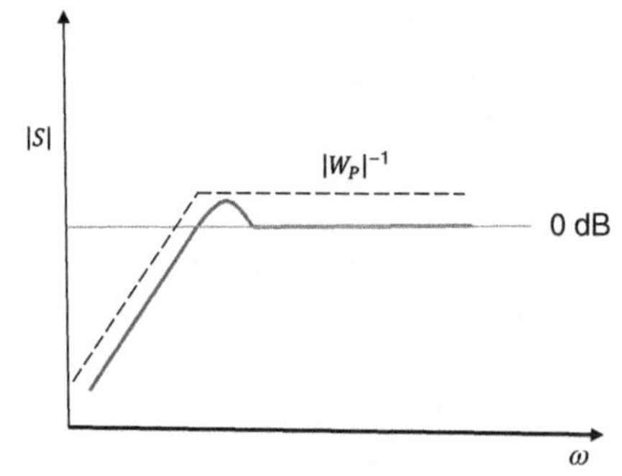

Figure 6.15 The $|W_P|^{-1}$ weighting function.

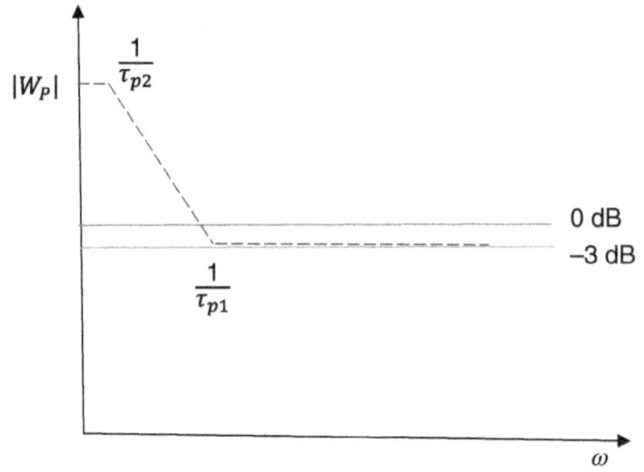

Figure 6.16 The $|W_P|$ weighting function.

When mixed sensitivity loop shaping algorithm is used, $|W_P|$ must be specified in place of $|W_P|^{-1}$, as shown in Figure 6.16.

It is not so difficult to see that $W_P = \frac{K(\tau_{p1}s+1)}{(\tau_{p2}s+1)}$, where $\frac{1}{\tau_{p2}}$ is selected to be a very small number so that W_P acts like an integrator at low frequency and $\frac{1}{\tau_{p1}}$ is selected to be approximately one decade after the closed-loop bandwidth frequency, so that T and S meet below 0 dB to enforce a good phase margin. The constant K is simply a tuning variable.

To obtain the desired shapes of T, the following inverse weighting transfer function, illustrated in Figure 6.17, could be utilized.

In this case, the optimization meets its objective if,

$$|T| < |W_\Delta|^{-1} \tag{6.77}$$

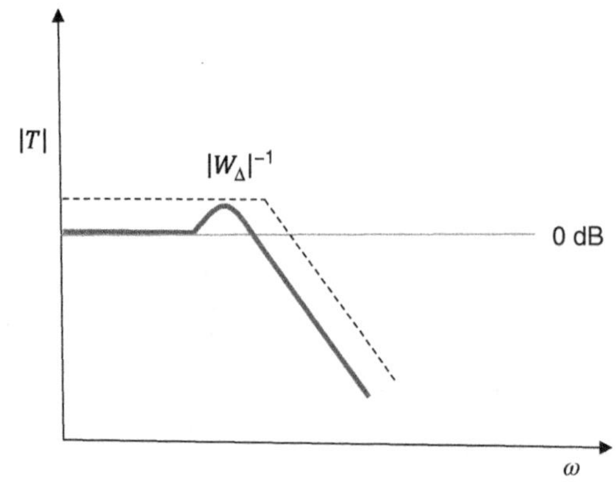

Figure 6.17 The $|W_\Delta|^{-1}$ weighting function.

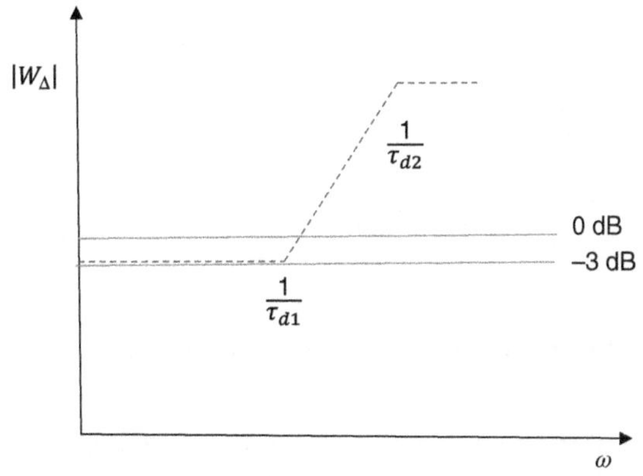

Figure 6.18 The $|W_\Delta|$ weighting function.

Again when mixed sensitivity loop shaping algorithm is used, $|W_\Delta|$ must be specified in place of $|W_\Delta|^{-1}$, as shown in Figure 6.18.

It is therefore straightforward to see that $W_\Delta = \frac{K_d(\tau_{d1}s+1)}{(\tau_{d2}s+1)}$, where $\frac{1}{\tau_{d2}}$ is selected to be a very large number so that magnitude of T to be very small at high frequency and $\frac{1}{\tau_{d1}}$ is selected to be approximately the closed-loop bandwidth frequency. The constant K_d is simply a tuning variable.

The weighting on Youla transfer function, $|W_Y|$, could be taken as a very small constant, e.g. ϵ, or an empty matrix, [...], so that no penalty on the control effort is enforced. The optimization usually computes the required controller for proving the shapes of S and T with a required control effort.

The robust control toolbox in Matlab over the years has evolved to provide the necessary tools to make this loop shaping effortless. The weighting function now could be defined with the following command,

```
makeweight(DCgain, cross-over frequency, High frequency gain)
```

where *makeweight*(100,1, 0.707) results in the Bode magnitude plot shown in (Figure 6.19),

Example code for using Matlab's Robust Control Toolbox can be found in Section 4.4.

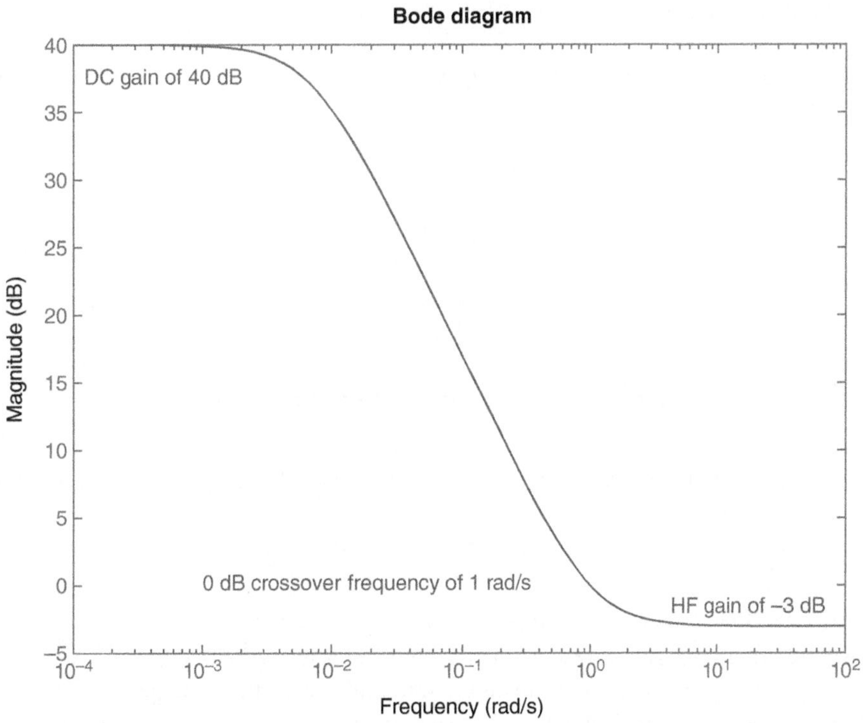

Figure 6.19 Bode magnitude plot of a typical W_p weighting function.

6.8 Problems

6.1 Consider the system shown in Figure 6.20.

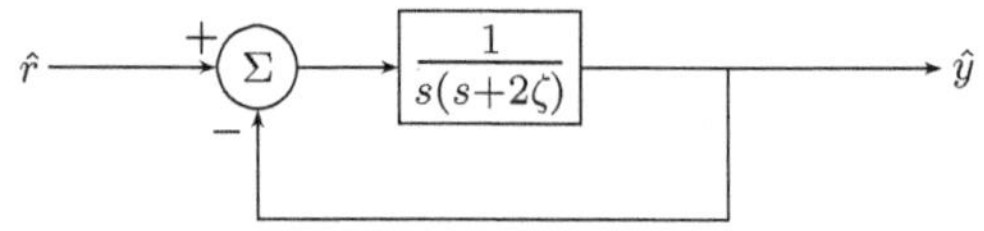

Figure 6.20 Block diagram for Problem 6.1.

Determine the optimal value of the damping ratio ζ so that when the system is subjected to a unit step input $r(t) = 1(t)$, the following performance index is minimized:

$$J = \int_0^\infty \underline{x}^T(t)Q\underline{x}(t)dt \tag{6.78}$$

where

$$\underline{x} = \begin{bmatrix} x_1 \\ x_2 \end{bmatrix} = \begin{bmatrix} x \\ \dot{x} \end{bmatrix}, \quad Q = \begin{bmatrix} 1 & 0 \\ 0 & M \end{bmatrix} \quad (M > 0) \tag{6.79}$$

and

$$x_1(0^-) = 0, \quad x_2(0^-) = 0 \tag{6.80}$$

Hint: You need to compute $\underline{P}$ from the Lyapunov function and minimize the performance index with respect to the damping ratio. You also need to note that

$$J = \int_0^\infty \underline{x}^T Q\underline{x}\ \mathrm{d}t = -\underline{x}^T P x\Big|_0^\infty = -\underline{x}^T(\infty)^T P\underline{x}^T(\infty) + \underline{x}^T(0)^T P\underline{x}^T(0) \tag{6.81}$$

Since all eigenvalues of A have negative real parts, we have $\underline{x}(\infty) \to 0$, therefore

$$J = \underline{x}^T(0)P\underline{x}(0) \tag{6.82}$$

This problem should be carried out by hand.

6.2 Consider the same quarter car model shown in Figure 6.21. Let $m = 250$ kg, $k = 16\ 000$ N/m, and $b = 1175$ Ns/m.

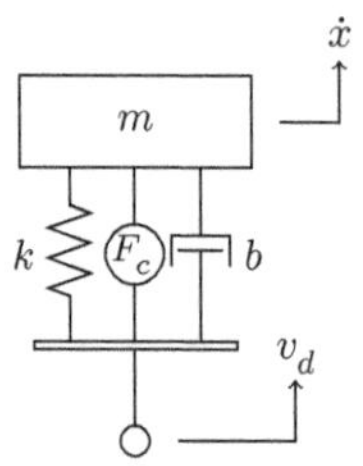

Figure 6.21 Active suspension for eighth-car model.

Your task is to design an optimal controller with control law $F_c = -k\underline{x}$ using the machinery of the linear quadratic regulator (LQR) by minimizing

$$J = \int_0^\infty \underline{x}^T Q\underline{x} + F_c^T R F_c\ \mathrm{d}t \tag{6.83}$$

Q must be selected so that $\underline{x}^T Q\underline{x}$ represents system energy. You are free to select R.
If you use the bond graph to derive your state model, your states would be $\underline{x} = \begin{bmatrix} p & q \end{bmatrix}^T$, where p is the momentum of the mass and q is the spring deflection. If you are unfamiliar with the bond graph method, you can obtain the equations of motion from a free body diagram. Use state variables $q = \int_0^t v_d\ \mathrm{d}t - x$ (so that $\dot{q} = v_d - \dot{x}$) and $p = m\dot{x}$ (so that $\dot{p} = m\ddot{x}$).

Use the following parameters:

$$m = 250 \text{ kg}, \ k = 16\ 000 \text{ N/m}, \ b = 1175 \text{ Ns/m} \tag{6.84}$$

The intent of an active control is to minimize the mass velocity response $\left|\frac{\dot{x}}{v_d}\right|$. You need to present your results the magnitude response of $\left|\frac{\dot{x}}{v_d}\right|$ as a function of frequency for the controlled and uncontrolled version. In addition, you need to plot $\left|\frac{F_c}{v_d}\right|$. Use 0.01–100 rad/s for you x-axis (frequency axis).

Hint: When you obtain your state-space model, your B matrix will have two columns – one for F_c and one for v_d (denoted B_F and B_v). You can ignore v_d and B_v when designing the LQR controller. To determine the frequency responses, let your closed-loop input matrix B_{cl} equal B_v and set C_{cl} so that $\dot{x}$ and F_c can be obtained as outputs (D_{cl} will be zero).

6.3 In this problem, you will explore the application of the LQR optimal controller for the linearized automobile suspension system shown in Figure 6.22 later.

The state vector is $x = \begin{bmatrix} p_s & q_s & p_u & q_t \end{bmatrix}^T$, where p_s is the momentum of the sprung mass m_s, q_s is the suspension deflection (positive in compression), p_u is the momentum of the unsprung mass m_u, and q_t is the tire deflection (positive in compression).

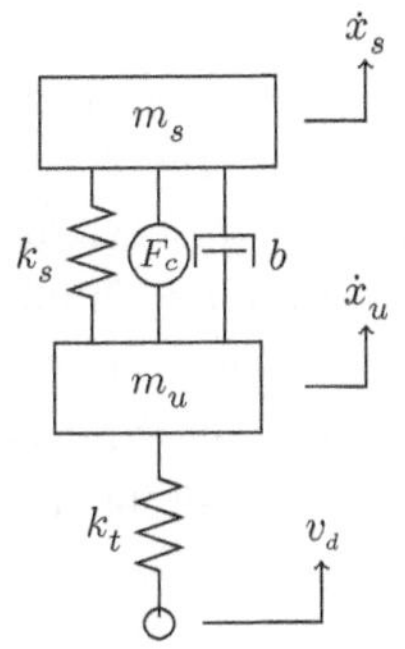

Figure 6.22 Active suspension for quarter-car model.

The following parameter values are to be used for this problem:

$$m_s = 240 \text{ kg}, \quad m_u = 36 \text{ kg}, \quad b = 1{,}175 \text{ Ns/m}, \quad k_s = 16\ 000 \text{ N/m}, \quad k_t = 160\ 000 \text{ N/m} \tag{6.85}$$

If you are unfamiliar with the bond-graph method, the state equations resulting from the bond graph are provided below.

$$\begin{bmatrix} \dot{p}_s \\ \dot{q}_s \\ \dot{p}_u \\ \dot{q}_t \end{bmatrix} = \begin{bmatrix} -\frac{b}{m_s} & k_s & \frac{b}{m_u} & 0 \\ -\frac{1}{m_s} & 0 & \frac{1}{m_u} & 0 \\ \frac{b}{m_s} & -k_s & -\frac{b}{m_u} & k_t \\ 0 & 0 & -\frac{1}{m_u} & 0 \end{bmatrix} \begin{bmatrix} p_s \\ q_s \\ p_u \\ q_t \end{bmatrix} + \begin{bmatrix} -1 & 0 \\ 0 & 0 \\ 1 & 0 \\ 0 & 1 \end{bmatrix} \begin{bmatrix} F_c \\ v_d \end{bmatrix} \tag{6.86}$$

The intent of the active suspension is to keep the motion of the sprung mass small (this is called comfort mode) while keeping the motions of the sprung mass not negatively impacted (in other words, preventing the loss of contact between the tire and ground, the security mode). These two optimization objectives need to be satisfied in the presence of the space constraint (rattle space) between the sprung mass and unsprung mass. The suspension cannot be allowed to bottom-out or over-extend, so the suspension deflection should be kept small.

You must select a weighting matrix Q and scalar R to obtain the best results. A good way to select Q is perhaps by including all the states you wish to minimize and your constraint state (rattle space). The scalar R will require some tuning as the active forces must be reasonable. You need to present your results by showing magnitude response plots for the following transfer functions:

(a) $\frac{\dot{x}_s}{v_d}(s)$

(b) $\frac{v_{rel}}{v_d}(s)$ (where $v_{rel} = \dot{x}_u - \dot{x}_s$)

(c) $\frac{\dot{x}_u}{v_d}(s)$
(d) $\frac{F_c}{v_d}(s)$
(e) $\frac{q_s}{v_d}(s)$
Show the magnitude responses for both the controlled and uncontrolled system (put the responses of the controlled and uncontrolled systems on the same figure). It is recommended you use the `bodemag` command to generate the magnitude plots:

```
w = logspace(0,2.3,140);
bodemag(your_tf_uncontrolled,'k-.',your_tf_controlled,'r-',w)
```

Comment on your findings.

6.4 Repeat Problem 6.3, but this time utilize the mixed sensitivity H_∞ method, where you augment your plant with W_1, W_2, and W_3 filters to derive a controller that meets the objectives and constraints stated above. The command in Matlab is `hinfsyn`.
The following filters are suggested:

- $W_1 = W_{1a} \cdot W_{1b}$, with $W_{1a} = G_{p,d}$ where $G_{p,d}$ is the open loop transfer function from v_d (disturbance input) to sprung mass velocity $\dot{x}_s$. You need to find a reasonable W_{1b}. Start simple, e.g. a first-order transfer function.
- For W_2, try a constant such as $\begin{bmatrix} 0 & 0.0005 \\ 0 & 1 \end{bmatrix}$.
- You also need to find a reasonable W_3. Like with W_1, start simple.

Keep in mind that this is not a tracking problem, so T (the closed-loop transfer function) does not need to be equal to 1 (0 dB) at very low frequencies. T needs to be closed to one at mid-frequencies (0.5–15 Hz). Make S smaller than one in this frequency range for better disturbance rejection capabilities. Note that the sprung mass natural frequency is around 1 Hz, while the wheel hop (unsprung mass) natural frequency is around 10 Hz.
You need to demonstrate your results by submitting the same plots as in Problem 6.3.
Hint: This is an output (not full-state) feedback H_∞ problem in which you have access to only the sprung mass velocity $\dot{x}_s$.

6.5 Repeat Problem 6.4, using the same filters, but minimize the H_2 norm instead. Present the same result plots as part (a). The command in Matlab is `h2lqg`.

6.6 Consider the second-order system described by

$$\begin{bmatrix} \dot{x}_1 \\ \dot{x}_2 \end{bmatrix} = \begin{bmatrix} 0 & 1 \\ -1 & -1 \end{bmatrix} \begin{bmatrix} x_1 \\ x_2 \end{bmatrix} \tag{6.87}$$

Clearly, the equilibrium state is the origin. Assuming that a Lyapunov function $V(x) = x^T Px$, determine the stability of the equilibrium state.
Hint: Assume $Q = I$ (identity matrix). Check for positive definiteness of P. You also need to compute $V(x)$ and its derivative.

6.7 Repeat Problem 6.6, but this time use the machinery of H_∞ optimization to compute the compensator, G_c. You need to provide $W_P(orW_1)$, $W_\delta(orW_3)$, in addition you may need to provide a weighting for the actuator, $W_u(orW_2)$. It is suggested to use a simple constant for your actuator weighting. Utilizing the H_∞ controller, compute T, Y, and S. Provide Bode plots of T, S, and Y, including your design phase and gain margin, and the step response of the closed-loop transfer function T. Compare your results to the results of the first problem and comment.
Hint: The structure W_3 filter, which is used for compensating delays, is discussed and provided to you in the class. You should use this filter for W_3 and then use a first-order filter to design W_1.

6.8 Consider the system show in Figure 6.23.
Assuming that the control signal to be

$$u(t) = -Kx(t) \tag{6.88}$$

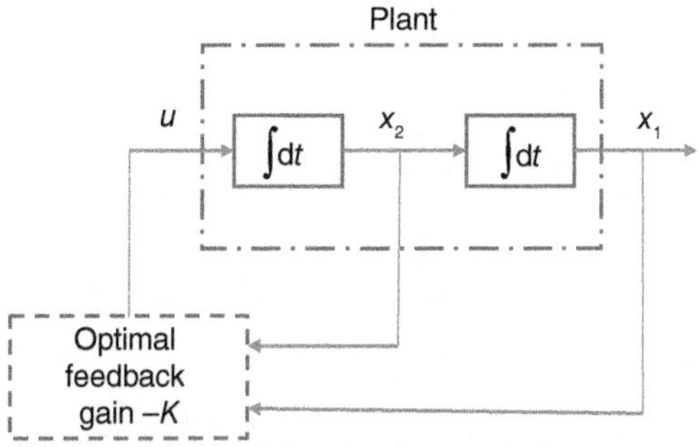

Figure 6.23 Block diagram for Problem 6.8.

design the optimal feedback gain matrix K such that the following performance index is minimized:

$$J = \int_0^{\infty} (x^T Qx + ru^2)dt \tag{6.89}$$

where

$$Q = \begin{bmatrix} 1 & 0 \\ 0 & \mu \end{bmatrix} \quad (\mu \geq 0) \tag{6.90}$$

6.9 A feedback control system can be rearranged as an LFT shown in Figure 6.24.

$$\begin{bmatrix} z \\ y \end{bmatrix} = \begin{bmatrix} G_{11} & G_{12} \\ G_{21} & G_{22} \end{bmatrix} \begin{bmatrix} w \\ u \end{bmatrix} \tag{6.91}$$

$$u = Ky \tag{6.92}$$

where w is all external inputs, u is the control inputs, z is the regulated outputs, and y is the measured outputs.

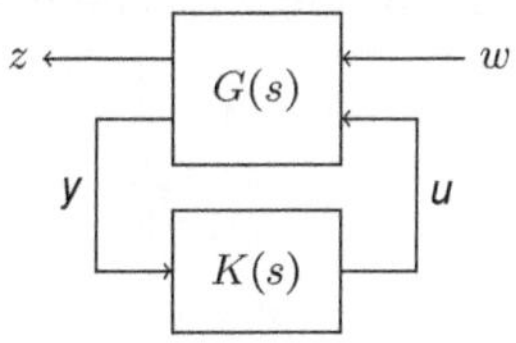

Figure 6.24 LFT.

Then,

$$T_{zw} = F_l(G, K) = G_{11} + G_{12}K(I - G_{22}K)^{-1}G_{21}$$

- LFT is a useful way to standardize block diagrams for robust control analysis and design
- $F_l(G, K)$ is the transfer function between the regulated signals and external inputs. In H_∞ control problems the objective is to minimize $\| F_l(G, K)\|_\infty$

Show the robust performance problem, as shown in Figure 6.25, has an equivalent infinity norm to the LFT shown above. That is, show that

$$\| F_l(G, K)\|_\infty = \left\| \begin{matrix} W_1 S \\ W_2 T \end{matrix} \right\|_\infty . \tag{6.93}$$

6.10 Consider the following transfer function.

$$G(s) = \frac{1 - s}{(2 - s)(1 + s)} \tag{6.94}$$

This system is both unstable and nonminimum phase, so we will have an **analytical constraint** on the maximum value of $\|S(j\omega)\|_\infty$.

$$\max_\omega |S(j\omega)| \geq \left| \frac{z + p}{z - p} \right| = \left| \frac{2 + 1}{2 - 1} \right| = 3 \approx 9.5\ dB \tag{6.95}$$

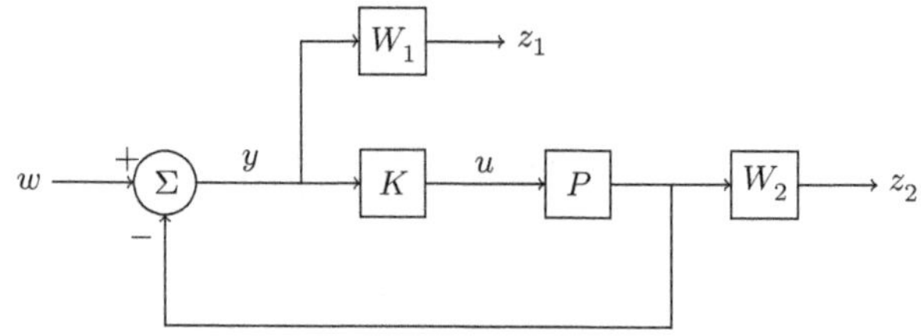

Figure 6.25 Robust performance.

$$\max_{\omega} |T(j\omega)| \geq \left|\frac{p+z}{p-z}\right| = \left|\frac{2+1}{2-1}\right| = 3 \approx 9.5 \text{ dB} \tag{6.96}$$

Use Youla parameterization to design a controller that will stabilize the system and provide command tracking. Provide the appropriate plots for the frequency response and the closed-loop step response. Do the analytical constraints on $||S||_\infty$ and $||T||_\infty$ hold true?

Hints:

(a) In addition to canceling the $(2 - s)$ pole, you will need to add a zero and three poles.
(b) Remember that your only goals here are (i) stability and (ii) command tracking.
(c) Although it is undesirable to have a controller with unstable poles, it may be unavoidable in this case. It is more important to ensure that internal stability is met.

6.11 Repeat Problem 3, but this time use the H_∞ method. Keep in mind the analytical constraints; you will need to make $W_1 \ll 0$ dB at high frequencies to keep the peak value of $|S|$ small. You could try shaping W_1 by approximating the inverse of the sensitivity transfer function from your Youla method controller. A single pole and zero should suffice. To make things easier for the H_∞ design, use the `hinfsyn` command in place of `hinf`. You can also choose the W_2 and W_3 filters to be empty vectors, such as (W_i = [ ]). *Use the uploaded m-file.* Provide the appropriate plots for the frequency response and the closed-loop step response. Do the analytical constraints on $||S||_\infty$ and $||T||_\infty$ hold true? Is there a difference in the form of your two controllers (i.e. number of poles and zeros)?

7

Estimation Design for SISO Using Parameterization Approach

As as been discussed so far, Youla parameterization is a well-established technique in deriving SISO (single-input single-output) and, to a lesser extent, MIMO (multiple-input multiple-output) controllers [1]. However, the utility of this methodology in estimation design, specifically in the framework of controller output observer (COO) [15], is not established. The fundamental question to be answered is as follows: is it possible to design a deterministic estimation technique using Youla parameterization with the same robust performance, or better, than well-established stochastic estimation techniques such as Kalman filtering? To prove this point, at this stage, a comparative analysis between Youla parameterization in estimation and Kalman filtering is performed through simulations only.

In this chapter, we provide an overview of Youla parameterization for estimation design. We develop a deterministic SISO Youla estimation technique in the framework of COO, and we investigate the utility of this method for two applications in the automotive domain.

7.1 Introduction

The COO framework is used for estimating system inputs and states that characterizes a physical system, as seen in Figure 7.1, through the use of other measurable outputs [15]. The estimation of unmeasurable quantities was elegantly introduced through the use of classical observers by Luenberger [16], one of the classical deterministic estimation approaches, as illustrated in Figure 7.2a. In this Figure, A_e, B_e, C_e matrices include the dynamics of the observer/estimator for a strictly proper system with feedthrough matrix D_e assumed to be zero without loss of generality.

This classical observer assumes that all system inputs are known, and hence, there are no distinctions between disturbance and control inputs. The observer gains, L, are computed by pole placement techniques. Information about the sensor dynamic response and its noise content are not considered in the design. These limitations drastically reduce the domain of applicability of these types of observers.

We will illustrate that by utilizing a COO with Youla parameterization, these limitations are directly addressed and that the pragmatic implementation of this method is feasible. In the Youla parameterization method, the closed-loop response is shaped directly and the Youla parameter is used to work backwards to form a controller or in this case, an estimator. (The Youla parameter is the control effort transfer function, relating a system input to the controller output, given that there is output feedback.) Because Youla parameterization allows direct shaping of the closed-loop response, the COO can be designed to be robust to plant uncertainties.

In Figure 7.2b, a simple transformation from the states to the input illustrates the invertability requirement of the input matrix, B_e. We will remove this required invertability by replacing it with a tunable transfer function, $G_c(s)$.

$$LB_e^{-1} \rightarrow G_c(s) \tag{7.1}$$

Now, Figure 7.2b is precisely equivalent to Figure 7.3, where LB_e^{-1} is replaced with $G_c(s)$.

With this type of observer structure, the inputs, as well as states, can be estimated without the need for matrix inversion. In fact, the estimation problem has been transformed to a control design problem and the dynamic controller, G_c, will be designed utilizing the Youla parameterization approach. As stated previously, all plant inputs have to be readily available to the classical observer for proper state estimation. However, there is no such requirement with the new observer structure in Figure 7.3. In fact, the input (from the plant to the observer) acts merely as a disturbance to the COO. It is a requirement

Robust Control: Youla Parameterization Approach, First Edition. Farhad Assadian and Kevin R. Mallon.

Companion Website: www.wiley.com/go/assadian/robust

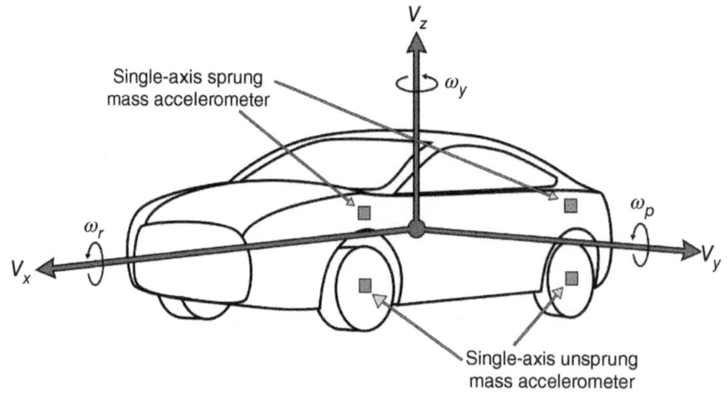

Figure 7.1 Automotive application for controller output observer.

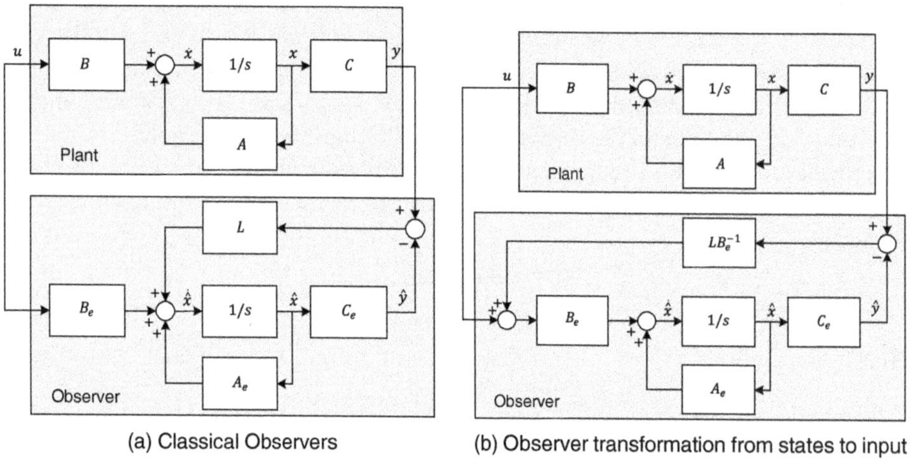

Figure 7.2 Classical observer design. (a) Classical observers. (b) Observer transformation from states to input.

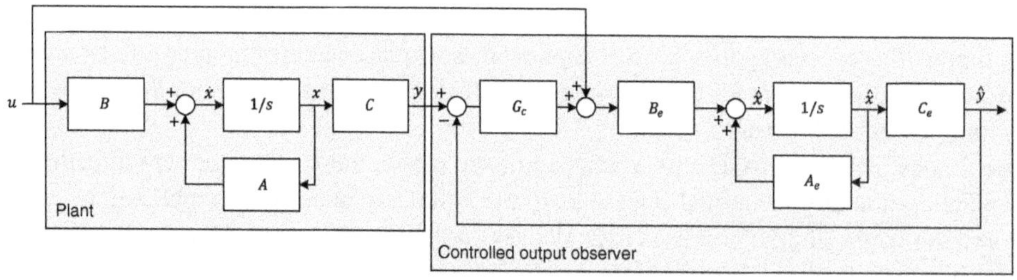

Figure 7.3 General controller output observer.

that this disturbance be properly rejected by the controller such that the observer output follows its input (plant measured output, y) for correct input (controller output, $\hat{u}$) and state estimation.

A final transformation of the COO, from Figure 7.3 to Figure 7.4, illustrates the equivalency of a feedback to a feed-forward controller using the Youla parameterization technique. We name this observer structure YCOO. The design of the feed forward controller using the YCOO structure for SISO systems will be discussed in this chapter, while its application to MIMO systems will be discussed in Chapter 16.

In comparison, a Kalman filter is formulated using the state-space approach, and its main function is to provide an estimate of the system states in the presence of both input and measurement uncertainties assuming zero-mean Gaussian white noise disturbances [17]. The YCOO makes no assumption on the statistics of the disturbances except that they must be bounded in terms of their energy or power, similar to the H_∞ control and filter design. Here lies the fundamental difference

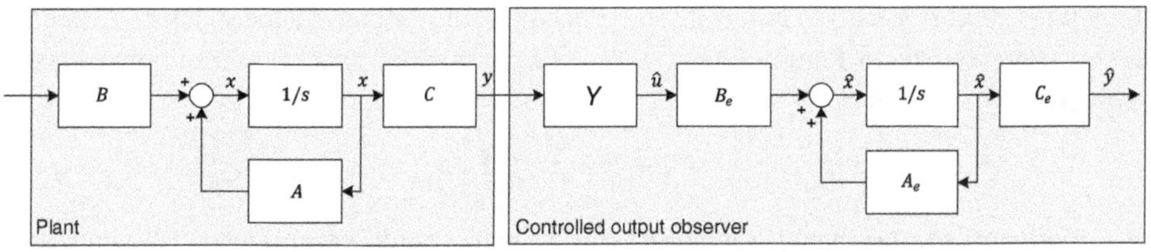

Figure 7.4 Youla controller output observer.

between YCOO and a Kalman filter as the stochastic processes are now transformed to deterministic ones. The procedure to design an estimator for the YCOO is similar and dual to its traditional feedback control loop counterpart.

In this chapter, we discuss the COO and its relation with the Luenberger Observer. We provide a simple SISO example of the COO using the Youla approach, and we compare the results with a discrete Kalman filter. The goal of this chapter is to demonstrate a deterministic Youla parameterization estimation concept utilizing the framework of the COO.

7.2 Youla Controller Output Observer Concept

It has been said that the only reason for using feedback in control design is to reduce the effects of uncertainty [18]. If there were no uncertainties, there would be no need for feedback control. In this sense, Youla parameterization is a powerful tool in compensating for the uncertainty and designing a balanced controller by shaping the closed-loop transfer functions.

When it comes to SISO control design, hand computation of the controller using Youla parameterization provides an insight into the design of a controller. Performance, internal stability, and robustness are the fundamental measures in control design that are enforced using this approach by directly shaping the closed-loop transfer functions [1]. Classical controls address these measures by indirectly shaping the open loop transfer function [19].

In a SISO estimator design, the classical feedback loop (as illustrated in Figure 7.5a) is transformed to an equivalent feedforward loop, Figure 7.5b, to directly shape complementary sensitivity, or closed-loop transfer function, by selecting a stable and proper Youla parameter, where $G_{p,e}(s) = C_e(sI - A_e)^{-1}B_e$, assuming the transfer function for the observer is strictly proper without loss of generality. This model is assumed to be always linear. Equivalently, $G_p(s) = C(sI - A)^{-1}B$ represents the actual linear plant model; however, this model could also be nonlinear. Therefore, we have transformed an estimator design to a neoclassical feedback control design utilizing Youla parameterization. It is important to note that the fundamental difference between designing an estimator using this technique and its dual control design is that in the case of an estimator design, the observer model (estimated model of the plant), $G_{p,e}(s)$, is an integral part of the feedback loop. While, in the case of the control design, this model is discarded and the controller is implemented on the actual plant.

From Figure 7.5b, the relation between Youla, $Y(s)$, and the controller, $G_c(s)$, is derived to be,

$$Y(s) = \frac{G_c(s)}{1 + G_c(s)G_{p,e}(s)} \tag{7.2}$$

or

$$G_c(s) = \frac{Y(s)}{1 - Y(s)G_{p,e}(s)} \tag{7.3}$$

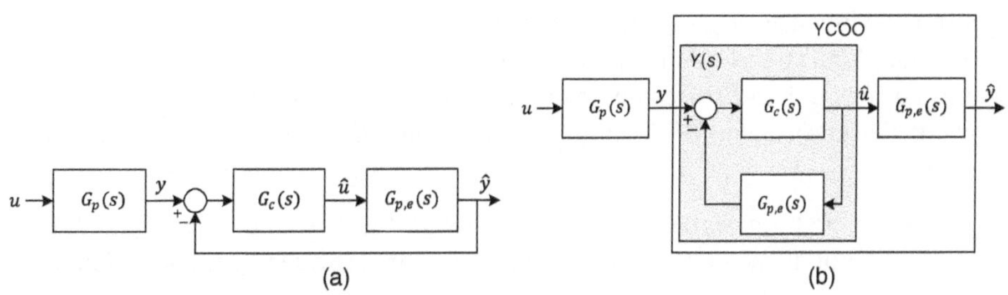

Figure 7.5 Comparison of classical feedback loop and Youla feedforward transformation. (a) Classical feedback loop. (b) Youla feedforward transformation.

For a SISO system, it is straightforward to illustrate the procedure for designing a feedback controller to meet the control loop objectives as stated previously. In Figure 7.5a, the closed-loop transfer function (complementary sensitivity transfer function), $T(s)$, is given by

$$T(s) = \frac{G_c(s)G_{p,e}(s)}{1 + G_c(s)G_{p,e}(s)} \tag{7.4}$$

From the transformation of the feedback to feedforward, by using Youla parameter as shown in Figure 7.5b, we can compute the closed-loop, or the complementary transfer function, as,

$$T(s) = Y(s)G_{p,e}(s) \tag{7.5}$$

Given a particular observer transfer function, $G_{p,e}(s)$, one typically fixes a preferred closed-loop transfer function, $T(s)$, as described later in this chapter, and then computes a proper and stable Youla, $Y(s)$ through Eq. (7.5). If the plant transfer function is stable and, in addition, does not include any non-minimum phase zeros, the internal stability is met by only selecting a proper and stable Youla. In the case of an unstable plant or a plant that includes non-minimum phase zeros [20], the following interpolation conditions must be satisfied to guarantee internal stability.

If "z" denotes a zero of the return ratio (open loop transfer function), L, of multiplicity a_z, and "p" denotes a pole of L of multiplicity a_p, then,

$$T(z) = 0 \qquad S(z) = 1 \tag{7.6}$$

and

$$\frac{d^kT}{ds^k}(z) = 0 = \frac{d^kS}{ds^k}(z) \qquad 1 \le k \le a_{z-1} \tag{7.7}$$

while

$$S(p) = 0 \qquad T(p) = 1 \tag{7.8}$$

and

$$\frac{d^kS}{ds^k}(p) = 0 = \frac{d^kT}{ds^k}(p) \qquad 1 \le k \le a_{p-1} \tag{7.9}$$

where $S(s)$ is the sensitivity transfer function. This transfer function specifies the ratio of plant parameter changes to its nominal parameter values in addition to the ratio of the control loop output to the output disturbance. The following algebraic constraint exists between the sensitivity and the complementary sensitivity transfer functions.

$$T(s) + S(s) = 1 \tag{7.10}$$

The relations, which are provided in Eqs. (7.6)–(7.9), are known as interpolation conditions. Due to these conditions, the closed-loop transfer functions, $S(s)$ and $T(s)$, are constrained to take on certain values in the complex plane, and these conditions influence these values on the imaginary axis, $S(j\omega)$ and $T(j\omega)$. These conditions cannot be removed by a pole–zero cancellation where "z" lies in the right half-plane or on the imaginary axis, where a factor of the return ratio is non-minimum phase. Constraining these transfer functions imposes irrevocable limits to feedback design.

From Eq. (7.3), the resulting controller can be written as

$$G_c(s) = \frac{Y(s)}{S(s)} \tag{7.11}$$

where $S(s) = 1 - Y(s)G_{p,e}(s)$.

The steps for designing a SISO estimator are given as follows:

1. Given an observer transfer function, $G_{p,e}(s)$, select a desired complementary transfer function, $T(s)$.
2. From Eq. (7.5), compute a Youla transfer function, $Y(s)$, which meets the aforementioned interpolation conditions.
3. From the constraint provided in Eq. (7.10), compute the sensitivity transfer function, $S(s)$.
4. Finally, from the expression given in Eq. (7.11), compute the controller, $G_c(s)$, which satisfies all the performance, robustness, and stability criterion.

To summarize, an estimation problem has been converted to a robust COO problem and the guarantee for robust stability and performance, as described in many robust control publications, are provided by the selection of the complementary sensitivity transfer function, $T(s)$, and sensitivity transfer function, $S(s)$ [14]. In addition, the well-posedness and the internal stability of the feedback loop are guaranteed by satisfying the interpolation conditions as discussed above.

It is important to note that the Youla parameterization approach discussed above cannot be casted in the internal model control (IMC) Paradigm [21], as in the case of the IMC, the controller output has to be simultaneously applied to the actual plant and the observer. In our case, as described above, the actual input to the plant merely acts as a disturbance input, as illustrated in Figure 6.10.

7.3 The SISO Case

This section provides a simple example of a linear time invariant (LTI) quarter car model of a vehicle suspension [22]. A comparison between the Kalman filter design and YCOO is made. The SISO Youla controller output observer (YCOO) concept is used to estimate the vertical road input velocity and all four states of a quarter car model while only requiring access to the sprung mass acceleration measurement. These results are then compared to the results from a discrete Kalman filter.

A simple quarter car model, as shown in Figure 7.6, is used to design the YCOO. The suspension stiffness, k_s, tire stiffness, k_t, and damper, b_s, are assumed linear and the masses, m_s and m_{us}, assumed known. A force actuator, F_a, represents the controllable input for semi-active damping or fully active suspension. However, in this example, this force is assumed zero.

The suspension stiffness can be obtained from the relationship between vertical wheel stroke, q_s, and spring force, and is therefore represented by a vertical spring. The passive damping force, F_d, is modified by a lever ratio. That is, the deflection is approximately proportional to (by $\frac{\partial q_s'}{\partial q_s}$), but not equal to, the wheel stroke. This is shown in Figure 7.7.

The plant is represented in the state space form as follows:

$$\dot{x} = Ax + Bu \qquad (7.12)$$
$$y = Cx + Du$$

The state and input matrices (Eqs. (7.13) and (7.14)) are shown below.

$$A = \begin{bmatrix} -b_s/m_s & b_s/m_s & k_s/m_s & 0 \\ b_s/m_{us} & -b_s/m_{us} & -k_s/m_{us} & k_t/m_{us} \\ -1 & 1 & 0 & 0 \\ 0 & -1 & 0 & 0 \end{bmatrix} \qquad (7.13)$$

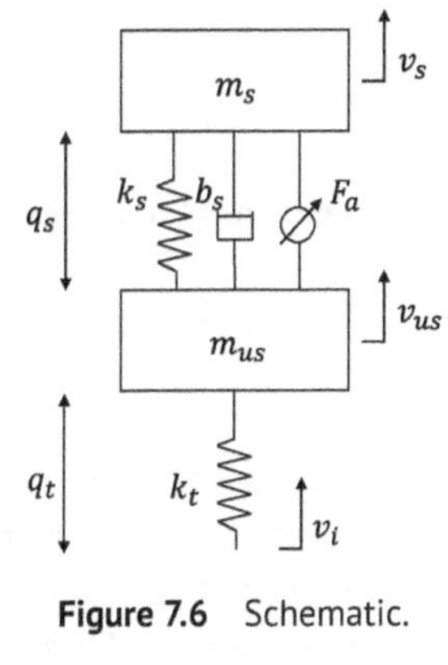

Figure 7.6 Schematic.

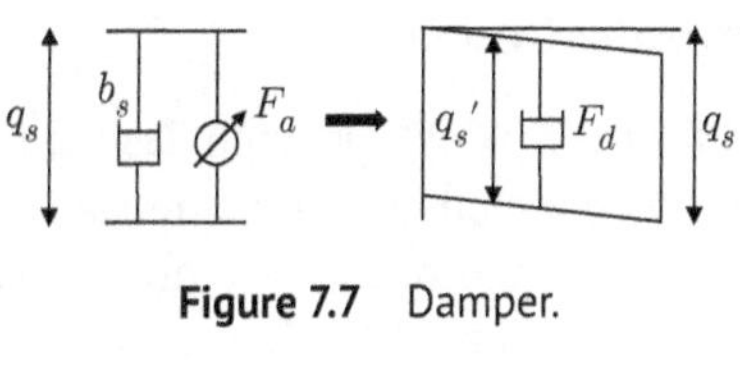

Figure 7.7 Damper.

The states, as depicted in Figure 7.7, are sprung mass velocity, v_s, unsprung mass velocity, v_{us}, suspension deflection, q_s, and tire vertical deflection, q_t.

$$B = \begin{bmatrix} 1/m_s & 0 \\ 1/m_{us} & 0 \\ 0 & 0 \\ 0 & 1 \end{bmatrix} \tag{7.14}$$

The two inputs are road velocity disturbance, v_i, and force input, F_a. Note that a physics-based modeling technique based on power is used – as a result, inputs to the system include velocities but not displacements.

7.3.1 Output and Feedthrough Matrices

The C and D matrices are shown in Eqs. (7.15) and (7.16) as the combination of individual output and feedthrough matrices. These individual output matrices are shown in Eqs. (7.17)–(7.24). The outputs, as provided by these matrices, are the sprung mass acceleration, the unsprung mass acceleration, the damper force, and the relative velocity between the sprung and unsprung mass.

$$C = \begin{bmatrix} C_{\dot{v}_s} & C_{\dot{v}_{us}} & C_{F_{de}} & C_{v_{rel}} \end{bmatrix}^T \tag{7.15}$$

$$D = \begin{bmatrix} D_{\dot{v}_s} & D_{\dot{v}_{us}} & D_{F_{de}} & D_{v_{rel}} \end{bmatrix}^T \tag{7.16}$$

We can obtain the following outputs by properly modifying the general C and D matrices as follows. Note that in Eqs. (7.17)–(7.20), we use the notation $M_{i,*}$ to indicate the entire ith row of a matrix M. Then,

- **Sprung Mass Acceleration**:

$$C_{\dot{v}_s} = A_{1,*} \tag{7.17}$$

$$D_{\dot{v}_s} = B_{1,*} \tag{7.18}$$

- **Unsprung Mass Acceleration**:

$$C_{\dot{v}_{us}} = A_{2,*} \tag{7.19}$$

$$D_{\dot{v}_{us}} = B_{2,*} \tag{7.20}$$

- **Damping Force**:

$$C_{F_{de}} = \left(\frac{\partial q_s'}{\partial q_s}\right)^{-1} \begin{bmatrix} -b_s & b_s & 0 & 0 \end{bmatrix} \tag{7.21}$$

$$D_{F_{de}} = \left(\frac{\partial q_s'}{\partial q_s}\right)^{-1} \begin{bmatrix} 1 & 0 \end{bmatrix} \tag{7.22}$$

- **Relative Velocity**:

$$C_{v_{rel}} = \begin{bmatrix} -1 & 1 & 0 & 0 \end{bmatrix} \tag{7.23}$$

$$D_{v_{rel}} = \begin{bmatrix} 0 & 0 \end{bmatrix} \tag{7.24}$$

It is important to note that the actual plant and the observer models in this example are assumed to be equal, such that $G_p = G_{p,e}, A = A_e, B = B_e, C = C_e$, and $D = D_e$.

7.3.2 SISO Estimator Design

The model used for the SISO YCOO includes the road velocity disturbance as an input and the sprung mass acceleration as an output. With this configuration, the states, inputs, and outputs are given by $\hat{x}$, $\hat{u}$, and $\hat{y}$, respectively, and are defined below. This method does not require access to the force input, F_a, to properly estimate the states and other unknown disturbance inputs. Referring to Figure 7.4, the input to the YCOO is the measured sprung mass acceleration including sensor noise, y_m (y in Figure 7.4), and the output of the YCOO is the road velocity disturbance, $\hat{v}_i$ ($\hat{u}$ in Figure 7.4). In the SISO configuration, the YCOO simultaneously estimates the road velocity disturbance input and the suspension states by only having access to the measured sprung mass acceleration. In the MIMO configuration presented later, road velocity disturbance and the force inputs, and the system states, are estimated while only having access to the measured sprung mass and unsprung mass accelerations. On the other hand, in the SISO configuration, the discrete Kalman filter has to have access to both measured noisy sprung mass acceleration and the force input, F_a, for proper estimation of the suspension states. Classical configuration of the Kalman filter does not lend itself to input estimation.

The plant input, output, and the states for this example are provided below.

$$\hat{x} = \begin{bmatrix} \hat{v}_s & \hat{v}_{us} & \hat{q}_s & \hat{q}_t \end{bmatrix}^T \tag{7.25}$$

$$\hat{u} = \hat{v}_i \tag{7.26}$$

$$\hat{y} = \dot{\hat{v}}_s \tag{7.27}$$

Below, we briefly outline the design of the SISO YCOO based on the design steps as provided above. It should be noted that the design of the discrete Kalman filter has been a topic of many papers [23] and is not discussed in this work. In designing the Kalman filter, the road disturbance is treated as process noise, and sensor noise is modeled using specifications of a known accelerometer. The selected values for the process noise and the sensor noise are not tuned or changed for different simulations. It should be noted that in practical applications, it is the discrete version of the Kalman filter which is always utilized. In the case of the YCOO, the discretization of the controller has no effect on its performance given a reasonable sampling rate. This is why the continuous version of the YCOO is included in the presented results.

The plant transfer function from the road velocity disturbance input to the sprung mass acceleration is derived to be

$$G_p(s) = \frac{(\beta_2 s^2 + \beta_1 s)}{(s^2 + \alpha_{22}s + \alpha_{21})(s^2 + \alpha_{21}s + \alpha_{11})} \tag{7.28}$$

$$= sG_p(s)^* \tag{7.29}$$

Asterisks (*) represent the portion of the plant with no zeros on the $j\omega$-axis or right-half plane, while αs and βs correspond to system natural frequencies and damping.

We select the Youla transfer function to be stable and proper based on the procedures given before, and furthermore, the zero of the plant is carried over to complementary/closed-loop transfer function for meeting the interpolation conditions (internal stability of the feedback loop), where this condition is given by

$$T(z) = 0 \tag{7.30}$$

where z is the zero of the plant on the $j\omega$-axis ($z = 0$).

We have the option to cancel all the stable poles of the plant and shape the closed-loop transfer function to closely match the second-order Butterworth filter. A Butterworth filter is chosen as it is an optimal, well damped filter with desirable response characteristics. This filter contains ζ and ω_n below and is tuned for good noise rejection. Additionally, a pole, p_1, is required for reasons that will be explained in the following paragraphs. Note that Youla must be proper. The resulting Youla transfer function is selected as follows:

$$Y(s) = \frac{\omega_n^2}{(s + p_1)(s^2 + 2\zeta\omega_n s + \omega_n^2)G_p(s)^*} \tag{7.31}$$

With this choice of Youla transfer function, the complementary transfer function is computed to be,

$$T(s) = Y(s)G_p(s) \tag{7.32}$$

Which results in the following:

$$T(s) = \frac{s\omega_n^2}{(s + p_1)(s^2 + 2\zeta\omega_n s + \omega_n^2)} \tag{7.33}$$

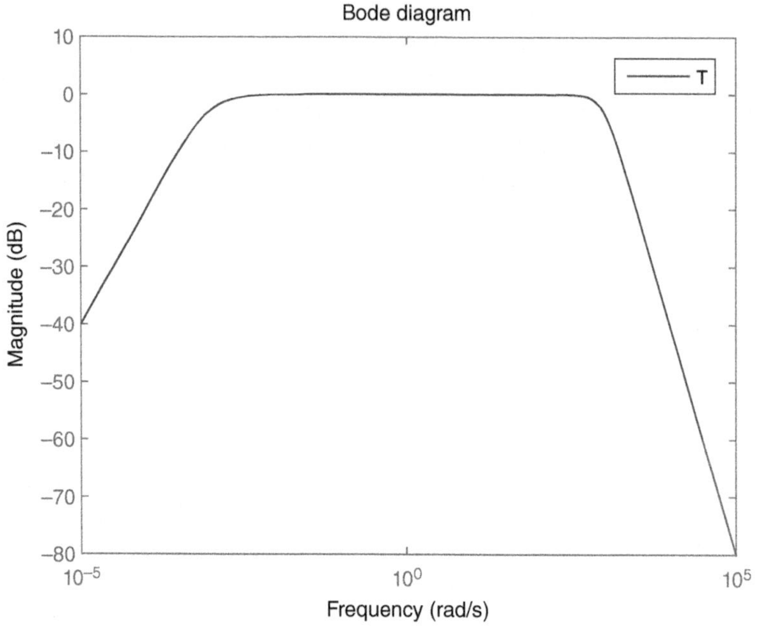

Figure 7.8 Closed-loop frequency response.

The $j\omega$-axis zero of the plant forces the complementary transfer function to be zero at low frequency to meet the interpolation conditions and therefore to have an internally stable feedback loop. This results in deterioration of the target following/estimation at low frequency as the closed-loop transfer function is required to be unity at low frequencies. The pole $(s + p_1)$ in the denominator of the Youla transfer function is added to force the closed-loop transfer function, $T(s)$, to act as a bandpass filter as shown in Figure 7.8. Due to selecting this pole to be at a very low frequency, the impact on the target following capability of the closed-loop transfer function is minimized.

The Youla transfer function is directly implemented using the configuration of Figure 7.4 to estimate the system states. Given that the models used for estimation are similar to the original plant model both Youla and Kalman estimation methods performed well. This is not shown in this chapter.

In order to reveal the robustness characteristics of this approach, we increased the observer model sprung mass by 50% relative to the actual plant model. The results for two states, wheel stroke and unsprung mass velocity, are depicted in Figure 7.9 for a random road input with a specific power spectral density of a typical D-road (farm road).

In addition, the RMS values of the unsprung mass velocities, for the actual plant, the Youla and the Kalman filter estimators, are provided in Table 7.1.

From Figure 7.9 and the results for the RMS values of unsprung mass velocities for the actual plant, and the Youla and the Kalman filter estimators provided in Table 7.1, we can conclude that the Youla estimation method is more robust than the Kalman filter.

It should also be noted that the Kalman filter has access to the input force and the noisy sprung mass acceleration signal while the Youla approach only has access to the noisy sprung mass acceleration signal. In addition, although not shown, the road disturbance input is also estimated by the Youla approach without any modifications to the base approach.

As the final comparative analysis for the SISO case, using the same quarter car model, Figure 7.10 presents the results for similar state estimates but this time with a prescribed speed bump as a road input. A similar conclusion can be drawn from these results.

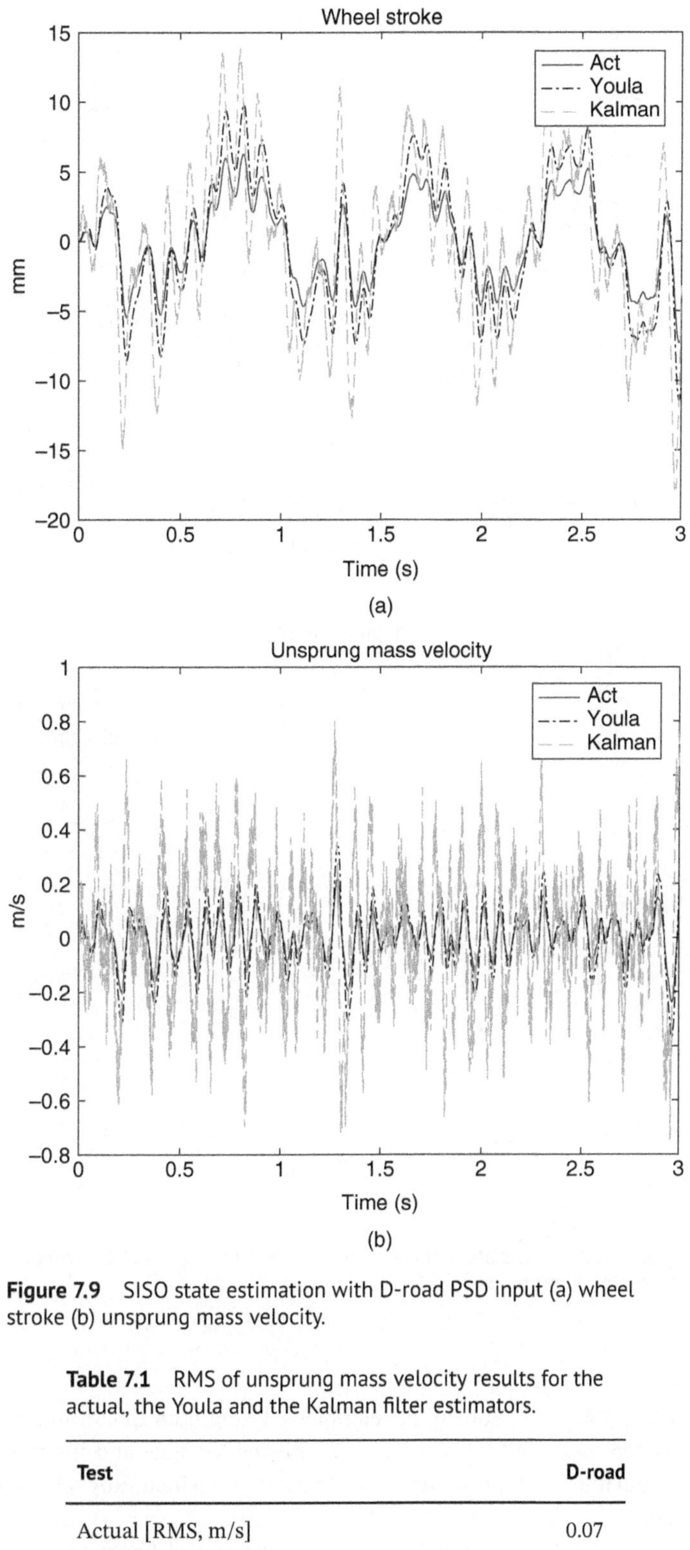

Figure 7.9 SISO state estimation with D-road PSD input (a) wheel stroke (b) unsprung mass velocity.

Table 7.1 RMS of unsprung mass velocity results for the actual, the Youla and the Kalman filter estimators.

Test	D-road
Actual [RMS, m/s]	0.07
Youla [RMS, m/s]	0.1
Kalman filter [RMS, m/s]	0.6

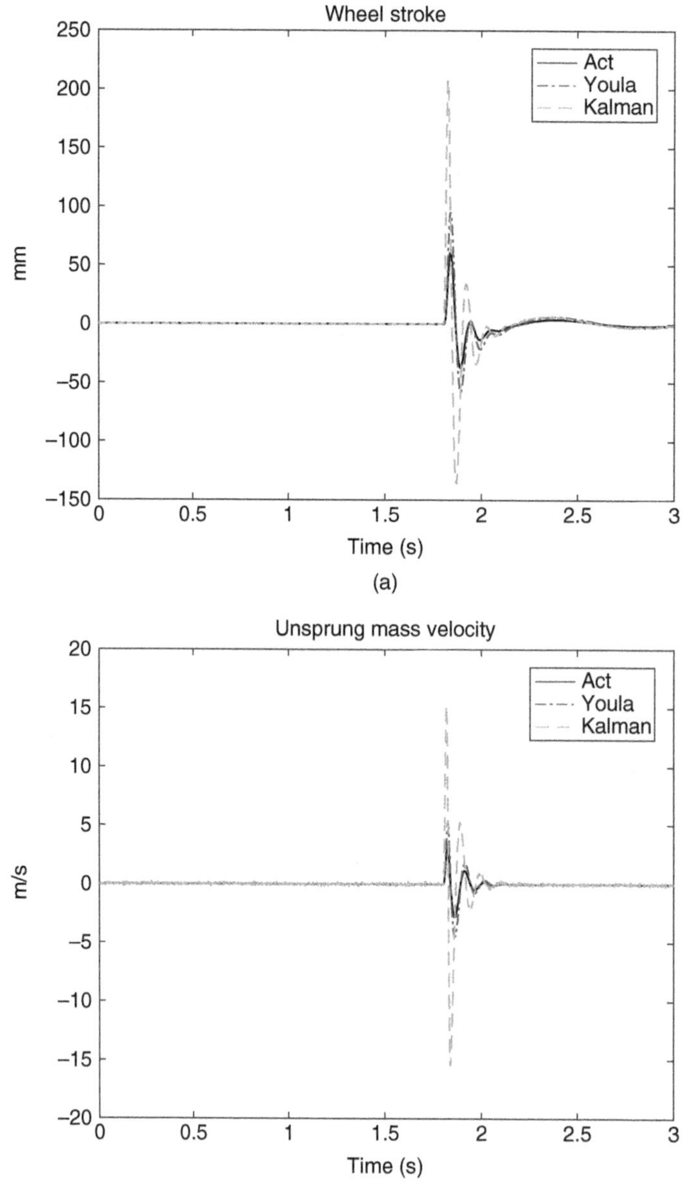

(a)

(b)

Figure 7.10 SISO state estimation with a speed bump input (a) Wheel stroke (b) Unsprung mass velocity.

7.4 Final Remarks

Youla parameterization has a long history in control development. Using both linear and nonlinear vehicle suspension models, it was shown that this approach could be equivalently utilized for state and input estimation. The comparative results between YCOO and Kalman filter were presented in the linear domain including robustness characteristics of YCOO versus Kalman filtering method.

The second volume of this book will introduce and develop the application of Youla parameterization to MIMO systems. Then in Chapter 16, we will then return to this system to develop MIMO state estimation using the YCOO approach.

8

Practical Applications

In this chapter, we present three practical applications for which we investigate SISO Youla control development using hand computation, SISO H_∞ control design, and finally, Youla loop-shaping control technique using internal model control (IMC). These applications include yaw stability control (YSC) development using active limited slip differential (ALSD), anti-lock braking system (ABS) control development using electric motors, and vehicle side-slip estimation using Youla controlled output observer and we compare the results with both LKF and non-LKF methods.

8.1 Yaw Stability Control with Active Limited Slip Differential

YSC systems have been and are addressed using various actuators in the automotive industry. The majority of current yaw stability systems, such as dynamic stability controller (DSC), utilize brake interventions at individual wheels to develop yaw moments to ensure vehicle stability at times when it would otherwise compromised. However, the focus of the premium vehicle manufacturers is not just to improve stability but also to increase driver enjoyment while driving both below and at the limits of adhesion. Brake-based stability systems tend to be somewhat intrusive, causing decelerations and hence, loss of vehicle speed. Alternative actuation systems are therefore being considered that may provide increased stability, without the intrusiveness of brake-based systems.

There are different versions of an active differential, here, an electric motor driven ALSD is considered, with a wet multi-disk friction clutch effectively providing a connection between the two driveshafts, as illustrated in Figure 8.1. By controlling the clamping force on the clutch, the torque transfer can be modulated. At maximum capacity, it is capable of providing 2500 Nm locking torque, which is equivalent to 1250 Nm of torque transfer from one driven wheel to the other.

One of the key limitations of an ALSD is that only control over the magnitude of torque transfer is possible. The direction of the torque transfer is determined by the wheel speed difference as torque is always transferred from faster to the slower wheel. As a result, the actuator is limited, in most cases, to generating more relative understeer and therefore, it is only capable of stabilizing an oversteering vehicle, as shown in Figure 8.2.

8.1.1 Model and Control Design

In this section, we describe the vehicle model (which is utilized for the control development), our simplified assumption for the actuator model, and our control development strategies including their comparison.

To reveal the basics of the vehicle dynamics with steady-state cornering and where the longitudinal velocity is assumed to be constant, one can use the so-called "linear bicycle model." This model is composed of one front and one rear tire linked by a rigid body, as illustrated in Figure 8.3. In this representation, the front and rear axles are squashed together. This model represents the yaw dynamics of a vehicle and its simplicity lends it to use for control design purposes.

The linear bicycle model includes two degrees of freedom. The degrees of freedom are yaw and side-slip velocity. In state-space form, this model may be written as:

$$\begin{bmatrix} \dot{V} \\ \dot{r} \end{bmatrix} = \begin{bmatrix} Y_V/M & Y_r/M - U \\ N_V/I_{zz} & N_r/I_{zz} \end{bmatrix} \begin{bmatrix} V \\ r \end{bmatrix} + \begin{bmatrix} 0 \\ 1/I_{zz} \end{bmatrix} [T_{yaw}] + \begin{bmatrix} Y_\delta/M \\ N_\delta/I_{zz} \end{bmatrix} [\delta] \tag{8.1}$$

or

$$\dot{X} = AX + Bu + Gw \tag{8.2}$$

Robust Control: Youla Parameterization Approach, First Edition. Farhad Assadian and Kevin R. Mallon.

Companion Website: www.wiley.com/go/assadian/robust

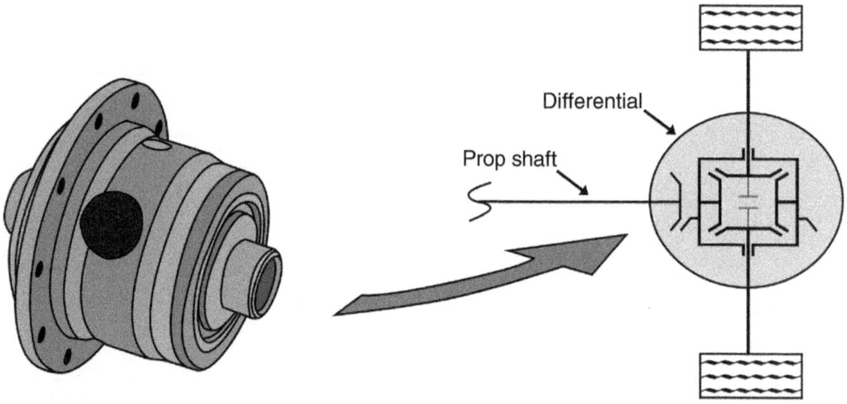

Figure 8.1 E-locker.

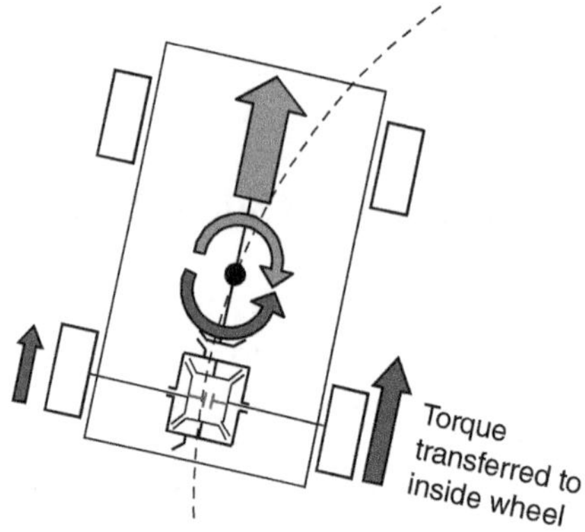

Figure 8.2 A schematic of the torque transfer capability and limitation of the E-locker.

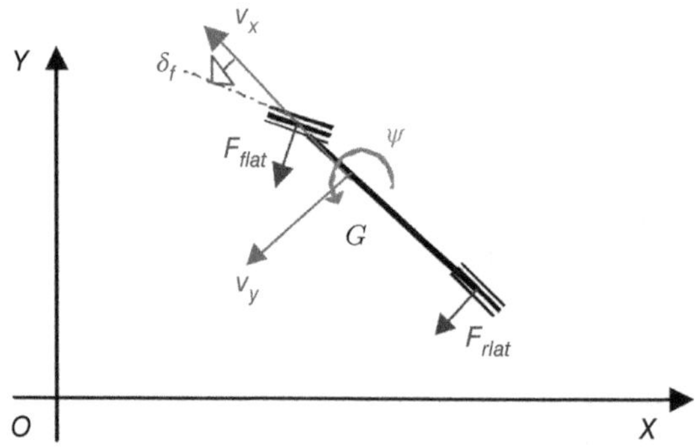

Figure 8.3 Bicycle model.

where

$$Y_V = \frac{-C_{\alpha f} - C_{\alpha r}}{U}, \quad Y_r = \frac{-aC_{\alpha f} + bC_{\alpha r}}{U}, \quad Y_\delta = C_{\alpha f}$$

$$N_V = \frac{-aC_{\alpha f} + bC_{\alpha r}}{U}, \quad N_r = \frac{-a^2C_{\alpha f} - b^2C_{\alpha r}}{U}, \quad N_\delta = aC_{\alpha f}$$

where M is the vehicle mass, I_{zz} is the mass moment of inertia around z-axis, $C_{\alpha f}$ is the front tire cornering stiffness, $C_{\alpha r}$ is the rear tire cornering stiffness, U is the longitudinal velocity, a is the distance from Center of Gravity (CG) to the front axle, and b is the distance from CG to the rear axle.

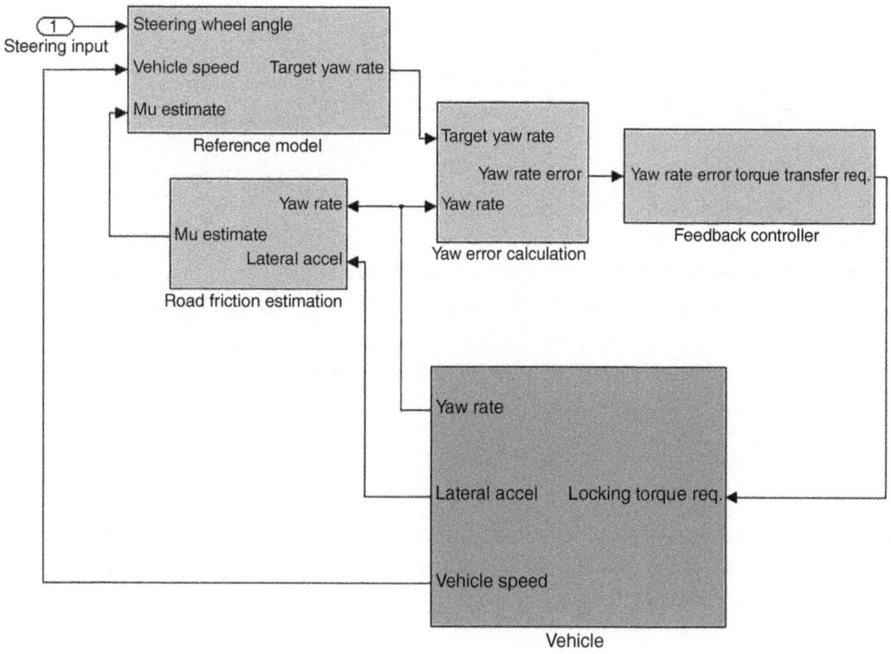

Figure 8.4 Controller structure.

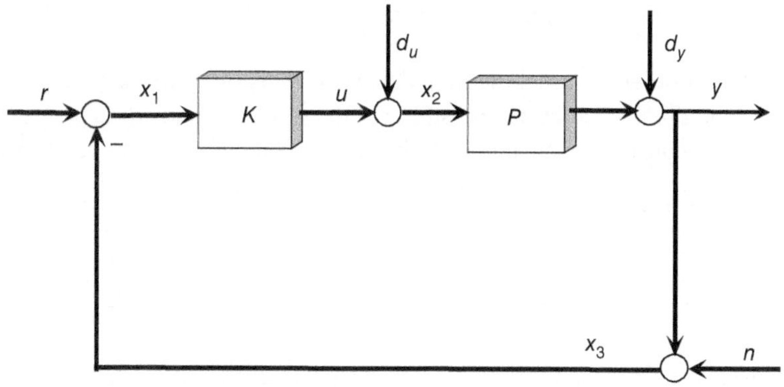

Figure 8.5 Feedback control loop.

The general structure of the controller implemented onboard vehicle is shown in Figure 8.4. To design the feedback controller using Youla parameterization and the techniques discussed in the previous chapters, we will utilize the feedback control loop shown in (Figure 8.5).

The feedback control design will control the yaw rate r by affecting the yaw torque T_{yaw}. From the above state-space representation, we can write the transfer function as:

$$\frac{r}{T_{yaw}} = \frac{\frac{1}{I_{zz}}\left(s + \frac{C_{af}+C_{ar}}{MU}\right)}{s^2 + D_1 s + D_0} \tag{8.3}$$

such that

$$D_1 = \frac{C_{af} + C_{ar}}{MU} + \frac{a^2 C_{af} + b^2 C_{ar}}{I_{zz} U} \tag{8.4}$$

$$D_1 = \frac{C_{af} + C_{ar}}{MU} + \frac{a^2 C_{af} + b^2 C_{ar}}{I_{zz} U} \tag{8.5}$$

$$D_0 = \left(\frac{-aC_{af} + bC_{ar}}{I_{zz}} + \frac{C_{af} C_{ar} (a+b)^2}{I_{zz} M U^2}\right) \tag{8.6}$$

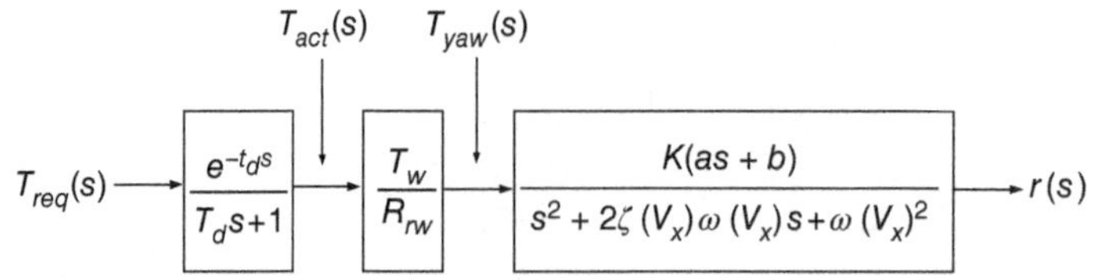

Figure 8.6 A third order plant model.

Additionally, we will consider actuator dynamics. Then, the feedback control design is performed after having obtained the plant transfer function from the actuator input (the locking torque request) to the vehicle yaw rate as illustrated in Figure 8.6.

Where t_d is the actuator model pure time delay, T_d is the actuator model first-order time constant, T_w is track width, and R_{rw} is the wheel radius. This means we have actuator dynamics:

$$\frac{T_{yaw}}{T_{req}} = \frac{T_w}{R_{rw}I_{zz}} \cdot \frac{e^{-t_d s}}{T_d s + 1} \tag{8.7}$$

The vehicle model from above can be rewritten as the second-order denominator where parameters ζ and ω are functions of vehicle forward velocity V_x. For the control design, V_x is assumed to be constant. Then, we have the complete transfer function from input torque request to output yaw rate.

$$\frac{r}{T_{req}} = \frac{T_w}{R_{rw}I_{zz}} \times \frac{e^{-t_d s}}{T_d s + 1} \times \frac{\left(s + \frac{C_{\alpha f} + C_{\alpha r}}{MU}\right)}{s^2 + \left(\frac{Caf + Car}{MU} + \frac{a^2C_{\alpha f} + b^2C_{\alpha r}}{IzzU}\right)s + \left(\frac{-aC_{\alpha f} + bC_{\alpha r}}{Izz} + \frac{C_{\alpha f}C_{\alpha r}(a+b)^2}{IzzMU^2}\right)} \tag{8.8}$$

It should be pointed out that in order to model the time delay associated with this actuator (electric motor and clutch plates), a pure time delay, t_d, and the first-order low-pass filter with a time constant T_d is utilized as shown in Figure 8.6. The pure time delay and the time constant associated with the actuator model are obtained from vehicle measurement as depicted in Figure 8.7.

```
%% Vehicle Data:
mf = 994.5260; % Total mass at front axle (kg)
mr = 882.1203; % Total mass at rear axle (kg)
i_zz = 3552;   % Yaw moment of inertia ($kg m^2$)
```

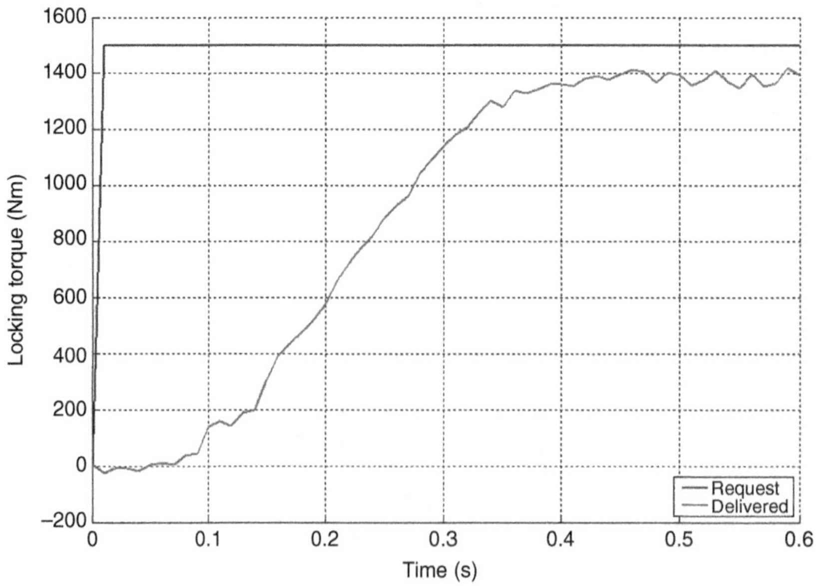

Figure 8.7 Actuator model.

```
l = 2.909;          % Wheelbase (m)
rr = 0.329;         % Rolling radius (m)
trr = 1.53;         % Front and Rear track (m)

% Understeer Gradient (rad/g):
K_under = (1.8*pi/180); % measured UG

% Cornering Stiffnesses:
Car = 1.5679e+005; % Rear (N/rad)
Caf = (mf*9.81) / (K_under + (mr*9.81/Car)); % Front (N/rad)

%% VEHICLE SPEED (m/s):
V_x = 25;

%% PRELIMINARY CALCULATIONS:
g = 9.81; % Acceleration due to gravity (m/s^2)
m = mf + mr;   % Total mass (kg)

% Weight distribution
a = mr*l/(mf+mr); % Distance from CG to front axle (m)
b = mf*l/(mf+mr); % Distance from CG to rear axle  (m)
```

The pure delay t_d is approximately 100 ms, while the time constant T_d is assumed to be 167 ms. Given the parameters above, the transfer function from the locking torque request $T_{req}(s)$ to the yaw rate $r(s)$ is given by:

$$\frac{r(s)}{T_{req}(s)} = \frac{-s+20}{s+20}\frac{1}{0.167s+1}\frac{4.65(0.000\,28s+0.0016)}{s^2+12.31s+60.6} \tag{8.9}$$

where the pure delay handled with a first-order Padè approximation.

Given the plant transfer function above, we derive three different SISO controllers using the techniques discussed in the previous chapters:

- Youla control design using hand computation
- H_∞ control design using loop-shaping technique.

8.1.2 Youla Control Design Using Hand Computation

The Youla transfer function is selected so that the closed-loop complementary transfer function would approach

$$T = \frac{\omega_n^2}{s^2+2\zeta\omega_n s+\omega_n^2} \tag{8.10}$$

where $\omega_n^2 = 60.6$ and $2\zeta\omega_n = 12.31$.

However, knowing that there is a pure delay, therefore, a non-minimum phase zero, and the Youla transfer function cannot be selected to cancel these non-minimum phase zero, the Youla transfer function is modified to be

$$Y = K\frac{\alpha s+\beta}{T_d} \tag{8.11}$$

where $K = 4.65$, $\alpha = 0.000\,28$, $\beta = 0.0016$, and $T_d = 1/0.167s + 1$.

This selection of Youla transfer function gives the following closed-loop complementary transfer function:

$$T = \frac{-s+\gamma_1}{s+\gamma_2}\frac{\omega_n^2}{s^2+2\zeta\omega_n s+\omega_n^2} \tag{8.12}$$

where $\gamma_1 = \gamma_2 = 20$ (it should be clear that $\gamma_2 \leq \gamma_1$ to limit the bandwidth due to the non-minimum phase zero), and ω_n and $\zeta\omega_n$ are the same as given above.

The following m-file is used to design this controller:

```
s=tf('s');

% Youla hand computation method

% Declaring plant model
```

```
td=(-s+20)/(s+20);
Td=1/(0.167*s+1);
G= (4.65*(0.000\,28*s+0.0016))/(s^2+12.31*s+60.6);

Gp= td*Td*G;

% Selecting Youla parameter
Y=(1/Td)*(60.6/(4.65*(0.000 28*s+0.0016)))*(1/(1/50*s+1)^2);

% Computing closed-loop transfer functions
T=minreal(Y*Gp);
S=minreal(1-T);
Gc=(Y/S);
Gc=minreal(Gc);

% Plotting results
figure(1)
bode(T);
hold on
bode(S);
bode(Y);
legend('T','S','Y')
grid on

figure(2)
step(Y)
legend('Actuator')
grid on


figure(3)
step(T)
legend('Yaw rate response')
grid on
```

Two high-frequency poles are added to the Youla transfer function to better shape its frequency response at high frequency as can be seen in the m-file above.

Figure 8.8 illustrates the yaw rate response to a locking torque step request. The non-minimum phase zero behavior is evident from this response. It should be clear that to control the non-minimum phase behavior, the constant γ_2 could have been taken to be much lower than γ_1. In this case, the impact of the non-minimum phase zero would be minimized while it decreases the closed-loop transfer function bandwidth; hence, it increases the rise time and settling time.

The locking torque capacity for this actuator is highly nonlinear and it depends on variables such as clutch plate and road coefficient of frictions, wheel speed difference, locking, and clutch plate temperature to name a few. This actuator is capable of providing a peak locking torque of 2500 Nm. Therefore, it is interesting to investigate the impact of the Youla linear control design on the actuator effort.

It should be noted that the longitudinal velocity for this linear model is assumed to be 25 m/s. Figure 8.9 shows that, given this longitudinal velocity, for good target following, the actuator requires almost more than three times its peak capacity at steady state. This means at high-speed (25 m/s or greater) stability maneuvers, the clutch pack of this actuator fully locks the two half shafts.

8.1.3 H_∞ Control Design Using Loop-shaping Technique

The loop-shaping approach as discussed in the previous chapters requires augmentation of the nominal plant by shaping filters, as illustrated in Figure 6.7. The filters that are used for the nominal plant augmentation are given as follows:

$$W_P = \frac{K_{W_P}\omega_n^2}{s^2 + 2\zeta\omega_n s + \omega_n^2} \tag{8.13}$$

where $K_{W_P} = 15$, $\omega_n = 0.1$, and $\zeta = 3.0$.

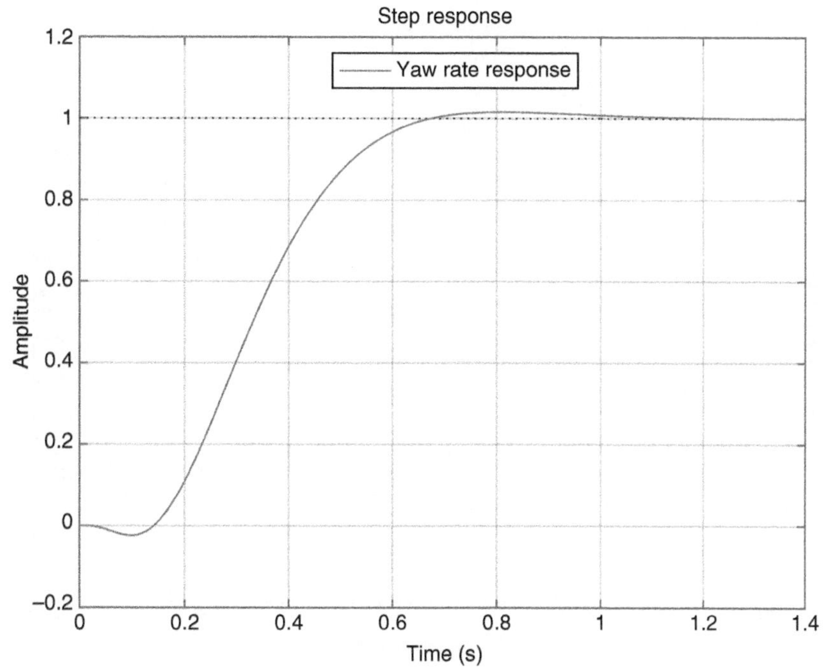

Figure 8.8 Yaw rate response to the locking torque step input request.

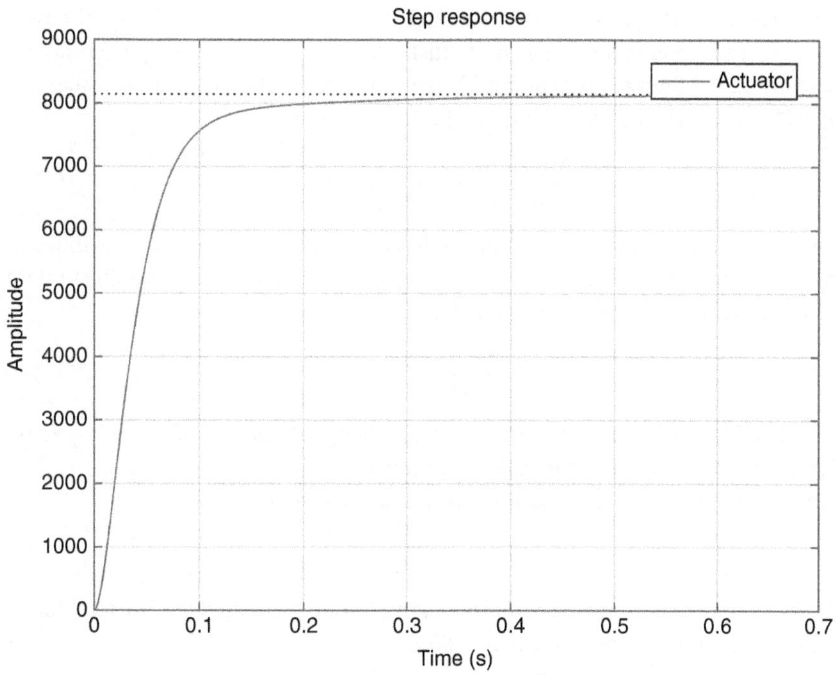

Figure 8.9 Actuator effort to the locking torque step input request.

As pointed out in Chapter 6, this filter weights the sensitivity transfer function. The shaping filter, W_Δ, for the closed-loop complementary transfer function is given as follows:

$$W_\Delta = \frac{K_{W_\Delta} s}{s + \frac{K_{W_\Delta}}{T_m ax}} \tag{8.14}$$

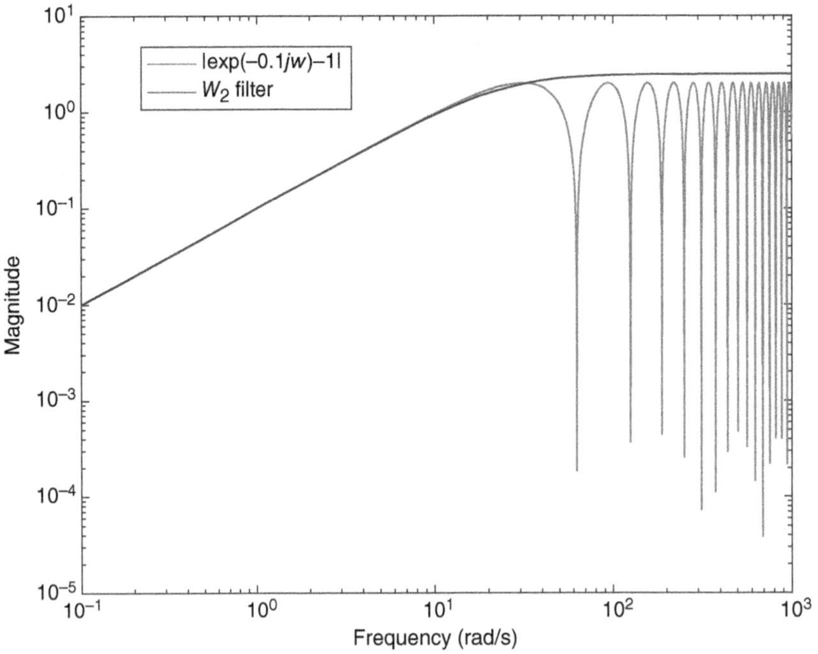

Figure 8.10 Bode plot of $|e^{-j0.1\omega} - 1|$ and W_Δ (W_2) filter.

This filter is used as weighting on the complementary transfer function so that the closed-loop stability is guaranteed in the presence of the pure time delay of T_{max} due to the actuator dynamics. In order to design the weighting filter W_Δ with parameter K_{W_Δ} and $T_{max} = 0.1$ seconds that will ensure robustness in the presence of a pure time delay ≤ 0.1 seconds, we utilize multiplicative perturbation.[1] Suppose that nominal plant transfer function is P and consider perturbed plant transfer function of the form $\hat{P} = (1 + \Delta W_\Delta)P$. Then knowing that $\|\Delta\|_\infty \leq 1$, then

$$\left|\frac{\hat{P}(j\omega)}{P(j\omega)} - 1\right| \leq |W_\Delta(j\omega)| \qquad \forall\, \omega \tag{8.15}$$

So $|W_\Delta(j\omega)|$ provides the uncertainty profile. This inequality describes a disk in the complex plane, where at each frequency the point $\hat{P}/P$ lies in the disk with center 1 and radius of $|W_\Delta|$. Typically, $|W_\Delta(j\omega)|$ is a increasing function of ω as uncertainty increases with increasing frequency. For the case of pure delay,

$$\frac{\hat{P}(j\omega)}{P(j\omega)} = e^{-jT_{max}\omega} \tag{8.16}$$

Therefore, the filter should be designed so that $|e^{-jT_{max}\omega} - 1| \leq |W_\Delta(j\omega)|$. A Bode plot of $|e^{-jT_{max}\omega} - 1|$ and $|W_\Delta(j\omega)|$ illustrates that the designed weighting filter satisfies the condition in Figure 8.10 when $K_{W_\Delta} = 2.5$,

The final weighting filter is used to limit the actuator effort and is given by:

$$W_u = \frac{1}{\rho} \tag{8.17}$$

where ρ is a constant and is equivalent to the maximum locking torque generated by the actuator (in this case, based on our previous analysis, this constant is taken to be equal to 8000).

The following m-file is used to design this controller:

```
s=zpk('s');

% Hinf method

% Declaring plant model
td=(-s+20)/(s+20);
```

1 We have shown in the previous chapters that multiplicative perturbation is equivalent to weighting the nominal complementary transfer function – small-gain theorem.

```
Td=1/(0.167*s+1);
G= Td*(4.65*(0.000\,28*s+0.0016))/(s^2+12.31*s+60.6);

Gp= td*Td*G;

% Shaping Filters
wn=0.10;
zeta=3.0;
Kw1=15.0;
Kw3=2.5;
Tmax=0.1;
W1=(Kw1*wn^2)/((s^2+2*zeta*wn*s+wn^2));
W2=1/8000; W3=(Kw3*s)/(s+Kw3/Tmax);

% Plant augmentation with Filters
P=augw(G,W1,W2,W3);
% Controller computation
[K,CL,GAM]=hinfsyn(P);
%
K=minreal(K);
Gc=minreal(tf(K));

% Computing closed-loop transfer functions
Yhinf=minreal(Gc/(1+Gc*Gp));
Thinf=minreal(Gp*Yhinf);
Shinf=minreal(1-Thinf);

% Plotting results
figure(1)
bode(Thinf);
hold on
bode(Shinf);
hold off
legend('sensitivity','complementary sensitivity')
grid on

figure(2)
bode(Gp)
hold on
bode(Yhinf)
bode(Gc)
hold off
legend('Plant','Youla','Gc')
grid on

figure(3)
step(Thinf)
legend('Yaw rate response ,Hinf')
grid on

figure(4)
bode(Shinf)
hold on
bode(W1)
bode(1/W1)
hold off
legend('sensitivity','Wp','1/Wp')
grid on

figure(5)
step(Yhinf)
legend('Actuator effort, Hinf' )
grid on
```

The yaw rate response to a locking torque step request is shown in Figure 8.11.

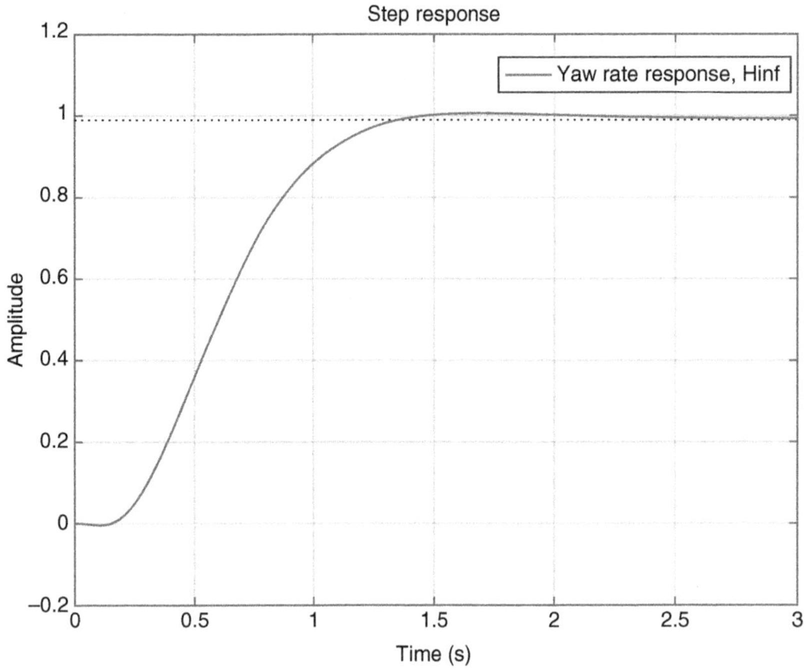

Figure 8.11 Yaw rate response to the locking torque step input request with Hinf controller.

Comparing the yaw rate responses of the H_∞ controller with the Youla hand computation method, it is obvious the results of the H_∞ controller is more conservative with respect to dealing with the pure time delay and response time. In addition, this response is more oscillatory than the yaw response from Youla hand computation results. Furthermore, Youla hand computation technique did not require any shaping filters and tuning these filters is not necessarily a straightforward task.

It would be informative to compare the actuator efforts of the H_∞ method with the Youla hand computation method.

Comparing the actuator effort results of the H_∞ controller with the Youla hand computation, it can be observed that the steady-state values are similar but the transient behaviors (rise time) are different. The response time for H_∞ is slower with slightly higher overshoot. This demonstrates that H_∞ method is conservative as it sacrifices performance (response time) for robust stability. It would be interesting to tune the Youla hand computation actuator effort transient behavior so that it is similar to H_∞ method, then compare yaw rate responses. However, before addressing this, we present the results of one final control method as discussed in Chapter 6; this method utilizes the structure of IMC for designing a Youla controller using H_∞ optimization. This will be presented next.

In Chapter 6, we discussed the method of designing a Youla controller with model matching using IMC structure. This method has significant benefits over loop-shaping technique as it does not require shaping filters. Furthermore, a Youla controller is computed directly using an optimization method rather than hand computation and only a target model is needed for this optimization. According to our previous discussion, this target model is selected to be

$$T_t = \frac{-s+\gamma_1}{s+\gamma_2}\frac{\omega_n^2}{s^2+2\zeta\omega_n s+\omega_n^2} \tag{8.18}$$

We have selected these constants to be the same as above.

In Figure 8.13, we have added the result of Youla optimization actuator effort to Figure 8.12. It is interesting to observe that the actuator effort response of the Youla optimization method differs from Youla hand computation technique. It should be noted that we included two high- frequency poles when selecting the Youla transfer function to better control the high-frequency response of the Youla parameter at high frequency when we hand-computed the Youla parameter. These two high-frequency poles had an impact on the closed-loop transfer function of the Youla controller by hand computation. However, when selecting the target model for the Youla optimization method, we did ignore these high-frequency poles. We will add these high-frequency poles to the target model and we will also limit the bandwidth of both Youla methodologies to the H_∞ method so that the actuator effort transient responses for all three strategies would be similar. Then we compare the results of these methods.

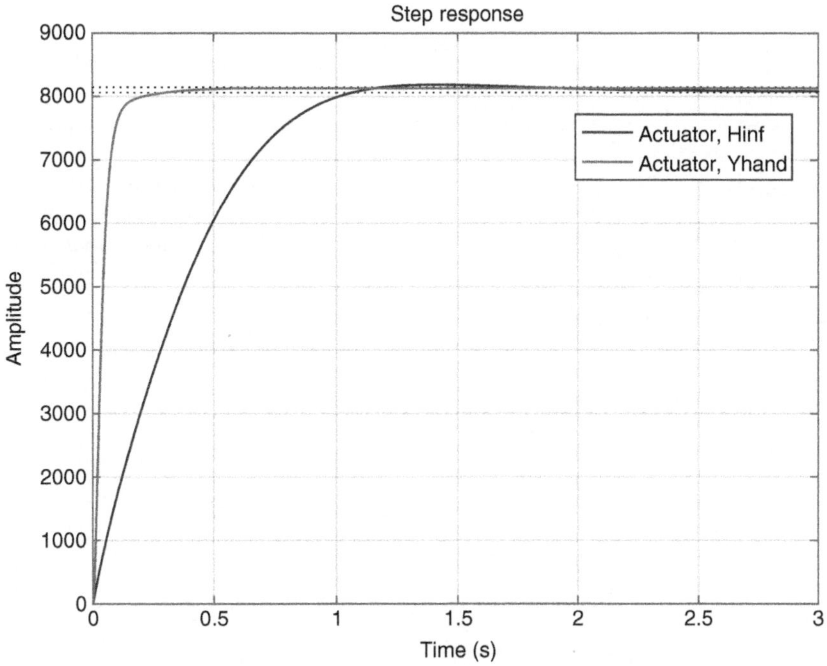

Figure 8.12 Actuator effort to the locking torque step input request comparison.

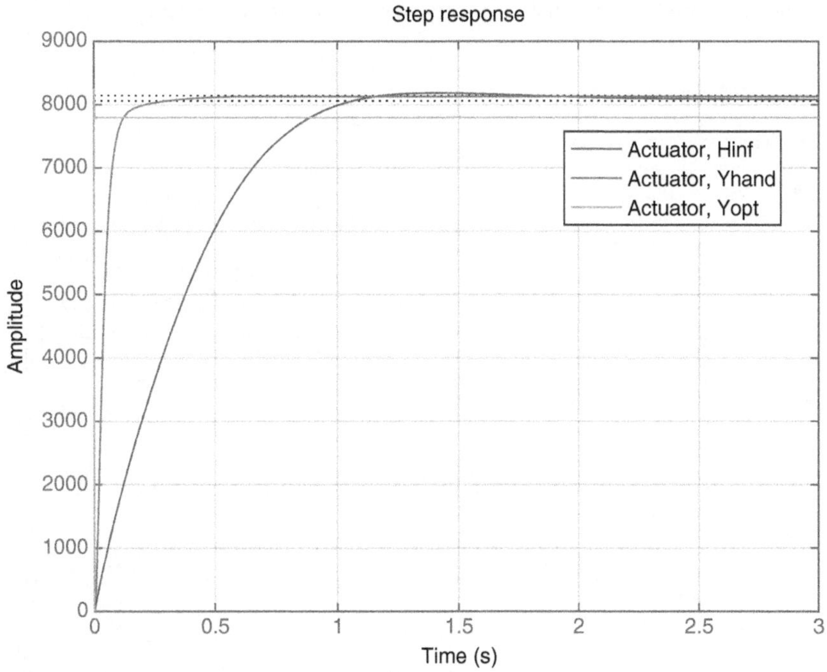

Figure 8.13 Actuator effort to the locking torque step input request comparison of all the proposed methods.

Let us assume a target model for the Youla optimization method as follows:

$$T_t = \left(\frac{\gamma_2}{\gamma_1}\right) \frac{-s+\gamma_1}{s+\gamma_2} \frac{\omega_n^2}{s^2+2\zeta\omega_n s+\omega_n^2} \frac{1}{(\alpha s+1)^2} \tag{8.19}$$

The constant γ_1 is the same as before but this time to limit the bandwidth $\gamma_2 = 3$. In addition, the constant $\alpha = 1/50$. The rest of the constant is the same as before. The ratio $\frac{\gamma_2}{\gamma_1}$ forces the closed-loop transfer function target to be one at low

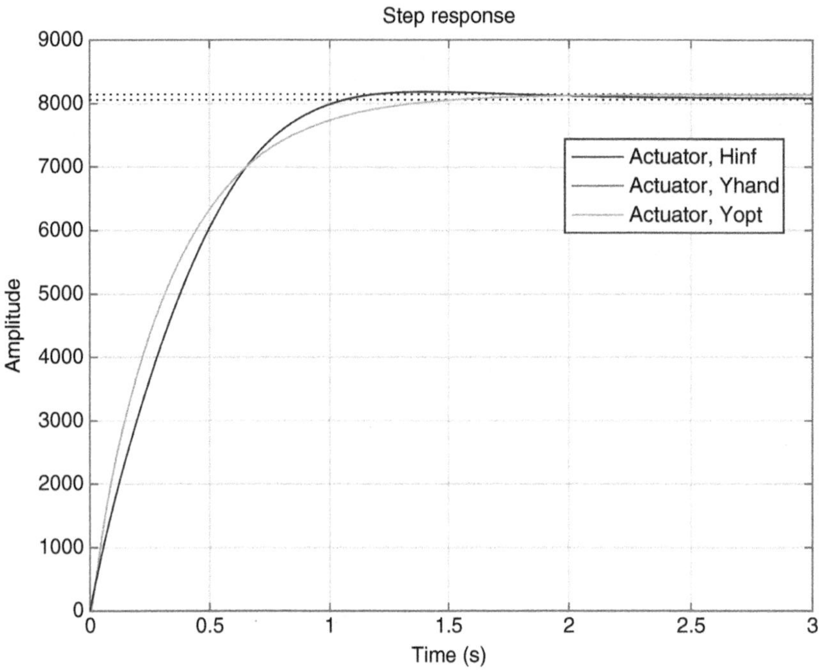

Figure 8.14 Actuator effort response to the locking torque request comparison of all the proposed methods with high-frequency poles and limited bandwidth.

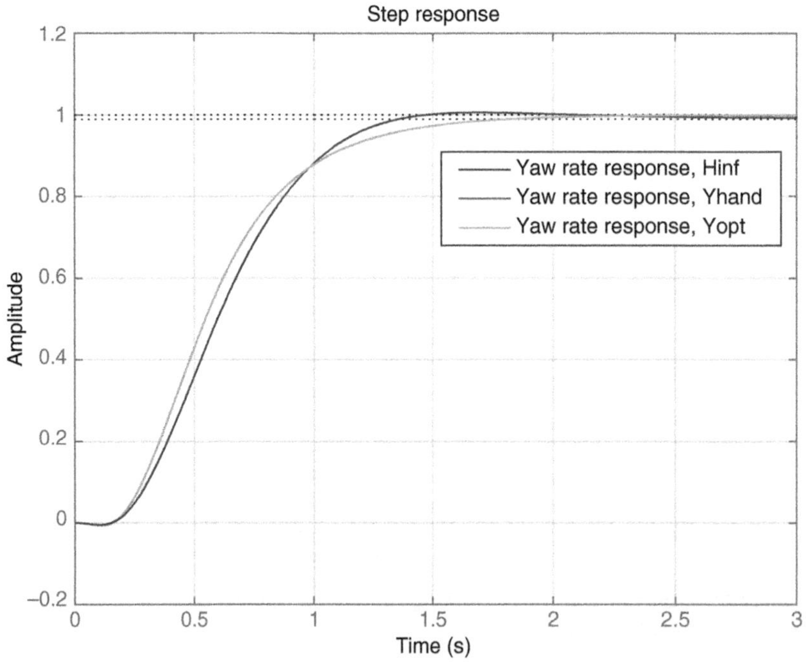

Figure 8.15 Yaw rate response to the locking torque request comparison of all the proposed methods with high-frequency poles and limited bandwidth.

frequency or DC. The Youla transfer function for hand computation is selected so that the closed-loop complementary transfer function to be precisely the same as this target model.

The actuator effort results are illustrated in Figure 8.14. The results for Youla hand computation and Youla optimization method are the same. This should be no surprise as we tried to force the closed-loop complementary transfer functions for both methods to be the same. We also limited the bandwidth so that the actuator effort for the H_∞ optimization would be similar to the Youla methods.

By observing Figure 8.15, two things are evident:

- Youla control design using hand computation perform exactly the same as Youla control design using IMC structure with H_∞ optimization.
- By limiting the bandwidth of Youla control design techniques, we can obtain very similar results to H_∞ loop-shaping control method.

These control methodologies could be implemented onboard vehicle for YSC with ALSD.

8.2 Vehicle Yaw Rate and Side-Slip Estimation

In this section, we will utilize the development of Chapter 7 in a vehicle dynamics application for yaw rate and side-slip estimation.

In the previous years, there are numerous articles for which many experts have attempted to estimate yaw rate and side-slip, e.g. see [24]. Particularly, vehicle side-slip estimation has important benefits YSC applications. The standard estimation techniques which have been pursued in the past to estimate these variables were nonlinear and unscented Kalman filtering techniques. Although the authors have shown successful estimation of these variables using these approaches, there is no discussion about robustness characteristics or calibration overhead, two most important implementation requirements, using these standard techniques.

Here, we will utilize Youla controller output observer (YCOO) as discussed in Chapter 7 to estimate vehicle yaw rate and side-slip. The YCOO turns out to have lower calibration overhead with better robustness characteristics in terms of many important vehicle dynamics parameters as will be investigated in this example.

Figure 8.16 illustrates a block diagram of COO with the acceleration measurement. The measured lateral acceleration including sensor noise, n, is used as a feedback loop reference target, and the output consists of lateral acceleration from state-space representation of the plant model.

8.2.1 Kalman Filters

(Linear) Kalman filtering (LKF) is an estimation technique which utilizes a statistical approach to estimate the mean and co-variance of the states. A plant model and sensor measurements are used to find the best estimation of the states. However, LKF uses linear state-space representation as a model; therefore, there are some limitations in the applications where plants are highly nonlinear or sensor measurements are of poor quality. For these applications, other types of Kalman filters are suitable, such as extended Kalman filter (EKF) and unscented Kalman filter (UKF). A linearization of the system equations at each time step is required for EKF and the linearized Jacobian matrices are utilized to estimate the variance matrix. This

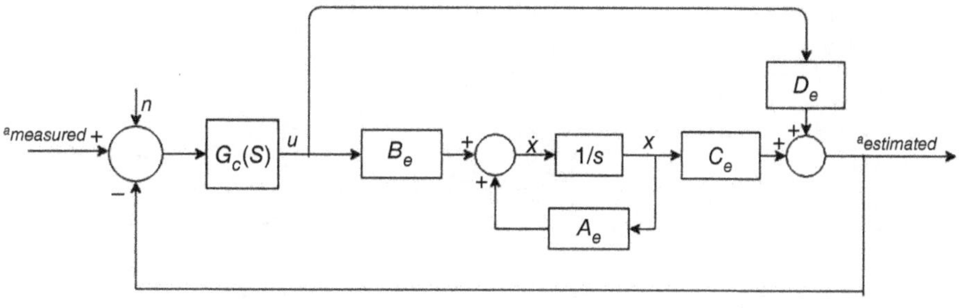

Figure 8.16 Block diagram of controller output observer (COO) using acceleration measurement.

technique is widely used in state, as well as certain parameter estimation. However, since the Jacobian matrices are required to be computed at each time step, the computational overhead can be high, especially for systems with large number of states. UKF uses unscented transformation to sample the statistics associated with the plant model and sensor uncertainties to better represent the mean and the co-variance at each time step without the need for linearization. Therefore, UKF requires less computational overhead than EKF and it also has wide applications.

In this example, we will make a comparison of SISO, YCOO, LKF, EKF, and UKF, mainly in terms of their robustness characteristics, in estimation of vehicle yaw rate and side-slip. However, before designing the estimators, we require a dynamic model of the vehicle and that is discussed next.

8.2.2 Vehicle Model – Nonlinear Bicycle Model with Pacejka Tire Model

One way to estimate the yaw rate is to use the lateral acceleration measurement at the vehicle center of gravity (c.g.) and using the dynamic relationship between the yaw rate and lateral acceleration. For a front steering vehicle, the system diagram of a simple vehicle dynamic model, using a two degrees of freedom bicycle model, is shown in Figure 8.17. Where U is the forward velocity, which is assumed to be constant. Parameters m and J are vehicle mass and moment of inertia. The distances from the vehicle c.g. to the front and rear axles are, respectively, shown as a and b. The front and rear tire lateral forces are indicated by F_f and F_r, respectively. The front wheel steering angle input is illustrated by δ. The model includes two states: the lateral velocity V and the yaw rate ω.

With body-fixed coordinates, the equations of motion of the bicycle model are written as:

$$\dot{V} = \frac{F_f \cos\delta + F_r}{m} - U\omega \tag{8.20}$$

$$\dot{\omega} = \frac{F_f \cos\delta \cdot a + F_r \cdot b}{J} \tag{8.21}$$

The lateral forces exerted on front and rear tires, F_f and F_r, are the functions of the slip angles of front and rear tires. The slip angles of front and rear are related to the system input and the states:

$$\alpha_f = \delta - \tan^{-1}\left(\frac{V + a\omega}{U}\right) \tag{8.22}$$

$$\alpha_r = \tan^{-1}\left(\frac{b\omega - V}{U}\right) \tag{8.23}$$

The tire slip angles result in generating lateral tire forces, and their relationship has been studied, for many decades, for developing an appropriate mathematical tire model. There are many tire models which relate the tire slip angles to the lateral tire forces. One commonly used and popular model is the Pacejka tire model, also known as the "Magic Formula" tire model. From [25], the basic formulas for lateral forces are

$$F_f = D_f \cdot \sin\left(C \cdot \arctan\left(B_f \cdot \left((1-E)\cdot\alpha_f + \frac{E}{B_f}\cdot\arctan(B_f\alpha_f)\right)\right)\right) \tag{8.24}$$

$$F_r = D_r \cdot \sin\left(C \cdot \arctan\left(B_r \cdot \left((1-E)\cdot\alpha_r + \frac{E}{B_r}\cdot\arctan(B_r\alpha_r)\right)\right)\right) \tag{8.25}$$

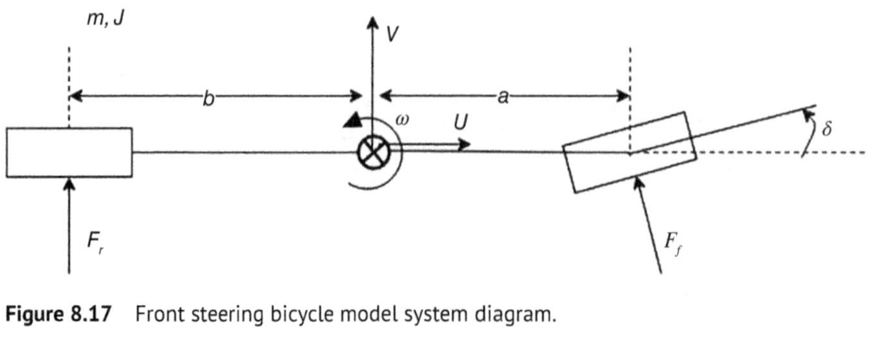

Figure 8.17 Front steering bicycle model system diagram.

where B, C, D, and E are various tire parameters, such as B is the stiffness factor, C is the shape factor, D is the peak factor which is related to the tire normal force (F_z) and tire–road friction coefficient (μ), and E is the curvature factor. These coefficients are extracted from various tire testing and measurement data. We have utilized a typical tire data in this chapter.

Given a steering angle input and system state initial conditions, tire slip angles are computed from Equations (8.24) and (8.25), where the relationship between slip angles and the system input and the states is defined. Knowing the tire slip angles, lateral tire forces are computed from the Pacejka tire model. Having the lateral tire forces, the equations of motion for lateral velocity and yaw rate are solved to obtain solutions of the new states. This cycle is then repeated at each time step.

The lateral acceleration at c.g. is related to the front and rear tire forces and is given by:

$$a_y = \frac{F_f \cos\delta + F_r}{m} \tag{8.26}$$

This SISO model then consists of one input, the front road wheel angle, δ, and one output, the vehicle lateral acceleration at the c.g., a_y.

This nonlinear model is used to represent the actual vehicle for testing the robustness and performance of the proposed estimators, while the linearized version of this model is used for developing YCOO and LKF estimators.

8.2.3 Linearizing the Bicycle Model

Both YCOO and LKF require a linearized model of the plant. Therefore, we need to linearize the vehicle and tire models to obtain a state-space representation of the plant to be used for YCOO and LKF estimators. For linearizing a bicycle model, all angles were assumed to be small, and the state equations are rewritten as:

$$\dot{V} = \frac{F_f + F_r}{m} - U\omega \tag{8.27}$$

$$\dot{\omega} = \frac{F_f \cdot a + F_r \cdot b}{J} \tag{8.28}$$

For linearizing a tire model, it is assumed that the lateral tire forces are proportional to slip angles with front and rear cornering stiffness coefficients C_f and C_r:

$$F_f = C_f \alpha_f \tag{8.29}$$

$$F_r = C_r \alpha_r \tag{8.30}$$

Then the state-space representation of the equations of motion is written as:

$$\begin{pmatrix} \dot{V} \\ \dot{\omega} \end{pmatrix} = \begin{pmatrix} -\dfrac{C_f + C_r}{mU} & -\left(\dfrac{aC_f - bC_r}{mU} + U\right) \\ -\dfrac{aC_f - bC_r}{JU} & -\dfrac{a^2C_f + b^2C_r}{JU} \end{pmatrix} \begin{pmatrix} V \\ \omega \end{pmatrix} + \begin{pmatrix} \dfrac{C_f}{m} \\ \dfrac{aC_f}{J} \end{pmatrix} \delta \tag{8.31}$$

The lateral acceleration at c.g. a_y can be expressed as:

$$a_y = \begin{pmatrix} -\dfrac{C_f + C_r}{mU} & -\dfrac{aC_f - bC_r}{mU} \end{pmatrix} \begin{pmatrix} V \\ \omega \end{pmatrix} + \left(\frac{C_f}{m}\right) \delta \tag{8.32}$$

8.2.4 Uncertainties

The acceleration sensor measurements in the simulation environment are made from a nonlinear model, a representation of an actual vehicle, as stated previously. While, the YCOO and LKF utilize a linear model to estimate the states of this nonlinear plant. Therefore, for these estimators, model uncertainties include the differences between linear and nonlinear plants used for state estimation and acceleration measurements, respectively. In addition, to this inherent uncertainty, we will examine parameter and environmental uncertainties, mass and tire–road coefficient of friction variations, for all estimator approaches.

For simulating sensor noise, data sheet of a single-axis accelerometer SCA830-D05 [26] is utilized. From the data sheet, we can assume that the sensor noise has a zero mean and a variance of $R = 2.5 \cdot 10^{-3}$.

8.2.5 State Estimation

In the state estimation for the SISO case, a front wheel steering vehicle with the front road wheel angle and one accelerometer at the center of gravity of the vehicle for acceleration measurement are utilized to estimate the vehicle yaw rate as shown in Figure 8.17. The input, in this case, is the front wheel steering angle (δ), and the output is the lateral acceleration at c.g. (a_y).

8.2.6 Youla Parameterization Estimator Design

The transfer function of the plant is the ratio between lateral acceleration and the road wheel angle:

$$G_p(s) = \frac{a_y}{\delta}(s) \tag{8.33}$$

There is no unstable poles and zeros, based on the Youla parameterization technique in [27], the Youla parameter can be computed by:[2]

$$Y = \frac{1}{Gp} \cdot T \tag{8.34}$$

where T is the desired closed-loop transfer function, and we select this desired transfer function to be the second-order well-damped filter with two additional poles at $s = 1/\tau$ for better shaping the Youla parameter at high frequency:

$$T = \frac{{\omega_n}^2}{(s^2 + 2\zeta\omega_n s + {\omega_n}^2)(\tau s + 1)^2} \tag{8.35}$$

Then the Youla parameter is written as:

$$Y = \frac{1}{Gp} \cdot \frac{{\omega_n}^2}{(s^2 + 2\zeta\omega_n s + {\omega_n}^2)(\tau s + 1)^2} \tag{8.36}$$

The sensitivity transfer function S of the feedback loop can be derived using the algebraic constraint between this transfer function and the closed-loop transfer function, T:

$$S = 1 - T \tag{8.37}$$

and finally the controller G_c to be used for state estimation using YCOO can be calculated by:

$$G_c = Y/S \tag{8.38}$$

With proper choice of ζ and ω_n, the controller, G_c, is designed so that the closed-loop transfer function, T, is at 0 dB at low frequency which indicates good target following capability, and the sensitivity transfer function, S, is small at low frequency which indicates good disturbance rejection capability. In selecting a bandwidth for the closed-loop transfer function, T, a trade-off between the transient response and the overshoot should be made.

Figure 8.18 shows the bode plot of the linear plant used in YCOO. The plant has high gain of 40 dB at lower frequency and the gain drops after frequency of 1 rad/s and reach the minimum at 20 dB around 8 rad/s. If the bandwidth of the controller/estimator is greater than 1 rad/sec, the system will have an overshoot due to the change in plant gain.

As shown in Figure 8.19, T is at 0 dB for a wide range of frequency. For the case of yaw rate estimation, due to designing a closed loop with large bandwidth, the yaw rate transient response will include a large overshoot. This overshoot can only be reduced or eliminated by reducing the closed-loop bandwidth. Regarding the bandwidth requirement/constraint, a set of objective requirements for the estimation results should be decided to reduce the subjectivity of our findings. For our comparative analysis in this chapter, we will use the results of the YCOO with the highest bandwidth.

It is important to point out that the bandwidth constraint in both control and estimation design has a significant importance. This constraint is directly addressed in the YCOO technique while in all Kalman filtering strategies, the bandwidth is dealt with as an afterthought.

2 It should be noted if the plant included unstable poles or non-minimum phase zeros, the interpolation conditions could be used to handle these types of plants.

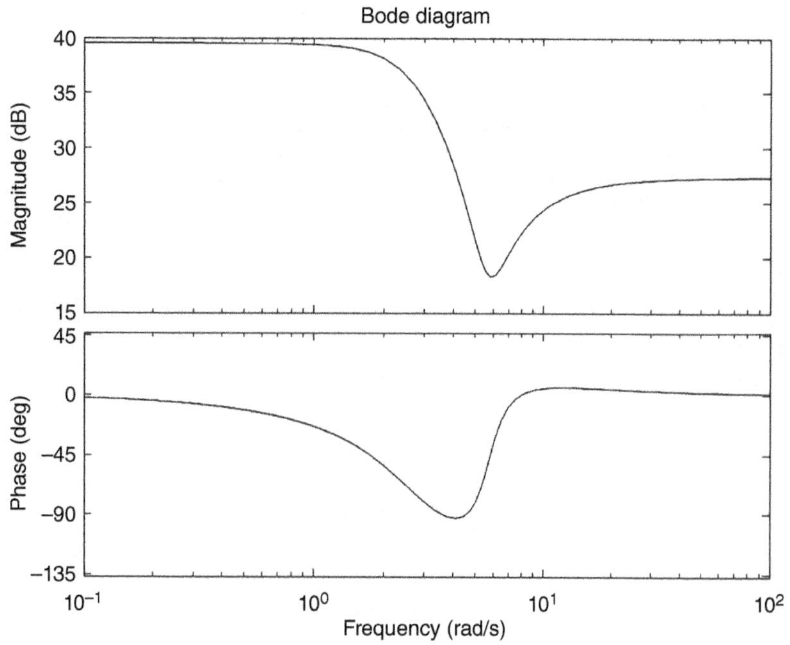

Figure 8.18 Bode plot of linear plant.

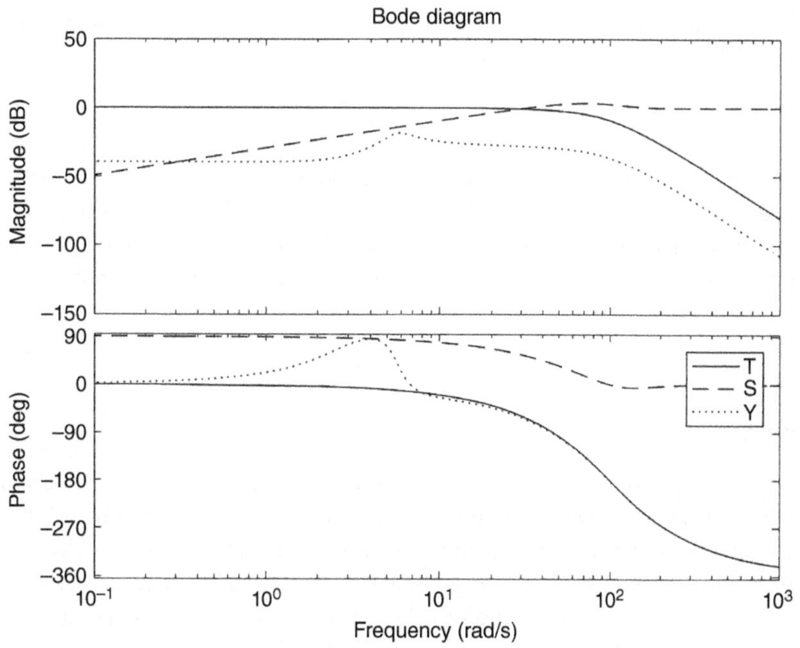

Figure 8.19 Bode plot for SISO case with large bandwidth.

Another way to look at the design is to look at the bode plot of the closed-loop transfer function of estimating the yaw rate T_ω. The transfer function of the yaw rate ω to the wheel angle input δ is

$$\begin{aligned} G_{p_\omega} &= \frac{\omega}{\delta}(s) \\ &= \frac{\dfrac{aC_f}{J}s + \dfrac{(a+b)C_fC_r}{mJU}}{s^2 + \left(\dfrac{C_f+C_r}{mU} + \dfrac{a^2C_f+b^2C_r}{JU}\right)s + \dfrac{(a+b)^2C_fC_r}{mJU^2} + \dfrac{bC_r - aC_f}{J}} \end{aligned} \tag{8.39}$$

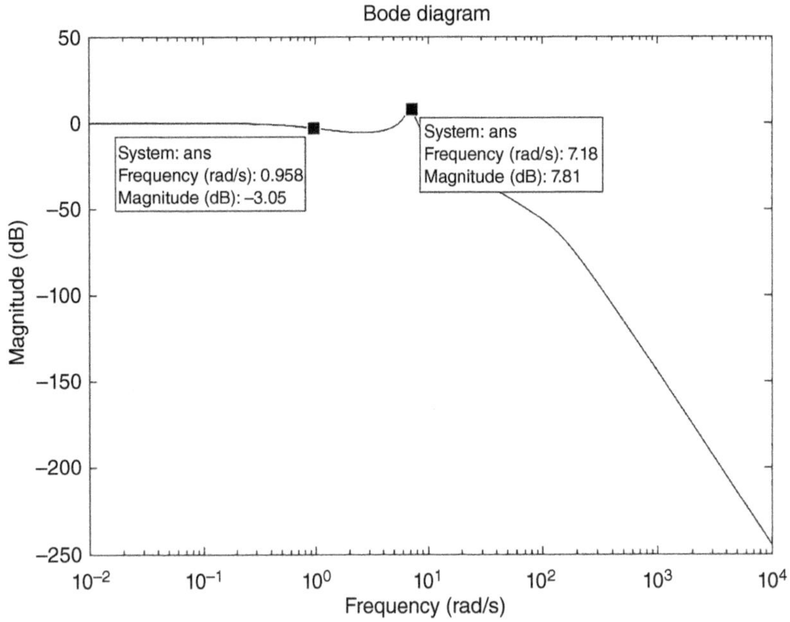

Figure 8.20 Bode plot of closed-loop transfer function of yaw rate estimation.

With the controller transfer function designed G_c, we can obtain the complementary closed-loop transfer function as:

$$T_{\omega} = \frac{G_c G_{p_{\omega}}}{1 + G_c G_{p_{\omega}}} \tag{8.40}$$

Looking at the bode plot of T_{ω} (Figure 8.20), the gain of T_{ω} is at 0 dB at low frequency, which means a time response to a step input with zero steady-state error. However, around 7 rad/s, the closed-loop transfer function has a peak magnitude of around 8 dB, which means a transient response with a large overshoot. If we decrease the bandwidth for T to 1 rad/s or even smaller, the yaw rate estimation will follow the target very well with no overshoot, but it will drastically slow down the estimation response time.

8.2.7 Simulation Results

Figure 8.21 and 8.22 show a comparison of the lateral acceleration at the c.g. and the yaw rate estimations using YCOO, LKF, EKF, and UKF with 2° road wheel angle step input. For both the lateral acceleration and the yaw rate estimates, all these strategies perform well in both sensor noise rejection and target following. Figures 8.23 and 8.24 illustrate the results of the yaw rate estimations with 6° and 10° road wheel angle step input. As we increase the wheel angle input, we can observe that LKF will have a steady-state error as the vehicle model (tire forces) starts operating in the nonlinear region. It should be noted that the lateral tire forces are linear at small tire slip angles, and as the slip angle increases, the tire forces begin to operate in the nonlinear region, and eventually at the very high tire slip angles, the lateral tire forces will saturate. As the tire slip angles increase, purely linear tire models will provide larger lateral forces than their nonlinear counterpart, and as a result, LKF overestimates the yaw rate. It is interesting to observe that in the case of YCOO, although this strategy utilizes a linear tire model, its robustness characteristics results in excellent steady-state target following. The robustness characteristics of YCOO will be further investigated in Section 8.2.8. However, as discussed in Section 8.2.6, the closed-loop bandwidth of YCOO is selected in such a way that its transient response due to the overshoot is not as good as the other proposed methods. The EKF and UKF perform well in both these cases as they are utilizing the same nonlinear vehicle model used for the reference generator (generating the lateral acceleration reference) as for the estimation. We will investigate the performance of the EKF and UKF in the robustness test of Section 8.2.8 when the model used for estimation differs from its nominal values.

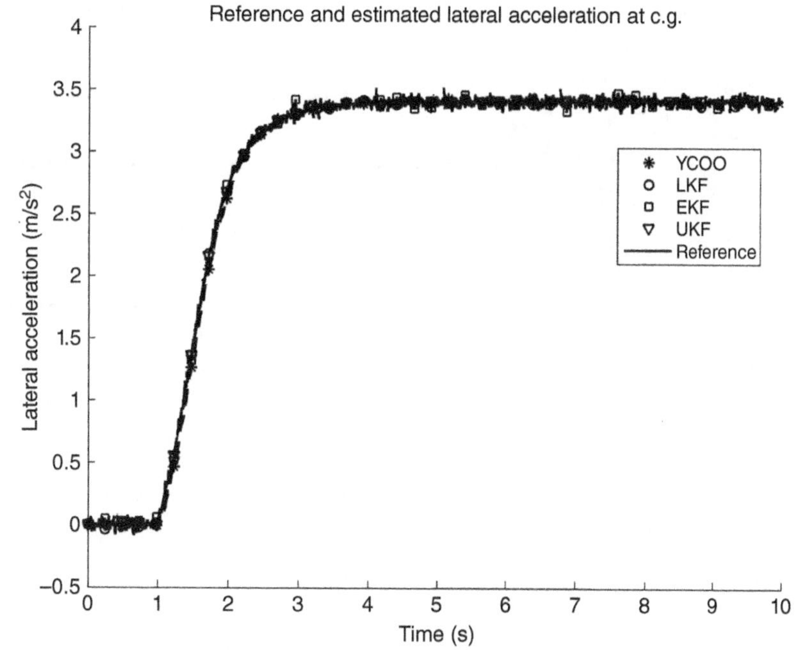

Figure 8.21 Lateral acceleration estimation using different strategies with 2° road wheel angle step input.

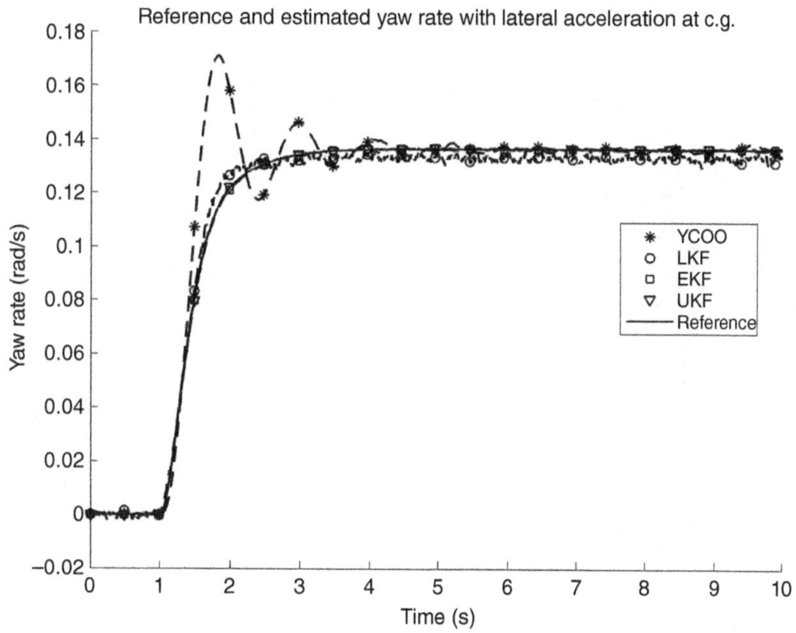

Figure 8.22 Yaw rate estimation using different strategies with 2° road wheel angle step input.

8.2.8 Robustness Test

8.2.8.1 Vehicle Mass Variation

Figure 8.25 shows the yaw rate estimation with different strategies with an addition of 30% change in the vehicle mass and a 2° road wheel angle step input. It illustrates that all Kalman filter approaches are sensitive to the vehicle mass variation and do not have good robustness characteristics with respect to this parameter. These results suggest that if the Kalman filters are tuned to have good sensor noise rejection capabilities and hence, more confidence is given to the model and if

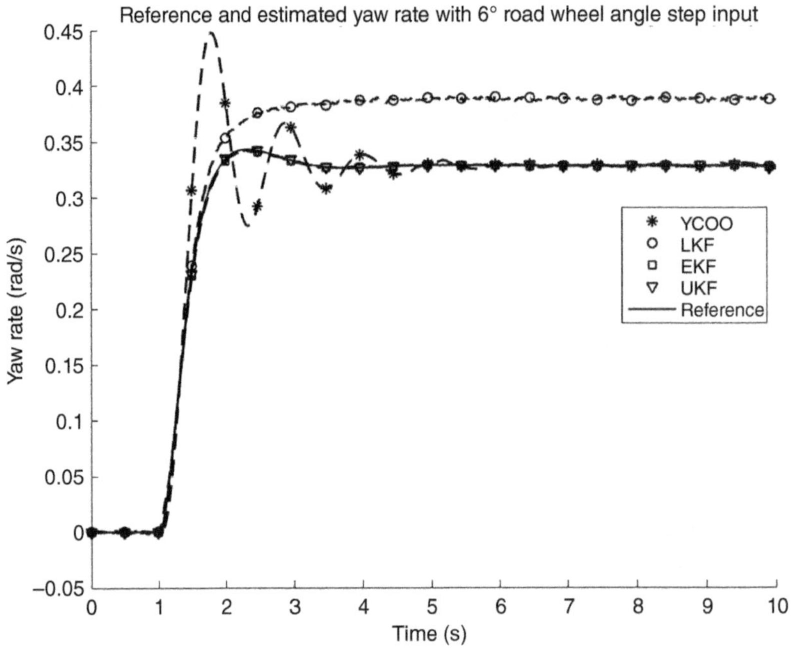

Figure 8.23 Yaw rate estimation using different strategies with 6° road wheel angle step input.

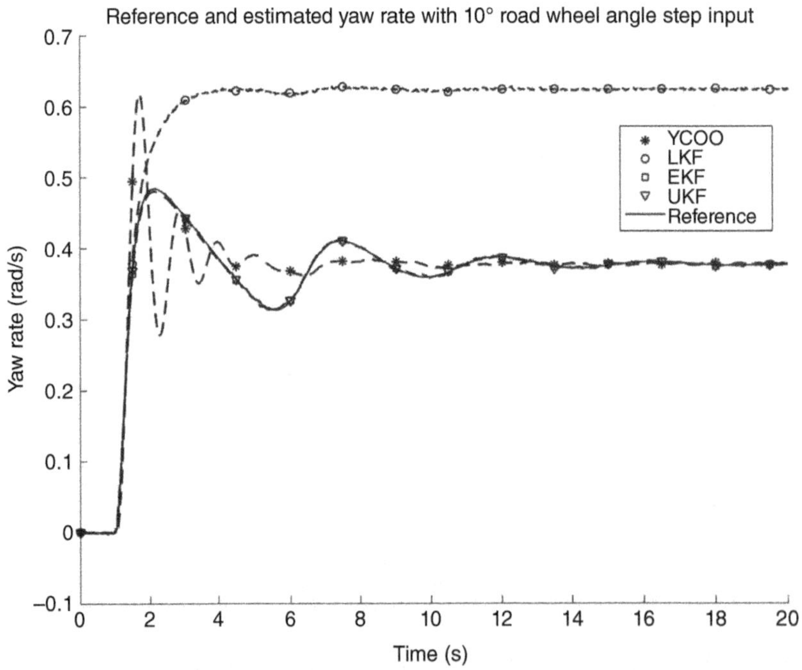

Figure 8.24 Yaw rate estimation using different strategies with 10° road wheel angle step input.

the sensitivities of the filters are high with respect to the model parameter changes, then this will negatively impact the estimation performance. But, it is interesting to observe that the YCOO performance has not been impacted with respect to this parameter variation. It could be argued that the plant parameter variations occur in the low-frequency region where the YCOO is designed to have a closed-loop sensitivity transfer function with a small magnitude, hence, resulting in a good robustness characteristic.

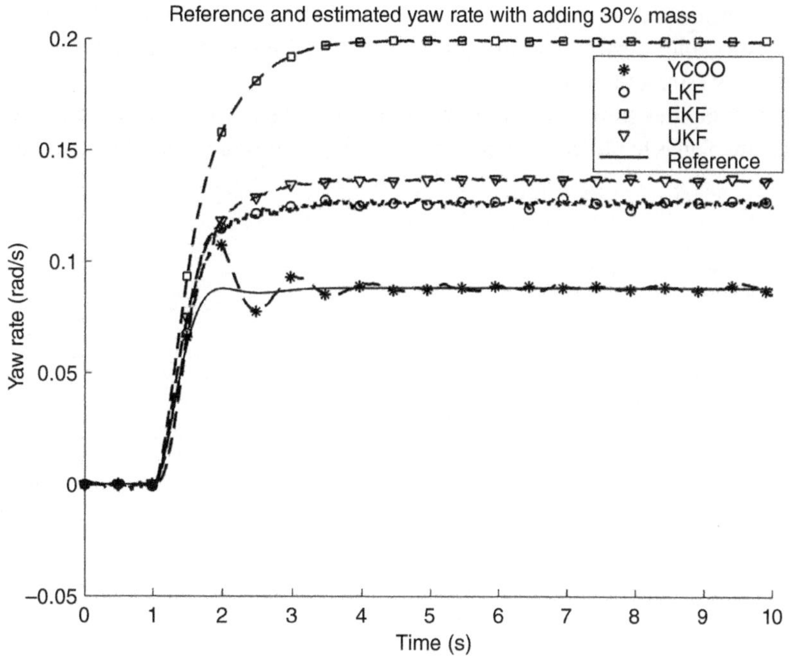

Figure 8.25 Yaw rate estimation using different strategies with +30% mass.

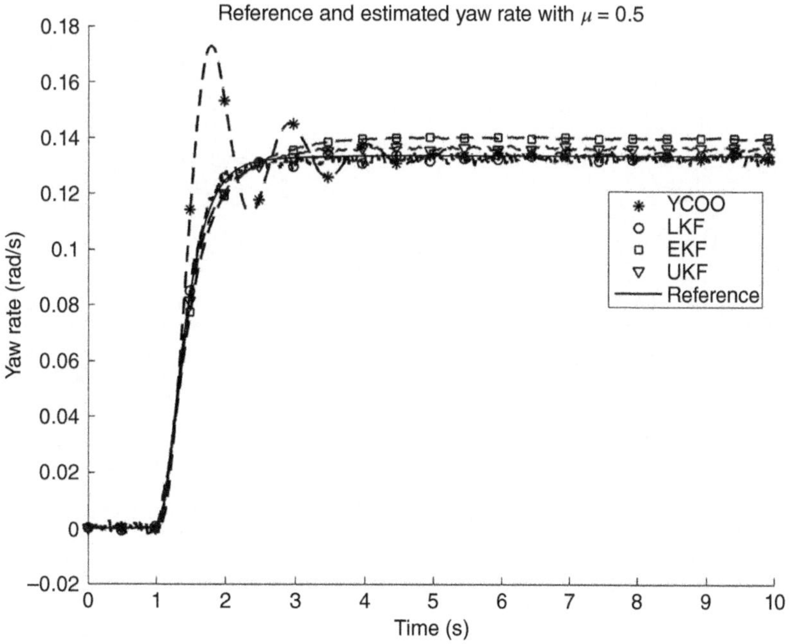

Figure 8.26 Yaw rate estimation using different strategies with $\mu = 0.5$.

8.2.8.2 Tire–road Coefficient of Friction

The yaw rate estimations with a reduced tire–road coefficient of friction $\mu = 0.5$ (the coefficient for dry asphalt is equal to 0.9–1) and a 2° road wheel angle step input are shown in Figure 8.26. The results indicate that all strategies are not very sensitive to the changes in μ as they are to the vehicle mass variations. Comparatively, EKF is more sensitive to μ than the other strategies.

The estimation results presented above suggest that the YCOO is an estimation method with both good performance and robustness characteristics. But, most importantly, it requires very little to no tuning when compared to the standard estimation techniques such as Kalman filtering.

It is worth mentioning that all the estimation approaches did show sensitivity to one important variable, the vehicle longitudinal velocity. Our future plan is to utilize reinforcement learning techniques with the YCOO to address this deficiency.

Part II

Control Design Using Youla Parametrization: Multi Input Multi Output (MIMO)

9

Introduction to Multivariable Feedback Control

Multivariable control design becomes sensible when there are more than one input and output that exist in a dynamic system and when there is relativity a strong coupling between these inputs and outputs.

Control system design, in the late 1800s, was presented in the form of differential equations [20]. Then, in 1940s, through the work of Bode and Nyquist, the frequency domain approach for feedback control design were introduced. In 1960s, with the advancement of the space program and through the work of Kalman and others, the benefits of differential equations through state-space representation of dynamic system was introduced [28]. Interestingly enough, once again in the late 1970s and early 80s through the work of Zames, Horowitz, Doyle and others, the importance of uncertainty quantification in control design was demonstrated through the frequency domain approach, and this approach once again find its way into the main stream control design community [13].

The multivariable or MIMO feedback control design includes a vast area of many different techniques. The original work for multivariable control design using frequency domain was pioneered by Ronsenbrock with the introduction of Nyquist-array method [2].

Multivariable control design methodologies in the frequency domain include methods such as sequential loop closures, the characteristic-locus, reversed-frame normalization, Nyquist-array, and quantitative feedback theory [2]. However, these techniques rely more on hand computation and engineering insights than deriving MIMO controllers directly using numerical optimization approaches. Furthermore, when it comes to MIMO control design, these aforementioned techniques could deviate from optimal design, and the multivariable design part is an extension of these techniques.

Optimal control techniques such as LQR (Linear Quadratic Regulators) and LQG (Linear Quadratic Gaussian) under the modern control domain introduced in the late 1950s and early 1960s are inherently multivariable strategies [2, 29]. These optimal control techniques are derived in the time domain. The equivalency between time and frequency domain is well understood. However, it begs the question that if optimal controllers are designed to not only stabilize the system but have a certain bandwidth, or satisfy any one of the desirable constraints associated with classical control (e.g. disturbance rejection, target following, robustness), and result in the best possible system of a particular type – hence, the word optimal, then why is it necessary to discuss and design robust control techniques?

In the rest of this chapter, we start by discussing nonoptimal and optimal control strategies, such as LQR and LQG, and robust control and attempt to answer the above question. Later, we introduce the Schur complement and provide an interpretation of poles and zeros. We end the chapter by introducing transfer function matrices.

9.1 Nonoptimal, Optimal, and Robust Control

In this section, we briefly review our choice of models for the model-based control development. We then review MIMO nonoptimal methods and in the last part of this section, we discuss optimal controllers such as LQR, LQG, and robust control including the fundamental reasons for designing controllers using robust control approach.

Given a MIMO complex system (G) as shown in Figure 9.1, and let's assume this system is represented by nonlinear partial differential equations (PDEs), we ask the following question: what are the choices to develop model-based controllers for this complex system?

Robust Control: Youla Parameterization Approach, First Edition. Farhad Assadian and Kevin R. Mallon.

Companion Website: www.wiley.com/go/assadian/robust

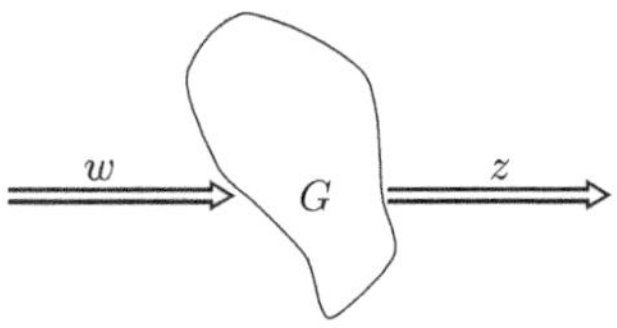

Figure 9.1 General Input–Output System.

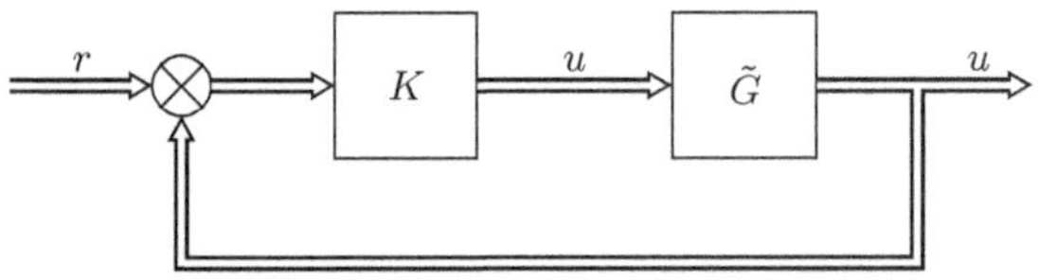

Figure 9.2 LTI Input–Output System.

We can globally divide these choices into two methods:

- **Direct**: control of the complex system using nonlinear PDEs. This is doable but suffers from robustness issues and complexity.
- **Indirect**: if the linearization of this complex system would be possible and PDE's could be transformed to ordinary differential equations (ODEs), then it is possible to write this model as linear time invariant (LTI) ODEs, and a controller could be designed using this simplified model, as shown in Figure 9.2. The tradeoff for this simplicity is additional uncertainty in the model; these uncertainties will be discussed in more depth in the robust control part of this section.

With respect to MIMO control design methods, these can be divided to two distinct approaches, nonoptimal and optimal methods:

9.1.1 Nonoptimal Control Methods

Nonoptimal MIMO control strategies have been thoroughly discussed in the references such as [2]. Below, we merely list these methods.

- Sequential Loop Closures
- Nyquist Array
- Characteristic Locus Method
- Quantitative Feedback Theory (QFT)

9.1.2 Optimal Control Methods

Classical control system design is generally a trial-and-error process in which various methods of analysis are used iteratively to determine the design parameters to satisfy the control requirements.

The objective of optimal control theory is to minimize/maximize a performance criterion while simultaneously satisfying the constraints (e.g. state and control signals physical limits) provided in the performance criterion. In a sense, optimal control replaces engineering judgment and a trial-and-error process with an organized mathematical process for solving a control system design problem.

Fundamentally, there are two methods of accomplishing the minimization of a given performance criterion, "the minimum Pontryagin principle" and "the Bellman dynamic programming". All optimal control laws start with either of these two methods. Interestingly, the minimum Pontryagin principle could be derived from calculus of variations or the dynamic programming [28]. The performance objective of this optimal control law could be posed as a quadratic cost function for a linear dynamic system. This is known as a "linear quadratic regulator" or "LQR". The infinite horizon solution (as time approaches infinity) of this deterministic problem results in a full state feedback controller with constant gains. In addition, LQR turns out to have very good classical phase (60°) and gain (≥6 dB) margins. As intuitive the development of the LQR is, it turns out that this method at times is neither practical nor accounts for model uncertainties and sensor noise (two very practical neoclassical disturbances for addressing robustness). In addition, the LQR requires having access to the entire system states, which is restrictive and at times costly or/and impractical.

One of the most popular and practical version of these optimal control methods is the "Linear Quadratic Gaussian" [2]. In a real feedback system, there are process and sensor noises, as addressed by Kalman in the 1960s [23]. Two additional inputs were added to the system resulting in Figure 9.2.

The LQG assumes that $w_{process}$ and v_{sensor} are statistically independent white Guassian noises. The idea of LQG optimal control is illustrated in Figure 9.4.

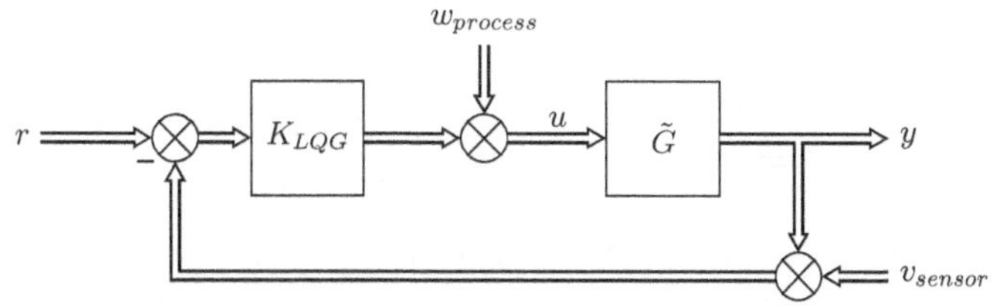

Figure 9.3 Full LTI Input-Output System.

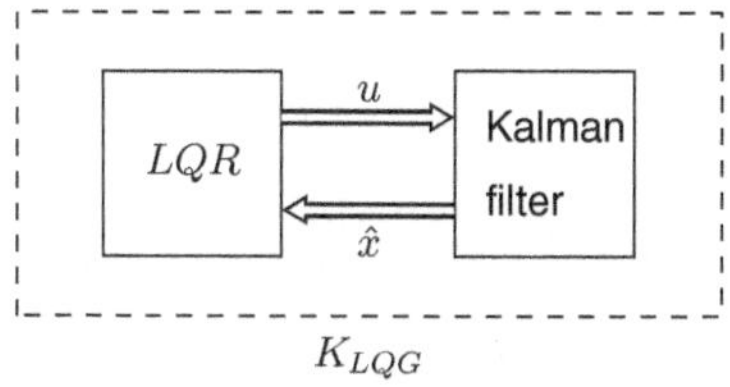

Figure 9.4 LTI Input–Output System.

We stated that the LQR is robust in a classical sense (60° phase margin and at least 6 dB-gain margin), and Kalman filter is also robust in a similar classical fashion as the LQR. The LQG optimal controller utilizes the LQR and Kalman filtering concepts as illustrated in both Figures 9.3 and 9.4 and the solution to the LQG problem is determined by what is known as **separation principle**, where unlike the LQR, the dynamic system states are not directly measured, they are estimated using Kalman filtering. Then the estimated states are utilized as the true states to synthesize a state feedback controller similar to the LQR technique. The LQG is a dynamic controller that could be posed as an output feedback as illustrated in Figure 9.3.

We have stated that both the LQR and Kalman filter have excellent classical robustness margins. However, it is surprising that as proposed by Doyle and Stein [30] through a simple example that LQG design can display poor stability margins. Fortunately, the stability margin could be recovered through an ad hoc method called **Loop Transfer Recovery**; this approach is discussed in many papers and books, such as [30] and [29], and will not be developed here. Simply put, this methodology indirectly re-shapes sensitivity transfer function to recover some of the lost stability margins, but one important issue that is either ignored or not mentioned in this development is the compromise between performance and stability. Interestingly enough, this compromise was directly addressed when utilizing frequency domain loop-shaping either through Youla parameterization technique or H_2 and H_∞ optimal robust control techniques as discussed in the first volume of this book.

Deriving an equivalency between the LQG (time domain) and H_2 (frequency domain), and the use of H_∞ norm and its role in shaping sensitivity introduced by G. Zames in [31], paved the way for understanding how these norms play a significant role in control system design. It became evident that the lack of stability margin in the LQG could be directly addressed by minimizing H_2 cost function and hence directly shaping sensitivity and complementary sensitivity transfer functions. In addition, it was shown H_∞ could be effectively utilized for plant uncertainty modeling. An era in robust control was started.

9.1.3 Optimal Robust Control

In this volume, we concentrate on designing MIMO optimal robust controllers. These controllers will be dynamic and constant gains.

Similar to the first volume, we start our development by introducing necessary mathematical tools so we would be able to uniquely design **Robust nonoptimal MIMO controllers using Youla parameterization** by hand computation.

We will then introduce H_2 and H_∞ norms for MIMO systems so that we would be able to design **Optimal robust MIMO controllers** by minimizing cost functions related to the aforementioned norms, using control optimization toolboxes.

One of the most important aspects of the feedback control design is how to manage uncertainties. We have discussed uncertainties in the first volume of this book but in this volume, and we will extend these ideas for MIMO control synthesis.

Before ending this section, it would be helpful to say a few words about the benefits of norm optimization, especially the H_∞ norm.

As illustrated in Figure 9.5, the power of H_∞ norm optimization comes from the following:

- Modeling uncertainties and revealing the trade-off between dealing with complexity (e.g. highly complex plant models) and uncertainty
- Dealing with uncertainties (e.g. disturbance signals) in a deterministic manner by assuming that disturbance signals are energy bounded

Our concentration in this book is to introduce a unique method to design MIMO controllers using Youla parameterization. Designing controllers using norm optimization is included only for comparative basis.

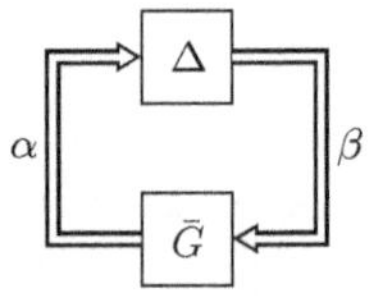

Figure 9.5 General plant configuration with structured uncertainty.

9.2 Review of the SISO Transfer Function

In this section, we start with reviewing deriving the SISO transfer function from its state-space matrices. We will introduce the Schur complement and Rosenbrock system matrix and their application in deriving transfer functions and transfer function matrices from state-space matrices. Then we provide an interpretation of poles, zeros, and transmission blocking zeros of transfer functions. We will end this section by introducing transfer function matrices including some of its basic aspects.

We derived the SISO transfer function from its state-space matrices in the first volume of this book. To refresh our memory, this derivation is provided below.

Given A, B, C, and D matrices as defined previously, the first-order LTI ODEs describing the system dynamics are written as

$$\dot{x} = Ax + Bu \tag{9.1}$$

$$y = C^T x + Du \tag{9.2}$$

where x are the system states. Transfer function from the input, u to the output, y is derived assuming zero state input (all ICs (initial conditions) are set to zero) using Laplace transform,

$$sX \equiv sIX = AX + DU \tag{9.3}$$

$$(sI - A)X = BU \tag{9.4}$$

$$X = (sI - A)^{-1} BU \tag{9.5}$$

where s is the Laplace operator, and X and U are functions of s in Laplace domain. Then the output is written as a function of the input,

$$Y = C^T X + DU \tag{9.6}$$

$$Y = (C^T (sI - A)^{-1} B + D)U \tag{9.7}$$

where

$$(sI - A)^{-1} = \frac{\text{adj}(sI - A)}{\det(sI - A)} \tag{9.8}$$

and finally, the transfer function describing the input/output relationship is written as,

$$G(s) = \frac{Y}{U} = \frac{\left(C^T \text{adj}(sI - A)B + \det(sI - A)D\right)}{\det(sI - A)} \tag{9.9}$$

where "zeros" are the zeros of the numerator polynomial and the "poles" are the zeros of the denominator.

9.2.1 Schur Complement

In this section, we introduce the Schur Complement and Rosenbrock's system matrix in computing system transfer function. Let's start by introducing the matrix Z with the given block dimensions,

$$Z = \left[\begin{array}{c|c} \underbrace{M}_{m\times m} & \underbrace{E}_{m\times n} \\ \hline \underbrace{H}_{n\times m} & \underbrace{N}_{n\times n} \end{array}\right] \tag{9.10}$$

Now, let's define,

$$U = M - EN^{-1}H \tag{9.11}$$

$$W = N - HM^{-1}E \tag{9.12}$$

The matrix U is defined to be the Schur Complement of N in Z and the matrix W is defined to be the Schur Complement of M in Z. The Schur Complement has many applications and properties in matrix manipulations and optimization [3].

One of the properties of the Schur Complement is as follows:

$$\det(N)\det(U) = \det \left[\begin{array}{c|c} \underbrace{M}_{m\times m} & \underbrace{E}_{m\times n} \\ \hline \underbrace{H}_{n\times m} & \underbrace{N}_{n\times n} \end{array} \right] = \det(M)\det(W) \tag{9.13}$$

Proof: Let's assume M is invertible, then

$$\begin{bmatrix} M & E \\ H & N \end{bmatrix} = \begin{bmatrix} I_m & 0 \\ HM^{-1} & I_n \end{bmatrix} \begin{bmatrix} M & 0 \\ 0 & N - HM^{-1}E \end{bmatrix} \begin{bmatrix} I_m & M^{-1}E \\ 0 & I_n \end{bmatrix} \tag{9.14}$$

Therefore

$$\det \begin{bmatrix} M & E \\ H & N \end{bmatrix} = \det M \det(N - HM^{-1}E) = \det M \det W \tag{9.15}$$

■

Rosenbrock's system matrix is given by

$$P(s) = \left[\begin{array}{c|c} \underbrace{sI - A}_{n\times n} & \underbrace{-B}_{n\times 1} \\ \hline \underbrace{C^T}_{1\times n} & \underbrace{D}_{1\times 1} \end{array} \right] \tag{9.16}$$

If we define the block dimension in the Rosenbrock's matrix as shown above (such as for SISO system), then we can write,

$$W = D - C^T(sI - A)^{-1}(-B) = G(s) \tag{9.17}$$

Using the property of the Schur Complement given above, the transfer function $G(s)$ can be written as,

$$G(s) = \frac{\det P(s)}{\det(sI - A)} \tag{9.18}$$

9.2.2 Interpretation of Poles and Zeros of a Transfer Function

Most students and control practitioners are familiar with the concept of poles and zeros of a transfer function. But, what might be an unfamiliar concept is the difference between a zero (what is also known as transmission zero) and transmission blocking zero. Therefore, it would be helpful to review these concepts.

9.2.2.1 Poles

An interpretation of a pole of a transfer function is acquired by the zero input or what is known as unforced response. The state-space equation is then written as

$$\dot{x} = Ax, \quad x(0) = x_0 \tag{9.19}$$

where the solution to this differential equation is

$$x(t) = x_0 e^{\lambda t} \tag{9.20}$$

and λ could be a real or complex number and it is called the eigenvalue. Inserting $x(t)$ into Equation (9.19) yields

$$\lambda x_0 e^{\lambda t} = Ax_0 e^{\lambda t} \tag{9.21}$$

Dividing out similar terms and moving everything to the left side yields

$$[\lambda I - A]x_0 = 0 \tag{9.22}$$

Knowing that the IC, x_0, is not equal to 0, then,

$$\det[\lambda I - A] = 0 \tag{9.23}$$

This equation allows you to solve for the eigenvalue, λ, which is equal to a pole of the transfer function or equivalently the roots of the characteristic equation as discussed in the example.

Example 9.1 Take a simple spring damper system in two different configurations as shown in Figures 9.6 and 9.7.

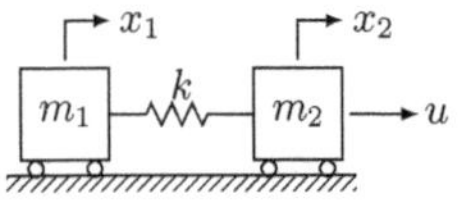

Figure 9.6 Example 1 – system configuration 1.

Controlling the position x_2 with the control force u in the first configuration as shown in Figure 9.6 results in the following transfer function,

$$\frac{X_2}{U} = \frac{\frac{1}{m_2}\left(s^2 + \frac{k}{m_1}\right)}{s^2\left(s^2 + \left(\frac{k}{m_1} + \frac{k}{m_2}\right)\right)} \tag{9.24}$$

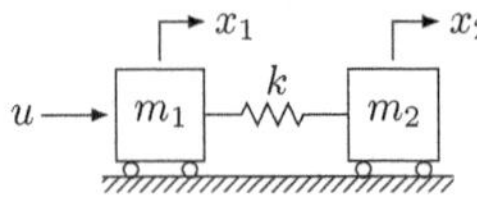

Figure 9.7 Example 1 – system configuration 2.

Controlling the position x_2 by moving the control force u in the second configuration as shown in Figure 9.7 results in the following transfer function,

$$\frac{X_2}{U} = \frac{\frac{k}{m_1 m_2}}{s^2\left(s^2 + \left(\frac{k}{m_1} + \frac{k}{m_2}\right)\right)} \tag{9.25}$$

This example illustrates that the location of the actuator input, and in the same manner, sensor location, does not impact the location of the eigenvalues or the roots of the characteristic equation. In both of these configurations, there are four eigenvalues, two are located at zero and two are complex conjugates and are located on the $j\omega$-axis. However, as briefly discussed in the first volume, the location of the actuator input, or sensor location, impacts the location of the zeros of the transfer function as can be seen in these two configurations.

9.2.2.2 Zeros

An interpretation of a zero of a transfer function is acquired by the zero state or what is known as the forced response. Let's select an input to the a dynamic system to be,

$$u(t) = u_0 e^{\mu t} \tag{9.26}$$

nonzero input. Then, we ask the following question: is it possible to have the output of this system precisely cancel this input?

The answer is yes, if the system transfer function has a zero at location equal to μ. This is simply called a zero of the system transfer function.

9.2.2.3 Transmission Blocking Zeros

However, a zero of a transfer function is different than a transmission blocking zero. We say a zero has a transmission blocking property when, assuming the same input as before, there is a specific IC associated with the zero input response,

$$x(t) = x_0 e^{\mu t} \quad \text{where} \quad x(0) = x_0 \tag{9.27}$$

so that the addition of the zero state response (forced response) and the zero input response (unforced or natural response) has precisely the following output response,

$$y(t) \equiv 0 \tag{9.28}$$

Given that the LTI ODE is,

$$\dot{x} = Ax + Bu \tag{9.29}$$

The total response becomes

$$\mu x_0 e^{\mu t} = Ax_0 e^{\mu t} + Bu_0 e^{\mu t} \tag{9.30}$$

Knowing that

$$y = C^T x + Du \tag{9.31}$$

Therefore,

$$0 \equiv C^T x_0 e^{\mu t} + Du_0 e^{\mu t} \tag{9.32}$$

From Eq. (9.30), we derive

$$(\mu I - A)x_0 = Bu_0 \tag{9.33}$$

Dividing Eq. (9.32) by $e^{\mu t}$, we can write

$$0 \equiv C^T x_0 + Du_0 \tag{9.34}$$

The last two equations can be written in the matrix form as

$$\left[\begin{array}{c|c} \mu I - A & -B \\ \hline C^T & D \end{array}\right] \begin{bmatrix} x_0 \\ u_0 \end{bmatrix} = 0 \tag{9.35}$$

which when x_0 and u_0 are not equal to zero then,

$$\det \left[\begin{array}{c|c} \mu I - A & -B \\ \hline C^T & D \end{array}\right] = 0 \tag{9.36}$$

from the Schur complement, the determinant of the matrix above can be written as,

$$\det \begin{bmatrix} \mu I - A & -B \\ C^T & D \end{bmatrix} = \det(\mu I - A) \det(C^T(\mu I - A)^{-1}B + D) = \det(\mu I - A) \det(G(\mu)) \tag{9.37}$$

If this determinant is precisely equal to zero, then a zero has transmission blocking property. An example below will help in shedding some light into this finding.

Example 9.2 Let

$$G(s) = \frac{s+2}{s^2 + 7s + 12} \tag{9.38}$$

It can be shown that in controllable canonical form, the system takes the form,

$$\dot{x} = \begin{bmatrix} 0 & 1 \\ -12 & -7 \end{bmatrix} x + \begin{bmatrix} 0 \\ 1 \end{bmatrix} u \tag{9.39}$$

$$y = \begin{bmatrix} 2 & 1 \end{bmatrix} x \tag{9.40}$$

Note: $D = 0$ in this example and therefore the system is strictly proper. If the system were just proper, i.e. $\deg(Num) = \deg(Den)$, use long division or partial fraction expansion to get the remainder which is used to obtain D for the canonical form.

The Rosenbrock system matrix (R) is used to define the response at a transmission zero (μ).

$$\left[\begin{array}{c|c} \mu I - A & -B \\ \hline C^T & D \end{array}\right] \begin{bmatrix} x_0 \\ u_0 \end{bmatrix} = 0 \tag{9.41}$$

Because we are assuming that x_0 and u_0 are nonzero, this leaves the condition

$$\det \left[\begin{array}{c|c} \mu I - A & -B \\ \hline C^T & D \end{array} \right] = 0 \tag{9.42}$$

This implies that

$$\det(\mu I - A)\det(C^T(\mu I - A)^{-1}B) = 0 \tag{9.43}$$

$$\det(\mu I - A)\det(G(\mu)) = 0 \tag{9.44}$$

$$\det(\mu I - A)\frac{\mu + 2}{\mu^2 + 7\mu + 12} = 0 \tag{9.45}$$

$$\mu + 2 = 0 \tag{9.46}$$

which implies that $\mu = -2$. Obviously this is not very interesting because we could have reached the same conclusion from $G(s)$. However, further information can be gained from this. Using Eq. (9.41), the following relationship exists

$$\begin{bmatrix} \mu & -1 & 0 \\ 12 & \mu + 7 & -1 \\ 2 & 1 & 0 \end{bmatrix} \begin{bmatrix} x_{0_1} \\ x_{0_2} \\ u_0 \end{bmatrix} = 0 \tag{9.47}$$

Letting $\mu = -2$

$$\begin{bmatrix} -2 & -1 & 0 \\ 12 & 5 & -1 \\ 2 & 1 & 0 \end{bmatrix} \begin{bmatrix} x_{0_1} \\ x_{0_2} \\ u_0 \end{bmatrix} = 0 \tag{9.48}$$

which results in the following,

$$2x_{0_1} + x_{0_2} = 0 \tag{9.49}$$

$$12x_{0_1} + 5x_{0_2} - u_0 = 0 \tag{9.50}$$

This results in the null space of the matrix in Eq. (9.48) to be,

$$\mathcal{N}(R) = k \begin{bmatrix} 1 \\ -2 \\ 2 \end{bmatrix} \tag{9.51}$$

where $k \in \mathbb{R}$.

Recall that for a linear system, the total response is a sum of two different parts,

$$Y = Y_1 + Y_2 \tag{9.52}$$

where Y_1 is the forced or zero-state response of the system and Y_2 is the IC or zero-input response of the system. Y_1 and Y_2 can be found with the following equations,

$$Y_1 = C^T(sI - A)^{-1}BU \tag{9.53}$$

$$Y_2 = C^T(sI - A)^{-1}x_0 \tag{9.54}$$

Taking the state-space system from Eqs. (9.39) to (9.40) and $u(t) = u_0 e^{\mu t} = 2e^{-2t}$, $U = \frac{2}{s+2}$ and Y_1 becomes

$$Y_1 = \frac{s+2}{s^2 + 7s + 12} \cdot \frac{2}{s+2} = \frac{2}{s^2 + 7s + 12} \tag{9.55}$$

now taking $x_0 = \begin{bmatrix} 1 & -2 \end{bmatrix}^T$ results in Y_2 being

$$Y_2 = \begin{bmatrix} 2 & 1 \end{bmatrix} \begin{bmatrix} s & -1 \\ 12 & s+7 \end{bmatrix}^{-1} \begin{bmatrix} 1 \\ -2 \end{bmatrix} = \frac{-2}{s^2 + 7s + 12} \tag{9.56}$$

Therefore $Y_1 + Y_2 = 0$ for the input $u(t) = u_0 e^{-2t}$ with x_{0_1}, x_{0_2}, and u_0 given in Eq. (9.51). This is a transmission blocking zero that will result in no output for the system.

9.3 Basic Aspects of Transfer Function Matrices

Define "r" as the sensor dimension, "m" as the actuator dimension, and "n" as the state dimension. This means that $A \in \mathbb{R}^{n\times n}$, $B \in \mathbb{R}^{n\times m}$, $C \in \mathbb{R}^{r\times n}$, and $D \in \mathbb{R}^{r\times m}$.

Recall, to obtain the transfer functions of the system, set all of the ICs to zero and use the following equation

$$Y = [C^T(sI - A)^{-1}B + D]U \tag{9.57}$$

Letting $C^T(sI - A)^{-1}B + D = G(s)$, $G(s)$ is now a rational transfer function matrix and with dimension $r \times m$. Further, $G(s)$ is either strictly proper and

$$\lim_{s\to\infty} G(s) = 0_{r\times m} \tag{9.58}$$

or $G(s)$ is proper and

$$\lim_{s\to\infty} G(s) = K_{r\times m} \tag{9.59}$$

where $k_{ij} \in \mathbb{R}$ and $k_{ij} < \infty$.

9.4 Problems

9.1 The Rosenbrock system matrix is given by

$$P(s) = \begin{bmatrix} sI - A & -b \\ C^T & D \end{bmatrix} \tag{9.60}$$

where $P(s)$ is $(n+1) \times (n+1)$, $sI - A$ is $n \times n$, $-b$ is $n \times 1$, C^T is $1 \times n$, and D is 1×1. Show that the transfer function $G(s)$ is equal to the following expression

$$G(s) = \frac{\det P(s)}{\det (sI - A)} \tag{9.61}$$

where "det" is the matrix determinant.

9.2 Given

$$G(s) = \frac{s+3}{(s+5)(s+2)} \tag{9.62}$$

(a) Using the Rosenbrock's system matrix, compute the transmission zero μ.
(b) Find the associated direction vector for the transmission blocking zero.
(c) Find the zero state response $Y_1(s)$, and the zero input response $Y_2(s)$.
(d) How do you show that your finding in (b) result in a transmission blocking zero?

9.3 Given

$$Y(s) = \begin{bmatrix} \frac{s+1}{s} & 0 \\ 0 & \frac{s+3}{s+1} \end{bmatrix} U(s) \tag{9.63}$$

(a) Will an input $u(t) = \begin{bmatrix} 1 \\ 0 \end{bmatrix} e^t$ contain an e^t term in the output $y(t)$?
(b) Will an input $u(t) = \begin{bmatrix} 0 \\ 1 \end{bmatrix} e^t$ contain an e^t term in the output $y(t)$?
(c) Will an input $u(t) = \begin{bmatrix} 1 \\ 0 \end{bmatrix} e^{3t}$ contain an e^{3t} term in the output $y(t)$?
(d) Will an input $u(t) = \begin{bmatrix} 0 \\ 1 \end{bmatrix} e^{3t}$ contain an e^{3t} term in the output $y(t)$?
(e) Will an input $u(t) = \begin{bmatrix} 1 \\ 1 \end{bmatrix} e^t$ contain an e^t term in the output $y(t)$?

10

Matrix Fractional Description

In this chapter, we introduce the required mathematical background necessary for designing MIMO control design using Youla parameterization.

10.1 Transfer Function Matrices

In this section, we discuss the matrix fraction description (MFD). Furthermore, we introduce the polynomial matrix properties and their relation to the MFDs.

Recall a MIMO dynamic system with n states, m actuators, and r sensors:

$$\underline{\dot{x}} = \underset{(n\times n)}{A}\,\underline{x} + \underset{(n\times m)}{B}\,\underline{u} \tag{10.1}$$

$$\underline{y} = \underset{(r\times n)}{C^T}\,\underline{x} + \underset{(r\times m)}{D}\,\underline{u} \tag{10.2}$$

As stated previously, we can follow the same process as for a SISO system to compute the transfer function:

$$s\underline{X} = sI\underline{X} = A\underline{X} + B\underline{U} \tag{10.3}$$

$$\underline{X} = (sI - A)^{-1}B\underline{U} \tag{10.4}$$

$$\underline{Y} = C^T\underline{X} + D\underline{U} \tag{10.5}$$

$$\underline{Y} = \left(C^T(sI - A)^{-1}B + D\right)\underline{U} \tag{10.6}$$

$$G(s) = C^T(sI - A)^{-1}B + D \tag{10.7}$$

$$G(s) = \frac{C^T\text{adj }(sI - A)\,B + D\det(sI - A)}{\det(sI - A)} \tag{10.8}$$

However, in this formulation, $G(s)$ is now a rational matrix (i.e. a matrix of rational functions):

$$\underset{(r\times 1)}{\underline{Y}} = \underset{(r\times m)}{G(s)}\ \underset{(m\times 1)}{\underline{U}} \tag{10.9}$$

$G(s)$ is either strictly proper:

$$\lim_{s\to\infty} G(s) = 0 \tag{10.10}$$

or proper,

$$\lim_{s\to\infty} G(s) < \infty \tag{10.11}$$

Robust Control: Youla Parameterization Approach, First Edition. Farhad Assadian and Kevin R. Mallon.

Companion Website: www.wiley.com/go/assadian/robust

10.1.1 Matrix Fraction Description

Every rational matrix can be written as a ratio of two polynomial matrices, N_R and D_R or N_L and D_L. So, we prefer to write

$$G(s) = N_R(s)D_R^{-1}(s) \tag{10.12}$$

This is known as a MFD. $N_R(s)$ is a right numerator polynomial matrix such that,

$$N_R(s) = (C^T \text{adj } (sI - A)B + D\det(sI - A)) \tag{10.13}$$

and D_R is a right denominator polynomial matrix such that,

$$D_R(s) = \det(sI - A)I_m \tag{10.14}$$

We could just as well write

$$G(s) = D_L^{-1}(s)N_L(s) \tag{10.15}$$

such that N_L is a left numerator polynomial matrix and D_L is a left denominator polynomial matrix. In this case,

$$N_L = N_R, \quad D_R(s) = \det(sI - A)I_r \tag{10.16}$$

Although the left and right numerator matrices are equal, the left and right denominator matrices are different sizes and are therefore *not* equal, however,

$$\underset{(r\times r)(r\times m)}{D_L^{-1}\,N_L} = \underset{(r\times m)(m\times m)}{N_R\,D_R^{-1}} \tag{10.17}$$

MFD's are never unique. Suppose F is any $m \times m$ polynomial matrix. Then,

$$G(s) = N_R D_R^{-1} \tag{10.18}$$

$$= N_R F F^{-1} D_R^{-1} \tag{10.19}$$

$$= (N_R F)(D_R F)^{-1} \quad \text{A new MFD!} \tag{10.20}$$

Example 10.1 Consider a left MFD as given below:

$$G(s) = D_L^{-1}(s)N_L(s), \tag{10.21}$$

where now the denominator polynomial matrix enters $G(s)$ from the left.

Consider the RLC circuit shown in Figure 10.1 below.

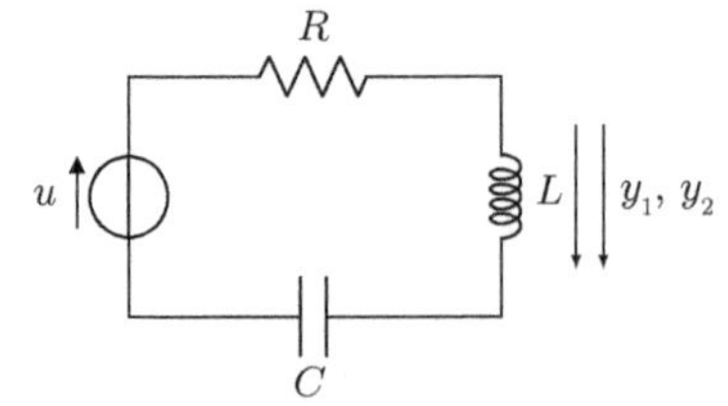

Figure 10.1 RLC circuit.

The state equations are given by:

$$\begin{bmatrix} \dot{y}_1 \\ \dot{y}_2 \end{bmatrix} = \begin{bmatrix} -\frac{R}{L} & -\frac{1}{C} \\ \frac{1}{L} & 0 \end{bmatrix} \begin{bmatrix} y_1 \\ y_2 \end{bmatrix} + \begin{bmatrix} 1 \\ 0 \end{bmatrix} u \tag{10.22}$$

where $\dot{y}_1$ is the inductor voltage and $\dot{y}_2$ is the current. y_1 and y_2 are both state variables and the outputs:

$$\begin{bmatrix} s + \frac{R}{L} & \frac{1}{C} \\ -\frac{1}{L} & s \end{bmatrix} \begin{bmatrix} Y_1 \\ Y_2 \end{bmatrix} = \begin{bmatrix} 1 \\ 0 \end{bmatrix} U \tag{10.23}$$

Assuming $\dot{y}_1$ and $\dot{y}_2$ are the outputs,

$$\begin{bmatrix} s+\frac{R}{L} & \frac{1}{C} \\ -\frac{1}{L} & s \end{bmatrix}\begin{bmatrix} Y_1 \\ Y_2 \end{bmatrix} = \begin{bmatrix} s \\ 0 \end{bmatrix} U \tag{10.24}$$

We can then write the above relationship in MFD as:

$$\begin{bmatrix} Y_1 \\ Y_2 \end{bmatrix} = \begin{bmatrix} s+\frac{R}{L} & \frac{1}{C} \\ -\frac{1}{L} & s \end{bmatrix}^{-1}\begin{bmatrix} s \\ 0 \end{bmatrix} \tag{10.25}$$

$$D_L = \det(sI - A)I_r = \left(s^2 + \frac{R}{L}s + \frac{1}{CL}\right)\begin{bmatrix} 1 & 0 \\ 0 & 1 \end{bmatrix} \tag{10.26}$$

$$N_L = \text{adj}(sI - A)\begin{bmatrix} s \\ 0 \end{bmatrix} = \begin{bmatrix} s^2 \\ \frac{s}{L} \end{bmatrix} \tag{10.27}$$

10.2 Polynomial Matrix Properties

Consider a polynomial matrix $P(s)$ that is $r \times m$ in size.

Definition 10.1 The degree of $P(s)$ (denoted $\deg P(s)$) is the highest degree of all nonzero minors of maximal order (usually, $\min(r, m)$). For a square matrix, this means $\deg\det P(s)$.

Definition 10.2 A polynomial matrix is unimodular if it is square and $\det P(s)$ is constant.

Definition 10.3 (Rank) The rank of a polynomial matrix, P, is the number of linearly independent columns or rows.

Definition 10.4 (Regular) A matrix, P, is regular if and only if $\text{rank}(P) = \min\{r, m\}$

Definition 10.5 $P(s)$ is a unit if its inverse is a polynomial.

Proposition 10.1 *A polynomial matrix is a unit if and only if it is unimodular.*

Proof: By example:

$$P_1 = \begin{bmatrix} s & s \\ 0 & 1 \end{bmatrix} \Rightarrow P_1^{-1} = \begin{bmatrix} \frac{1}{s} & -1 \\ 0 & 1 \end{bmatrix} \quad \text{Is this a polynomial?} \tag{10.28}$$

P_1^{-1} is not a polynomial and therefore, P_1 is not a unit as it is not unimodular. ■

Definition 10.6 (Zero) $s = z$ is a zero of a polynomial matrix $P(s)$ if
$\text{rank}(P(z)) < \mathit{rank}(P(s))$.

Example 10.2 Consider the following polynomial matrix,

$$P(s) = \begin{bmatrix} 1 & s \\ 0 & 1 \end{bmatrix} \tag{10.29}$$

$\det P(s) = 1$, so $P(s)$ is unimodular and $\deg P(s) = 0$. In fact, all unimodular polynomials have degree $= 0$.

Example 10.3 In this example, we examine the meaning of the "highest degree of all nonzero minors of maximal order", consider the following polynomial matrix,

$$P(s) = \begin{bmatrix} 1 & s & 1 \\ 0 & 1 & 1 \end{bmatrix} \tag{10.30}$$

Identify all minors of maximal order for this matrix,

$$\text{minors of maximal order: } \begin{bmatrix} 1 & s \\ 0 & 1 \end{bmatrix}, \begin{bmatrix} s & 1 \\ 1 & 1 \end{bmatrix}, \begin{bmatrix} 1 & 1 \\ 0 & 1 \end{bmatrix} \tag{10.31}$$

Compute the degree of this polynomial matrix,

$$\deg P(s) = \max\left(\deg\det\begin{bmatrix} 1 & s \\ 0 & 1 \end{bmatrix}, \deg\det\begin{bmatrix} s & 1 \\ 1 & 1 \end{bmatrix}, \deg\det\begin{bmatrix} 1 & 1 \\ 0 & 1 \end{bmatrix}\right) \tag{10.32}$$

$$\deg P(s) = \max\left(\deg(1), \deg(s-1), \deg(1)\right) \tag{10.33}$$

So, $\deg P(s) = 1$.

Example 10.4 If $P(s)$ is scalar, then unimodular polynomials are the scalars.

Example 10.5 Let

$$P_1 = \begin{bmatrix} 1 & 0 \\ 0 & s \end{bmatrix} \text{ and } P_2 = \begin{bmatrix} s & s \\ 1 & 1 \end{bmatrix} \tag{10.34}$$

$\text{rank}(P_1) = 2$ and $\text{rank}(P_2) = 1$ therefore, P_1 is regular and P_2 is not regular.

Example 10.6 Using P_1 from above with rank $= 2$,

$$P_1(0) = \begin{bmatrix} 1 & 0 \\ 0 & 0 \end{bmatrix} \tag{10.35}$$

and $\text{rank}(P_1(0)) = 1$. Therefore $s = 0$ is a zero. Defining

$$P_3(s) = \begin{bmatrix} 1 & s \\ 0 & 1 \end{bmatrix} \tag{10.36}$$

rank $(P_3(s))$ =rank$(P_3(0)) = 2$ so therefore $s = 0$ is not a zero of P_3.

10.2.1 Minimum-Degree Factorization

Every rational transfer function matrix can be written as a ratio of a polynomial matrix numerator and an identity matrix times the transfer function matrix's characteristic polynomial. For example, consider the following matrix $G(s)$,

$$G(s) = \begin{bmatrix} \frac{1}{s+1} & 1 \\ 0 & \frac{1}{s+2} \end{bmatrix} = \frac{\begin{bmatrix} (s+2) & (s+1)(s+2) \\ 0 & (s+1) \end{bmatrix}}{(s+1)(s+2)} \tag{10.37}$$

$$= \underbrace{\begin{bmatrix} (s+2) & (s+1)(s+2) \\ 0 & (s+1) \end{bmatrix}}_{N_R} \underbrace{\begin{bmatrix} (s+1)(s+2) & 0 \\ 0 & (s+1)(s+2) \end{bmatrix}^{-1}}_{D_R} \tag{10.38}$$

For this example, $\deg N_R = 2$ and $\deg D_R = 4$. But can we obtain a lower-degree factorization (analogous to checking for pole–zero cancellations in a SISO system)?

A different factorization for $G(s)$ is

$$G(s) = \begin{bmatrix} \frac{1}{s+1} & 1 \\ 0 & \frac{1}{s+2} \end{bmatrix} = \begin{bmatrix} 1 & s+2 \\ 0 & 1 \end{bmatrix} \begin{bmatrix} \frac{1}{s+1} & 0 \\ 0 & \frac{1}{s+2} \end{bmatrix} \tag{10.39}$$

$$= \underbrace{\begin{bmatrix} 1 & s+2 \\ 0 & 1 \end{bmatrix}}_{N_R,\ \text{deg}=0} \underbrace{\begin{bmatrix} s+1 & 0 \\ 0 & s+2 \end{bmatrix}^{-1}}_{D_R,\ \text{deg}=2} \tag{10.40}$$

This factorization results in the deg $N_R = 0$ and deg $D_R = 2$, so we can firmly state that the original factorization was not minimum degree.

In the following sections, we will discuss the conditions for the minimum-degree factorization. But, first, in Section 10.3, we introduce the equivalency condition between polynomial matrices.

10.3 Equivalency of Polynomial Matrices

Definition 10.7 (Equivalent) Two matrices are equivalent if one can be converted to the other using a combination of

1. The interchange of columns or rows
2. The multiplication of one column or row by a scalar
3. The addition of a polynomial multiple of a row or column to another row or column

---

Example 10.7 Interchange

$$P = \begin{bmatrix} 1 & 2 & 3 \\ 4 & 5 & 6 \\ 7 & 8 & 9 \end{bmatrix} \sim \begin{bmatrix} 2 & 1 & 3 \\ 5 & 4 & 6 \\ 8 & 7 & 9 \end{bmatrix} = Q \tag{10.41}$$

or

$$P = \begin{bmatrix} 1 & 2 & 3 \\ 4 & 5 & 6 \\ 7 & 8 & 9 \end{bmatrix} \sim \begin{bmatrix} 4 & 5 & 6 \\ 1 & 2 & 3 \\ 7 & 8 & 9 \end{bmatrix} = R \tag{10.42}$$

Let

$$D = \begin{bmatrix} 0 & 1 & 0 \\ 1 & 0 & 0 \\ 0 & 0 & 1 \end{bmatrix} \tag{10.43}$$

Then

$$PD = Q \tag{10.44}$$

$$DP = R \tag{10.45}$$

Example 10.8 Scalar Multiplication

$$P = \begin{bmatrix} 1 & 2 & 3 \\ 4 & 5 & 6 \\ 7 & 8 & 9 \end{bmatrix} \sim \begin{bmatrix} 1 & 4 & 3 \\ 4 & 10 & 6 \\ 7 & 16 & 9 \end{bmatrix} = Q \tag{10.46}$$

or

$$P = \begin{bmatrix} 1 & 2 & 3 \\ 4 & 5 & 6 \\ 7 & 8 & 9 \end{bmatrix} \sim \begin{bmatrix} 1 & 2 & 3 \\ 8 & 10 & 12 \\ 7 & 8 & 9 \end{bmatrix} = R \tag{10.47}$$

Let

$$D = \begin{bmatrix} 1 & 0 & 0 \\ 0 & 2 & 0 \\ 0 & 0 & 1 \end{bmatrix} \tag{10.48}$$

Then

$$PD = Q \tag{10.49}$$

$$DP = R \tag{10.50}$$

Example 10.9 Addition of Polynomial Product
Using P from above, (1st column)–s×(2nd column)

$$P = \begin{bmatrix} 1 & 2 & 3 \\ 4 & 5 & 6 \\ 7 & 8 & 9 \end{bmatrix} \sim \begin{bmatrix} 1-2s & 2 & 3 \\ 4-5s & 5 & 6 \\ 7-8s & 8 & 9 \end{bmatrix} = Q \tag{10.51}$$

or (2nd row)–s×(1st row)

$$P = \begin{bmatrix} 1 & 2 & 3 \\ 4 & 5 & 6 \\ 7 & 8 & 9 \end{bmatrix} \sim \begin{bmatrix} 1 & 2 & 3 \\ 4-s & 5-2s & 6-3s \\ 7 & 8 & 9 \end{bmatrix} = R \tag{10.52}$$

Let

$$D = \begin{bmatrix} 1 & 0 & 0 \\ -s & 1 & 0 \\ 0 & 0 & 1 \end{bmatrix} \tag{10.53}$$

Then

$$PD = Q \tag{10.54}$$

$$DP = R \tag{10.55}$$

In this section, we introduce the concept of Smith form.

10.4 Smith Canonical Form

In the analysis and design of SISO controllers, the locations of zeros and poles of a transfer function model are critical. Therefore, when we are dealing with MIMO dynamic systems and, hence, transfer function matrices one can ask how we could obtain the location of zeros and poles. Before, introducing generalized transfer function matrices and their associated poles and zeros, we start with introducing Smith canonical form. One benefit of Smith canonical form is easy identification of the zeros of polynomial matrices.

We generalize the polynomial matrices equivalency by the following statement: Matrices $P(s)$ and $Q(s)$ (both $r \times m$) are in an equivalence relationship if and only if there exist unimodular U_R and U_L such that

$$\underset{(r\times m)}{P} = \underset{(r\times r)}{U_L}\ \underset{(r\times m)}{Q}\ \underset{(m\times m)}{U_R} \tag{10.56}$$

Let's suppose that $r = m$ and P is regular. Then,

$$\det P = \underbrace{\det U_L}_{\text{Constant}} \det Q \underbrace{\det U_R}_{\text{Constant}} \tag{10.57}$$

So, P and Q have the same degree and same zeros (this is also true for $r \neq m$).

Theorem 10.1 Smith Canonical Form *Any $r \times m$ polynomial matrix P is in an equivalence relationship to a diagonal matrix S whose diagonal elements are*

$$S_{kk} = \begin{cases} \epsilon_k(s), & k \leq rank\ \ P \\ 0, & k > rank\ \ P \end{cases} \tag{10.58}$$

where each $\epsilon_k(s)$ is monic and divides into $\epsilon_{k+1}(s)$ without remainder.

Proof: See [2]. ■

Definition 10.8 A polynomial is monic if the coefficient of the highest degree is equal to 1.

For square, full rank $P(s)$ with rank $P(s) = p$,

$$P \sim S = \begin{bmatrix} \epsilon_1(s) & 0 & \cdots & 0 \\ 0 & \epsilon_2(s) & & \\ \vdots & & \ddots & \\ 0 & & & \epsilon_p(s) \end{bmatrix} \tag{10.59}$$

Add zeros as needed for $P(s)$ not full rank or $P(s)$ not square.

Example 10.10 Consider the matrix P.

$$P = \begin{bmatrix} s & 1 \\ 0 & s \end{bmatrix} \tag{10.60}$$

We will put P into Smith canonical form by performing four transformations:

1. Exchanging columns 1 and 2
2. Subtract s times the first row from the second row into second row
3. Subtract s times the first column from the second column into second column
4. Multiply the second row by -1

$$P(s) = \begin{bmatrix} s & 1 \\ 0 & s \end{bmatrix} \overset{\text{Swap Columns}}{\sim} \begin{bmatrix} 1 & s \\ s & 0 \end{bmatrix} \overset{\text{2nd Row}-s\times\text{1st Row}}{\sim} \begin{bmatrix} 1 & s \\ 0 & -s^2 \end{bmatrix} \overset{\text{2nd Col.}-s\times\text{1st Col.}}{\sim} \begin{bmatrix} 1 & 0 \\ 0 & -s^2 \end{bmatrix} \overset{-1\times\text{2nd Col.}}{\sim} \begin{bmatrix} 1 & 0 \\ 0 & s^2 \end{bmatrix} = N(s) \tag{10.61}$$

For every action or matrix manipulation, there is an associated unimodular matrix, such that

1. Exchanging columns 1 and 2,

$$U_R(s) = \begin{bmatrix} 0 & 1 \\ 1 & 0 \end{bmatrix} \tag{10.62}$$

2. Subtract s times the first row from the second row,

$$U_L(s) = \begin{bmatrix} 1 & 0 \\ -s & 1 \end{bmatrix} \tag{10.63}$$

3. Subtract s times the first column from the second column,

$$U_R(s) = \begin{bmatrix} 1 & -s \\ 0 & 1 \end{bmatrix} \tag{10.64}$$

4. Multiply the second row by -1,

$$U_L(s) = \begin{bmatrix} 1 & 0 \\ 0 & -1 \end{bmatrix} \tag{10.65}$$

Therefore, we can multiply these unimodular matrices by $P(s)$ in one operation as follows:

$$\begin{bmatrix} 1 & 0 \\ 0 & -1 \end{bmatrix}\left(\left(\begin{bmatrix} 1 & 0 \\ -s & 1 \end{bmatrix}\left(P(s)\begin{bmatrix} 0 & 1 \\ 1 & 0 \end{bmatrix}\right)\right)\begin{bmatrix} 1 & -s \\ 0 & 1 \end{bmatrix}\right) = \begin{bmatrix} 1 & 0 \\ 0 & s^2 \end{bmatrix} \tag{10.66}$$

$P(s)$ includes two zeros at $s = 0$ as given by $N(s)$, this was not obvious from $P(s)$, unless, we compute the determinant of this matrix, but, this would not be possible if $P(s)$ is not a square matrix.

As shown in the example above, $P(s) \sim N(s)$ means that there exist right and left unimodular matrices U_R and U_L such that

$$U_L P U_R = N \tag{10.67}$$

But, what is the procedure to find U_R and U_L? To answer this question, we will set up the $P(s)$ and $N(s)$ matrices as shown below:

$$\left[\begin{array}{c:c} I_r & P \\ \hdashline & I_m \end{array}\right] \sim \left[\begin{array}{c:c} U_L & N \\ \hdashline & U_R \end{array}\right] \tag{10.68}$$

(the bottom left quadrant can be neglected).

Let's now repeat the same example as above by assuming P to be

$$P = \begin{bmatrix} s & 1 \\ 0 & s \end{bmatrix} \tag{10.69}$$

We know how to compute $N(s)$ from $P(s)$, by performing the same row and column operations. But this time, every time we perform a row or column operation, we will keep track of them in I_r and I_m.

So we must simply carry out the same steps on the larger matrix. we need to write down the operation steps on the top of each matrix:

$$\left[\begin{array}{cc:cc} 1 & 0 & s & 1 \\ 0 & 1 & 0 & s \\ \hdashline & & 1 & 0 \\ & & 0 & 1 \end{array}\right] \sim \left[\begin{array}{cc:cc} 1 & 0 & 1 & s \\ 0 & 1 & s & 0 \\ \hdashline & & 0 & 1 \\ & & 1 & 0 \end{array}\right] \sim \left[\begin{array}{cc:cc} 1 & 0 & 1 & s \\ -s & 1 & 0 & -s^2 \\ \hdashline & & 0 & 1 \\ & & 1 & 0 \end{array}\right] \tag{10.70}$$

$$\sim \left[\begin{array}{cc:cc} 1 & 0 & 1 & 0 \\ -s & 1 & 0 & -s^2 \\ \hdashline & & 0 & 1 \\ & & 1 & -s \end{array}\right] \sim \left[\begin{array}{cc:cc} 1 & 0 & 1 & 0 \\ s & -1 & 0 & s^2 \\ \hdashline & & 0 & 1 \\ & & 1 & -s \end{array}\right] \tag{10.71}$$

Verify the result:

$$U_L = \begin{bmatrix} 1 & 0 \\ s & -1 \end{bmatrix}, \quad U_R = \begin{bmatrix} 0 & 1 \\ 1 & -s \end{bmatrix} \tag{10.72}$$

Both U_R and U_L are unimodular.

$$\begin{bmatrix} 1 & 0 \\ s & -1 \end{bmatrix}\begin{bmatrix} s & 1 \\ 0 & s \end{bmatrix}\begin{bmatrix} 0 & 1 \\ 1 & -s \end{bmatrix} = \begin{bmatrix} 1 & 0 \\ s & -1 \end{bmatrix}\begin{bmatrix} 1 & 0 \\ s & -s^2 \end{bmatrix}$$

$$= \begin{bmatrix} 1 & 0 \\ 0 & s^2 \end{bmatrix}$$

$$= N(s) \ \checkmark$$

Example 10.11 Consider the matrix

$$P = \begin{bmatrix} s & s^2+s \\ 1 & s \end{bmatrix} \tag{10.73}$$

We will find the equivalent Smith canonical form matrix.

$$P(s) = \begin{bmatrix} s & s^2+s \\ 1 & s \end{bmatrix} \overset{\text{Swap rows}}{\sim} \begin{bmatrix} 1 & s \\ s & s^2+s \end{bmatrix} \overset{\substack{\text{2nd Row}-\\ s\times\text{1st Row}}}{\sim} \begin{bmatrix} 1 & s \\ 0 & s \end{bmatrix} \overset{\substack{\text{1st Row}-\\ \text{2nd row}}}{\sim} \begin{bmatrix} 1 & 0 \\ 0 & s \end{bmatrix} = N(s) \tag{10.74}$$

1. Swap the first and second rows
2. Subtract $s\times$ the first row from the second row
3. Subtract the second row from the first row

There is one zero at $s = 0$. Computation of the unimodular matrices are left as an exercise.

Example 10.12 Consider the matrix,

$$P = \begin{bmatrix} s+3 & -1 \\ -4 & 1 \end{bmatrix} \tag{10.75}$$

We will find the equivalent Smith canonical form matrix:

$$P(s) = \begin{bmatrix} s+3 & -1 \\ -4 & 1 \end{bmatrix} \overset{\text{Swap columns}}{\sim} \begin{bmatrix} -1 & s+3 \\ 1 & -4 \end{bmatrix} \overset{\substack{\text{Add 1st row} \\ \text{to 2nd row}}}{\sim} \begin{bmatrix} -1 & s+3 \\ 0 & s-1 \end{bmatrix} \overset{\substack{\text{Add } (s+3)\times \\ \text{1st Col. to} \\ \text{2nd Col.}}}{\sim} \begin{bmatrix} 1 & 0 \\ 0 & s-1 \end{bmatrix} = N(s) \tag{10.76}$$

1. Swap the first and second columns
2. Add the first row to the second row
3. Add $(s + 3)\times$ the first column to the second column

There is one zero at $s = 1$. Computation of the unimodular matrices are left as an exercise.

Matlab command "smithForm (P)" could be utilized to compute the unimodular and diagonal matrices. However, this command only works for square matrices.

10.5 Smith–McMillan Form

In Section 10.4, we introduced Smith form of matrices and discussed that one of the advantages of this form is determining the zeros of polynomial matrices.

In this section, we will introduce Smith–McMillan form, this form of transfer function matrices is beneficial in determining the location of both poles and zeros of these matrices, and hence it will be utilized in later in this book for MIMO control design. Furthermore, we discuss the relation between the MFD's and Smith–McMillan form.

10.5.1 Smith–McMillan Form

Theorem 10.2 *Every transfer function matrix $G(s)$ of a general size of $r \times m$ can be transformed to a pseudo-diagonal matrix through row and column operations:*

$$\underset{(r\times m)}{G(s)} \sim \left[\begin{array}{cccc|c} \frac{\eta_1}{\delta_1} & 0 & \cdots & 0 & 0 \\ 0 & \frac{\eta_2}{\delta_2} & & & \\ \vdots & & \ddots & & \\ 0 & & & \frac{\eta_p}{\delta_p} & \\ \hline 0 & & & & 0 \end{array}\right] = M \tag{10.77}$$

M is an $r \times m$ rational matrix whose first p (where $p = \min(r, m)$) diagonal elements ratios of polynomials and all other elements are 0. η_k divides into η_{k+1} without remainder, while δ_k divides into δ_{k-1} without remainder.

Proof: Assuming $r = m$ without loss of generality, we can always write $G(s) = \frac{p(s)}{\delta(s)}$, where $p(s)$ is a polynomial matrix and $\delta(s)$ is the greatest common multiple of the denominators $G_{ij}(s)$. Using row and column operations, $p(s)$ can be reduced to Smith form:

$$G(s) \sim \frac{1}{\delta}\begin{bmatrix} \varepsilon_1 & 0 & \cdots & 0 \\ 0 & \varepsilon_2 & \cdots & 0 \\ \vdots & \ddots & \ddots & \vdots \\ 0 & 0 & \cdots & \varepsilon_p \end{bmatrix} \tag{10.78}$$

After canceling any factors between δ and ε_i and rearranging to meet the division requirements,

$$\frac{1}{\delta}\begin{bmatrix} \varepsilon_1 & 0 & \cdots & 0 \\ 0 & \varepsilon_2 & \cdots & 0 \\ \vdots & \ddots & \ddots & \vdots \\ 0 & 0 & \cdots & \varepsilon_p \end{bmatrix} = \begin{bmatrix} \frac{n_1}{\delta_1} & 0 & \cdots & 0 \\ 0 & \frac{n_2}{\delta_2} & \cdots & 0 \\ \vdots & \ddots & \ddots & \vdots \\ 0 & 0 & \cdots & \frac{n_p}{\delta_p} \end{bmatrix} = M \tag{10.79}$$

By transitivity, $G(s) \sim M$. ■

There are different approaches to compute Smith–McMillan form. Two of the most popular approaches are listed below:

1) Factor $G(s) = \frac{p(s)}{\delta(s)}$ then transform $p(s)$ into Smith form by row and column manipulations.
2) Define a determinantal divisor D_k = greatest common divisor (gcd) of all $k \times k$ minors and setting $D_0 = 1$ with the following steps:
   a) Factor $G(s) = \frac{p(s)}{\delta(s)}$,
   b) Find all minors of $p(s)$,
   c) Compute D_k for each $k \times k$ minor,
   d) Calculate $\varepsilon_k = \frac{D_k}{D_{k-1}}$ for each $i \in \{1, \ldots, p\}$,
   e) Find M using $M = \frac{1}{\delta(s)}\text{diag}(\varepsilon_k)$.

We demonstrate the first approach by the following example:

Example 10.13 Consider the dynamic system defined in state-space form by:

$$A = \begin{bmatrix} -2 & 1 & 1 \\ 1 & -2 & 1 \\ 1 & 1 & -2 \end{bmatrix}, \quad B = \begin{bmatrix} 1 & 0 \\ 0 & 1 \\ 0 & 0 \end{bmatrix} \tag{10.80}$$

$$C^T = \begin{bmatrix} 1 & 0 & 0 \\ 0 & 0 & 1 \end{bmatrix}, \quad D = \begin{bmatrix} 0 & 0 \\ 0 & 0 \end{bmatrix} \tag{10.81}$$

$$\begin{aligned} G(s) &= \frac{1}{\det(sI - A)}\begin{bmatrix} (s+3)(s+1) & (s+3) \\ (s+3) & (s+3) \end{bmatrix} \\ &= \frac{1}{s(s+3)^2}\underbrace{\begin{bmatrix} (s+3)(s+1) & (s+3) \\ (s+3) & (s+3) \end{bmatrix}}_{p(s)} \end{aligned}$$

Transform $p(s)$ into Smith form by the following row and column manipulations:

1. Swap the first and second rows.
2. Subtract $(s+1)$ times the first row from the second row.
3. Subtract the first column from the second column.

$$p(s) = \begin{bmatrix} (s+3)(s+1) & (s+3) \\ (s+3) & (s+3) \end{bmatrix} \sim \begin{bmatrix} (s+3) & (s+3) \\ (s+3)(s+1) & (s+3) \end{bmatrix} \tag{10.82}$$

$$\sim \begin{bmatrix} (s+3) & (s+3) \\ 0 & s(s+3) \end{bmatrix} \sim \begin{bmatrix} (s+3) & 0 \\ 0 & s(s+3) \end{bmatrix} \tag{10.83}$$

Then, the Smith–McMillan form of $G(s)$ is

$$G(s) \sim \frac{1}{s(s+3)^2} \begin{bmatrix} (s+3) & 0 \\ 0 & s(s+3) \end{bmatrix} = \begin{bmatrix} \frac{1}{s(s+3)} & 0 \\ 0 & \frac{1}{s+3} \end{bmatrix} \tag{10.84}$$

We will now look at the second or a more procedural approach to compute the Smith–McMillan form. Let determinantal divisors $D_0 = 1$, D_k = the greatest common divisor of all $k \times k$ minors. It can be shown that

$$\epsilon_k = \frac{D_k}{D_{k-1}} \tag{10.85}$$

We repeat the example above but this time, we use the second approach to compute the Smith–McMillan form.

Example 10.14

$$p(s) = \begin{bmatrix} (s+3)(s+1) & (s+3) \\ (s+3) & (s+3) \end{bmatrix} \tag{10.86}$$

$$1 \times 1 \text{ Minors:} (s+3)(s+1),\ (s+3),\ (s+3),\ (s+3)$$

$$2 \times 2 \text{ Minors: } \det \begin{bmatrix} (s+3)(s+1) & (s+3) \\ (s+3) & (s+3) \end{bmatrix}$$

$$\begin{aligned} D_0 &= 1 \\ D_1 &= \gcd\,((s+3)(s+1), (s+3)) \\ &= s+3 \\ D_2 &= s(s+3)^2 \end{aligned}$$

$$\Rightarrow\ \epsilon_1 = \frac{D_1}{D_0} = s+3, \quad \epsilon_2 = \frac{D_2}{D_1} = s(s+3) \tag{10.87}$$

Therfore, with ϵ_1 and ϵ_2, it is straightforward to show that we will obtain the same Smith–McMillan form as the first approach.

Example 10.15 Let's repeat the Example 10.14 but this time with a different B matrix. Select B to be

$$B = \begin{bmatrix} 1 & 0 \\ 0 & \frac{1}{2} \\ 0 & \frac{1}{2} \end{bmatrix} \quad \Rightarrow \quad G(s) = \frac{1}{s(s+3)^2} \underbrace{\begin{bmatrix} (s+1)(s+3) & (s+3) \\ (s+3) & \frac{1}{2}(s+2)(s+3) \end{bmatrix}}_{p(s)} \tag{10.88}$$

$$1 \times 1 \text{ Minors:} (s+3)(s+1),\ (s+3),\ (s+3),\ \frac{1}{2}(s+3)(s+2)$$

$$2 \times 2 \text{ Minors: } \det \begin{bmatrix} (s+1)(s+3) & (s+3) \\ (s+3) & \frac{1}{2}(s+2)(s+3) \end{bmatrix}$$

$$\begin{aligned}
D_0 &= 1\\
D_1 &= \gcd\left((s+3)(s+1), (s+3), \frac{1}{2}(s+3)(s+2)\right)\\
&= s+3\\
D_2 &= \frac{1}{2}s(s+3)^3\\
\Rightarrow\ \epsilon_1 &= \frac{D_1}{D_0} = s+3, \quad \epsilon_2 = \frac{D_2}{D_1} = \frac{1}{2}s(s+3)^2
\end{aligned} \tag{10.89}$$

$$G(s) \sim \frac{1}{s(s+3)^2}\begin{bmatrix} (s+3) & 0 \\ 0 & \frac{1}{2}s(s+3)^2 \end{bmatrix} = \begin{bmatrix} \frac{1}{s(s+3)} & 0 \\ 0 & \frac{1}{2} \end{bmatrix} \tag{10.90}$$

It is not as obvious as the previous example, but in this example δ_k divides into δ_{k-1}, this means, $s(s+3) = 2 \times \frac{1}{2} \times s(s+3)$ is a factor.

10.5.2 MFD's and Their Relations to Smith–McMillan Form

Recall left and right MFDs for MIMO systems,

$$G(s) = D_L^{-1}N_L \ = N_R D_R^{-1} \tag{10.91}$$

We discussed the idea of coprimeness for SISO transfer functions. Here, we extend this idea to MIMO systems,

Definition 10.9 (Right Common Factor) W is a right common factor of N_R and D_R if $N_R = N_R'W$ and $D_R = D_R'W$ where N_R', D_R', and W are polynomial matrices. Left common factors are equivalently defined.

Definition 10.10 (Coprimeness) N_R and D_R are right coprime if all right common factors, W, are unimodular. Left coprime is equivalently defined.

Definition 10.11 (Irreducible MFD) $N_R D_R^{-1}$ is an irreducible MFD of $G(s)$ if N_R and D_R are right coprime. Irreducibility is equivalently defined for a left MFD.

We will now consider finding a MFD from the Smith–McMillan form of matrix $G(s)$:

$$G(s) \sim M \quad \Rightarrow \quad U_L G(s) U_R = M = \operatorname{diag}\left(\frac{\eta_1}{\delta_1}, \frac{\eta_2}{\delta_2}, \ldots, \frac{\eta_p}{\delta_p}, 0, \ldots, 0\right) \tag{10.92}$$

So,

$$G(s) = U_L^{-1}\left[\begin{array}{cccc|c} \eta_1 & 0 & \cdots & 0 & 0 \\ 0 & \eta_2 & & & \\ \vdots & & \ddots & & \\ 0 & & & \eta_p & \\ \hline 0 & & & & 0 \end{array}\right]\left[\begin{array}{cccc|c} \delta_1 & 0 & \cdots & 0 & 0 \\ 0 & \delta_2 & & & \\ \vdots & & \ddots & & \\ 0 & & & \delta_p & \\ \hline 0 & & & & 1 \end{array}\right]^{-1} U_R^{-1} \tag{10.93}$$

Therefore,

$$G(s) = N_R D_R^{-1} \quad \text{where} \quad \begin{aligned} N_R &= U_L^{-1}\operatorname{diag}(\eta_1, \ldots, \eta_p, 0, \ldots, 0) \\ D_R &= U_R \operatorname{diag}(\delta_1, \ldots, \delta_p, 1, \ldots, 1) \end{aligned} \tag{10.94}$$

However,

$$\left[\begin{array}{cccc|c} \eta_1 & 0 & \cdots & 0 & 0 \\ 0 & \eta_2 & & & \\ \vdots & & \ddots & & \\ 0 & & & \eta_p & \\ \hline 0 & & & & 0 \end{array}\right] \quad \text{and} \quad \left[\begin{array}{cccc|c} \delta_1 & 0 & \cdots & 0 & 0 \\ 0 & \delta_2 & & & \\ \vdots & & \ddots & & \\ 0 & & & \delta_p & \\ \hline 0 & & & & 1 \end{array}\right] \tag{10.95}$$

are coprime because each η_k/δ_k is irreducible. Hence, this MFD, N_R and D_R, will be coprime.

Poles of $G(s)$ are the zeros of D_R, so the characteristic polynomial is

$$\delta_1(s)\delta_2(s)\ldots\delta_p(s) \tag{10.96}$$

The degree of the characteristic polynomial is called the Smith–McMillan (SM) degree. **It is important to note that the SM form is the minimum realization of** $G(s)$. All poles are the eigenvalues of A matrix in the state-space representation.

Zeros of $G(s)$ are zeros of N_R. The zero polynomial is therefore,

$$\eta_1(s)\eta_2(s)\ldots\eta_p(s) \tag{10.97}$$

It is also interesting to note if one could compute a coprime MFD and Smith–MacMillan form, then from relationship above, the left and right unimodular matrices could be computed. This will be presented in an example later in this section. Computation of the unimodular matrices become important during MIMO control design computation as will be presented in the later chapters.

10.5.3 Computing an Irreducible (Coprime) Matrix Fraction Description

Given N_R and D_R (right MFD in this case, left MFD is equivalent) which are not coprime, we ask the following question, how can we find the common divisor of greatest degree for computing a coprime MFD? Imagine row operations on,

$$\begin{bmatrix} \underset{}{\overset{(m\times m)}{D_R}} \\ ----- \\ \underset{(r\times m)}{N_R} \end{bmatrix} \sim \begin{bmatrix} \overset{(m\times m)}{W} \\ ----- \\ \underset{(r\times m)}{0} \end{bmatrix} \tag{10.98}$$

Now, we can consider a matrix V such that,

$$\begin{bmatrix} D_R \\ ---- \\ N_R \end{bmatrix} = V \begin{bmatrix} W \\ ---- \\ 0 \end{bmatrix} = \left[\begin{array}{c|c} V_{11} & V_{12} \\ \hline V_{21} & V_{22} \end{array}\right] \begin{bmatrix} W \\ ---- \\ 0 \end{bmatrix} \tag{10.99}$$

So,

$$D_R = V_{11}W, \quad N_R = V_{21}W \tag{10.100}$$

or conversely,

$$V_{11} = D_R W^{-1}, \quad V_{21} = N_R W^{-1} \tag{10.101}$$

Then,

$$G(s) = V_{21}V_{11}^{-1} \tag{10.102}$$

is an irreducible factorization of $G(s)$, where V_{11} and V_{21} are coprime.

In the next two examples, we will illustrate the procedure above for computing a coprime MFD. The first example is for the SISO case and the second example is for the MIMO case.

Example 10.16 Consider the transfer function,

$$G(s) = \frac{s^2+s}{s^2+3s+2} = \frac{s(s+1)}{(s+1)(s+2)} \tag{10.103}$$

which we know is not coprime due to the common factor of $(s+1)$ but we will verify this below. We select the numerator and the denominator of this transfer function to be $N_R = s(s+1)$ and $D_R = (s+1)(s+2)$. Then given the procedure above, we can write

$$\begin{bmatrix} D_R \\ ---- \\ N_R \end{bmatrix} = \underset{(a)}{\begin{bmatrix} s^2+3s+2 \\ ---------- \\ s^2+s \end{bmatrix}} \sim \underset{(b)}{\begin{bmatrix} 2s+2 \\ ------ \\ s^2+s \end{bmatrix}} \sim \underset{(c)}{\begin{bmatrix} 2s+2 \\ ------ \\ 0 \end{bmatrix}} \sim \underset{(d)}{\begin{bmatrix} s+1 \\ ----- \\ 0 \end{bmatrix}} = \begin{bmatrix} W \\ ---- \\ 0 \end{bmatrix} \tag{10.104}$$

$a \rightarrow b$ Subtract the second row from the first.

$b \rightarrow c$ Subtract $\frac{1}{2}s$ times the first row from the second row.

$c \rightarrow d$ Divide the first row by 2.

We find that $W = s+1$. W is not unimodular (scalar), so N_R and D_R are not coprime. Now, we convert N_R and D_R to a coprime MFD,

$$N_R = V_{21}W \quad \Rightarrow \quad N_R' = V_{21} = N_R W^{-1} = s \tag{10.105}$$

$$D_R = V_{11}W \quad \Rightarrow \quad D_R' = V_{11} = D_R W^{-1} = s+2 \tag{10.106}$$

$$G(s) = \underbrace{N_R' D_R'^{-1}}_{\text{Coprime}} = \frac{s}{s+2} \tag{10.107}$$

Example 10.17 Consider the following non-coprime MFD,

$$G(s) = N_R D_R^{-1} \tag{10.108}$$

With the following N_R and D_R,

$$N_R = \begin{bmatrix} s+2 & (s+1)(s+2) \\ 0 & s(s+1) \end{bmatrix}, \quad D_R = \begin{bmatrix} (s+1)(s+2) & 0 \\ 0 & (s+1)(s+2) \end{bmatrix} \tag{10.109}$$

Now, we compute the common divisor of greatest degree, W matrix, using the procedure above as follows:

$$\begin{bmatrix} D_R \\ \hline N_R \end{bmatrix} = \underbrace{\begin{bmatrix} (s+1)(s+2) & 0 \\ 0 & (s+1)(s+2) \\ \hline s+2 & (s+1)(s+2) \\ 0 & s(s+1) \end{bmatrix}}_{(a)} \sim \underbrace{\begin{bmatrix} (s+1)(s+2) & 0 \\ 0 & 2(s+1) \\ \hline s+2 & 2(s+1) \\ 0 & s(s+1) \end{bmatrix}}_{(b)} \sim \tag{10.110}$$

$$\underbrace{\begin{bmatrix} (s+1)(s+2) & 0 \\ 0 & (s+1) \\ \hline s+2 & 2(s+1) \\ 0 & 0 \end{bmatrix}}_{(c)} \sim \underbrace{\begin{bmatrix} (s+1)(s+2) & 0 \\ 0 & (s+1) \\ \hline s+2 & 0 \\ 0 & 0 \end{bmatrix}}_{(d)} \sim \tag{10.111}$$

$$\underbrace{\begin{bmatrix} (s+2) & 0 \\ 0 & (s+1) \\ \hline s+2 & 0 \\ 0 & 0 \end{bmatrix}}_{(e)} \sim \underbrace{\begin{bmatrix} s+2 & 0 \\ 0 & s+1 \\ \hline 0 & 0 \\ 0 & 0 \end{bmatrix}}_{(f)} \tag{10.112}$$

$a \to b$ Subtract the last row from the second and third rows.
$b \to c$ Divide the second row by 2, subtract s times the second row from the last row.
$c \to d$ Add -2 times the second row to the third row.
$d \to e$ Subtract s times the third row from the first row.
$e \to f$ Multiply the first row by -1 and then subtract the first row from the third row.

$$W = \begin{bmatrix} s+2 & 0 \\ 0 & s+1 \end{bmatrix} \quad \Rightarrow \quad W^{-1} = \begin{bmatrix} \frac{1}{s+2} & 0 \\ 0 & \frac{1}{s+1} \end{bmatrix} \tag{10.113}$$

$$N_R' = V_{21} = N_R W^{-1} = \begin{bmatrix} s+2 & (s+1)(s+2) \\ 0 & s(s+1) \end{bmatrix} \begin{bmatrix} \frac{1}{s+2} & 0 \\ 0 & \frac{1}{s+1} \end{bmatrix} = \begin{bmatrix} 1 & s+2 \\ 0 & s \end{bmatrix} \tag{10.114}$$

$$D'_R = V_{11} = D_R W^{-1} = \begin{bmatrix} (s+1)(s+2) & 0 \\ 0 & (s+1)(s+2) \end{bmatrix} \begin{bmatrix} \frac{1}{s+2} & 0 \\ 0 & \frac{1}{s+1} \end{bmatrix}$$

$$= \begin{bmatrix} s+1 & 0 \\ 0 & s+2 \end{bmatrix}$$

$$\Rightarrow G(s) = \underbrace{\begin{bmatrix} 1 & s+2 \\ 0 & s \end{bmatrix} \begin{bmatrix} s+1 & 0 \\ 0 & s+2 \end{bmatrix}^{-1}}_{\text{Coprime}} \tag{10.115}$$

In the next example, we compute the Smith–McMillan form of a transfer function matrix. In the last example, we demonstrate how to compute the unimodular matrices from MFDs and Smith–McMillan forms. We accomplish this by first computing a coprime MFD and then use this MFD with its Smith–McMillan form to compute the unimodular matrices.

Example 10.18 Consider the following transfer function matrix,

$$G(s) = \begin{bmatrix} \frac{1}{(s+1)^2} & \frac{1}{(s+1)(s+2)} \\ \frac{1}{(s+1)(s+2)} & \frac{s+3}{(s+2)^2} \end{bmatrix}$$
$$= \begin{bmatrix} (s+2)^2 & (s+1)(s+2) \\ (s+1)(s+2) & (s+1)^2(s+3) \end{bmatrix} \frac{1}{(s+1)^2(s+2)^2}$$

then using the second approach for computing the Smith–McMillan form, the determinantal divisors are written as:

$$D_0 = 1, \quad D_1 = 1, \quad D_2 = (s+1)^2(s+2)^3 \tag{10.116}$$

$$\Longrightarrow \epsilon_1 = \frac{D_1}{D_0} = 1, \quad \epsilon_2 = \frac{D_2}{D_1} = (s+1)^2(s+2)^3 \tag{10.117}$$

Therefore, the SM form can be written as:

$$G(s) \sim \begin{bmatrix} \frac{1}{(s+1)^2(s+2)^2} & 0 \\ 0 & s+2 \end{bmatrix} \tag{10.118}$$

The zero polynomial of $G(s)$ is therefore $s+2=0$, and the characteristic polynomial is $(s+1)^2(s+2)^2 = 0$. $G(s)$ has a Smith–McMillan degree of 4.

Therefore, all other MFD's are irreducible if and only if their deg $= 4$.

Note:

(1) Not all the entries of Smith–McMillan form are proper transfer functions. It is important to know that there is no restriction on the properness of the SM forms, and they simply reveal the poles and zeros of MIMO systems.
(2) The poles and zeros of det $(G(s))$ are not similar to the poles and zeros of the Smith–McMillan form of $G(s)$. A MIMO system is directional-based, so zeros and poles of the same magnitude but different direction do not cancel each other. But this is not the same as when one computes the determinant of a matrix as the act of computing the determinant eliminates directional sensitivity, and hence all the poles and zeros with the same magnitude are then canceled.

Example 10.19 Consider the same transfer function matrix as the above example, $G(s)$ can be written as multiplication of a numerator, N_R and denominator, D_R matrices as follows:

$$G(s) = \begin{bmatrix} \frac{1}{(s+1)^2} & \frac{1}{(s+1)(s+2)} \\ \frac{1}{(s+1)(s+2)} & \frac{s+3}{(s+2)^2} \end{bmatrix}$$

$$= \begin{bmatrix} (s+2) & (s+2) \\ (s+1) & (s+1)(s+3) \end{bmatrix} \begin{bmatrix} \overset{\text{GCD of 1st Col.}}{\frac{1}{(s+1)^2(s+2)}} & 0 \\ 0 & \underset{\text{GCD of 2nd Col.}}{\frac{1}{(s+1)(s+2)^2}} \end{bmatrix}$$

$$= \underbrace{\begin{bmatrix} (s+2) & (s+2) \\ (s+1) & (s+1)(s+3) \end{bmatrix}}_{N_R} \underbrace{\begin{bmatrix} (s+1)^2(s+2) & 0 \\ 0 & (s+1)(s+2)^2 \end{bmatrix}^{-1}}_{D_R:\ \deg=6}$$

It is obvious that this MFD is not coprime as the degree of D_R is greater than 4. Hence, we can compute a coprime factorization by calculating the W matrix:

$$\begin{bmatrix} D_R \\ \hline N_R \end{bmatrix} = \underset{(a)}{\begin{bmatrix} (s+1)^2(s+2) & 0 \\ 0 & (s+1)(s+2)^2 \\ \hline s+2 & s+2 \\ s+1 & (s+1)(s+3) \end{bmatrix}} \sim \underset{(b)}{\begin{bmatrix} 0 & -(s+1)^2(s+2) \\ 0 & (s+1)(s+2)^2 \\ \hline s+2 & s+2 \\ s+1 & (s+1)(s+3) \end{bmatrix}} \sim$$

$$\underset{(c)}{\begin{bmatrix} 0 & -(s+1)^2(s+2) \\ 0 & (s+1)(s+2) \\ \hline s+2 & s+2 \\ s+1 & (s+1)(s+3) \end{bmatrix}} \sim \underset{(d)}{\begin{bmatrix} 0 & 0 \\ 0 & (s+1)(s+2) \\ \hline s+2 & s+2 \\ s+1 & (s+1)(s+3) \end{bmatrix}} \sim$$

$$\underset{(e)}{\begin{bmatrix} 0 & 0 \\ 0 & (s+1)(s+2) \\ \hline s+2 & s+2 \\ s+1 & s+1 \end{bmatrix}} \sim \underset{(f)}{\begin{bmatrix} 0 & 0 \\ 0 & (s+1)(s+2) \\ \hline 1 & 1 \\ s+1 & s+1 \end{bmatrix}} \sim$$

$$\underset{(g)}{\begin{bmatrix} 0 & 0 \\ 0 & (s+1)(s+2) \\ \hline 1 & 1 \\ 0 & 0 \end{bmatrix}} \sim \underset{(h)}{\begin{bmatrix} 1 & 1 \\ 0 & (s+1)(s+2) \\ \hline 0 & 0 \\ 0 & 0 \end{bmatrix}} \sim$$

$a \to b$ Subtract $(s+1)^2$ times the 3rd row from the 1st row.
$b \to c$ Add the 1st row to the 2nd row.
$c \to d$ Add $(s+1)$ times the 2nd row to the 1st row.
$d \to e$ Subtract the 2nd row from the 4th row.
$e \to f$ Subtract the 4th row from the 3rd row.
$f \to g$ Subtract $(s+1)$ times the 3rd row from the 4th row.
$g \to h$ Swap the 1st and 3rd row.

So, the right common factor, W, is

$$W = \begin{bmatrix} 1 & 1 \\ 0 & (s+1)(s+2) \end{bmatrix} \quad \Rightarrow \quad W^{-1} = \begin{bmatrix} 1 & -\dfrac{1}{(s+1)(s+2)} \\ 0 & \dfrac{1}{(s+1)(s+2)} \end{bmatrix} \tag{10.119}$$

$$\begin{aligned} N_R' = N_R W^{-1} &= \begin{bmatrix} (s+2) & (s+2) \\ (s+1) & (s+1)(s+3) \end{bmatrix} \begin{bmatrix} 1 & -\dfrac{1}{(s+1)(s+2)} \\ 0 & \dfrac{1}{(s+1)(s+2)} \end{bmatrix} \\ &= \begin{bmatrix} s+2 & 0 \\ s+1 & 1 \end{bmatrix} \end{aligned}$$

$$\begin{aligned} D_R' = D_R W^{-1} &= \begin{bmatrix} (s+1)^2(s+2) & 0 \\ 0 & (s+1)(s+2)^2 \end{bmatrix} \begin{bmatrix} 1 & -\dfrac{1}{(s+1)(s+2)} \\ 0 & \dfrac{1}{(s+1)(s+2)} \end{bmatrix} \\ &= \begin{bmatrix} (s+1)^2(s+2) & -(s+1) \\ 0 & s+2 \end{bmatrix} \end{aligned}$$

So, an irreducible realization of $G(s)$ is

$$G(s) = N_R' D_R'^{-1} = \underbrace{\begin{bmatrix} s+2 & 0 \\ s+1 & 1 \end{bmatrix}}_{\text{Zeros}=-2} \underbrace{\begin{bmatrix} (s+1)^2(s+2) & -(s+1) \\ 0 & s+2 \end{bmatrix}^{-1}}_{\text{Poles}=-2,-2,-1,-1} \tag{10.120}$$

Note that the degree of this denominator polynomial matrix, D_R', is 4, precisely equal to the SM degree.

Next, we must find U_R and U_L, by knowing both the SM form and a coprime MFD:

$$G(s) = \begin{bmatrix} s+2 & 0 \\ s+1 & 1 \end{bmatrix} \begin{bmatrix} (s+1)^2(s+2) & -(s+1) \\ 0 & s+2 \end{bmatrix}^{-1} \sim \underbrace{\begin{bmatrix} \dfrac{1}{(s+1)^2(s+2)^2} & 0 \\ 0 & s+2 \end{bmatrix}}_{\text{SM form}} \tag{10.121}$$

we know

$$N_R' = U_L^{-1}\operatorname{diag}(\eta_1, \eta_2, \ldots, \eta_p) \tag{10.122}$$

$$\begin{bmatrix} s+2 & 0 \\ s+1 & 1 \end{bmatrix} = U_L^{-1} \begin{bmatrix} 1 & 0 \\ 0 & s+2 \end{bmatrix} \tag{10.123}$$

therefore,

$$U_L = \begin{bmatrix} 1 & 0 \\ 0 & s+2 \end{bmatrix} \begin{bmatrix} s+2 & 0 \\ s+1 & 1 \end{bmatrix}^{-1} = \begin{bmatrix} \frac{1}{s+2} & 0 \\ -(s+1) & s+2 \end{bmatrix} \tag{10.124}$$

and

$$D_R = U_R \operatorname{diag}(\delta_1, \delta_2, \ldots, \delta_p) \tag{10.125}$$

$$\begin{bmatrix} (s+1)^2(s+2) & -(s+1) \\ 0 & s+2 \end{bmatrix} = U_R \begin{bmatrix} (s+1)^2(s+2)^2 & 0 \\ 0 & 1 \end{bmatrix} \tag{10.126}$$

so,

$$\begin{aligned} U_R &= \begin{bmatrix} (s+1)^2(s+2) & -(s+1) \\ 0 & s+2 \end{bmatrix} \begin{bmatrix} (s+1)^2(s+2)^2 & 0 \\ 0 & 1 \end{bmatrix}^{-1} \\ &= \begin{bmatrix} \dfrac{1}{s+2} & -(s+1) \\ 0 & s+2 \end{bmatrix} \end{aligned}$$

Knowing the unimodular matrices, U_L and U_R, we now verify if we could obtain the same Smith–McMillan form computed above, directly from the transfer function matrix, $G(s)$, as follows:

$$U_L G(s) U_R = M \tag{10.127}$$

$$\begin{bmatrix} \frac{1}{s+2} & 0 \\ -s-1 & s+2 \end{bmatrix} \begin{bmatrix} \frac{1}{(s+1)^2} & \frac{1}{(s+1)(s+2)} \\ \frac{1}{(s+1)(s+2)} & \frac{s+3}{(s+2)^2} \end{bmatrix} \begin{bmatrix} \frac{1}{s+2} & -s-1 \\ 0 & s+2 \end{bmatrix} = \begin{bmatrix} \frac{1}{(s+1)^2(s+2)^2} & 0 \\ 0 & s+2 \end{bmatrix}$$

This is the same SM form computed above.

Note:

1) U_L and U_R in the example above are not polynomial matrices, but their determinants are constants and are equal to 1, as some of the entries are rational functions.
2) Unimodular matrices U_L and U_R can be computed using hermite form (as is done in Matlab and not included here) to obtain the following matrices:

$$U_L = \begin{bmatrix} 1 & -(s^2+4s+5) \\ s+1 & -(s^3+5s^2+9s+6) \end{bmatrix}$$

$$U_R = \begin{bmatrix} s+2 & -(s^5+9s^4+32s^3+55s^2+44s+13) \\ -1 & s^4+7s^3+18s^2+19s+6 \end{bmatrix}$$

Multiplication of these unimodular matrices with $G(s)$ results in the same M matrix! (Try it).

10.6 MIMO Controllability and Observability

In this section, we discuss controllability and observability as they relate to coprime factorization. We then briefly present the SISO state-space realization before discussing the steps required for the computation of MIMO state-space realization in controllable form (observable form).

Consider the state equations for a MIMO system:

$$\left.\begin{aligned} \overset{(n)}{\dot{\underline{x}}} &= A\underline{x} + B\overset{(m)}{\underline{u}} \\ \underset{(r)}{\underline{y}} &= C^T\underline{x} + D\underline{u} \end{aligned}\right\} \underset{(r\times m)}{G(s)} = C^T(sI-A)^{-1}B + D = \underset{(r\times m)(m\times m)}{N_R \; D_R^{-1}} = \underset{(r\times r)(r\times m)}{D_L^{-1} \; N_L} \tag{10.128}$$

These two realizations are strictly comparable if and only if deg D_L or deg $D_R = n$. We present the controllability and observability of MIMO dynamic systems as follows: The realization is controllable (inputs can effect all states) if

$$\text{rank}\,\underset{(n\times nm)}{\begin{bmatrix} B & AB & \cdots & A^{n-1}B \end{bmatrix}} = n \tag{10.129}$$

The realization is observable (outputs are affected by all states) if

$$\text{rank}\,\underset{(rn\times n)}{\begin{bmatrix} C^T & C^TA & \cdots & C^TA^{n-1} \end{bmatrix}^T} = n \tag{10.130}$$

We have previously discussed the idea of comprimeness of an MFD or factorization. We can show observability and controllability if and only if N_R and D_R (or N_L and D_L) are coprime. We know that eigenvalues of A are poles of $G(s)$ (and zeros of D_R or D_L) and the transmission zeros of (A, B, C^T, D) are zeros of $G(s)$ (zeros of N_R or N_L).

If the realization is uncontrollable or unobservable, then there is at least one transmission zero on an eigenvalue. A common factor in D and N will also have a pole and zero at the same location.

The minimum/irreducible realization is controllable and observable if deg D_R or deg $D_L = n$ and where N_R and D_R (or N_L and D_L) are coprime.

10.6.1 State-Space Realization

We have shown how to convert a state-space system representation to a transfer function (or transfer function matrix), but how do we convert a transfer function (or transfer function matrix) to state-space form? Before, we discuss state-space representation for MIMO systems, we present this realization for SISO systems which has been covered in various texts but we present this material for the sake of completeness.

10.6.1.1 SISO System

We can easily express a transfer function in state-space form by first expressing the transfer function as a differential equation and then rewriting it as a set of first-order differential equations. We will consider three cases for the transfer function $G(s)$:

1. $G(s)$ is strictly proper and has a numerator equal to unity.
2. $G(s)$ is strictly proper.
3. $G(s)$ is proper.

A third-order system will be used for this example, but the method can be generalized to systems of any order.

Case 1:

$$G(s) = \frac{1}{s^3 + a_1 s^2 + a_2 s + a_3} = \frac{Y}{U}(s) \tag{10.131}$$

First, write as a differential equation,

$$\dddot{y} + a_1 \ddot{y} + a_2 \dot{y} + a_3 y = u \tag{10.132}$$

Then, express in matrix form as a system of first-order equations:

$$\begin{bmatrix} \dot{y} \\ \ddot{y} \\ \dddot{y} \end{bmatrix} = \underbrace{\begin{bmatrix} 0 & 1 & 0 \\ 0 & 0 & 1 \\ -a_3 & -a_2 & -a_1 \end{bmatrix}}_{A} \cdot \begin{bmatrix} y \\ \dot{y} \\ \ddot{y} \end{bmatrix} + \underbrace{\begin{bmatrix} 0 \\ 0 \\ 1 \end{bmatrix}}_{B} \cdot u \tag{10.133}$$

$$y = \underbrace{\begin{bmatrix} 1 & 0 & 0 \end{bmatrix}}_{C^T} \cdot \begin{bmatrix} y \\ \dot{y} \\ \ddot{y} \end{bmatrix} + \underbrace{\begin{bmatrix} 0 \end{bmatrix}}_{D} \cdot u \tag{10.134}$$

Case 2:

$$G(s) = \frac{b_1 s^2 + b_2 s + b_3}{s^3 + a_1 s^2 + a_2 s + a_3} = \frac{Y}{U}(s) \tag{10.135}$$

First, we define a new transfer function Y/V, where V is a pseudo-state, such that,

$$\frac{Y}{U} = \frac{Y}{V} \cdot \frac{V}{U} \tag{10.136}$$

We define Y/V to be the numerator of the transfer function and V/U to be the denominator. Then,

$$\frac{Y}{V} = b_1 s^2 + b_2 s + b_3, \quad \frac{V}{U} = \frac{1}{s^3 + a_1 s^2 + a_2 s + a_3} \tag{10.137}$$

Clearly, V/U is identical to the problem shown in Case 1. However, we now have a new differential equation to define the system output, which will be realized as a new C^T matrix:

$$\dddot{v} + a_1 \ddot{v} + a_2 \dot{v} + a_3 v = u \tag{10.138}$$

$$y = b_1 \ddot{v} + b_2 \dot{v} + b_3 v \tag{10.139}$$

Expressed in matrix form:

$$\begin{bmatrix} \dot{v} \\ \ddot{v} \\ \dddot{v} \end{bmatrix} = \begin{bmatrix} 0 & 1 & 0 \\ 0 & 0 & 1 \\ -a_3 & -a_2 & -a_1 \end{bmatrix} \cdot \begin{bmatrix} v \\ \dot{v} \\ \ddot{v} \end{bmatrix} + \begin{bmatrix} 0 \\ 0 \\ 1 \end{bmatrix} \cdot u \tag{10.140}$$

$$y = \begin{bmatrix} b_3 & b_2 & b_1 \end{bmatrix} \cdot \begin{bmatrix} v \\ \dot{v} \\ \ddot{v} \end{bmatrix} + \begin{bmatrix} 0 \end{bmatrix} \cdot u \tag{10.141}$$

Case 3:

$$G(s) = \frac{b_0 s^3 + b_1 s^2 + b_2 s + b_3}{s^3 + a_1 s^2 + a_2 s + a_3} = \frac{Y}{U}(s) \tag{10.142}$$

We can rewrite G(s) as:

$$G(s) = \frac{c_1 s^2 + c_2 s + c_3}{s^3 + a_1 s^2 + a_2 s + a_3} + d \tag{10.143}$$

where $d = G(\infty)$.

The problem is then analogous to Case 2, but with an additional feed-through element, d, expressed in the output equation.

Expressed in matrix form:

$$\begin{bmatrix} \dot{v} \\ \ddot{v} \\ \dddot{v} \end{bmatrix} = \begin{bmatrix} 0 & 1 & 0 \\ 0 & 0 & 1 \\ -a_3 & -a_2 & -a_1 \end{bmatrix} \cdot \begin{bmatrix} v \\ \dot{v} \\ \ddot{v} \end{bmatrix} + \begin{bmatrix} 0 \\ 0 \\ 1 \end{bmatrix} \cdot u \tag{10.144}$$

$$y = \begin{bmatrix} c_3 & c_2 & c_1 \end{bmatrix} \cdot \begin{bmatrix} v \\ \dot{v} \\ \ddot{v} \end{bmatrix} + [d] \cdot u \tag{10.145}$$

The state-space representations for the above cases have a special form known as ***controllable-form*** *or* ***controllable canonical form****, where controllability in this representation is guaranteed.*

Now, we are in the position to discuss this realization for MIMO systems.

10.6.1.2 MIMO System

It would be informative to first illustrate how we can quickly convert a transfer function matrix to its equivalent state-space representation using two approaches before discussing its controllable form.

Consider the following general transfer function matrix, $G(s)$:

$$G(s) = \begin{bmatrix} g_{11}(s) & \cdots & g_{1m}(s) \\ \vdots & \ddots & \vdots \\ g_{r1}(s) & \cdots & g_{rm}(s) \end{bmatrix} \tag{10.146}$$

The simplest approach to transform this transfer function matrix to its equivalent state-space representation would be to develop a state-space model for each element of $G(s)$, similar to the SISO case, in the form of $(A_{ij}, B_{ij}, C_{ij}, D_{ij})$ and then assemble the complete state-space representation. If $G(s)$ is $r \times m$, then

$$A = \left[\begin{array}{cc|c|cc} \begin{matrix} A_{11} & & 0 \\ & \ddots & \\ 0 & & A_{1m} \end{matrix} & & & & 0 \\ \hline & & \ddots & & \\ \hline & 0 & & \begin{matrix} A_{r1} & & 0 \\ & \ddots & \\ 0 & & A_{rm} \end{matrix} & \end{array}\right] \tag{10.147}$$

$$B = \left[\begin{array}{c} \begin{matrix} B_{11} & & 0 \\ & \ddots & \\ 0 & & B_{1m} \end{matrix} \\ \hline \vdots \\ \hline \begin{matrix} B_{r1} & & 0 \\ & \ddots & \\ 0 & & B_{rm} \end{matrix} \end{array}\right] \tag{10.148}$$

$$C = \left[\begin{array}{ccc|c|ccc} C_{11} & \cdots & C_{1m} & & 0 & & 0 \\ & & & \cdots & & & \\ 0 & & 0 & & C_{r1} & \cdots & C_{rm} \end{array}\right] \tag{10.149}$$

$$D = \begin{bmatrix} D_{11} & \cdots & D_{1m} \\ \vdots & \ddots & \vdots \\ D_{r1} & \cdots & D_{rm} \end{bmatrix} \tag{10.150}$$

However, this method is inefficient and most definitely not a minimal realization.

Another approach, called Gilbert's realization and in [32], requires rewriting the transfer function matrix as:

$$G(s) = \frac{p(s)}{\delta(s)} \tag{10.151}$$

Then, finding the Smith–McMillan form and the SM degree, and finally, applying partial fraction expansion to each of the elements of $G(s)$ and collect residues for each distinct pole:

$$G(s) = \sum_{i}^{m} \frac{R_i}{s - p_i}, \quad \text{where } R_i = \lim_{s \to p_i} (s - p_i)G(s) \tag{10.152}$$

The residues, R_i, are matrices in the MIMO case, and the sum of the ranks of matrices R_i gives the SM degree.

The steps for forming the state-space representations are detailed below:

1. Set $R_i = C_i B_i$ and find appropriate B_i and C_i.
2. Form A_i by placing the poles on the diagonal as many times as needed (determined by rank R_i).

The minimal state-space realization is then written as:

$$\dot{\underline{x}} = \begin{bmatrix} A_1 & & 0 \\ & \ddots & \\ 0 & & A_m \end{bmatrix} \underline{x} + \begin{bmatrix} B_1 \\ \vdots \\ B_m \end{bmatrix} \underline{u} \tag{10.153}$$

$$\underline{y} = \begin{bmatrix} C_1 & \cdots & C_m \end{bmatrix} \underline{x} + D\underline{u} \tag{10.154}$$

Example 10.20

$$G(s) = \begin{bmatrix} \dfrac{1}{s+2} & \dfrac{1}{s+2} \\ \dfrac{1}{s-2} & \dfrac{s-2}{(s+1)(s+2)} \end{bmatrix} = \frac{1}{(s+1)(s+2)(s-2)} \begin{bmatrix} (s+1)(s-2) & (s+1)(s-2) \\ (s+1)(s+2) & (s-2)^2 \end{bmatrix}$$

We then find the SM form (left here as an exercise) and realize the characteristic equation is a fourth-order system with poles at $s = -2, -2, -1, 2$.

The transfer function matrix, $G(s)$ is written as a ratio of $\dfrac{p(s)}{\delta(s)}$,

$$G(s) = \frac{\begin{bmatrix} (s+1)(s-2) & (s+1)(s-2) \\ (s+1)(s+2) & (s-2)^2 \end{bmatrix}}{(s+1)(s+2)(s-2)}$$

and using partial fraction expansion, $G(s)$ is written as follows:

$$G(s) = \frac{R_1}{s+1} + \frac{R_2}{s-2} + \frac{R_3}{s+2}$$

then the residue matrices are computed:

$$R_1 = \lim_{s \to -1} (s+1)G(s) = \begin{bmatrix} 0 & 0 \\ 0 & -3 \end{bmatrix} \tag{10.155}$$

$$R_2 = \lim_{s \to 2} (s-2)G(s) = \begin{bmatrix} 0 & 0 \\ 1 & 0 \end{bmatrix} \tag{10.156}$$

$$R_3 = \lim_{s \to -2} (s+2)G(s) = \begin{bmatrix} 1 & 1 \\ 0 & 4 \end{bmatrix} \tag{10.157}$$

now, $G(s)$ can be written as:

$$G(s) = \frac{1}{s+1}\begin{bmatrix} 0 & 0 \\ 0 & -3 \end{bmatrix} + \frac{1}{s-2}\begin{bmatrix} 0 & 0 \\ 1 & 0 \end{bmatrix} + \frac{1}{s+2}\begin{bmatrix} 1 & 1 \\ 0 & 4 \end{bmatrix} \tag{10.158}$$

Note that the last residue matrix is rank 2, so there will be two poles at $s = -2$ to make the system order 4, equal to the SM degree.

It is evident from $G(s)$ that the number of inputs and outputs are equal 2. In addition, knowing that the system order is 4, we can then state that the dimensions for matrices, A is 4×4, B is 4×2, C is 2×4, and D is 2×2. Next, we find the elements of the B and C matrices from the previously computed residue matrices. And easy way to form B and C is to set $C_i = 1$ or 0 as follows:

$$\begin{bmatrix} 0 & 0 \\ 0 & -3 \end{bmatrix} = \begin{bmatrix} C_{11} \\ C_{21} \end{bmatrix}\begin{bmatrix} B_{11} & B_{12} \end{bmatrix} \Rightarrow \begin{matrix} B_{11}C_{11} = 0 & B_{12}C_{11} = 0 \\ B_{11}C_{21} = 0 & B_{12}C_{21} = -3 \end{matrix} \tag{10.159}$$

$$\Rightarrow \begin{matrix} B_{11} = 0 & C_{11} = 0 \\ B_{12} = -3 & C_{21} = 1 \end{matrix} \Rightarrow B_1 = \begin{bmatrix} 0 & -3 \end{bmatrix}, \quad C_1 = \begin{bmatrix} 0 \\ 1 \end{bmatrix} \tag{10.160}$$

$$\begin{bmatrix} 0 & 0 \\ 1 & 0 \end{bmatrix} = \begin{bmatrix} C_{12} \\ C_{22} \end{bmatrix}\begin{bmatrix} B_{21} & B_{22} \end{bmatrix} \Rightarrow \begin{matrix} B_{21}C_{12} = 0 & B_{22}C_{12} = 0 \\ B_{21}C_{22} = 1 & B_{22}C_{22} = 0 \end{matrix} \tag{10.161}$$

$$\Rightarrow \begin{matrix} B_{21} = 1 & C_{12} = 0 \\ B_{22} = 0 & C_{22} = 1 \end{matrix} \Rightarrow B_2 = \begin{bmatrix} 1 & 0 \end{bmatrix}, \quad C_2 = \begin{bmatrix} 0 \\ 1 \end{bmatrix} \tag{10.162}$$

$$\begin{bmatrix} 1 & 1 \\ 0 & 4 \end{bmatrix} = \begin{bmatrix} C_{13} & C_{14} \\ C_{23} & C_{24} \end{bmatrix}\begin{bmatrix} B_{31} & B_{32} \\ B_{41} & B_{42} \end{bmatrix} \tag{10.163}$$

$$\Rightarrow \text{Set } C_3 = \begin{bmatrix} C_{13} & C_{14} \\ C_{23} & C_{24} \end{bmatrix} = \begin{bmatrix} 1 & 0 \\ 0 & 1 \end{bmatrix} \Rightarrow B_3 = \begin{bmatrix} 1 & 1 \\ 0 & 4 \end{bmatrix} \tag{10.164}$$

The A matrices can then be formed by placing poles on the diagonal:

$$A_1 = -1, \quad A_2 = 2, \; A_3 = \begin{bmatrix} -2 & 0 \\ 0 & -2 \end{bmatrix} \tag{10.165}$$

The D matrix is equal to $G(\infty)$. So, the complete state-space system is

$$A = \begin{bmatrix} -1 & 0 & 0 & 0 \\ 0 & 2 & 0 & 0 \\ 0 & 0 & -2 & 0 \\ 0 & 0 & 0 & -2 \end{bmatrix}, \quad B = \begin{bmatrix} 0 & -3 \\ 1 & 0 \\ 1 & 1 \\ 0 & 4 \end{bmatrix} \tag{10.166}$$

$$C = \begin{bmatrix} 0 & 0 & 1 & 0 \\ 1 & 1 & 0 & 1 \end{bmatrix}, \quad D = \begin{bmatrix} 0 & 0 \\ 0 & 0 \end{bmatrix} \tag{10.167}$$

10.6.2 Controllable Form of State-Space Realization of MIMO System

We start by introducing the concept of column reduced matrix as applied to D_R matrix in the right fraction.

Let $K_{c,i}$ be the maximum degree in column i. We say D_R is column-reduced if

$$\deg(D_R) = \deg(\det D_R) = \sum_i K_{c,i} \tag{10.168}$$

We will show in the mathematical details section below that the column reduced D_R matrix results in the controllable form of state-space realization. Equivalently, if D_L in the left fraction is row reduced, this left fraction results in the observable form realization. The proof of this is not discussed here; see [32] for proofs and other details.

A proper transfer function matrix when expressed as a right **coprime column reduced** fraction, then its state-space realization will automatically be both **controllable** and **observable**.

Example 10.21

$$D_R = \begin{bmatrix} s^2+s & s+1 \\ s+2 & 1 \end{bmatrix}: \quad \text{where } K_{c,1} = 2, \quad K_{c,2} = 1, \quad K_{c,1} + K_{c,2} = 3 \tag{10.169}$$

$$\det D_R = s^2 + s - s^2 - 2s - 2 = -2s - 2 \quad \Rightarrow \quad \deg(\det D_R) = 1 \tag{10.170}$$

Therefore, D_R is not column-reduced, as $\deg(D_R) > \deg(\det D_R)$.

We can make it column-reduced by using column operations. Subtract s times the second column from the first column:

$$\begin{bmatrix} s^2+s & s+1 \\ s+2 & 1 \end{bmatrix} \begin{bmatrix} 1 & 0 \\ -s & 1 \end{bmatrix} = \begin{bmatrix} 0 & s+1 \\ 2 & 1 \end{bmatrix} \tag{10.171}$$

$$K_{c,1} = 0, \quad K_{c,2} = 1, \quad K_{c,1} + K_{c,2} = 1 : \text{ column reduced} \tag{10.172}$$

Let D' be the reduced form and U be the appropriate transformation matrix. Then, we say

$$D_R U = D' \quad \Longrightarrow \quad D_R = D'U^{-1} \tag{10.173}$$

Similarly,

$$N_R U = N' \quad \Longrightarrow \quad N_R = N'U^{-1} \tag{10.174}$$

Then,

$$G(s) = N_R D_R^{-1} = N'U^{-1}(D'U^{-1})^{-1} = N'D'^{-1} \tag{10.175}$$

10.6.2.1 Mathematical Details

To obtain the state-space realization in controllable form, we, first, introduce a pseudo-state ζ with length $m \times 1$, where m is the number of column of D_R. These developments are compared to the general SISO state-space realization for comparative and clarification purposes.

SISO. First, we consider the single-input, single-output state-space realization:

$$G(s) = \frac{NUM}{DEN} \tag{10.176}$$

$$\frac{Y}{U} = \frac{NUM}{DEN} \tag{10.177}$$

$$\frac{V}{U} = \frac{1}{DEN} \tag{10.178}$$

$$\frac{Y}{V} = NUM \tag{10.179}$$

where V is the Laplace transform of v, the pseudo-state:

$$DEN\ V = U \tag{10.180}$$

$$Y = NUM\ V \tag{10.181}$$

The general nth order characteristic equation is written as,

$$DEN = s^n + a_1 s^{n-1} + a_2 s^{n-2} \cdots + a_{n-1}s + a_n \tag{10.182}$$

Then, (10.180) is written in differential equation form as follows:

$$\frac{d^n v}{dt^n} = -a_1 \frac{d^{n-1}v}{dt^{n-1}} - a_2 \frac{d^{n-2}v}{dt^{n-2}} \cdots - a_{n-1}\frac{dv}{dt} - a_n v + u \tag{10.183}$$

setting,

$$\begin{aligned} x_1 &= \frac{d^{n-1}v}{dt^{n-1}} = \dot{x}_2 \\ x_2 &= \frac{d^{n-2}v}{dt^{n-2}} = \dot{x}_3 \\ &\vdots \\ x_{n-1} &= \frac{dv}{dt} = \dot{x}_n \\ x_n &= v \end{aligned} \tag{10.184}$$

and, therefore, $\dot{x}_1 = \dfrac{d^n v}{dt^n}$, we can write

$$\dot{x}_1 = -a_1x_1 - a_2x_2 \ldots - a_{n-1}x_{n-1} - a_nx_n + u \tag{10.185}$$

Finally,

$$\Psi(s)_{\text{SISO}} V = X \tag{10.186}$$

where $\Psi(s)_{\text{SISO}} = [s^{n-1}; s^{n-2}; \ldots ; s^2; s; 1]$.

Let's finally define

$$\Gamma(s)_{\text{SISO}} = s^n \tag{10.187}$$

Then, we can always write

$$DEN = \Delta_{h\text{SISO}}\Gamma_{\text{SISO}} + \Delta_{p\text{SISO}}\Psi_{\text{SISO}}, \tag{10.188}$$

where $\Delta_{h\text{SISO}}$ is non-zero if and only if DEN is nth order.

MIMO. The multi-input, multi-output state-space realization will logically follow from the SISO realization:

$$G(s) = N_R D_R^{-1} \tag{10.189}$$

$$Y = N_R D_R^{-1} U \tag{10.190}$$

$$Y = N_R Z (D_R Z)^{-1} U \tag{10.191}$$

where Z is the Laplace transform of ζ, the pseudo-state, and

$$D_R Z = U \tag{10.192}$$

$$Y = N_R Z \tag{10.193}$$

Next we define

$$\Gamma(s) = \text{diag}\ \left(s^{K_{c1}},\ s^{K_{c2}},\ \ldots,\ s^{K_{cm}}\right) \tag{10.194}$$

where K_{ck} is the column degrees of D_R. Then,

$$\Psi(s) = \left[\begin{array}{c|c|c} s^{K_{c1}-1} & 0 & \\ s^{K_{c1}-2} & 0 & \\ \vdots & \vdots & \\ s & 0 & \\ & & \\ 1 & 0 & \\ \hline 0 & s^{K_{c2}-1} & \\ 0 & s^{K_{c2}-2} & \text{etc.} \\ \vdots & \vdots & \\ 0 & s & \\ & & \\ 0 & 1 & \\ \hline \vdots & \vdots & \\ \hline 0 & 0 & \end{array}\right] = \textbf{block diag}\left\{\begin{bmatrix} s^{K_{ci}-1} \\ s^{K_{ci}-2} \\ \vdots \\ s \\ 1 \end{bmatrix}, i = 1, \ldots, m\right\}$$

where

$$\Psi(s)Z = X \tag{10.195}$$

and X are the system states in Laplace domain. We can always write

$$D_R = \Delta_h\Gamma + \Delta_p\Psi, \tag{10.196}$$

where Δ_h is non-singular if and only if D_R is column reduced.

The above developments should manifest the equivalency between the MIMO and SISO realizations. Next, we illustrate this mathematical development through a simple example.

Example

$$D_R = \begin{bmatrix} s^2+3s & 0 \\ 0 & s^2+3s \end{bmatrix} \tag{10.197}$$

$$K_{c1} = 2, \quad K_{c2} = 2 \tag{10.198}$$

$$D_R = \underbrace{\begin{bmatrix} 1 & 0 \\ 0 & 1 \end{bmatrix}}_{\Delta_h} \underbrace{\begin{bmatrix} s^2 & 0 \\ 0 & s^2 \end{bmatrix}}_{\Gamma} + \underbrace{\left[\begin{array}{cc|cc} 3 & 0 & 0 & 0 \\ 0 & 0 & 3 & 0 \end{array}\right]}_{\Delta_p} \underbrace{\left[\begin{array}{cc} s & 0 \\ 1 & 0 \\ \hline 0 & s \\ 0 & 1 \end{array}\right]}_{\Psi} \tag{10.199}$$

To complete this mathematical development, next, we define the state variables ($X = \Psi(s)Z$) as:

$$\begin{aligned} x_1 &= \frac{d^{K_{c1}-1}}{dt^{K_{c1}-1}}\zeta_1 \\ x_2 &= \frac{d^{K_{c1}-2}}{dt^{K_{c1}-2}}\zeta_1 \\ &\vdots \\ x_{K_{c1}} &= \zeta_1 \\ x_{K_{c1}+1} &= \frac{d^{K_{c2}-1}}{dt^{K_{c2}-1}}\zeta_2 \\ &\vdots \\ x_{K_{c1}+K_{c2}} &= \zeta_2 \end{aligned} \tag{10.200}$$

How many states are there?

$$K_{c1} + K_{c2} + \cdots + K_{cm} = \deg D_R \tag{10.201}$$

so, the total number of states are equal to the deg D_R.

Recall that we want

$$\underline{\dot{x}} = A\underline{x} + B\underline{u} \tag{10.202}$$

$$\underline{y} = C^T\underline{x}, \tag{10.203}$$

considering $G(s)$ to be strictly proper without loss of generality,

$$D_R Z = \left(\Delta_h \Gamma + \Delta_p \Psi\right) Z = U \tag{10.204}$$

So,

$$\Gamma Z = -\Delta_h^{-1}\Delta_p \Psi Z + \Delta_h^{-1} U \tag{10.205}$$

We know $\Psi Z = X$, so we make the substitution into the above equation,

$$\Gamma Z = -\Delta_h^{-1}\Delta_p X + \Delta_h^{-1} U \tag{10.206}$$

Let's use the previous example for revealing the utility of this equation for deriving a state-space realization for a MIMO system.

Example 10.22

$$\underbrace{\begin{bmatrix} s^2 & 0 \\ 0 & s^2 \end{bmatrix}}_{\Gamma} Z = -\Delta_h^{-1}\Delta_p \Psi Z + \Delta_h^{-1} U \tag{10.207}$$

where,

$$X = \underbrace{\begin{bmatrix} s & 0 \\ 1 & 0 \\ \hdashline 0 & s \\ 0 & 1 \end{bmatrix}}_{\Psi} \begin{bmatrix} Z_1 \\ Z_2 \end{bmatrix} \tag{10.208}$$

and therefore,

$$\begin{bmatrix} X_1 \\ X_2 \\ X_3 \\ X_4 \end{bmatrix} = \begin{bmatrix} sZ_1 \\ Z_1 \\ sZ_2 \\ Z_2 \end{bmatrix} \tag{10.209}$$

furthermore,

$$\underbrace{\begin{bmatrix} s^2 & 0 \\ 0 & s^2 \end{bmatrix}}_{\Gamma} \begin{bmatrix} Z_1 \\ Z_2 \end{bmatrix} = \mathscr{L} \begin{bmatrix} \dot{x}_1 \\ \dot{x}_3 \end{bmatrix} \tag{10.210}$$

we then can write

$$\begin{bmatrix} \dot{x}_1 \\ \dot{x}_3 \end{bmatrix} = \underbrace{\begin{bmatrix} -3 & 0 & 0 & 0 \\ 0 & 0 & -3 & 0 \end{bmatrix}}_{-\Delta_h^{-1}\Delta_p} \begin{bmatrix} x_1 \\ x_2 \\ x_3 \\ x_4 \end{bmatrix} + \underbrace{\begin{bmatrix} 1 & 0 \\ 0 & 1 \end{bmatrix}}_{\Delta_h^{-1}} u \tag{10.211}$$

It is evident for the example above that not all the states could be included in this derivation as the state derivatives include only the highest-order derivative of each column and not all state derivatives. Knowing this, we can then write

$$\begin{bmatrix} \dot{x}_1 \\ \dot{x}_{K_{c1}+1} \\ \dot{x}_{K_{c1}+K_{c2}+1} \\ \vdots \end{bmatrix} = \underbrace{-\Delta_h^{-1}\Delta_p}_{m \text{ rows of } A} x + \underbrace{\Delta_h^{-1}}_{m \text{ rows of } B} u \tag{10.212}$$

while, in order to complete this realization, we need to add other rows which are,

$$\frac{\mathrm{d}x_k}{\mathrm{d}t} = x_{k-1} \tag{10.213}$$

Next we can write the outputs (Y) as:

$$\begin{aligned} Y &= N_R Z \\ &= N_l \Psi X \end{aligned} \tag{10.214}$$

Therefore,

$$y = N_l x \tag{10.215}$$

and we can then write the output matrix as:

$$C^T = N_l \tag{10.216}$$

Example 10.23

$$G = \begin{bmatrix} \dfrac{4s-10}{2s+1} & \dfrac{3}{s+2} \\ \dfrac{1}{(2s+1)(s+2)} & \dfrac{s+1}{(s+2)^2} \end{bmatrix} = G(\infty) + G_{sp} \tag{10.217}$$

$$= \begin{bmatrix} 2 & 0 \\ 0 & 0 \end{bmatrix} + \begin{bmatrix} \dfrac{-12}{2s+1} & \dfrac{3}{s+2} \\ \dfrac{1}{(2s+1)(s+2)} & \dfrac{s+1}{(s+2)^2} \end{bmatrix} \tag{10.218}$$

where "sp" stands for strictly proper. The strictly proper part of G is written as a MFD as follows:

$$G_{sp} = \underbrace{\begin{bmatrix} -6s-12 & -9 \\ 0.5 & 1 \end{bmatrix}}_{N_R} \underbrace{\begin{bmatrix} s^2+2.5s+1 & 2s+1 \\ 0 & s+2 \end{bmatrix}}_{D_R}^{-1} \tag{10.219}$$

$$K_{c1} = 2, \quad K_{c2} = 1 \tag{10.220}$$

so D_R is column reduced.

$$D_R = \underbrace{\begin{bmatrix} 1 & 2 \\ 0 & 1 \end{bmatrix}}_{\Delta_h} \underbrace{\begin{bmatrix} s^2 & 0 \\ 0 & s \end{bmatrix}}_{\Gamma} + \underbrace{\left[\begin{array}{cc|c} 2.5 & 1 & 1 \\ 0 & 0 & 2 \end{array}\right]}_{\Delta_p} \underbrace{\left[\begin{array}{cc} s & 0 \\ 1 & 0 \\ \hline 0 & 1 \end{array}\right]}_{\Psi} \tag{10.221}$$

$$-\Delta_h^{-1}\Delta_p = \left[\begin{array}{cc|c} -2.5 & -1 & 3 \\ 0 & 0 & -2 \end{array}\right] \tag{10.222}$$

We know that there are total of three states and from Equation (10.212):

$$\begin{bmatrix} \dot{x}_1 \\ \dot{x}_3 \end{bmatrix} = \left[\begin{array}{cc|c} -2.5 & -1 & 3 \\ 0 & 0 & -2 \end{array}\right] \begin{bmatrix} x_1 \\ x_2 \\ x_3 \end{bmatrix} + \begin{bmatrix} 1 & -2 \\ 0 & 1 \end{bmatrix} u \tag{10.223}$$

From Equation (10.213),

$$\dot{x}_2 = x_1 \tag{10.224}$$

Therefore,

$$\begin{bmatrix} \dot{x}_1 \\ \dot{x}_2 \\ \dot{x}_3 \end{bmatrix} = \underbrace{\begin{bmatrix} -2.5 & -1 & 3 \\ 1 & 0 & 0 \\ 0 & 0 & -2 \end{bmatrix}}_{A} \begin{bmatrix} x_1 \\ x_2 \\ x_3 \end{bmatrix} + \underbrace{\begin{bmatrix} 1 & -2 \\ 0 & 0 \\ 0 & 1 \end{bmatrix}}_{B} u \tag{10.225}$$

From Equation (10.214),

$$N_R = \underbrace{\begin{bmatrix} -6 & -12 & -9 \\ 0 & 0.5 & 1 \end{bmatrix}}_{N_l} \underbrace{\left[\begin{array}{cc} s & 0 \\ 1 & 0 \\ \hline 0 & 1 \end{array}\right]}_{\Psi} \tag{10.226}$$

So we can write the output equation as:

$$y = \underbrace{\begin{bmatrix} -6 & -12 & -9 \\ 0 & 0.5 & 1 \end{bmatrix}}_{C^T} \begin{bmatrix} x_1 \\ x_2 \\ x_3 \end{bmatrix} + \underbrace{\begin{bmatrix} 2 & 0 \\ 0 & 0 \end{bmatrix}}_{D} u \tag{10.227}$$

10.7 Straightforward Computational Procedures

Previous mathematical approach for the computation of the state-space matrices can be made easier by defining the following matrices:

$$A_0 = \textbf{block diag}\left\{\begin{bmatrix} 0 & & 0 \\ 1 & \ddots & \\ 0 & 1 & 0 \end{bmatrix}, K_{ci} \times K_{ci}, i = 1, \ldots, m\right\} \tag{10.228}$$

such when $K_{ci} = 2$, and $i = 1,2$, as given previously, for two blocks, we have

$$A_0 = \left[\begin{array}{cc|cc} 0 & 0 & & \\ 1 & 0 & & \\ \hline & & 0 & 0 \\ & & 1 & 0 \end{array}\right] \tag{10.229}$$

and when $K_{ci} = 3$ for two blocks, we can write,

$$A_0 = \left[\begin{array}{ccc|ccc} 0 & 0 & 0 & & & \\ 1 & 0 & 0 & & & \\ 0 & 1 & 0 & & & \\ \hline & & & 0 & 0 & 0 \\ & & & 1 & 0 & 0 \\ & & & 0 & 1 & 0 \end{array}\right] \tag{10.230}$$

and

$$B_0 = \textbf{block diag}\left\{\begin{bmatrix} 1 \\ 0 \\ \vdots \\ 0 \end{bmatrix}, K_{ci} \times 1, i = 1, \ldots, m\right\} \tag{10.231}$$

Then, it can be shown that the state-space matrices can be written as:

$$A = A_0 - B_0\Delta_h^{-1}\Delta_p \tag{10.232}$$

$$B = B_0\Delta_h^{-1} \tag{10.233}$$

$$C^T = N_l \tag{10.234}$$

$$D = G(\infty) \tag{10.235}$$

Example 10.24

$$G = \begin{bmatrix} s+1 & 1 \\ 1 & 1 \end{bmatrix}\begin{bmatrix} s^2+3s & 0 \\ 0 & s^2+3s \end{bmatrix}^{-1} \tag{10.236}$$

$$K_{c1} = 2, \quad K_{c2} = 2 \tag{10.237}$$

$$D_R = \underbrace{\begin{bmatrix} 1 & 0 \\ 0 & 1 \end{bmatrix}}_{\Delta_h}\underbrace{\begin{bmatrix} s^2 & 0 \\ 0 & s^2 \end{bmatrix}}_{\Gamma} + \underbrace{\left[\begin{array}{cc|cc} 3 & 0 & 0 & 0 \\ 0 & 0 & 3 & 0 \end{array}\right]}_{\Delta_p}\underbrace{\left[\begin{array}{cc} s & 0 \\ 1 & 0 \\ \hline 0 & s \\ 0 & 1 \end{array}\right]}_{\Psi} \tag{10.238}$$

$$-\Delta_h\Delta_p = \left[\begin{array}{cc|cc} -3 & 0 & 0 & 0 \\ 0 & 0 & -3 & 0 \end{array}\right] \tag{10.239}$$

$$N_R = \begin{bmatrix} s+1 & 1 \\ 1 & 1 \end{bmatrix} = \underbrace{\begin{bmatrix} 1 & 1 & 0 & 1 \\ 0 & 1 & 0 & 1 \end{bmatrix}}_{N_l}\underbrace{\left[\begin{array}{cc} s & 0 \\ 1 & 0 \\ \hline 0 & s \\ 0 & 1 \end{array}\right]}_{\Psi} \tag{10.240}$$

Using the equations above

$$A_0 = \begin{bmatrix} 0 & 0 & 0 & 0 \\ 1 & 0 & 0 & 0 \\ 0 & 0 & 0 & 0 \\ 0 & 0 & 1 & 0 \end{bmatrix}, \quad B_0 = \begin{bmatrix} 1 & 0 \\ 0 & 0 \\ 0 & 1 \\ 0 & 0 \end{bmatrix} \tag{10.241}$$

Therefore,

$$A = \begin{bmatrix} -3 & 0 & 0 & 0 \\ 1 & 0 & 0 & 0 \\ 0 & 0 & -3 & 0 \\ 0 & 0 & 1 & 0 \end{bmatrix}, \quad B = \begin{bmatrix} 1 & 0 \\ 0 & 0 \\ 0 & 1 \\ 0 & 0 \end{bmatrix}, \quad C^T = \begin{bmatrix} 1 & 1 & 0 & 1 \\ 0 & 1 & 0 & 1 \end{bmatrix}, \quad D = 0 \tag{10.242}$$

Example 10.25

$$G = \begin{bmatrix} s+2 & 0 \\ s+1 & 1 \end{bmatrix} \begin{bmatrix} (s+1)^2(s+2) & -(s+1) \\ 0 & s+2 \end{bmatrix}^{-1} \tag{10.243}$$

$$K_{c1} = 3, \quad K_{c2} = 1 \tag{10.244}$$

$$D_R = \underbrace{\begin{bmatrix} 1 & -1 \\ 0 & 1 \end{bmatrix}}_{\Delta_h} \underbrace{\begin{bmatrix} s^3 & 0 \\ 0 & s \end{bmatrix}}_{\Gamma} + \underbrace{\left[\begin{array}{ccc|c} 4 & 5 & 2 & -1 \\ \hline 0 & 0 & 0 & 2 \end{array}\right]}_{\Delta_p} \underbrace{\left[\begin{array}{c|c} s^2 & 0 \\ s & 0 \\ 1 & 0 \\ \hline 0 & 1 \end{array}\right]}_{\Psi} \tag{10.245}$$

$$-\Delta_h \Delta_p = \left[\begin{array}{ccc|c} -4 & -5 & -2 & -1 \\ \hline 0 & 0 & 0 & -2 \end{array}\right] \tag{10.246}$$

$$N_R = \underbrace{\left[\begin{array}{ccc|c} 0 & 1 & 2 & 0 \\ \hline 0 & 1 & 1 & 1 \end{array}\right]}_{N_l} \underbrace{\left[\begin{array}{c|c} s^2 & 0 \\ s & 0 \\ 1 & 0 \\ \hline 0 & 1 \end{array}\right]}_{\Psi} = \begin{bmatrix} s+2 & 0 \\ s+1 & 1 \end{bmatrix} \tag{10.247}$$

Then,

$$A_0 = \begin{bmatrix} 0 & 0 & 0 & 0 \\ 1 & 0 & 0 & 0 \\ 0 & 1 & 0 & 0 \\ 0 & 0 & 0 & 0 \end{bmatrix}, \quad B_0 = \begin{bmatrix} 1 & 0 \\ 0 & 0 \\ 0 & 0 \\ 0 & 1 \end{bmatrix} \tag{10.248}$$

Therefore,

$$A = \begin{bmatrix} -4 & -5 & -2 & -1 \\ 1 & 0 & 0 & 0 \\ 0 & 1 & 0 & 0 \\ 0 & 0 & 0 & -2 \end{bmatrix}, \quad B = \begin{bmatrix} 1 & 1 \\ 0 & 0 \\ 0 & 1 \\ 0 & 0 \end{bmatrix}, \quad C^T = \begin{bmatrix} 0 & 1 & 2 & 0 \\ 0 & 1 & 1 & 1 \end{bmatrix}, D = 0 \tag{10.249}$$

10.8 Problems

10.1 Given

$$G(s) = \begin{bmatrix} \dfrac{1}{s+1} & \dfrac{1}{(s+1)(s+2)} \\ \dfrac{s}{(s+1)(s+2)} & \dfrac{2s+1}{(s+1)(s+2)} \end{bmatrix} \tag{10.250}$$

find the Smith–McMillan form and the location of the poles and zeros of this transfer function matrix.

10.2 Show that

$$G(s) = \frac{1}{s(s+3)^2} \begin{bmatrix} (s+3)(s+1) & s+3 \\ s+3 & s+3 \end{bmatrix} \tag{10.251}$$

can be written as:

$$\begin{bmatrix} s+1 & 1 \\ 1 & 1 \end{bmatrix} \begin{bmatrix} s(s+3) & 0 \\ 0 & s(s+3) \end{bmatrix}^{-1} \tag{10.252}$$

where the first transfer function matrix is N_R and the second transfer function matrix is D_R. Provide the locations of the poles and zeros. Verify that

$$\begin{bmatrix} 1 & 1 \\ 0 & s \end{bmatrix} \tag{10.253}$$

is a common right factor of N_R and D_R. Is this a coprime factorization? Verify that

$$\begin{bmatrix} s+1 & -1 \\ 1 & 1 \end{bmatrix} \begin{bmatrix} s(s+3) & -(s+3) \\ 0 & s+3 \end{bmatrix}^{-1} \tag{10.254}$$

is a coprime factorization. Provide the locations of the poles and zeros:

10.3

$$G_c(s) = \begin{bmatrix} \dfrac{1}{(s+1)(s+2)} & \dfrac{-1}{(s+1)(s+2)} \\ \dfrac{s^2+s-4}{(s+1)(s+2)} & \dfrac{2s^2-s-8}{(s+1)(s+2)} \\ \dfrac{s-2}{s+1} & \dfrac{2s-4}{s+1} \end{bmatrix} \tag{10.255}$$

Find the Smith–McMillan form for this compensator. Provide the locations of the poles and zeros of this transfer function matrix. What is the McMillan degree?

10.4 Given the following plant TFM,

$$G_p = 50 \begin{pmatrix} s^2+1 & -1 \\ 1 & 0 \end{pmatrix} \underbrace{\begin{pmatrix} \dfrac{1}{s^2+1} & 0 \\ 0 & \dfrac{s+1}{s} \end{pmatrix}}_{G_p^D} \begin{pmatrix} 1 & 0 \\ 1 & 1 \end{pmatrix} \tag{10.256}$$

Find if G_p^D is the Smith–McMillan form of G. (What is the Smith–McMillan form?)

10.5 Control of a vectored-thrust ducted-fan aircraft is a challenging task. After linearization, the transfer function matrix is given as:

$$\begin{bmatrix} \alpha_1 \\ \alpha_2 \\ \alpha_3 \end{bmatrix} = \begin{bmatrix} \dfrac{a_1 s^2 + a_2}{s^2(s^2+b_1)} & 0 \\ 0 & \dfrac{a_4}{s^2+b_2} \\ \dfrac{a_3}{s^2+b_1} & 0 \end{bmatrix} \begin{bmatrix} f_1 \\ f_2 \end{bmatrix} \tag{10.257}$$

where the inputs f_1 and f_2 are the forces and α_1, α_2, and α_3 are the yaw, roll, and pitch angles. a_1, a_2, a_3, a_4, b_1, and b_2 are all constants.

(a) Derive the Smith–McMillan form of this TFM.
(b) Provide a right matrix fractional description (RMFD) of this TFM and state if this factorization is coprime. No computation is necessary, but you need to provide a reason for this based on your finding in (a).

10.6 Given the following controlled object $G_p(s)$,

$$G_p(s) = \begin{bmatrix} 0 & \dfrac{-1}{s+1} \\ 1 & \dfrac{1}{s(s+1)} \end{bmatrix} \tag{10.258}$$

11

Eigenvalues and Singular Values

In this chapter, we discuss some of the properties (eigenvectors and eigenvalues) of matrices and our concentration will be on square matrices. We will then discuss singular value decomposition (SVD) and generalize the results for both square and nonsquare matrices. There are several texts and technical publications that discuss these topics in details. Our hope here is to provide some insight into these topics that are very relevant in MIMO control design. We start this chapter by discussing eigenvalues and eigenvectors next.

11.1 Eigenvalues and Eigenvectors

Suppose we have a square matrix A. Matrices act on vectors like a function: in goes vector $\underline{x}$ and out comes vector $A\underline{x}$ (like $T(\underline{x})$, but multidimensional). In particular, we are interested in vectors that come out parallel to the vector that went in.

$$A\underline{x} = \lambda\underline{x} \quad \text{such that} \quad A\underline{x} \parallel \underline{x} \tag{11.1}$$

Such vectors are known as <u>eigenvectors</u>, while the values λ are known as <u>eigenvalues</u>. So, if the multiplication of a vector by a general matrix results in simultaneously stretching/shrinking and rotating the vector, then if this vector is an eigenvector, the outcome would be equal to only stretching/shrinking this vector by a scalar called an eigenvalue.

We can state the above in a different way by first assuming the following transformation, $T : \mathbb{R}^n \to \mathbb{R}^n$ then,

$$T(\underline{x}) = \lambda\underline{x} \tag{11.2}$$

if and only if $\underline{x}$ is an eigenvector. if $T(\underline{x}) = A\underline{x}$, where A is any matrix, then,

$$A\underline{x} = \lambda\underline{x} \tag{11.3}$$

However, we have two unknowns (the scalar λ, eigenvalue, and the eigenvector $\underline{x}$) but only one equation. So how can we find both the eigenvalues and eigenvectors?

First, let's consider the case where $\lambda = 0$: $A\underline{x} = 0$, so A is singular (meaning it takes some vector and maps it to zero) and the eigenvectors form the basis of the null space of A, denoted $\eta(A)$.

Let's now consider another example, this time a permutation matrix. Let

$$A = \begin{bmatrix} 0 & 1 \\ 1 & 0 \end{bmatrix} \tag{11.4}$$

We are looking for vectors $\underline{x}$ and scalars λ such that $A\underline{x} = \lambda\underline{x}$. Intuitively,

$$\underline{x}_1 = \begin{bmatrix} 1 \\ 1 \end{bmatrix}, \quad \lambda_1 = 1, \quad \Rightarrow \quad \begin{bmatrix} 0 & 1 \\ 1 & 0 \end{bmatrix}\begin{bmatrix} 1 \\ 1 \end{bmatrix} = 1 \cdot \begin{bmatrix} 1 \\ 1 \end{bmatrix} \tag{11.5}$$

$$\underline{x}_2 = \begin{bmatrix} -1 \\ 1 \end{bmatrix}, \quad \lambda_2 = -1, \quad \Rightarrow \quad \begin{bmatrix} 0 & 1 \\ 1 & 0 \end{bmatrix}\begin{bmatrix} -1 \\ 1 \end{bmatrix} = -1 \cdot \begin{bmatrix} -1 \\ 1 \end{bmatrix} \tag{11.6}$$

It is important to note that for two-dimensional matrices, we have shortcuts for finding the eigenvalues and eigenvectors. The eigenvalues can be computed using the followings:

$$\lambda_1 + \lambda_2 = \text{trace}\ (A) = \sum \text{diag}\ (A), \quad \det (A) = \lambda_1\lambda_2 \tag{11.7}$$

Robust Control: Youla Parameterization Approach, First Edition. Farhad Assadian and Kevin R. Mallon.

Companion Website: www.wiley.com/go/assadian/robust

While the eigenvectors are solved by rearranging Eq. (11.3) and writing,

$$A\underline{x} - \lambda I\underline{x} = 0 \quad \Rightarrow \quad \underbrace{(A - \lambda I)}_{\text{Singular}}\underline{x} = 0 : \quad \underline{x} \in \eta(A - \lambda I) \tag{11.8}$$

We will demonstrate the above computations for eigenvalues and eigenvectors through the following three examples:

Example 11.1

$$A = \begin{bmatrix} 3 & 1 \\ 1 & 3 \end{bmatrix} \tag{11.9}$$

$$\text{trace}(A) = \lambda_1 + \lambda_2 = 6, \quad \det(A) = \lambda_1\lambda_2 = 8 \quad \Rightarrow \quad \lambda_1 = 2, \quad \lambda_2 = 4 \tag{11.10}$$

$$\lambda_1 = 2 : \quad \left(\overbrace{\begin{bmatrix} 3 & 1 \\ 1 & 3 \end{bmatrix} - \begin{bmatrix} 2 & 0 \\ 0 & 2 \end{bmatrix}}^{A-\lambda_1 I} \right) \underline{x}_1 = \begin{bmatrix} 1 & 1 \\ 1 & 1 \end{bmatrix} \underline{x}_1 = \begin{bmatrix} 0 \\ 0 \end{bmatrix} \quad \Rightarrow \quad \underline{x}_1 = \begin{bmatrix} 1 \\ -1 \end{bmatrix} \tag{11.11}$$

$$\lambda_2 = 4 : \quad \left(\begin{bmatrix} 3 & 1 \\ 1 & 3 \end{bmatrix} - \begin{bmatrix} 4 & 0 \\ 0 & 4 \end{bmatrix} \right) \underline{x}_2 = \begin{bmatrix} -1 & 1 \\ 1 & -1 \end{bmatrix} \underline{x}_2 = \begin{bmatrix} 0 \\ 0 \end{bmatrix} \Rightarrow \underline{x}_2 = \begin{bmatrix} 1 \\ 1 \end{bmatrix} \tag{11.12}$$

Example 11.2 Rotation matrix:

$$Q = \begin{bmatrix} \cos\theta & -\sin\theta \\ \sin\theta & \cos\theta \end{bmatrix} \tag{11.13}$$

Let $\theta = 90°$ (90° rotation).

$$Q = \begin{bmatrix} 0 & -1 \\ 1 & 0 \end{bmatrix} \quad \Rightarrow \quad \lambda_1 + \lambda_2 = 0, \quad \lambda_1\lambda_2 = 1 \quad \Rightarrow \quad \lambda_1 = i, \quad \lambda_2 = -i \tag{11.14}$$

It is interesting to note that for rotation matrices, the eigenvalues are imaginary and there are no real eigenvectors.

Example 11.3

$$A = \begin{bmatrix} 3 & 1 \\ 0 & 3 \end{bmatrix} \quad \Rightarrow \quad \lambda_1 + \lambda_2 = 6, \quad \lambda_1\lambda_2 = 9 \quad \Rightarrow \quad \lambda_1 = 3, \quad \lambda_2 = 3 \tag{11.15}$$

$$\lambda_1 = 3 : \quad \left(\begin{bmatrix} 3 & 1 \\ 0 & 3 \end{bmatrix} - \begin{bmatrix} 3 & 0 \\ 0 & 3 \end{bmatrix} \right) \underline{x}_1 = \begin{bmatrix} 0 & 1 \\ 0 & 0 \end{bmatrix} \underline{x}_1 = \begin{bmatrix} 0 \\ 0 \end{bmatrix} \quad \Rightarrow \quad \underline{x}_1 = \begin{bmatrix} 1 \\ 0 \end{bmatrix} \tag{11.16}$$

What happens for λ_2? There is no other linearly independent eigenvector, so $\underline{x}_2$ does not exist. This is a degenerate case. This should be evident by looking at the matrix A as the vector (first column) $\begin{bmatrix} 3 & 0 \end{bmatrix}^T$ is one dimensional and as it stretches or shrinks, the vector remains on its own span.

11.2 Matrix Diagonalization

In the previous section, we discussed the concepts of eigenvalues and eigenvectors. In this section, we briefly introduce the application of eigenvalues and eigenvectors in matrix diagonalization including some important properties of matrices.

We are interested in diagonalizing a matrix (or alternately, eigenvalue decoupling).

Suppose a matrix A has n independent eigenvectors/eigenvalues. Then, $A\underline{x}_i = \lambda_i \underline{x}_i$ for all $i = 1, \ldots, n$. We can express this equation for all eigenvalues and eigenvectors by using a matrix representation,

$$\Lambda = \text{diag}\left(\lambda_1, \ \ldots, \ \lambda_n\right), \quad S = \left[\underline{x}_1, \ \ldots, \ \underline{x}_n\right] \tag{11.17}$$

where Λ is a diagonal matrix of all eigenvalues and S is a matrix consists of all eigenvectors. Then,

$$\begin{aligned} AS &= A\begin{bmatrix} | & | & | \\ \underline{x}_1, & \ldots, & \underline{x}_n \\ | & | & | \end{bmatrix} \\ &= \left[\lambda_1\underline{x}_1, \;\ldots,\; \lambda_n\underline{x}_n\right] \\ &= \left[\underline{x}_1, \;\ldots,\; \underline{x}_n\right]\begin{bmatrix} \lambda_1 & & 0 \\ & \ddots & \\ 0 & & \lambda_n \end{bmatrix} = S\Lambda \end{aligned} \tag{11.18}$$

To diagonalize matrix A, S must be invertible. If so, then we can premultiply both sides by S^{-1},

$$S^{-1}AS = \Lambda \tag{11.19}$$

Diagonalization of matrix A with dimension of $n \times n$ is possible if matrix S consists of n independent eigenvectors, in addition, matrix A must be square. There are some matrices that do not contain n independent eigenvectors (consider the degenerate case example). Therefore, we can ask the following question: **Which matrices are diagonalizable?**

Matrix A must have n linearly independent eigenvectors. If there are no repeated eigenvalues, A is diagonalizable. If A has repeated eigenvalues, then A may or may not have sufficient linearly independent eigenvectors.

If matrix A is diagonalizable, its exponential can be computed as follows:

$$\begin{aligned} A\underline{x} &= \lambda\underline{x} \\ A^2\underline{x} &= \lambda A\underline{x} = \lambda^2\underline{x} \\ A^2 &= (S\Lambda S^{-1})(S\Lambda S^{-1}) = S\Lambda^2 S^{-1} \\ A^k &= S\Lambda^k S^{-1} \end{aligned}$$

It can be shown that for symmetric real matrices, where $A = A^T$, their eigenvectors are orthogonal ($Q^{-1} = Q^T$, where Q is the normalized eigenvector matrix, in addition, the eigenvalues of symmetric matrices are all real. We could always then diagonalize a symmetric matrix $A = Q\Lambda Q^T$, as shown in the following example,

---

Example 11.4

$$A = \begin{bmatrix} 3 & 1 \\ 1 & 3 \end{bmatrix} \quad \Rightarrow \quad \lambda_1 + \lambda_2 = 6, \quad \lambda_1\lambda_2 = 8 \quad \Rightarrow \quad \lambda_1 = 4, \quad \lambda_2 = 2 \tag{11.20}$$

$$\lambda_1 = 4: \quad \left(\begin{bmatrix} 3 & 1 \\ 1 & 3 \end{bmatrix} - \begin{bmatrix} 4 & 0 \\ 0 & 4 \end{bmatrix}\right)\underline{x}_1 = \begin{bmatrix} -1 & 1 \\ 1 & -1 \end{bmatrix}\underline{x}_1 = \begin{bmatrix} 0 \\ 0 \end{bmatrix} \Rightarrow \underline{x}_1 = \begin{bmatrix} 1 \\ 1 \end{bmatrix} \tag{11.21}$$

$$\lambda_2 = 2: \quad \left(\begin{bmatrix} 3 & 1 \\ 1 & 3 \end{bmatrix} - \begin{bmatrix} 2 & 0 \\ 0 & 2 \end{bmatrix}\right)\underline{x}_2 = \begin{bmatrix} 1 & 1 \\ 1 & 1 \end{bmatrix}\underline{x}_2 = \begin{bmatrix} 0 \\ 0 \end{bmatrix} \Rightarrow \underline{x}_2 = \begin{bmatrix} 1 \\ -1 \end{bmatrix} \tag{11.22}$$

$$\Lambda = \begin{bmatrix} 4 & 0 \\ 0 & 2 \end{bmatrix}, \quad S = \begin{bmatrix} 1 & 1 \\ 1 & -1 \end{bmatrix} \quad \Rightarrow \quad Q = \begin{bmatrix} \frac{1}{\sqrt{2}} & \frac{1}{\sqrt{2}} \\ \frac{1}{\sqrt{2}} & -\frac{1}{\sqrt{2}} \end{bmatrix} \tag{11.23}$$

$$\begin{bmatrix} \frac{1}{\sqrt{2}} & \frac{1}{\sqrt{2}} \\ \frac{1}{\sqrt{2}} & -\frac{1}{\sqrt{2}} \end{bmatrix}\begin{bmatrix} 3 & 1 \\ 1 & 3 \end{bmatrix}\begin{bmatrix} \frac{1}{\sqrt{2}} & \frac{1}{\sqrt{2}} \\ \frac{1}{\sqrt{2}} & -\frac{1}{\sqrt{2}} \end{bmatrix} = \begin{bmatrix} 4 & 0 \\ 0 & 2 \end{bmatrix} \tag{11.24}$$

We can generalize the diagonalizability criterion for a matrix A, but before we do that, we need to introduce the **algebraic and geometric multiplicities**.

Definition 11.1 Algebraic Multiplicity. Let

$$(A - \lambda I)\underline{x} = 0 \tag{11.25}$$

where $\underline{x} \neq 0$. Then,

$$\det(A - \lambda I) = 0 = \underbrace{\delta(\lambda)}_{\text{Polyn of deg. } n} \tag{11.26}$$

where

$$\delta(\lambda) = (\lambda - \lambda_1)^{a_1}(\lambda - \lambda_2)^{a_2} \dots (\lambda - \lambda_d)^{a_d}, \quad \lambda_1, \lambda_2, \dots, \lambda_d \tag{11.27}$$

are distinct eigenvalues and size(A) = n. Then,

$$a_k, \quad k \in 1, \cdots, d$$

is the **Algebraic Multiplicity**.

Definition 11.2 Geometric Multiplicity. Find the eigenvector $\underline{x}_k$ from

$$(A - \lambda I)\underline{x} = 0, \ \underline{x} \neq 0 \tag{11.28}$$

Then, compute

$$g_k = n - \text{rank}\ (A - \lambda_k I) \tag{11.29}$$

where "g_k" is the **Geometric Multiplicity**.

Geometric Multiplicity is the null space of $(A - \lambda_k I)$ or $g_k \in \eta(A - \lambda_k I)$. Independent solutions can show $1 \leq g_k \leq a_k$. There are classes of matrices that $g_k \equiv a_k$, these classes are called "nondefective", and it can always be diagonalized, i.e. symmetric matrices.

Example 11.5

$$A = \begin{bmatrix} 2 & 1 \\ 0 & 2 \end{bmatrix} \quad \Rightarrow \quad \lambda_1 + \lambda_2 = 4, \quad \lambda_1\lambda_2 = 4 \quad \Rightarrow \quad \lambda_1 = 2, \quad \lambda_2 = 2 \tag{11.30}$$

$$\lambda_1 = 2: \quad \left(\begin{bmatrix} 2 & 1 \\ 0 & 2 \end{bmatrix} - \begin{bmatrix} 2 & 0 \\ 0 & 2 \end{bmatrix} \right) \underline{x}_1 = \begin{bmatrix} 0 & 1 \\ 0 & 0 \end{bmatrix} \underline{x}_1 = \begin{bmatrix} 0 \\ 0 \end{bmatrix} \quad \Rightarrow \quad \underline{x}_1 = \begin{bmatrix} 1 \\ 0 \end{bmatrix} \tag{11.31}$$

But, there is no second linearly independent eigenvector. So, the algebraic multiplicity is 2 ($a_k = 2$, number of repeated eigenvalues), but the geometric multiplicity is 1 ($g_k = 1$, number of independent eigenvectors). Note that

$$1 \leq g_k \leq a_k \tag{11.32}$$

When $g_k = a_k$, the matrix is diagonalizable.

11.2.1 Classes of Diagonalizable Matrices

If a matrix A is nondefective, then

$$T = [\underline{x}_1 | \underline{x}_2 | \dots | \underline{x}_n] \tag{11.33}$$

is non-singular.

An important class is the set of "normal" matrices which is nondefective.

Let's first introduce some notations.

We say $\underline{W}$ and $\underline{V}$ are orthogonal if

$$< \underline{W}^*, \underline{V} >= \sum_K \overline{W}_K V_K = 0 \tag{11.34}$$

And pseudo-orthogonal if

$$< \underline{W}^T, \underline{V} >= \sum_K W_K V_K = 0 \tag{11.35}$$

where the bracket represents the inner product, the star represents complex conjugate transpose, the bar represents complex conjugate, and T represents transpose.

Example 11.6 Let $\underline{W} = \begin{bmatrix} 1 \\ j \end{bmatrix}$ then $W^T W = \begin{bmatrix} 1 & j \end{bmatrix} \begin{bmatrix} 1 \\ j \end{bmatrix} = 0$ where this is pseudo-orthogonal.

Definition 11.3 Matrix $\underline{N}$ is **Normal** if it commutes with its own adjoint (complex conjugate transpose),

$$\underline{N}\,\underline{N}^* = \underline{N}^*\underline{N} \tag{11.36}$$

This class has the following properties:

(1) It is nondefective, meaning $g_k \equiv a_k$.
(2) One can select $\underline{x}_k$ such that $\underline{x}_k^*\underline{x}_i = 0$, $i \neq k$.

Example 11.7 Assume $\underline{Q}$ is real and symmetric then $\underline{Q} = \underline{Q}^T$ and

$$\underline{Q}\,\underline{Q}^* = \underline{Q}\,\underline{Q}^T = \underline{Q}^2 = \underline{Q}^*\underline{Q} = \underline{Q}^T\underline{Q} \tag{11.37}$$

Therefore, $\underline{Q}$ is normal. Additionally, eigenvalues are real. What if $\underline{Q}$ is complex and symmetric? Then it could be shown that $\underline{Q}$ is not normal. This is left as an exercise.

Definition 11.4 A complex matrix $\underline{H}$ is **Hermitian** if

$$\underline{H}\,\underline{H}^* = \underline{H}^2 = \underline{H}^*\underline{H} \tag{11.38}$$

Then, $\underline{H}$ is normal. Additionally, the eigenvalues of $\underline{H}$ is real as is illustrated in the example below.

Example 11.8 A **Hermitian matrix** is a matrix for which $H = H^*$. For instance,

$$A = \begin{bmatrix} 3 & 2-i & -3i \\ 2+i & 0 & 1-i \\ 3i & 1+i & 0 \end{bmatrix}, \quad \bar{A} = \begin{bmatrix} 3 & 2+i & 3i \\ 2-i & 0 & 1+i \\ -3i & 1-i & 0 \end{bmatrix} \tag{11.39}$$

$$\Rightarrow \quad A^* = \bar{A}^T = \begin{bmatrix} 3 & 2-i & -3i \\ 2+i & 0 & 1-i \\ 3i & 1+i & 0 \end{bmatrix} \tag{11.40}$$

So, A is Hermitian.

Example 11.9 In this example, we will show that the eigenvalues of a Hermitian matrix are real. We write,

$$\underline{H}\,\underline{x} = \lambda\underline{x} \tag{11.41}$$

We obtain the complex conjugate of the Eq. (11.41),

$$\underline{x}^*\underline{H}^* = \bar{\lambda}\underline{x}^* \tag{11.42}$$

then as H is Hermitian, we can write,

$$\underline{x}^*\underline{H} = \bar{\lambda}\underline{x}^* \tag{11.43}$$

From Eq. (11.41), it follows:

$$\underline{x}^*\underline{H}\,\underline{x} = \lambda\underline{x}^*\underline{x} \tag{11.44}$$

From Eq. (11.43), it follows,

$$\underline{x}^*\underline{H}\,\underline{x} = \bar{\lambda}\underline{x}^*\underline{x} \tag{11.45}$$

Subtract Eq. (11.45) from (11.44),

$$0 = (\lambda - \bar{\lambda})\underline{x}^*\underline{x} \tag{11.46}$$

Knowing that $\underline{x}^*\underline{x} = \sum_K \bar{x}_k x_K = \sum(x_K^2) > 0$, then, $(\lambda = \bar{\lambda})$ which means eigenvalues are real

---

Example 11.10 Let's assume matrix A is given as

$$A = \begin{bmatrix} 1 & 0 \\ 0 & 1 \end{bmatrix} \tag{11.47}$$

then

$$\delta(\lambda) = (1 - \lambda)^2 \tag{11.48}$$

where $\lambda = 1$ and the algebraic multiplicity $a = 2$.

This matrix is diagonal with algebraic multiplicity of 2. Therefore, the geometric multiplicity of this matrix has to be equal to 2 and this is both evident from the matrix A and can also be computed from the null space of $([A - \lambda_k I])$.

Let's check to find out if A is normal,

$$A^* = A^T \text{ and } A^T A = AA^T \quad \rightarrow \text{Normal} \tag{11.49}$$

In addition, the eigenvectors are

$$\underline{x}_1 = \begin{bmatrix} 1 \\ 0 \end{bmatrix} \tag{11.50}$$

and

$$\underline{x}_2 = \begin{bmatrix} 0 \\ 1 \end{bmatrix} \tag{11.51}$$

which are orthogonal.

Example 11.11 Let's assume matrix A is given as

$$A = \begin{bmatrix} 0 & 1 \\ 0 & 0 \end{bmatrix} \tag{11.52}$$

then

$$\delta(\lambda) = (\lambda)^2 \tag{11.53}$$

where $\lambda = 0$ and the algebraic multiplicity $a = 2$.

Let's check to find out if A is normal,

$$AA^T = \begin{bmatrix} 0 & 1 \\ 0 & 0 \end{bmatrix} \begin{bmatrix} 0 & 0 \\ 1 & 0 \end{bmatrix} = \begin{bmatrix} 1 & 0 \\ 0 & 0 \end{bmatrix} \tag{11.54}$$

$$A^T A = \begin{bmatrix} 0 & 0 \\ 1 & 0 \end{bmatrix} \begin{bmatrix} 0 & 1 \\ 0 & 0 \end{bmatrix} = \begin{bmatrix} 0 & 0 \\ 0 & 1 \end{bmatrix} \neq AA^T \tag{11.55}$$

therefore, A is not normal. In addition, the eigenvector is

$$\underline{x}_1 = \begin{bmatrix} 1 \\ 0 \end{bmatrix} \tag{11.56}$$

and there is no second linearly independent eigenvector, so the geometric multiplicity is 1.

Definition 11.5 A complex matrix $\underline{U}$ is Unitary if

$$\underline{U}^* = \underline{U}^{-1} \tag{11.57}$$

For this matrix, we can write,

$$\underline{U}^* \underline{U} = \underline{U}^{-1} \underline{U} = \underline{I} = \underline{U}\, \underline{U}^* \tag{11.58}$$

In addition, unitary matrices are Normal, nondefective, eigenvectors are orthogonal, and

$$|\lambda| = 1 \tag{11.59}$$

Example 11.12 For instance,

$$A = \begin{bmatrix} \frac{1+i}{2} & \frac{1-i}{2} \\ \frac{1-i}{2} & \frac{1+i}{2} \end{bmatrix} \tag{11.60}$$

where $A^{*}=A^{-1}$.

Example 11.13 A real matrix $\underline{T}$ is Unitary if

$$\underline{T}^{T} = \underline{T}^{-1} \tag{11.61}$$

For this matrix, we can write,

$$\underline{T}^{*}\underline{T} = \underline{T}^{T}\underline{T} = \underline{T}^{-1}\underline{T} = \underline{I} = \underline{T}\,\underline{T}^{*} \tag{11.62}$$

In addition, $\underline{T}$ is Normal, nondefective, eigenvectors are orthogonal, and

$$|\lambda_k| = 1 \tag{11.63}$$

There are many other important matrix sub-classes. Suppose that $\underline{H}$ is Hermitian so that,

$$\underline{x}_i^{*}\underline{x}_k = 0 \quad \text{for all} \quad i \neq k \tag{11.64}$$

suppose we normalize $\underline{x}_k$ such that,

$$\underline{x}_k^{*}\underline{x}_k = \sum_{k} |\underline{x}_k|^2 = 1 \tag{11.65}$$

and

$$\underline{U} = [\underline{x}_1 | \underline{x}_2 | \underline{x}_3 \cdots | \underline{x}_n] \tag{11.66}$$

where $\underline{U}$ is known as the Modal matrix. Then,

$$\underline{U}^{*}\underline{U} = \begin{bmatrix} \underline{x}_1^{*} \\ \underline{x}_2^{*} \\ \vdots \\ \underline{x}_n^{*} \end{bmatrix} [\, \underline{x}_1 | \underline{x}_2 \cdots | \underline{x}_n \,] = \underline{I} \tag{11.67}$$

then $\underline{U}^{*} = \underline{U}^{-1}$, hence, $\underline{U}$ is unitary. It should be noted that the modal matrix of any normal matrix is unitary.

11.3 Singular Value Decomposition

One may question that why is it necessary to decouple/diagonalize a matrix? Diagonalization of a matrix generally reduces computational overhead, e.g. numerical optimization. But, most importantly, as will be pointed out in the rest of this book, it plays a significant role in MIMO control system design. But, not all matrices can be diagonalized using the approaches presented in the last section.

In this section, we present an approach that is general enough to be applied for diagonalization of all matrices.

MIMO systems often do not result in square matrices, nor have independent eigenvectors, and so on. So, we turn our attention to SVD to analyze these systems. The origin of SVD comes from eigenvector analysis of symmetric matrices $A = A^T$, which have orthogonal eigenvectors. If we have a nonsymmetric matrix, is it possible to transform this matrix to a symmetric one? For instance,

$$A = \begin{bmatrix} 1 & 2 \\ 3 & 4 \end{bmatrix} \quad \Rightarrow \quad AA^T = \begin{bmatrix} 5 & 11 \\ 11 & 25 \end{bmatrix} \quad \text{Symmetric!} \tag{11.68}$$

So, any real nonsymmetric matrices could be transformed to a symmetric one by multiplying it with its transpose. Furthermore, AA^T or A^TA is always a square matrix regardless of the A matrix. Therefore, we can always find an eigenvalue decomposition for AA^T or A^TA.

11.3.1 What is a Singular Value Decomposition?

All square or nonsquare matrices can be decomposed into a diagonal Σ and an orthonormal basis (normalized orthogonal eigenvectors.)[1] U, V.

$$\underset{(r\times m)}{A} = \underset{(r\times r)}{U}\underset{(r\times m)}{\Sigma}\underset{(m\times m)}{V^T} \tag{11.69}$$

where the left singular vectors are the column vectors of U, the right singular vectors are the row vectors of V^T, and Σ is a diagonal matrix of singular values. The singular values are always real if the matrix A is real.

The SVD and the eigenvalue decomposition have similarities which we investigate next. Because U and V are orthonormal, we have

$$AV = U\Sigma \underbrace{V^TV}_{=1} = U\Sigma, \quad U^TA = \underbrace{U^TU}_{=1} \Sigma V^T = \Sigma V^T \tag{11.70}$$

That is to say, $V^T = V^{-1}$ and $U^T = U^{-1}$; therefore, these matrices are unitary. For a singular value σ_i, V_i is defined as the right singular vector and U_i is defined as the left singular vector.

$$AV_i = \sigma_i U_i \tag{11.71}$$

$$(\text{Analogous to } A\underline{x} = \lambda\underline{x}) \tag{11.72}$$

This mapping is illustrated in Figure 11.1. The size of the matrix Σ is the same as the matrix A, while U and V are square. The null space of A presents itself as zeros on the diagonal of the matrix Σ, as shown below,

$$A\begin{bmatrix} V_1 & V_2 & \cdots & V_m \end{bmatrix} = \begin{bmatrix} U_1 & U_2 & \cdots & U_r \end{bmatrix}\Sigma \tag{11.73}$$

$$\text{if } r = m \text{ then } \Sigma = \begin{bmatrix} \sigma_1 & & \\ & \ddots & \\ & & \sigma_r \end{bmatrix} \tag{11.74}$$

$$\text{if } r < m \text{ then } \Sigma = \begin{bmatrix} \sigma_1 & & & & & \\ & \ddots & & & & \\ & & \sigma_r & 0 & \cdots & 0 \end{bmatrix} \tag{11.75}$$

$$\text{if } r > m \text{ then } \Sigma = \begin{bmatrix} \sigma_1 & & \\ & \ddots & \\ & & \sigma_m \\ & & 0 \\ & & \vdots \\ & & 0 \end{bmatrix} \tag{11.76}$$

Typically, singular values are ordered from the largest to the smallest. That is, $\sigma_1 \geq \cdots \geq \sigma_{\min(r,m)}$.

How do singular values relate to eigenvalues and singular vectors relate to eigenvectors? We can write the following relationships:

$$\begin{aligned} AV &= U\Sigma \\ A &= U\Sigma V^{-1} = U\Sigma V^T \\ A^TA &= V\Sigma^T U^T U\Sigma V^T \\ \underbrace{A^TA}_{\text{Square}} &= V\Sigma^T\Sigma V^T \end{aligned} \tag{11.77}$$

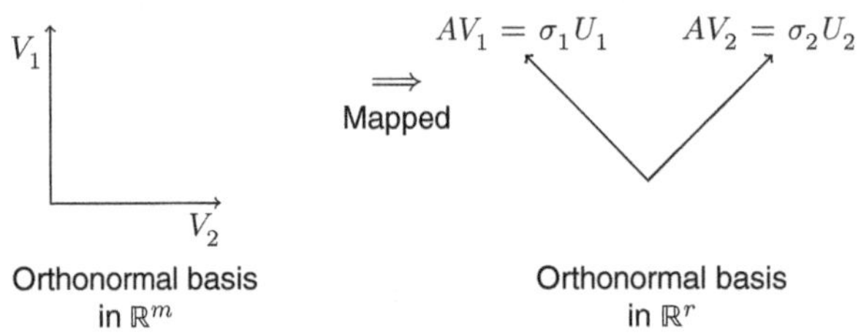

Figure 11.1 SVD orthonormal basis.

1 Normalized orthogonal eigenvectors always form an orthonormal basis, but basis vectors do not always form eigenvectors.

From the above development, we can state that the right singular vectors V are exactly the eigenvectors of A^TA and the square of the singular values $\Sigma^T\Sigma$ are exactly the eigenvalues of A^TA. Similarly, we can write relate the left singular vectors U to the eigenvectors by writing the following expressions:

$$\begin{aligned} A^TU &= V\Sigma^T \\ A^T &= V\Sigma^TU^{-1} = V\Sigma^TU^T \\ AA^T &= U\Sigma V^TV\Sigma^TU^T \\ AA^T &= U\Sigma\Sigma^TU^T \end{aligned} \tag{11.78}$$

Therefore, the left singular vectors, U are the eigenvectors of AA^T and the square of the singular values $\Sigma\Sigma^T$ are the eigenvalues of AA^T.

Example 11.14

$$A = \begin{bmatrix} 4 & 4 \\ -3 & 3 \end{bmatrix} \tag{11.79}$$

V_1, V_2 are in row space $\mathbb{R}^2$. U_1, U_2 are in column space $\mathbb{R}^2$. $\sigma_1 > 0, \sigma_2 > 0$.

$$A^TA = \begin{bmatrix} 4 & -3 \\ 4 & 3 \end{bmatrix}\begin{bmatrix} 4 & 4 \\ -3 & 3 \end{bmatrix} = \begin{bmatrix} 25 & 7 \\ 7 & 25 \end{bmatrix} \tag{11.80}$$

$$\lambda_1 = 32 \quad \Rightarrow \quad V_{1n} = \begin{bmatrix} 1 \\ 1 \end{bmatrix} \Rightarrow V_1 = \begin{bmatrix} \frac{1}{\sqrt{2}} \\ \frac{1}{\sqrt{2}} \end{bmatrix} \tag{11.81}$$

$$\lambda_2 = 18 \quad \Rightarrow \quad V_{2n} = \begin{bmatrix} -1 \\ 1 \end{bmatrix} \Rightarrow V_2 = \begin{bmatrix} -\frac{1}{\sqrt{2}} \\ \frac{1}{\sqrt{2}} \end{bmatrix} \tag{11.82}$$

$$AA^T = \begin{bmatrix} 4 & 4 \\ -3 & 3 \end{bmatrix}\begin{bmatrix} 4 & -3 \\ 4 & 3 \end{bmatrix} = \begin{bmatrix} 32 & 0 \\ 0 & 18 \end{bmatrix} \tag{11.83}$$

$$\lambda_1 = 32 \quad \Rightarrow \quad U_1 = \begin{bmatrix} 1 \\ 0 \end{bmatrix}, \quad \lambda_2 = 18 \quad \Rightarrow \quad U_2 = \begin{bmatrix} 0 \\ 1 \end{bmatrix} \tag{11.84}$$

$$\sigma_1 = \sqrt{\lambda_1} = \sqrt{32}, \quad \sigma_2 = \sqrt{\lambda_2} = \sqrt{18} \tag{11.85}$$

$$\begin{bmatrix} 4 & 4 \\ -3 & 3 \end{bmatrix} = \left[\begin{array}{c:c} 1 & 0 \\ 0 & 1 \end{array}\right]\begin{bmatrix} \sqrt{32} & 0 \\ 0 & \sqrt{18} \end{bmatrix}\left[\begin{array}{cc} \frac{1}{\sqrt{2}} & \frac{1}{\sqrt{2}} \\ \hdashline -\frac{1}{\sqrt{2}} & \frac{1}{\sqrt{2}} \end{array}\right] \tag{11.86}$$

11.3.2 Orthonormal Vectors

A set of vectors are normal if

$$q_i^T q_j = \begin{cases} 0 & \text{if } i \neq j \\ 1 & \text{if } i = j \end{cases} \tag{11.87}$$

for

$$Q = \begin{bmatrix} q_1 & \cdots & q_n \end{bmatrix} \tag{11.88}$$

so that

$$Q^TQ = \begin{bmatrix} q_1^T \\ \vdots \\ q_n^T \end{bmatrix}\begin{bmatrix} q_1 & \cdots & q_n \end{bmatrix} = \begin{bmatrix} 1 & & 0 \\ & \ddots & \\ 0 & & 1 \end{bmatrix} = I_n \tag{11.89}$$

If Q is square, then $Q^TQ = I$, which tells us that $Q^T = Q^{-1}$.

For any independent vectors a and b, Gram–Schmidt process [28] could be utilized to find orthonormal basis vectors.

Example 11.15

$$A = \begin{bmatrix} 4 & 3 \\ 8 & 6 \end{bmatrix} \tag{11.90}$$

The rows and columns of A are not linearly independent, so, in this example, there will be a singular value (square root of an eigenvalue) at 0, and only one linearly independent right and one linearly independent left singular vectors. The second left and right singular vectors, as will be shown in this example, will be computed from the null space of A or $\eta(A)$. The row and column spaces of A are visualized in Figure 11.2.

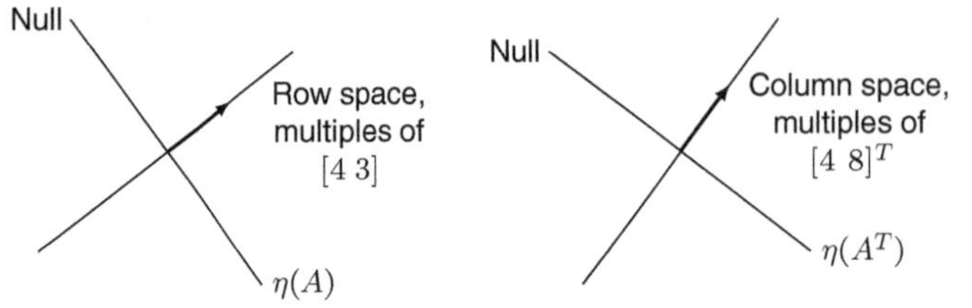

Figure 11.2 Row and column spaces of A.

What should V be? We can define V_1 as basis of the row space of A.

$$V_1 = \begin{bmatrix} \frac{4}{5} \\ \frac{3}{5} \end{bmatrix} \tag{11.91}$$

Then, the only orthogonal vector V_1 is the basis of $\eta(A)$.

$$V_2 = \begin{bmatrix} \frac{3}{5} \\ -\frac{4}{5} \end{bmatrix} \tag{11.92}$$

Similarly, we choose U to contain the basis of the column space of A and the basis of $\eta(A^T)$.

$$U_1 = \begin{bmatrix} \frac{1}{\sqrt{5}} \\ \frac{2}{\sqrt{5}} \end{bmatrix}, \quad U_2 = \begin{bmatrix} \frac{2}{\sqrt{5}} \\ -\frac{1}{\sqrt{5}} \end{bmatrix} \tag{11.93}$$

We can find the singular values from either A^TA or AA^T.

$$AA^T = \begin{bmatrix} 4 & 3 \\ 8 & 6 \end{bmatrix}\begin{bmatrix} 4 & 8 \\ 3 & 6 \end{bmatrix} = \begin{bmatrix} 25 & 50 \\ 50 & 100 \end{bmatrix} \quad \Rightarrow \quad \sigma_1^2 = \lambda_1 = 125, \quad \sigma_2^2 = \lambda_2 = 0 \tag{11.94}$$

So,

$$\begin{bmatrix} 4 & 3 \\ 8 & 6 \end{bmatrix} = \begin{bmatrix} \frac{1}{\sqrt{5}} & \frac{2}{\sqrt{5}} \\ \frac{2}{\sqrt{5}} & -\frac{1}{\sqrt{5}} \end{bmatrix}\begin{bmatrix} \sqrt{125} & 0 \\ 0 & 0 \end{bmatrix}\begin{bmatrix} \frac{4}{5} & \frac{3}{5} \\ \frac{3}{5} & -\frac{4}{5} \end{bmatrix} \tag{11.95}$$

Example 11.16 In this example, we illustrate a different approach in the calculation of the left singular vectors U after having obtained the right singular vectors V.

Consider a matrix C.

$$C = \begin{bmatrix} 5 & 5 \\ -1 & 7 \end{bmatrix} \tag{11.96}$$

We want to find U and V such that $C = U\Sigma V^T$, we can then write,

$$C^TC = V\Sigma^T\Sigma V^T \tag{11.97}$$

and then use this equation to compute right singular vectors V, furthermore,

$$CV = U\Sigma \tag{11.98}$$

So, U can be computed from Eq. (11.98), by multiplying both sides of (11.98) with Σ^{-1},

$$CV\Sigma^{-1} = U \tag{11.99}$$

This computation is provided below,

$$C^T C = \begin{bmatrix} 26 & 18 \\ 18 & 74 \end{bmatrix} \tag{11.100}$$

$$\Rightarrow \quad \det(C^T C - \lambda I) = \lambda^2 - 100\lambda + 1600 = (\lambda - 20)(\lambda - 80) \tag{11.101}$$

$$\lambda_1 = 80, \quad \lambda_2 = 20 \tag{11.102}$$

$$C^T C - 80I = \begin{bmatrix} -54 & 18 \\ 18 & -6 \end{bmatrix} \quad \Rightarrow \quad V_1 = \frac{1}{\sqrt{10}} \begin{bmatrix} 1 \\ 3 \end{bmatrix} \tag{11.103}$$

$$C^T C - 20I = \begin{bmatrix} 6 & 18 \\ 18 & 54 \end{bmatrix} \quad \Rightarrow \quad V_2 = \frac{1}{\sqrt{10}} \begin{bmatrix} -3 \\ 1 \end{bmatrix} \tag{11.104}$$

$$V = \begin{bmatrix} \frac{1}{\sqrt{10}} & -\frac{3}{\sqrt{10}} \\ \frac{3}{\sqrt{10}} & \frac{1}{\sqrt{10}} \end{bmatrix}, \quad \Sigma = \begin{bmatrix} 4\sqrt{5} & 0 \\ 0 & 2\sqrt{5} \end{bmatrix} \tag{11.105}$$

So, with C, Σ, and V known, we can work backwards to find U with the correct signs.

$$CV = U\Sigma \quad \Rightarrow \quad CV\Sigma^{-1} = U \tag{11.106}$$

$$U = \begin{bmatrix} 5 & 5 \\ -1 & 7 \end{bmatrix} \begin{bmatrix} \frac{1}{\sqrt{10}} & -\frac{3}{\sqrt{10}} \\ \frac{3}{\sqrt{10}} & \frac{1}{\sqrt{10}} \end{bmatrix} \begin{bmatrix} 4\sqrt{5} & 0 \\ 0 & 2\sqrt{5} \end{bmatrix}^{-1} \tag{11.107}$$

$$\Rightarrow \quad U = \frac{1}{\sqrt{2}} \begin{bmatrix} 1 & -1 \\ 1 & 1 \end{bmatrix} \tag{11.108}$$

We will present a summary of our development about SVD including generalizing SVD for all square and nonsquare, real and imaginary matrices in the next section.

11.4 Singular Value Decomposition Properties

Let M be a $r \times m$ matrix that may have complex elements, k will denote the kth singular value, left singular vector, or right singular vector, as appropriate. M^* denotes the complex conjugate transpose of M. Without loss of generality, to be more specific, we will change our notation from V_k for the right singular vector to r_k and similarly, for the left singular vector from U_k to l_k.

Definition 11.6 σ_k is a singular value, $\underline{r}_k$ is a right singular vector, and $\underline{l}_k$ is a left singular vector of M if

1. σ_k is real and ≥ 0.
2. $M\underline{r}_k = \begin{cases} \underline{l}_k \sigma_k, & k \leq \text{rank } M \\ 0, & k > \text{rank } M \end{cases}$.
3. $M^* \underline{l}_k = \begin{cases} \underline{r}_k \sigma_k, & k \leq \text{rank } M \\ 0, & k > \text{rank } M \end{cases}$.

Consider M^* times $M\underline{r}_k$.

$$M^* M \underline{r}_k = M^* \underline{l}_k \sigma_k \tag{11.109}$$

$$M^* M \underline{r}_k = \sigma_k^2 \underline{r}_k \tag{11.110}$$

similar to the eigenvalue problem.

Consider a matrix M and vectors $\underline{x}$ and $\underline{y}$ such that

$$\underline{y} = M\underline{x} \tag{11.111}$$

We can define a scalar q such that

$$q = \underline{x}^* M^* M \underline{x} \tag{11.112}$$

$$q = \underline{y}^* \underline{y} = \sum_{k=1}^{m} |y_k|^2 \geq 0 \tag{11.113}$$

So, q is positive definite or positive semi-definite (nonnegative definite) and

$$\lambda_k(M^*M) \geq 0 \tag{11.114}$$

Therefore, we can always find real singular values,

$$\sigma_k(M^*M) = +\sqrt{\lambda_k(M^*M)} \tag{11.115}$$

We can now define the procedure to find singular values and singular vectors:

1. From M, form M^*M and find the eigenvalues λ_k and normalized eigenvectors $\underline{\epsilon}_k$.
2. Set $\sigma_k = \sqrt{\lambda_k}$, $\underline{r}_k = \underline{\epsilon}_k$ and generate $\underline{l}_k = M\underline{r}_k\sigma_k^{-1}$ and, if necessary, $\underline{l}_k$ from $M^*\underline{l}_k = 0$.

Given matrix, M with the right singular vector, $\underline{r}_k$, the left singular vector, $\underline{l}_k$, and the singular value, σ_k, then we can write, $M\underline{r}_k = \underline{l}_k\sigma_k$, then knowing this relationship, we can write the following equivalencies,

- $M^*\underline{l}_k = \sigma_k\underline{r}_k$
- $\sigma(M) = \sigma(M^*)$
- $\underline{r}_k(M^*) = \underline{l}_k(M)$ (and $\underline{l}_k$ of M is an $\underline{r}_k$ of M^*)
- $\underline{l}_k(M^*) = \underline{r}_k(M)$ (and $\underline{r}_k$ of M is an $\underline{l}_k$ of M^*)

Suppose that M is square with $M\underline{r}_k = \underline{l}_k\sigma_k$, then,

$$M^{-1}M\underline{r}_k = \sigma_k M^{-1}\underline{l}_k \quad \Rightarrow \quad M^{-1}\underline{l}_k = \frac{1}{\sigma_k}\underline{r}_k \tag{11.116}$$

This implies that

- $\sigma(M^{-1}) = (\sigma(M))^{-1}$
- $\underline{r}_k(M^{-1}) = \underline{l}_k(M)$
- $\underline{l}_k(M^{-1}) = \underline{r}_k(M)$

11.5 Comparison of Eigenvalue and Singular Value Decompositions

So far, we have discussed both eigenvalue and SVDs. In this section, we will further discuss and compare these two decompositions to develop a better understanding and insight in how to use these methodologies in practical applications.

The following development is provided in [3]; we reiterate this here for completeness. Consider a static system with m inputs and r outputs.

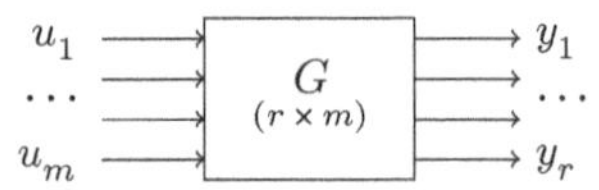

Let $g_{ij}(\omega)$ represent the harmonic (sinusoidal) response from input j to output i.

$$u_j(t) = u_{j0}\sin(\omega t + \alpha_j) \quad \Rightarrow \quad y_i(t) = y_{i0}\sin(\omega t + \beta_i) \tag{11.117}$$

where the amplification (gain) and the phase shift may be obtained from the complex number $g_{ij}(j\omega)$ as follows.

$$\left|g_{ij}(j\omega)\right| = \frac{y_{i0}}{u_{j0}}, \quad \angle g_{ij}(j\omega) = \beta_i - \alpha_j \tag{11.118}$$

In phasor notation,

$$y_i(\omega) = g_{ij}(j\omega)u_j(\omega) \tag{11.119}$$

where

$$u_j(\omega) = u_{j0}e^{j\alpha_j}, \quad y_i(\omega) = y_{i0}e^{i\beta_i} \tag{11.120}$$

The overall response to simultaneous input signals, of the same frequency, in several channels is

$$y_i(\omega) = g_{i1}(j\omega)u_1(\omega) + g_{i2}(\omega)u_2(\omega) + \cdots = \sum_{j=1}^{m} g_{ij}(j\omega)u_j(\omega) \tag{11.121}$$

In matrix form,

$$y(\omega) = G(j\omega)u(\omega) \tag{11.122}$$

$$u(\omega) = \begin{bmatrix} u_1(\omega) \\ u_2(\omega) \\ \vdots \\ u_m(\omega) \end{bmatrix}, \quad y(\omega) = \begin{bmatrix} y_1(\omega) \\ y_2(\omega) \\ \vdots \\ y_r(\omega) \end{bmatrix} \tag{11.123}$$

It can be stated that eigenvalues $|\lambda\,(G(j\omega))|$ are a poor measure of gain as they do not provide a useful means of generalizing the SISO gain. As mentioned, eigenvalues can only be computed for square systems, and even then they can be very misleading. For instance, take $y = Gu$ with,

$$G = \begin{bmatrix} 0 & 10 \\ 0 & 0 \end{bmatrix} \tag{11.124}$$

which has both eigenvalues equal to zero. However, to conclude from the eigenvalues that the system gain is zero is clearly misleading.

$$\text{Input:} \quad u = \begin{bmatrix} 0 \\ 1 \end{bmatrix} \quad \Rightarrow \quad \text{Output:} \quad y = \begin{bmatrix} 10 \\ 0 \end{bmatrix} \tag{11.125}$$

We can ask, what seems to be the problem? The eigenvalues measure the gain for the special case when the inputs and the outputs are in the same direction, namely in the direction of the eigenvectors.

Let t_i be the eigenvector of G and consider $u = t_i$. Then,

$$y = Gt_i = \lambda_i t_i \tag{11.126}$$

where λ_i is the corresponding eigenvalue. We can write,

$$\frac{\|y\|}{\|u\|} = \frac{\|\lambda_i t_i\|}{\|t_i\|} = |\lambda_i| \tag{11.127}$$

The eigenvalues are a good measure for system stability but not for system performance.

11.5.1 System Gain

The magnitude of the input/output, as illustrated in Figure 11.3, is the Euclidean distance.

$$\|\underline{u}\|_E = \sqrt{\sum_k u_k^2}, \quad \|\underline{y}\|_E = \sqrt{\sum_k y_k^2} \tag{11.128}$$

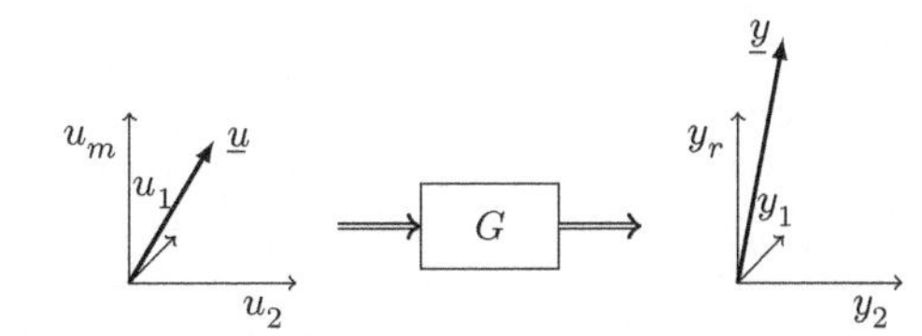

Figure 11.3 Multidimensional input/output map.

"Gain", in linear systems, is independent of input magnitude (Euclidean distance) and depends on the input frequency (similar to SISO systems),

$$\text{Gain} = \frac{\|\underline{y}\|_E}{\|\underline{u}\|_E} = \frac{\|G(j\omega)\underline{u}(\omega)\|_E}{\|\underline{u}(\omega)\|_E} \tag{11.129}$$

but now for a MIMO system, there are additional degrees of freedom and the gain depends on the direction of each input u_k or the relative magnitude of each input, e.g. u_1/u_2, as well as the input frequency.

Let's get a better understanding of what we mean by input direction through analyzing a static system in the next example.

Example 11.17 For a system with two inputs

$$u = \begin{bmatrix} u_{10} \\ u_{20} \end{bmatrix} \tag{11.130}$$

the gain is in general different for the following five inputs.

$$u_1 = \begin{bmatrix} 1 \\ 0 \end{bmatrix}, \quad u_2 = \begin{bmatrix} 0 \\ 1 \end{bmatrix}, \quad u_3 = \begin{bmatrix} 0.707 \\ 0.707 \end{bmatrix} \tag{11.131}$$

$$u_4 = \begin{bmatrix} 0.707 \\ -0.707 \end{bmatrix}, \quad u_5 = \begin{bmatrix} 0.6 \\ -0.8 \end{bmatrix} \tag{11.132}$$

In each of these cases, the input magnitude $\|\underline{u}\|_E = 1$, but with a different relative magnitude or direction, as illustrated in Figure 11.4.

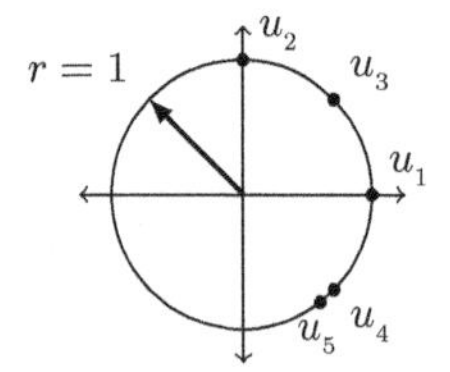

Figure 11.4 Projection of a vector onto a plane.

We can determine the outputs for the following static system G.

$$G = \begin{bmatrix} 5 & 4 \\ 3 & 2 \end{bmatrix} \tag{11.133}$$

Then,

$$y_1 = \begin{bmatrix} 5 \\ 3 \end{bmatrix}, \quad y_2 = \begin{bmatrix} 4 \\ 2 \end{bmatrix}, \quad y_3 = \begin{bmatrix} 6.36 \\ 3.53 \end{bmatrix} \tag{11.134}$$

$$y_4 = \begin{bmatrix} 0.707 \\ 0.707 \end{bmatrix}, \quad y_5 = \begin{bmatrix} -0.2 \\ 0.2 \end{bmatrix} \tag{11.135}$$

$$\|\underline{y_1}\|_E = 5.83, \quad \|\underline{y_2}\|_E = 4.47, \quad \|\underline{y_3}\|_E = 7.30 \tag{11.136}$$

$$\|\underline{y_4}\|_E = 1, \quad \|\underline{y_5}\|_E = 0.28 \tag{11.137}$$

The resulting gains are plotted in Figure 11.5.

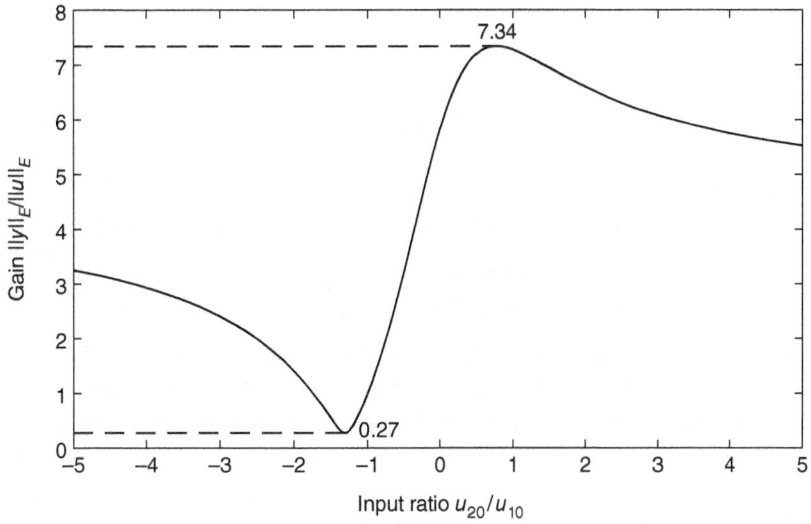

Figure 11.5 Gain as a function of input direction.

So, the singular values (σ) are precisely the system gains given an input direction, and they can be written as

$$\max_{\underline{u}\neq 0} \frac{\|G\underline{u}\|_E}{\|\underline{u}\|_E} = \max_{\|\underline{u}\|_E=1} \|G\underline{u}\|_E = \overline{\sigma}(G) \tag{11.138}$$

$$\min_{\underline{u}\neq 0} \frac{\|G\underline{u}\|_E}{\|\underline{u}\|_E} = \min_{\|\underline{u}\|_E=1} \|G\underline{u}\|_E = \underline{\sigma}(G) \tag{11.139}$$

and hence,

$$\underline{\sigma}(G) \leq \frac{\|G\underline{u}\|_E}{\|\underline{u}\|_E} \leq \overline{\sigma}(G) \tag{11.140}$$

In MIMO systems, input directions are determined by right singular vectors and the gains by singular values. We stated that singular vectors are unitary therefore they have unit gains, so they only affect the direction while the singular values affect the magnitude.

Let $\underline{u} = u_0\underline{r}_k$, where u_0 is the input magnitude and $\underline{r}_k$ is the direction, a right singular vector of G. We know the singular vectors are unitary,

$$\underline{r}_k^*\underline{r}_k = 1 = \underline{l}_k^*\underline{l}_k \tag{11.141}$$

What is $\|\underline{u}\|_E$?

$$\|\underline{u}\|_E = \sqrt{\underline{u}^*\underline{u}} = \sqrt{u_0^2\underline{r}_k^*\underline{r}_k} = u_0 = |u| \tag{11.142}$$

What is $\|\underline{y}\|_E$?

$$\underline{y} = G\underline{u} = u_0 G\underline{r}_k \tag{11.143}$$

But, $G\underline{r}_k = \underline{l}_k\sigma_k$, from definition of SVD in the previous section. So,

$$\underline{y} = u_0\underline{l}_k\sigma_k, \quad \left\|\underline{y}\right\|_E = \sqrt{\sigma_k^2 u_0^2 \underline{l}_k^* \underline{l}} = \sigma_k |u| \tag{11.144}$$

So, for $\underline{u} = u_0\underline{r}_k$,

$$\frac{\left\|\underline{y}\right\|_E}{\left\|\underline{u}\right\|_E} = \sigma_k \tag{11.145}$$

This again illustrates that singular values determine the system gains.

In general, any real matrix can always be written, in this case, 2×2, in the following form of,

$$G = \begin{bmatrix} \cos\theta_1 & -\sin\theta_1 \\ \sin\theta_1 & \cos\theta_1 \end{bmatrix} \begin{bmatrix} \sigma_1 & 0 \\ 0 & \sigma_2 \end{bmatrix} \begin{bmatrix} \cos\theta_2 & \sin\theta_2 \\ -\sin\theta_2 & \cos\theta_2 \end{bmatrix} \tag{11.146}$$

where the matrix in the right is the right singular vectors and the matrix in the left is the left singular vectors, and the middle matrix are the singular values. So, the left and right singular vectors purely rotate the input vector, while, the singular values stretch the input vector. We will generalize our findings for nonsquare matrices in the next section.

11.6 Generalized Singular Value Decomposition

Below, we will generalize the SVD for any nonsquare matrix M with dimension $r \times m$. We will start by writing the following expression,

$$\underbrace{\begin{bmatrix} \underline{l}_1^* \\ \vdots \\ \underline{l}_r^* \end{bmatrix}}_{(r\times r)} \underset{(r\times m)}{M} \underbrace{\left[\, \underline{r}_1 \mid \cdots \mid \underline{r}_m \,\right]}_{(m\times m)} \tag{11.147}$$

where,

$$\underbrace{\begin{bmatrix} \underline{l}_1^* \\ \vdots \\ \underline{l}_r^* \end{bmatrix}}_{(r\times r)} = U^* \tag{11.148}$$

the left singular column vectors, and

$$\underbrace{\left[\, \underline{r}_1 \mid \cdots \mid \underline{r}_m \,\right]}_{(m\times m)} = V \tag{11.149}$$

the right singular row vectors. Then we can write,

$$U^*MV = U^*\left[\, M\underline{r}_1 \mid M\underline{r}_2 \mid \cdots \mid M\underline{r}_m \,\right] \tag{11.150}$$

where we learned that $M\underline{r}_k = \sigma_k\underline{l}_k$, hence, we can write the following relationship,

$$\begin{aligned} U^*\left[\, M\underline{r}_1 \mid M\underline{r}_2 \mid \cdots \mid M\underline{r}_m \,\right] &= U^*\left[\, \sigma_1\underline{l}_1 \mid \sigma_2\underline{l}_2 \mid \cdots \mid \sigma_m\underline{l}_m \,\right] \\ &= \begin{bmatrix} \underline{l}_1^* \\ \underline{l}_2^* \\ \vdots \\ \underline{l}_r^* \end{bmatrix} \left[\, \sigma_1\underline{l}_1 \mid \sigma_2\underline{l}_2 \mid \cdots \mid \sigma_m\underline{l}_m \,\right] \end{aligned} \tag{11.151}$$

We know: $U^* = U^{=1}$ and $V^* = V^{-1}$, therefore,

$$\begin{bmatrix} \underline{l}_1^* \\ \underline{l}_2^* \\ \vdots \\ \underline{l}_r^* \end{bmatrix} \left[\, \sigma_1\underline{l}_1 \mid \sigma_2\underline{l}_2 \mid \cdots \mid \sigma_m\underline{l}_m \,\right] = \mathrm{diag}\left(\sigma_1,\ \sigma_2,\ \ldots,\ \sigma_m,\ 0,\ \ldots\right) = \Sigma \tag{11.152}$$

So,

$$U^*MV = \Sigma \Leftrightarrow M = U\Sigma V^* \tag{11.153}$$

A reminder note on dimensions:

- If $m > r$, columns of Σ (let's call it j) $j > r$ and Σ is going to be a wide matrix with $j - r$ column zeroed.
- If $m < r$, rows of Σ (let's call it i) $i > m$ and Σ is going to be a tall matrix with $i - m$ row zeroed.

Example 11.18 In the field of integrated vehicle dynamics and controls (IVDC), it is well known that control integration of active-chassis systems has many benefits, such as an increase in vehicle agility and stability (to name a few). The following transfer function matrix (TFM) is computed at low frequency when active differential and active roll control are integrated.

$$G = \begin{bmatrix} -0.0342 & -0.1195 \\ 0.0058 & 0.0214 \end{bmatrix} \tag{11.154}$$

The following singular vectors and singular values are calculated for this TFM.

$$G = \underbrace{\begin{bmatrix} \overbrace{\begin{matrix} -0.9845 \\ 0.1756 \end{matrix}}^{l_1} & \overbrace{\begin{matrix} 0.1756 \\ 0.9845 \end{matrix}}^{l_2} \end{bmatrix}}_{U} \underbrace{\begin{bmatrix} 0.1263 & 0 \\ 0 & 0.0003 \end{bmatrix}}_{\Sigma} \underbrace{\begin{bmatrix} \overbrace{0.2747 \quad 0.9615}^{r_1} \\ \underbrace{-0.9615 \quad 0.2747}_{r_2} \end{bmatrix}}_{V^*} \tag{11.155}$$

This system has two inputs $\underline{u} = [\ u_1,\ u_2\ ]^T$ and two outputs, $\underline{y} = [\ y_1,\ y_2\ ]^T$. In particular, u_1 is the roll torque distribution γ, and u_2 is the yaw torque T_ψ, y_1 is the vehicle lateral velocity V_y, and y_2 is the vehicle yaw rate V_ψ.

This turns out to be not an easy plant to control, let's first investigate the directional properties of this plant. From the first right singular vector r_1 (the first vector in V^*), we conclude that if both inputs are in the same directions, then the gain is 0.1263. However, from the second right singular vector r_2 (the second vector in V^*), we conclude that the gain is only 0.0003 when the inputs are in different directions. We will define the condition number ($\kappa(G)$) as follows:

$$\text{Condition number, } \kappa(G) = \frac{\overline{\sigma}(G)}{\underline{\sigma}(G)} = \frac{0.1263}{0.0003} = 469 \tag{11.156}$$

A large condition number means that the matrix is ill-conditioned. Control of ill-conditioned plants is difficult, as some combinations of the inputs have a strong effect on the outputs, whereas other combinations have a weak effect on the outputs. Hence, based on the aforementioned information, with our current actuators, under certain conditions (low adherence conditions), it becomes extremely difficult to satisfy both lateral velocity and yaw rate desired targets.

Consider the following possible inputs:

$$\underline{u} = \begin{bmatrix} 1 \\ 0 \end{bmatrix} \Rightarrow \underline{y} = \begin{bmatrix} -0.0342 \\ 0.0058 \end{bmatrix}, \quad \underline{u} = \begin{bmatrix} 0 \\ 1 \end{bmatrix} \Rightarrow \underline{y} = \begin{bmatrix} -0.1195 \\ 0.0214 \end{bmatrix} \tag{11.157}$$

It is obvious from that the first input, roll torque distribution, has much less authority over the outputs relative to the second input, the yaw torque. Now let's consider simultaneous change of both inputs in the same and opposite directions as follows:

$$\underline{u} = \begin{bmatrix} 1 \\ 1 \end{bmatrix} \Rightarrow \underline{y} = \begin{bmatrix} -0.1537 \\ 0.0272 \end{bmatrix}, \quad \underline{u} = \begin{bmatrix} -1 \\ 1 \end{bmatrix} \Rightarrow \underline{y} = \begin{bmatrix} -0.0853 \\ 0.0156 \end{bmatrix} \tag{11.158}$$

So, if we apply both inputs in the same directions, then the authority on the outputs are much greater than when we apply the inputs in different directions, as stated above. This also has a physical significance, if both actuators are working together to generate understeer or oversteer, their impacts are greater, versus when one actuator generates understeer and the other oversteer [33].

Finally, the direction of the left singular vectors impacts the output direction such that the direction of the first left singular vector, l_1, pulls the outputs in different direction, while, the direction of the second left singular vector, l_2, results in forcing the outputs to be in the same direction. This says that the actuators could have a bigger impact on the outputs with the decrease of lateral velocity and the increase of yaw rate or vice versa.

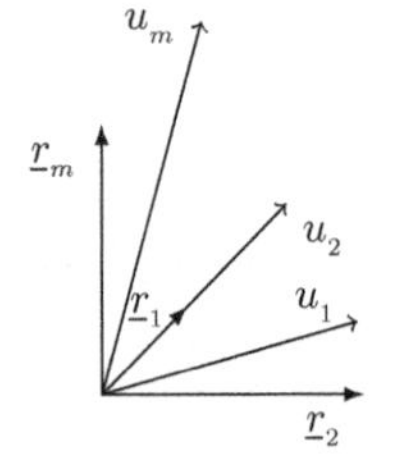

Figure 11.6 Altered view of the input space.

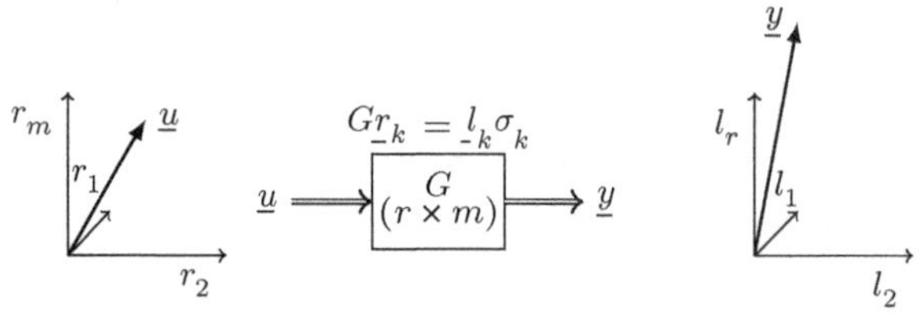

Figure 11.7 Input and output spaces of a dynamic system.

11.6.1 The Scalar Case

Consider the scalar case such that M is a complex number, then the singular value of M is

$$\sigma(M) = \sqrt{\lambda(M^*M)} = \sqrt{\lambda(\overline{M}M)} = \sqrt{\overline{M}M} = |M| \tag{11.159}$$

In other words, in the scalar case, the singular value of M is the modulus (used in the Bode magnitude plot).

11.6.2 Input and Output Spaces

We say that $\{\underline{r}_1, \ldots, \underline{r}_m\}$, the right singular vectors, is a "basis" for the input space. Similarly, $\{\underline{l}_1, \ldots, \underline{l}_r\}$, the left singular vectors, is a basis for the output space. In other words, the input $\underline{u}$ can be written as some linear combination of the right singular vectors and the output $\underline{y}$ can equivalently be written as some linear combination of the left singular vectors. This concept is illustrated in Figures 11.6 and 11.7.

Example 11.19 Suppose matrix G is $r \times m$, where $r = 2$, $m = 3$. Then, the right singular vectors consist of r_1, r_2 and r_3 but r_3 is in the null space of G.

$$\begin{bmatrix} y_1 \\ y_2 \end{bmatrix} = \begin{bmatrix} g_{11} & g_{12} & g_{13} \\ g_{21} & g_{22} & g_{23} \end{bmatrix} \begin{bmatrix} u_1 \\ u_2 \\ u_3 \end{bmatrix} \tag{11.160}$$

The right and left singular vectors can be written as

$$V = [\, r_1 \mid r_2 \mid r_3 \,], \quad U = [\, l_1 \mid l_2 \,] \tag{11.161}$$

While the singular value matrix, Σ is a wide matrix with the last column equal to zero,

$$GV = \begin{bmatrix} & & 0 \\ & & 0 \end{bmatrix} \quad \Rightarrow \quad U^*GV = \begin{bmatrix} \sigma_1 & 0 & 0 \\ 0 & \sigma_2 & 0 \end{bmatrix} \tag{11.162}$$

Any component along $\underline{r}_3$ is annihilated, and hence the output space is two dimensional, as shown in Figure 11.8,

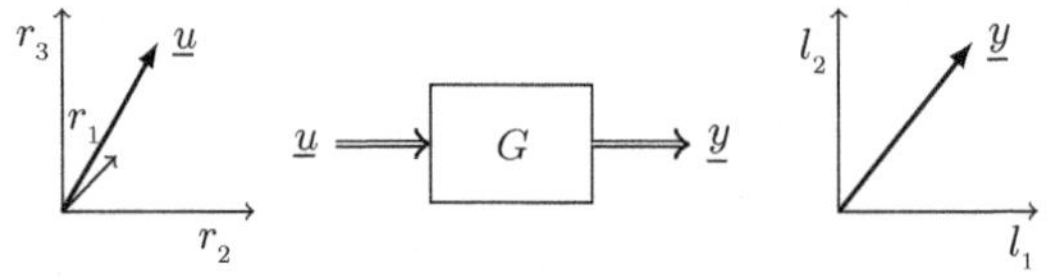

Figure 11.8 A system with 3 inputs and 2 outputs.

Example 11.20 Consider matrix G with $r \times m$ size, but this time, suppose $r = 3$, $m = 2$, then,

$$\begin{bmatrix} y_1 \\ y_2 \\ y_3 \end{bmatrix} = \begin{bmatrix} g_{11} & g_{12} \\ g_{21} & g_{22} \\ g_{31} & g_{32} \end{bmatrix} \begin{bmatrix} u_1 \\ u_2 \end{bmatrix} \tag{11.163}$$

The right and left singular vectors can be written as

$$V = \left[\, r_1 \,\vdots\, r_2 \,\right], \quad U = \left[\, l_1 \,\vdots\, l_2 \,\vdots\, l_3 \,\right] \tag{11.164}$$

While, this time, the singular value matrix, Σ is a tall matrix with the last row equal to zero,

$$GV = \begin{bmatrix} \; & \vdots & \; \\ \hline \; & \vdots & \; \\ \hline 0 & \vdots & 0 \end{bmatrix} \quad \Rightarrow \quad U^*GV = \begin{bmatrix} \sigma_1 & 0 \\ 0 & \sigma_2 \\ 0 & 0 \end{bmatrix} \tag{11.165}$$

The input space is two dimensional and eventhough, the output space is three dimensional, as shown in Figure 11.9, all outputs rest along $\underline{l}_1$ and $\underline{l}_2$.

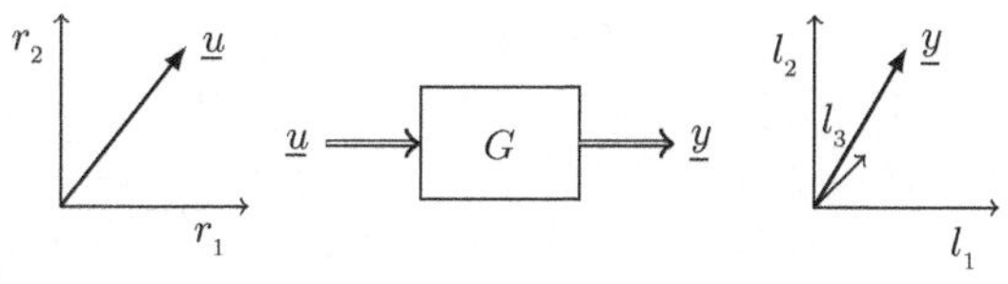

Figure 11.9 A system with 2 inputs and 3 outputs.

11.7 Norms

How *big* is a vector or matrix? We can use norms to find out. For example, consider the vector (or euclidean) norm:

$$||\underline{x}|| = \sqrt{\underline{x}^*\underline{x}} = \sqrt{\sum_k |x_k|^2} \tag{11.166}$$

In more general terms, norms obey the following rules. For vector norms,

- $||\underline{x}|| \geq 0$ (with equality only where $\underline{x} = \underline{0}$)
- $||\alpha\underline{x}|| = |\alpha| \cdot ||\underline{x}||$ (where α is any real or complex number)
- $||\underline{x}+\underline{y}|| \leq ||\underline{x}|| + ||\underline{y}||$ (triangle inequality)

While for matrix norms,

- $||M|| \geq 0$ (with equality only where $M = 0$)
- $||\alpha M|| = |\alpha| \cdot ||M||$ (where α is any real or complex number)
- $||M + N|| \leq ||M|| + ||N||$ (triangle inequality)

11.7.1 The Spectral Norm

Consider a dynamic system with input $\underline{x}$, output $\underline{y}$, shown in Figure 11.10.

The size of the input is $||\underline{x}||_E$, while the size of the output is given by $||\underline{y}||_E = ||M\underline{x}||_E$. We define the spectral norm of a matrix M to be

$$||M||_S = \max_{\text{all } \underline{x}} \frac{||M\underline{x}||_E}{||\underline{x}||_E} \tag{11.167}$$

It can be shown that

$$||M||_S = \max_{\text{all } \underline{x}} \frac{||M\underline{x}||_E}{||\underline{x}||_E} = \sqrt{\lambda_{max}(M^*M)} = \sigma_{max} \tag{11.168}$$

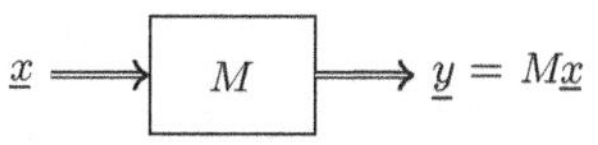

Figure 11.10 A multivariable dynamic system.

Also,

$$\sigma_{min} \leq \frac{\|M\underline{x}\|_E}{\|\underline{x}\|_E} \leq \sigma_{max} \tag{11.169}$$

If M is square, then

$$\sigma_{min} \leq |\lambda(M)| \leq \sigma_{max} \tag{11.170}$$

11.8 Problems

11.1 Classify the following matrices; that is, indicate whether each matrix is Symmetric, Hermitian, Orthogonal, and/or Unitary.

$$\text{(a)} \begin{bmatrix} 0 & 1 \\ -1 & -2 \end{bmatrix} \quad \text{(b)} \begin{bmatrix} 0 & 1 \\ 1 & 0 \end{bmatrix} \quad \text{(c)} \begin{bmatrix} 0 & i \\ i & 0 \end{bmatrix} \quad \text{(d)} \begin{bmatrix} 0 & i \\ -i & 0 \end{bmatrix} \tag{11.171}$$

$$\text{(e)} \begin{bmatrix} 1/5 & -2/5 \\ 2/5 & 1/5 \end{bmatrix} \quad \text{(f)} \begin{bmatrix} i/5 & -2i/5 \\ 2i/5 & i/5 \end{bmatrix} \quad \text{(g)} \begin{bmatrix} 1/2 & -i/2 \\ i/5 & 1/2 \end{bmatrix} \tag{11.172}$$

Which of these matrices are normal?

11.2 Use MATLAB to determine the eigenvalues and eigenvectors of matrices (e)–(g) from Problem 1. Check the eigenvectors for orthogonality.

11.3 Find the eigenvalues, eigenvectors, and geometric multiplicities of the matrix

$$\begin{bmatrix} 4 & 0 & -2 \\ 2 & 5 & 4 \\ 0 & 0 & 5 \end{bmatrix} \tag{11.173}$$

11.4 Suppose that a certain static system has the input–output relation

$$y = \begin{bmatrix} 1 & 2 & 1 \\ -1 & 3 & 6 \end{bmatrix} u \tag{11.174}$$

Find the singular values and the singular vectors. Calculate y where u is equal to each of the right singular vectors. Is $|y|/|u|$ the "directional gain" and is y pointed in the direction of a left singular vector? Are there any directional gains equal to zero? What would be the meaning of that?

12

MIMO Feedback Principals

In this chapter, we discuss MIMO feedback principals before introducing how to design MIMO controllers using Youla parameterization.

12.1 Mutlivariable Closed-Loop Transfer Functions

In this section, we derive relationships for MIMO return ratio and return difference, complementary sensitivity, sensitivity, and Youla transfer function matrices. We finish this section by introducing several state-space realization short cuts.

Consider the multivariable feedback control loop, with their associated vector signal and matrices dimensions, as shown in Figure 12.1. In a SISO closed-loop system, the controller and plant were both transfer functions, but now they are represented by transfer function matrices. This means we will have to rethink how we derive the closed-loop transfer function matrices, as G_cG_p is no longer equal to G_pG_c.

Consider the feedback system shown in Figure 12.2. Let $G_c(s)$ denote an $m \times r$ matrix of controller transfer functions and let $G_p(s)$ denote an $r \times m$ matrix of plant transfer functions.

We want to derive the return ratio (or loop gain) transfer function matrix (TFM), $L(s)$. One way to measure the loop gain is to break the loop at the output or the input of the plant. A good question to ask is if this return ratio matrix stays the same no matter where we break the loop. Let's investigate this by first breaking the loop at the output of the plant as shown in Figure 12.3.

We can then derive the return ratio matrix taken from the plant input, $\underline{i}$, to the plant output, $\underline{v}$ denoted by $L_y(s)$ using the following relationships:

$$\underline{v} = G_p(s)\underline{u} = -G_p(s)G_c(s)\underline{i} \tag{12.1}$$

$$\underset{(r\times r)}{L_y(s)} = \underset{(r\times m)}{G_p(s)}\underset{(m\times r)}{G_c(s)} \tag{12.2}$$

Similarly, we can break the loop at the plant input, shown in Figure 12.4.

We can then derive the return ratio taken from the plant input to output, similar to our previous derivation; however, this time, we denote the return ratio by $L_u(s)$ and using the following relationships:

$$\underline{v} = -G_c(s)\underline{y} = -G_c(s)G_p(s)\underline{i} \tag{12.3}$$

$$\underset{(m\times m)}{L_u(s)} = \underset{(m\times r)}{G_c(s)}\underset{(r\times m)}{G_p(s)} \tag{12.4}$$

We stated that for MIMO transfer function matrices, G_cG_p is no longer equal to G_pG_c, hence the return ration, $L_y \neq L_u$, so in MIMO feedback control loop, unlike the SISO counterpart, we have two distinct return ratios.

We can now form the closed-loop transfer function matrices for these two distinct return ratios. These will all be based on the block diagram shown in Figure 12.1.

Robust Control: Youla Parameterization Approach, First Edition. Farhad Assadian and Kevin R. Mallon.

Companion Website: www.wiley.com/go/assadian/robust

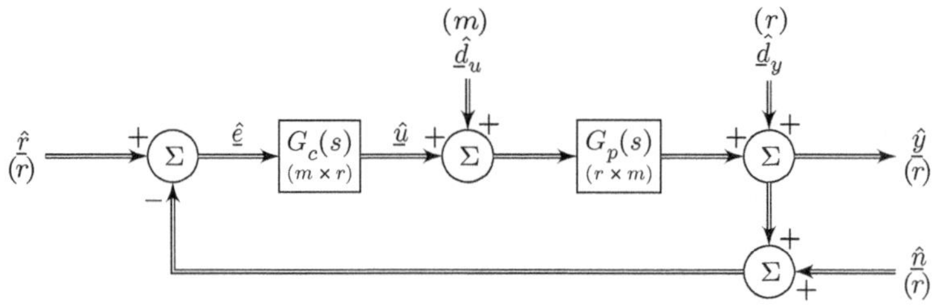

Figure 12.1 Block diagram of a multivariable control system.

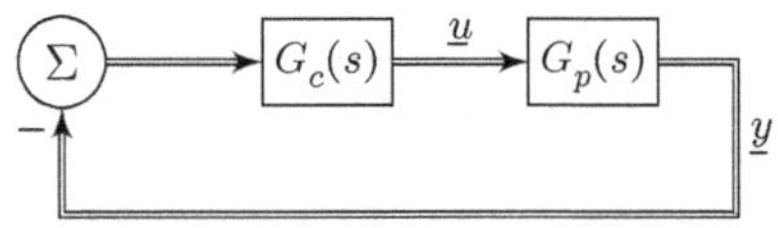

Figure 12.2 State feedback control system.

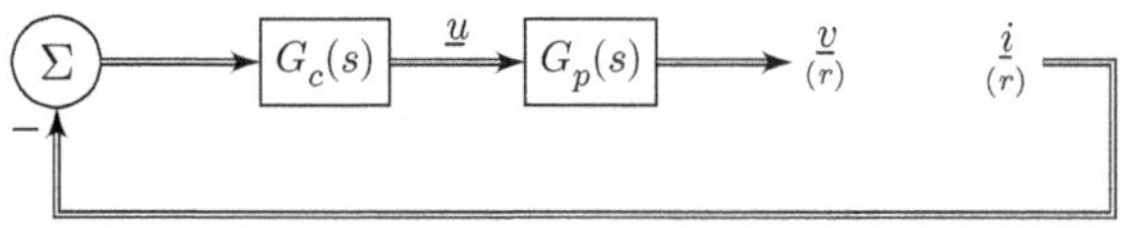

Figure 12.3 Open-loop gain from the plant output.

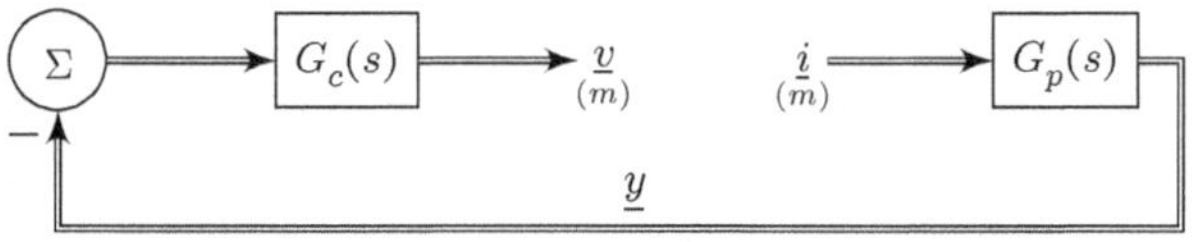

Figure 12.4 Open-loop gain from the plant input.

12.1.1 Transfer Function Matrix, From $\underline{r}$ to $\underline{y}$

$$\underline{y} = \underbrace{G_p G_c}_{L_y} \underline{r} - \underbrace{G_p G_c}_{L_y} \underline{y} \tag{12.5}$$

$$\begin{aligned} (I + L_y)\underline{y} &= L_y \underline{r} \\ \underline{y} &= (I + L_y)^{-1} L_y \underline{r} = T_y \underline{r} \\ \Rightarrow \underset{(r\times r)}{T_y} &= (I + L_y)^{-1} L_y \end{aligned} \tag{12.6}$$

The TFM T_y is known as the output complementary sensitivity or closed-loop TFM.

12.1.2 Transfer Function Matrix From $\underline{d}_y$ to $\underline{y}$ As Shown in Figure 12.1

$$\underline{y} = \underline{d}_y - L_y \underline{y} \tag{12.7}$$

$$\begin{aligned} (I + L_y)\underline{y} &= \underline{d}_y \\ \underline{y} &= (I + L_y)^{-1} \underline{d}_y = S_y \underline{d}_y \\ \Rightarrow \underset{(r\times r)}{S_y} &= (I + L_y)^{-1} \end{aligned} \tag{12.8}$$

where the TFM S_y is known as the output sensitivity TFM.

Note:

$$\begin{aligned} S_y + T_y &= (I + L_y)^{-1} + (I + L_y)^{-1} L_y \\ &= (I + L_y)^{-1}(I + L_y) \\ \Rightarrow S_y + T_y &= I \end{aligned} \tag{12.9}$$

This is similar to the SISO case where $S + T = 1$.

12.1.3 Transfer Function Matrix From $\underline{r}$ to $\underline{e}$

$$\underline{e} = \underline{r} - \underbrace{G_p G_c}_{L_y} \underline{e} \tag{12.10}$$

$$\underline{e} = (I + L_y)^{-1}\underline{r} = S_y\underline{r} \tag{12.11}$$

12.1.4 Transfer Function From $\underline{r}$ to $\underline{u}$

$$\underline{u} = G_c\underline{r} - \underbrace{G_c G_p}_{L_u} \underline{u} \tag{12.12}$$

$$(I + L_u)\underline{u} = G_c\underline{r}$$

$$\underline{u} = (I + L_u)^{-1}G_c\underline{r} = Y\underline{r}$$

$$\Rightarrow \underset{(m\times r)}{Y} = (I + L_u)^{-1}G_c \tag{12.13}$$

This is known as the Youla TFM. We can use this TFM to derive a relationship for the input closed-loop TFM, T_u:

$$T_u = (I + L_u)^{-1}L_u \tag{12.14}$$

$$= (I + L_u)^{-1}G_c G_p$$

$$\Rightarrow \underset{(m\times m)}{T_u} = \underset{(m\times r)}{Y}\ \underset{(r\times m)}{G_p} \tag{12.15}$$

A similar definition for T_y can be obtained by using the following identity:

$$G_p(I + G_c G_p) = G_p + G_p G_c G_p$$

$$= (I + G_p G_c)G_p \tag{12.16}$$

Now, multiplying both sides first by $(I + G_p G_c)^{-1}$ on the left, then by $(I + G_c G_p)^{-1}$ on the right,

$$\Rightarrow (I + L_y)^{-1}G_p = G_p(I + L_u)^{-1} \tag{12.17}$$

Now we use this property in deriving T_y as a function of Youla TFM as follows:

$$T_y = (I + L_y)^{-1}L_y \tag{12.18}$$

$$= (I + L_y)^{-1}G_p G_c$$

$$= G_p(I + L_u)^{-1}G_c$$

$$\Rightarrow \underset{(r\times r)}{T_y} = \underset{(r\times m)}{G_p}\ \underset{(m\times r)}{Y} \tag{12.19}$$

Additionally,

$$Y = (I + G_c G_p)^{-1}G_c$$

$$\Rightarrow Y = S_u G_c \tag{12.20}$$

where the TFM S_u is known as the input sensitivity TFM. We can also write

$$Y = (I + G_c G_p)^{-1}G_c \tag{12.21}$$

$$= G_c(I + G_p G_c)^{-1}$$

$$\Rightarrow Y = G_c S_y \tag{12.22}$$

12.1.5 Realization Tricks

Since we know that the general MIMO sensitivity TFM $S = I - T$ where T is the complementary sensitivity TFM and has the following state-space realization:

$$\dot{x} = A_T x + B_T u \tag{12.23}$$

$$y = C_T^T x + D_T u \tag{12.24}$$

where the TFM for this realization can be written as:

$$G_T(s) = C_T^T(sI - A_T)^{-1}B_T + D_T \tag{12.25}$$

Inserting this into the equation $S = I - T$ yields the following:

$$G_S(s) = I - [C_T^T(sI - A_T)^{-1}B_T + D_T] \tag{12.26}$$

which yields an equivalent state-space realization for the sensitivity TFM as given below:

$$A_S = A_T \tag{12.27}$$

$$B_S = B_T \tag{12.28}$$

$$C_S = -C_T \tag{12.29}$$

$$D_S = I - D_T \tag{12.30}$$

Therefore, knowing the state-space realization for either complementary sensitivity or sensitivity TFM is sufficient to obtain the state-space realization for the other using the equivalency provided above.

In this section, we start by introducing the concept of well-posedness for MIMO systems.

12.2 Well-Posedness of MIMO Systems

Let's refresh our memory about the definition and meaning of well-posedness for SISO systems. When a SISO feedback system is well posed, all the closed-loop transfer functions are either proper or strictly proper. In this case, the mathematical condition for well-posedness was stated to be $1 + L(\infty) \neq 0$. Let's pause for a moment and discuss why such a condition as $1 + L(\infty) \neq 0$. Consider the Heaviside expansion of the return ratio:

$$L(s) = L(\infty) + L'(s) \tag{12.31}$$

where $L'(s)$ is strictly proper and has the same denominator as $L(s)$. It follows that all closed-loop transfer functions, such as,

$$\frac{e(s)}{r(s)} = \frac{1}{1 + L(s)} = S(s) \tag{12.32}$$

include $1 + L(s)$ as their denominators. It is clear that if the denominator, $1 + L(s)$, is strictly proper (if $1 + L(\infty) = 0$), then $\frac{1}{1+L(s)}$ is improper. Hence, for the properness of all closed-loop transfer functions, we have to enforce the mathematical condition $1 + L(\infty) \neq 0$.

Now we are in a position to derive a mathematical condition for well-posedness of MIMO systems. Consider the system shown in Figure 12.5,

The state-space representation of the compensator G_c is written as:

$$\dot{\underline{x}}_c = A_c \underline{x}_c + B_c \underline{e} \tag{12.33}$$

$$\underline{u} = C_c \underline{x}_c + D_c \underline{e} \tag{12.34}$$

and the controlled object state-space representation is given by,

$$\dot{\underline{x}}_p = A_p \underline{x}_p + B_p \bar{u} \tag{12.35}$$

$$\underline{y} = C_p \underline{x}_p + D_p \bar{u} \tag{12.36}$$

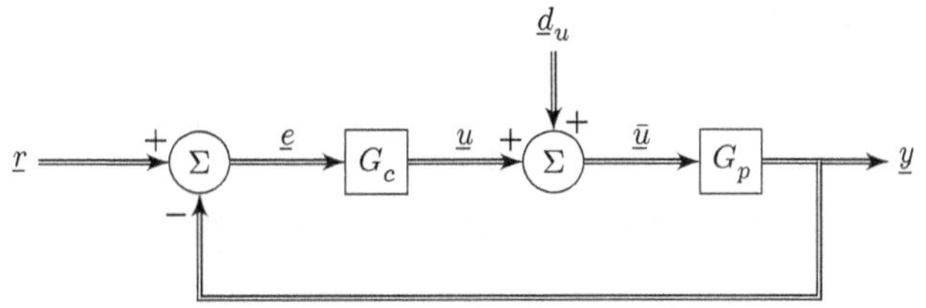

Figure 12.5 Block diagram of a multivariable control system.

The relationship between the inputs, $\underline{u}$ and $\underline{d}_u$, and the independent nodes, $\underline{\bar{u}}$ and $\underline{e}$, can be written as:

$$\underline{e} = \underline{r} - \underline{y} \tag{12.37}$$

$$\underline{\bar{u}} = \underline{u} + \underline{d}_u \tag{12.38}$$

Substituting the node equations into the output equations, $\underline{u}$ and $\underline{y}$ above, gives us

$$\underline{e} = \underline{r} - C_p\underline{x}_p - D_p\underline{\bar{u}} \tag{12.39}$$

$$\underline{\bar{u}} = C_c\underline{x}_c + D_c\underline{e} + \underline{d}_u \tag{12.40}$$

We can reorganize and rewrite these equations as,

$$\underline{e} + D_p\underline{\bar{u}} = \underline{r} - C_p\underline{x}_p \tag{12.41}$$

$$\underline{\bar{u}} - D_c\underline{e} = \underline{d}_u + C_c\underline{x}_c \tag{12.42}$$

These equations can be rewritten in the matrix form:

$$\underbrace{\left[\begin{array}{c|c} I & D_p \\ \hline -D_c & I \end{array}\right]\left[\begin{array}{c} \underline{e} \\ \hline \underline{\bar{u}} \end{array}\right]}_{\text{(Nodes)}} = \left[\begin{array}{c|c} -C_p & 0 \\ \hline 0 & C_c \end{array}\right]\underbrace{\left[\begin{array}{c} \underline{x}_p \\ \hline \underline{x}_c \end{array}\right]}_{\text{(States)}} + \underbrace{\left[\begin{array}{c} \underline{r} \\ \hline \underline{d}_u \end{array}\right]}_{\text{(Inputs)}} \tag{12.43}$$

Definition 12.1 We say that a MIMO feedback is well-posed if the nodes are uniquely determined by the inputs and the states. The feedback is ill-posed otherwise. That is to say, the feedback is well-posed if and only if,

$$\det \left[\begin{array}{c|c} I & D_p \\ \hline -D_c & I \end{array}\right] \neq 0 \tag{12.44}$$

$$\det(I_m + D_cD_p) = \det(I_r + D_pD_c) \neq 0 \tag{12.45}$$

If either D_p or D_c is zero, then the return ratio is strictly proper, and the feedback is well posed.

This condition is very similar to the well-posedness condition for a SISO feedback given above where $1 + L(\infty) \neq 0$.

12.3 State Variable Compositions

In this section, we introduce state-space representation for the return ratio, $L(s)$, for MIMO systems. This representation plays an important role when presenting MIMO well-posedness and stability concepts.

Given the matrices, A, B, C, and D of a MIMO dynamic system, we know that the easiest way to obtain the plant TFM $G_p(s)$ is

$$G_p(s) = C(sI - A)^{-1}B + D \tag{12.46}$$

But the question is how do we obtain a state-space realization for the return ratio $L(s)$, where we computed the return ratio in the last section? We start with $L(s) = G_p(s)G_c(s)$ and state the realizations for G_c and G_p as follows:

$$\left.\begin{array}{l} \dot{\underline{x}}_c = A_c\underline{x}_c + B_c\underline{e} \\ \underline{u} = C_c\underline{x}_c + D_c\underline{e} \end{array}\right\} \quad G_c \text{ Realization} \tag{12.47}$$

where $\underline{e}$ is the error signal in the feedback loop, and $\underline{x}_c$ is the controller, G_c is the state variable, and $\underline{u}$ is its output:

$$\left.\begin{aligned}\dot{\underline{x}}_p &= A_p\underline{x}_p + B_p\underline{u}\\ \underline{y} &= C_p\underline{x}_p + D_p\underline{u}\end{aligned}\right\} \quad G_p \text{ Realization} \tag{12.48}$$

where $\underline{x}_p$ is the plant, G_p is the state variable, and $\underline{y}$ is the plant output. Then, by substituting $\underline{u}$, $\underline{u} = C_c\underline{x}_c + D_c\underline{e}$, into G_p state-space realization, we obtain the followings:

$$\dot{\underline{x}}_p = A_p\underline{x}_p + B_pC_c\underline{x}_c + B_pD_c\underline{e} \tag{12.49}$$

$$\underline{y} = C_p\underline{x}_p + D_pC_c\underline{x}_c + D_pD_c\underline{e} \tag{12.50}$$

So, this can be written in the matrix form as:

$$\frac{\mathrm{d}}{\mathrm{d}t}\begin{bmatrix}\underline{x}_c\\ \underline{x}_p\end{bmatrix} = \begin{bmatrix}A_c & 0\\ B_pC_c & A_p\end{bmatrix}\begin{bmatrix}\underline{x}_c\\ \underline{x}_p\end{bmatrix} + \begin{bmatrix}B_c\\ B_pD_c\end{bmatrix}\underline{e} \tag{12.51}$$

$$\underline{y} = \begin{bmatrix}D_pC_c & C_p\end{bmatrix}\begin{bmatrix}\underline{x}_c\\ \underline{x}_p\end{bmatrix} + \begin{bmatrix}D_pD_c\end{bmatrix}\underline{e} \tag{12.52}$$

We can write this as:

$$\dot{\underline{x}}_L = A_L\underline{x}_L + B_L\underline{i} \tag{12.53}$$

$$-\underline{v} = C_L\underline{x}_L + D_L\underline{i} \tag{12.54}$$

where i and v are given in the last section. The loop is closed for $\underline{v} = \underline{i}$. Then,

$$-\underline{i} = C_L\underline{x}_L + D_L\underline{i} \tag{12.55}$$

We can write,

$$(-I - D_L)\underline{i} = C_L\underline{x}_L \tag{12.56}$$

and

$$\underline{i} = -(I + D_L)^{-1}C_L\underline{x}_L \tag{12.57}$$

We have shown in the previous section of this chapter that $(I + D_L)^{-1}$, where $D_L = D_pD_c$, always exist if and only if the feedback loop is well posed. Substituting for $\underline{i}$ into the differential equation for $\underline{x}_L$, we obtain the closed-loop form:

$$\dot{\underline{x}}_L = (A_L - B_L(I + D_L)^{-1}C_L)\underline{x}_L \tag{12.58}$$

So, the closed-loop characteristic polynomial (C.P.), $\Delta(s)$, can be written as:

$$\Delta(s) = \det\left(sI - A_L + B_L(I + D_L)^{-1}C_L\right) \tag{12.59}$$

where the closed-loop C.P. is

$$\Delta(s) = \underbrace{\det\left(sI - A_L\right)}_{\text{Open-loop C.P.}}\det\left(I + \left(sI - A_L\right)^{-1}B_L(I + D_L)^{-1}C_L\right) \tag{12.60}$$

where we set the open-loop C.P. $(\det(sI - A_L))$ to be $\delta(s)$, and knowing that $\det(I + (sI - A_L)^{-1}B_L(I + D_L)^{-1}C_L) = \det(I + C_L(sI - A_L)^{-1}B_L(I + D_L)^{-1})$ or,

$$\Delta(s) = \delta(s)\det\left(I + C_L(sI - A_L)^{-1}B_L(I + D_L)^{-1}\right) \tag{12.61}$$

The ratio of closed-loop C.P. to the open-loop C.P. is given by:

$$\frac{\Delta(s)}{\delta(s)} = \frac{(\det(I + D_L + C_L(sI - A_L)^{-1}B_L)}{\det(I + D_L)} \tag{12.62}$$

where $\det(I + D_L)^{-1} = \frac{1}{\det(I+D_L)}$. Similar to the state-space representation of $G_p(s)$, the state-space representation of the return ratio, $L(s)$ is given by:

$$L(s) = D_L + C_L(sI - A_L)^{-1}B_L \tag{12.63}$$

Therefore,

$$\frac{\Delta(s)}{\delta(s)} = \frac{\det(I + L(s))}{\det(I + D_L)} \tag{12.64}$$

If the system is well posed, then $\det(I + D_L)$ exist and it is a constant which has no effect on the system stability and stability is only determined by:

$$\det(I + L(s)) \tag{12.65}$$

which will be discussed in the next section.

This derivation for MIMO stability is again analogous to SISO systems, where

$$\frac{\Delta(s)}{\delta(s)} = 1 + L(s) \tag{12.66}$$

where the zeros of $1 + L(s)$ are the poles of the closed loop and the poles of $1 + L(s)$ are the poles of the open loop.

12.4 Nyquist Criterion for MIMO Systems

Similar to SISO systems if the feedback system is well posed and stable, then this does not necessarily guarantee *system internally stability*, such as if there is a pole-zero cancellation on the $j\omega$-axis or in the right half-plane by the controller and the plant (the description of internal stability for SISO systems is provided in Chapter 3). Our interest in internal stability here lies in the input–output maps rather than system state-space representation as will be discussed in the last section of this chapter.

We start with the general relationship for Nyquist criterion, avoiding the details of derivations and noting that $\det(I + L(j\omega))$ is a transfer function with zeros and poles, then according to Nyquist criterion,

$$N_c = N_o + N \tag{12.67}$$

where:

- N_c is the closed-loop poles in the right half-plane (right half-plane zeros of $\det(I + L(j\omega))$).
- N_o is the open-loop poles in the right half-plane (right half-plane poles of $\det(I + L(j\omega))$).
- N is the number of clockwise encirclements of the origin by the Nyquist plot of $\det(I + L(j\omega))$.

A Nyquist plot for an arbitrary MIMO system is given in Figure 12.6.

For a closed-loop system to be stable, $N_c = 0$, N_0, number of unstable open-loop poles, must be precisely equal to $-N$, number of counterclockwise encirclements of the origin by the Nyquist plot of $\det(I + L(j\omega))$.

12.4.1 Characteristic Gains

Another intuitive way to compute $\det(I + L(j\omega))$ is with the use of *Characteristic Gains*. The characteristic gains or characteristic gain functions, $\lambda_k(j\omega)$, of a square TFM M are the eigenvalues of M as a function of frequency.

Note: The characteristic gains can be found in MATLAB with the command `cgloci (M)` *where* `M` *is a TFM.*

The characteristic gains have the following property:

$$\sigma_{\min}(M) \leq |\lambda_k| \leq \sigma_{\max}(M) \tag{12.68}$$

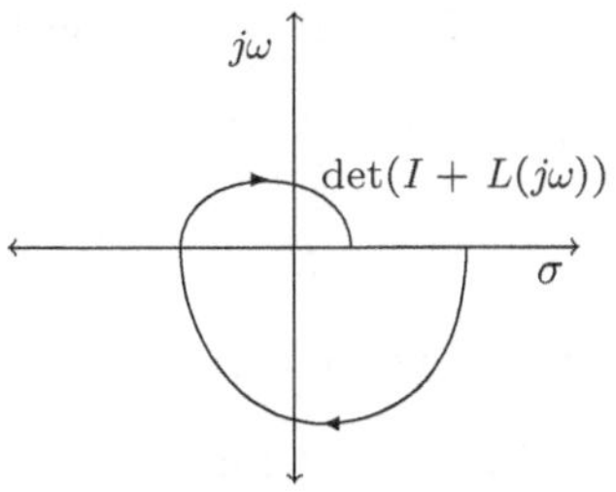

Figure 12.6 Nyquist plot for a MIMO system.

Next, we will discuss two identities which could be used to illustrate the use of characteristic gains in computing $\det(I + L(j\omega))$.

First, for any general square matrix M,

$$\det(M) = \prod_{k=1}^{n} \lambda_k \tag{12.69}$$

where λ_k is the eigenvalue of matrix M.

Secondly, knowing that $\lambda(I + M) = 1 + \lambda(M)$.

Now applying these two eigenvalue identities, we can write

$$\det[I + L(s)] = \prod_k \lambda[I + L(s)] = \prod_k (1 + \lambda[L(s)]) \tag{12.70}$$

This yields a Nyquist array and by applying the Nyquist criterion across all of the resulting characteristic gains,

$$N_c = N_o + \sum_k n_k \tag{12.71}$$

where n_i is the number of clockwise encirclements of -1 by λ_i. Now, this could be utilized to compute system stability using Nyquist criterion, such as for stable open-loop system where $N_o = 0$, for the closed-loop system to be stable, $\sum_k n_k = 0$.

12.4.2 Poles and Zeros

In this section, we discuss the important topic of *internal stability* for MIMO systems. Before, addressing this topic, we will discuss poles and zeros of MIMO systems using matrix factorization to provide some insight into the interpretation of poles and zeros of MIMO systems.

Recall that the complementary sensitivity and sensitivity transfer function matrices can be written as:

$$T = (I + L)^{-1}L = L(I + L)^{-1} \tag{12.72}$$

$$S = (I + L)^{-1} \tag{12.73}$$

Using the right factorization of L,

$$L = ND^{-1} \tag{12.74}$$

Then, T and S can be written as:

$$\begin{aligned} T &= ND^{-1}(I + ND^{-1})^{-1} \\ &= ND^{-1}((D + N)D^{-1})^{-1} \\ &= N(D + N)^{-1} \end{aligned} \tag{12.75}$$

Equivalently, it can be shown that

$$S = D(D + N)^{-1} \tag{12.76}$$

Similar development for the left factorization is not going to be repeated here. Remember if this is the coprime factorization, it is the minimal realization of the system, i.e. there is no pole/zero cancellation.

Definition 12.2 (Coprime) Suppose W is a right factor of maximal degree such that

$$N = N'W \tag{12.77}$$

$$D = D'W \tag{12.78}$$

We know if W is unimodular, then (N, D) is coprime. In general we can write

$$L = N'W(D'W)^{-1} \tag{12.79}$$

$$= N'(D')^{-1} \tag{12.80}$$

and then,

$$T = N'(D' + N')^{-1} \tag{12.81}$$

$$S = D'(D' + N')^{-1} \tag{12.82}$$

Therefore, uncanceled zeros of L are the zeros of T and uncanceled poles of L are the zeros of S.

Recall the SISO interpolation conditions:

$$S + T = 1 \tag{12.83}$$

$$T(z) = 0 \rightarrow S(z) = 1 \tag{12.84}$$

$$S(p) = 0 \rightarrow T(p) = 1 \tag{12.85}$$

For a MIMO system, these interpolation conditions become

$$S + T = I \tag{12.86}$$

$$\det\ T(z) = 0 \text{ if } z \text{ is an uncanceled zero of } L \tag{12.87}$$

$$\det\ S(p) = 0 \text{ if } p \text{ is an uncanceled pole of } L \tag{12.88}$$

However, note that $S + T = I$ does not yield a similar condition for MIMO systems, i.e.

$$S + T = I \tag{12.89}$$

$$\det(T(z)) = 0 \nrightarrow \det(S(z)) = 1 \tag{12.90}$$

$$\det(S(p)) = 0 \nrightarrow \det(T(p)) = 1 \tag{12.91}$$

These MIMO interpolation conditions are for single unstable poles and zeros and repeated poles and zeros are not discussed above. For repeated poles or zeros, use the same derivative constraints as a SISO system. This works because derivatives are linear operators, i.e. $\det[\frac{d^k}{ds^k}T(z)] = 0$.

However, we will present a more straightforward and computationally less intensive method than the MIMO interpolation conditions given above. This method takes advantage of matrix diagonalization and SM form to enforce internal stability but first let's provide an overview of the internal stability of MIMO systems.

12.4.3 Internal Stability

It is important to remember that internal stability refers to the initial condition response in the absence of generalized inputs; in this regard, it coincides with the standard notion of *asymptotic stability* of dynamic systems. However, it has immediate implications on the input–output properties of the system. For the SISO feedback system, we pointed out that all closed-loop transfer functions can be reduced to only four to address internal stability. We will extend this concept to the MIMO system, by finding the closed-loop maps from r and d_u to y and $\bar{u}$ as illustrated in Figure 12.7,[1]

Recall the following transfer function matrices:

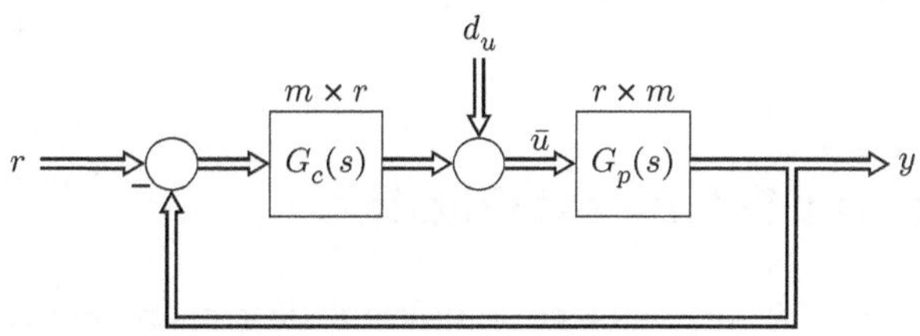

Figure 12.7 MIMO feedback system without noise and output disturbance injection.

1 In place of the input r, we could equivalently use the disturbance input d_y or the noise input n (not shown); furthermore, in place of the output y, we could use the error signal e, the conclusion would have been the same.

- **Output Complementary Sensitivity**

$$y = (I + G_p G_c)^{-1} G_p G_c r \tag{12.92}$$
$$= T_y r \tag{12.93}$$

- **Youla Parameter**

$$\bar{u} = (I + G_p G_c)^{-1} G_c r \tag{12.94}$$
$$= Yr \tag{12.95}$$

- **Output Sensitivity**

$$y = (I + G_p G_c)^{-1} G_p d_u \tag{12.96}$$
$$= S_y G_p d_u \tag{12.97}$$

- **Input Sensitivity**

$$\bar{u} = (I + G_c G_p)^{-1} d_u \tag{12.98}$$
$$= S_u d_u \tag{12.99}$$

If these four transfer function matrices are BIBO stable, then the system is internally stable. Note, that well-posedness is a necessary condition for BIBO stability:

$$M = \begin{bmatrix} Y & S_u \\ T_y & S_y G_p \end{bmatrix} \tag{12.100}$$

These four transfer function matrices can be written in terms of Youla parameter, Y. Recall that

$$\bar{u} = Yr \tag{12.101}$$
$$\bar{u} = S_u d_u = (I - T_u) d_u = (I - YG_p) d_u \tag{12.102}$$
$$y = T_y r = G_p Y r \tag{12.103}$$
$$y = G_p S_u d_u = G_p (I - YG_p) d_u \tag{12.104}$$

Then

$$\begin{bmatrix} \bar{u} \\ y \end{bmatrix} = \underbrace{\begin{bmatrix} Y & I - YG_p \\ G_p Y & G_p(I - YG_p) \end{bmatrix}}_{M} \begin{bmatrix} r \\ d_u \end{bmatrix} \tag{12.105}$$

The closed-loop TFM is parameterized by Y and G_p. Each entry of matrix M has to be BIBO stable to guarantee the *internal stability* of the closed-loop system.

Further, if G_p is stable, then the closed loop is internally stable if and only if Y is stable. All internally stabilizing compensators are parameterized by:

$$G_c = (I - YG_p)^{-1} Y \tag{12.106}$$

We will further discuss the topic of internal stability and the criteria for enforcing internal stability in the following chapter.

12.5 MIMO Performance and Robustness Criteria

For the full MIMO system in Figure 12.8, we are unable to utilize Bode gain and phase plots, as were done in the SISO frequency domain control method. Instead, we use singular values of the relevant closed-loop transfer function matrices, such as T, S, and Y as will be discussed in this section.

It is important to discuss and develop objective measures to aid in both MIMO control design and analysis using frequency-shaping techniques, similar to the SISO case, before fully present the MIMO Youla parameterization control design method.

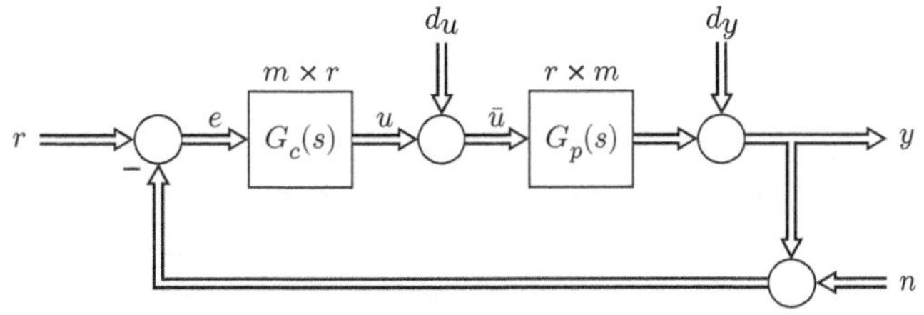

Figure 12.8 MIMO feedback system.

Note: The MATLAB command `sigma (A, B, C, D, ω)`, *where ω is a frequency vector, gives the singular values as a function of ω.*

We start with the transfer function matrices which map the input signals, r, n, and d_y, to the actuator inputs, u:

$$u = -Yd_y \tag{12.107}$$

$$u = Yr \tag{12.108}$$

$$u = -Yn \tag{12.109}$$

Typically, we ideally design a controller for which $\sigma_{\max}(Y) \ll 1$ at all frequencies to protect the actuator and keep it from saturating, i.e. exceeding the maximum level of efforts/flows it could produce. The frequency shape of Y along with all the existing constraints will be discussed later in this section.

For the output sensitivity TFM, S_y, which maps the input signals r and d_y to the error signals e and the output signals y as follows:

$$e = S_y r \tag{12.110}$$

$$y = S_y d_y \tag{12.111}$$

We design a controller so that $\sigma_{\max}(S_y) \ll 1$. This will result in good disturbance rejection and good target following. These are the control design performance measures.

Finally to ensure robust stability in the presence of high-frequency disturbances such as sensor noise and unstructured model uncertainties, as discussed for SISO control design, the following measures are imposed at high frequencies:

$$\lim_{\omega\to\infty} T_y(j\omega) \to 0_{r\times r} \Rightarrow \sigma_{\max}(T_y) \ll 1 \tag{12.112}$$

but, as $T_y + S_y = I$ (algebraic constraint), this means,

$$\lim_{\omega\to\infty} S_y(j\omega) \to I_{r\times r} \Rightarrow \sigma_{\max}(S_y) \to 1 \tag{12.113}$$

as illustrated in Figure 12.9.

Therefore, here a clear trade-off between performance and robustness (robust stability) is given. As we try to make the loop robustly stable to uncertainties at high frequency, such that $\sigma_{\max}(T_y) \ll 1$, we are forced to compromise disturbance rejection and target following capabilities or the loop performance in this frequency range, such that $\sigma_{\max}(S_y) \to 1$.

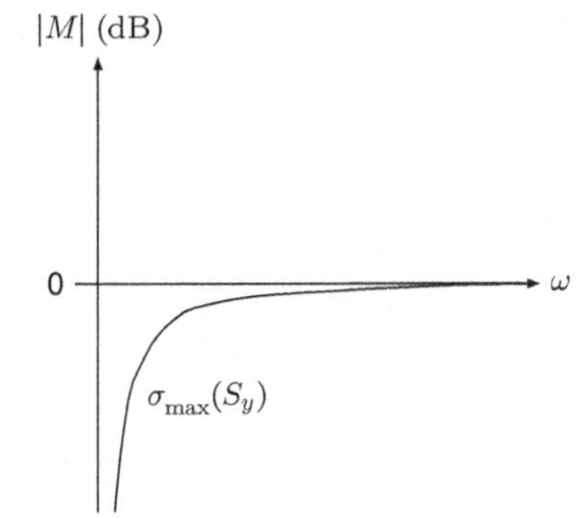

Figure 12.9 Frequency plot of $\sigma_{\max}(S_y)$

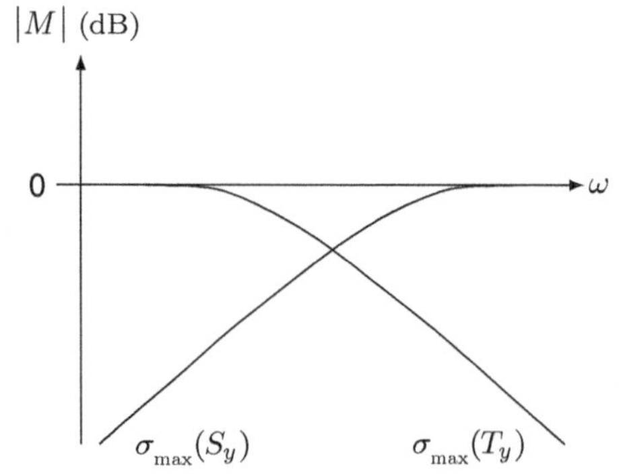

Figure 12.10 Frequency plot of $\sigma_{\max}(S_y)$ and $\sigma_{\max}(T_y)$.

Let's look at the same measures (performance and robust stability) in terms of the transfer function matrices which map the feedback loop reference and disturbance inputs to the output, we can then write

$$y = T_y r \tag{12.114}$$

$$y = -T_y n \tag{12.115}$$

$$y = S_y d_y \tag{12.116}$$

$$y = S_y G_p d_u \tag{12.117}$$

MIMO control design using frequency-shaping technique is similar to the SISO case. However, as mentioned above, in the MIMO case, we shape the singular values rather than Bode gains.

The fundamental question is how do we want to shape the transfer function matrices T_y and S_y given the equations above. The secondary question, which will be discussed later in this section, is the frequency shape of Y and the role of T_u and S_u in this design.

First, how do we want to frequency-shape the singular values of T_y and S_y? If $y = T_y r$, then $T_y = I$ yields perfect tracking but by looking at Eq. (12.115), we are unable to keep $T_y = I$ at all frequencies, as if this was the case the sensor noise at higher frequencies will not be rejected at the output. Here again lies the fundamental trade-off, as discussed above, between performance and robustness. Therefore, we shape $T_y(0) = I$ (means T_y at zero frequency and without abuse of notation, we also imply T_y at very low frequency) for good target following. By enforcing $T_y(0) = I$, this results $\sigma_{\max}(T_y(0))) = \sigma_{\min}(T_y(0)) = 1$. [2]

Knowing the algebraic constraint, $S + T = I$, Figure 12.10 provides our target frequency shapes of T_y and S_y singular values.

12.6 Open-Loop Singular Values

In this section, we map the closed-loop performance and robust stability measures to the equivalent open-loop (return ratio) measures. This way, by plotting the singular values of the return ratio, we could decide if our design measures would be acceptable. Recall that

$$S = (I + L)^{-1} \tag{12.118}$$

Figure 12.11 illustrates an acceptable frequency shape of the sensitivity TFM maximum singular value. From previous sections, we learned that

$$\sigma_{\max}(M^{-1}) = \frac{1}{\sigma_{\min}(M)} \tag{12.119}$$

Therefore,

$$\sigma_{\max}(S) = \sigma_{\max}((I + L)^{-1}) = \frac{1}{\sigma_{\min}(I + L)} \tag{12.120}$$

2 We will discuss the benefits of having minimum and maximum singular values of T_y to be equal in the following chapters.

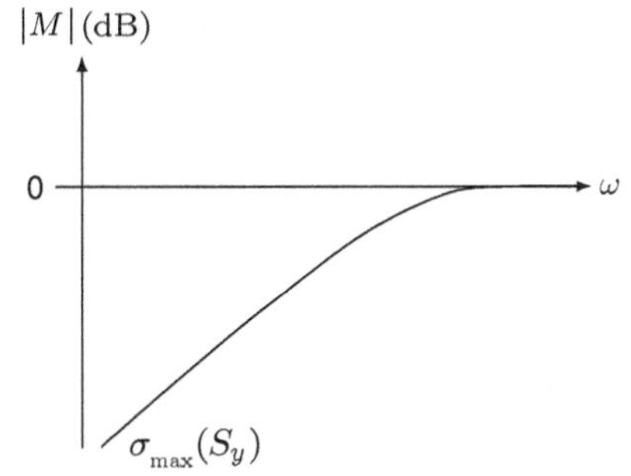

Figure 12.11 Frequency plot of $\sigma_{\max}(S_y)$.

Another property of SVD (discussed in the Appendix B) is,

$$|\sigma_{min}(M) - \sigma_{\min}(N)| \leq \sigma_{\min}(M+N) \leq \sigma_{\min}(M) + \sigma_{\min}(N) \tag{12.121}$$

Substituting I (identity matrix) and L in place of M and N yields,

$$|\sigma_{\min}(I) - \sigma_{\min}(L)| \leq \sigma_{\min}(I+L) \leq \sigma_{\min}(I) + \sigma_{\min}(L) \tag{12.122}$$

$$|1 - \sigma_{\min}(L)| \leq \sigma_{\min}(I+L) \leq 1 + \sigma_{\min}(L) \tag{12.123}$$

Therefore,

$$\sigma_{\max}(S) = \frac{1}{\sigma_{\min}(I+L)} \leq \frac{1}{|1 - \sigma_{\min}(L)|} \tag{12.124}$$

The relationship in Eq. (12.124) provides a mapping from the closed-loop measures to the open-loop measures.

For example, at low frequencies, the performance measure enforces $\sigma_{\max}(S) \ll 1$ which would imply that $|1 - \sigma_{\min}(L)| \gg 1$ or $\sigma_{\min}(L) \gg 1$. Therefore, at low frequency, we can write

$$\sigma_{\max}(S) \leq \frac{1}{\sigma_{\min}(L)} \tag{12.125}$$

Similarly, at high frequency, where $\sigma_{\max}(S) \to 1$, this implies $\sigma_{\max}(L) \ll 1$. The proof of this is straightforward and left as an exercise.

So, in terms of the return ratio TFM, our performance and robust stability measures dictate at low frequency, $\sigma_{\min}(L) \gg 1$, while at high frequency, $\sigma_{\max}(L) \ll 1$. Next, we discuss how the maximum and minimum singular values of L should behave at the crossover frequency.

12.6.1 Crossover Frequency

It is important to note that at the crossover frequency, we start to lose the benefits of feedback control. Therefore, at crossover frequency, when shaping the closed-loop matrices or equivalently the return ratio, the gap between $\sigma_{\min}(L)$ and $\sigma_{\max}(L)$ should be avoided. The large gaps between minimum and maximum singular values of L make the we ideally design a to be vulnerable to noise, i.e. it can be shown that the closed-loop complementary TFM singular values do not decrease fast enough below 0 dB when large gaps exist. There are other benefits for having this gap reduced and this will be discussed when presenting the condition numbers and their role in MIMO control design approach. An example problem is provided in the problem set to provide a better understanding of the impact of minimizing this gap.

Therefore, for a MIMO system, we want $\sigma_{\min}(L) \cong \sigma_{\max}(L)$ around the crossover frequency, to reduce the impact of noise in the feedback loop.

Recall the condition number of a matrix, where for L we can write

$$\frac{\sigma_{\max}(L)}{\sigma_{\min}(L)} = \mathscr{K}(L) \tag{12.126}$$

where $\mathscr{K}(L) \approx 1$ around the crossover frequency as illutrated in Figures 12.12 and 12.13. Next, we discuss the important topic of bandwidth and its constraints for MIMO systems.

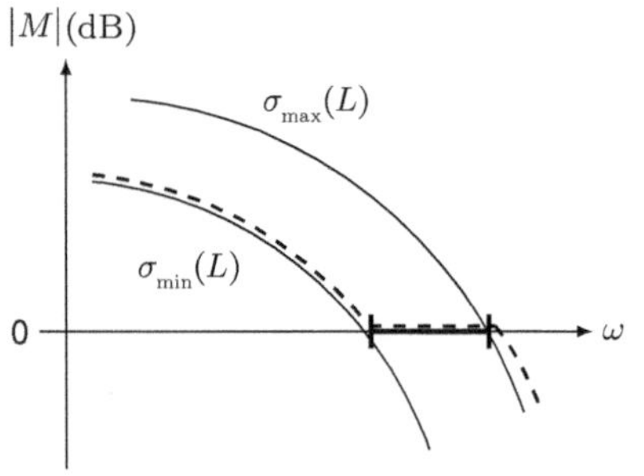

Figure 12.12 Comparison of plot of $\sigma_{max}(L)$ and $\sigma_{min}(L)$ for poorly conditioned L.

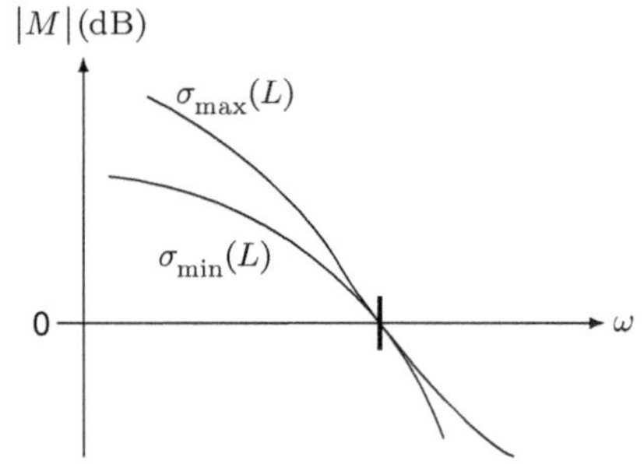

Figure 12.13 Comparison of plot of $\sigma_{max}(L)$ and $\sigma_{min}(L)$ for well-conditioned L.

12.6.2 Bandwidth Constraints

As discussed previously, ideally, we want $\sigma_{max}(Y)$ to be low at all frequencies for minimizing actuator effort and avoiding actuator saturation. But, as we will find next, this is usually not possible for all frequencies.

We know the following relationships:

$$T_y = G_p Y \tag{12.127}$$

$$T_u = Y G_p \tag{12.128}$$

We will use the following singular value property:

$$\sigma_{min}(M)\sigma_{max}(N) \leq \sigma_{max}(MN) \leq \sigma_{max}(M)\sigma_{max}(N) \tag{12.129}$$

using this property, we can write the following inequality,

$$\sigma_{min}(G_p)\sigma_{max}(Y) \leq \sigma_{max}(T_y) \tag{12.130}$$

$$\Rightarrow$$

$$\sigma_{max}(Y) \leq \frac{\sigma_{max}(T_y)}{\sigma_{min}(G_p)} \tag{12.131}$$

From this relationship, we can state that it is undesirable to have $\sigma_{max}(T_y)$ to be much greater than $\sigma_{min}(G_p)$, as this results in large maximum singular values of Y. Let's investigate this maximum singular value at different frequency ranges, such as low and high frequencies.

At low frequencies, a good target following is obtained by $\sigma_{max}(T_y) \cong 1$ and combining this with Eq. (12.131) yields

$$\sigma_{max}(Y) \leq \frac{1}{\sigma_{min}(G_p)} \tag{12.132}$$

This is an interesting result, similar to the SISO case, as it states that we are unable to shape the control effort Y at low frequencies! If the plant is low gain, it will then a large actuator effort would be required and the actuator will be vulnerable to saturation and there is nothing we could do to avoid this without compromising a good target following requirement!

Note: Recall that for SISO systems, this changes to $T \cong 1$ and $Y \cong \frac{1}{G_p}$ at low frequencies.

At high frequencies, $\sigma_{max}(S_y) \cong 1$ due to the algebraic constraint $T + S = I$, where $\sigma_{max}(T_y) \ll 1$, we have the following relationship:

$$Y = G_c S_y \tag{12.133}$$

Combining this with the singular value property above,

$$\sigma_{max}(Y) = \sigma_{max}(G_c S_y) \tag{12.134}$$

Then,

$$\sigma_{max}(S_y)\sigma_{min}(G_c) \leq \sigma_{max}(Y) \leq \sigma_{max}(S_y)\sigma_{max}(G_c) \tag{12.135}$$

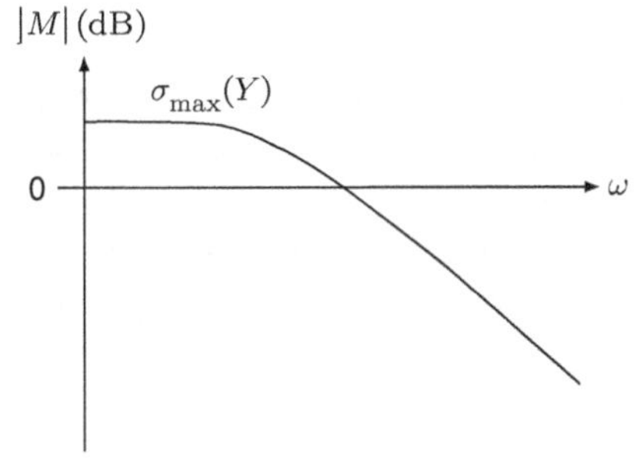

Figure 12.14 Frequency plot of $\sigma_{max}(Y)$.

so at high frequencies with $\sigma_{max}(S_y) \cong 1$, the following expression is obtained:

$$\sigma_{min}(G_c) \leq \sigma_{max}(Y) \leq \sigma_{max}(G_c) \tag{12.136}$$

Hence, at high frequencies, the control effort Y could be shaped, unlike at the low frequencies, by assuring the singular values of G_c are not large. A typical plot of $\sigma_{max}(Y)$ will follow something similar to Figure 12.14, if both $\sigma_{min}(G_p)$ and $\sigma_{max}(G_c)$ are below 0 dB at low and high frequencies.

Next, we discuss the condition number and its role in MIMO control design.

12.7 Condition Number and its Role in MIMO Control Design

The condition number of a general matrix M is given by:

$$\mathscr{K}(M) = \frac{\sigma_{max}(M)}{\sigma_{min}(M)} \tag{12.137}$$

if M is well conditioned, then, $\mathscr{K} \sim 1$ and if ill conditioned, then, $\mathscr{K} \gg 1$.

12.7.1 Condition Numbers and Decoupling

We stated that for good target following at low frequencies, we are required to design a MIMO controller so that $T(0) = I$, that is, $\sigma_{max}(T(0)) = \sigma_{min}(T(0))$, which in turn results in closed-loop input–output decoupling. Having $T(0) = I$ provides perfect reference tracking and zero steady-state error and input–output decoupling in low frequencies. In MIMO control design, decoupling could be very desirable as every reference input is only affecting one output, and this could make controller tuning and calibration an easier task.

The question becomes, how do we recognize this decoupling on a MIMO singular value plot? $\sigma_{max}(T_y) = \sigma_{min}(T_y)$ at low frequencies, as shown in Figure 12.15.

It is important to note that it could be shown when $\sigma_{min}(L)$ is large at low frequencies with having $\mathscr{K}(L) \cong 1$, these conditions also result in input–output decoupling.

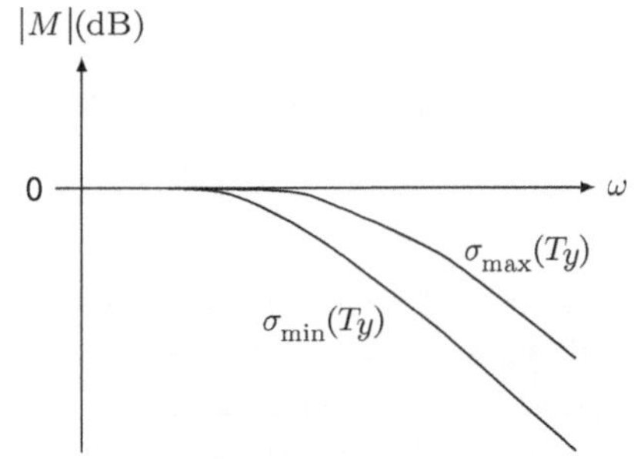

Figure 12.15 Frequency plot of $\sigma(T_y)$.

12.7.2 Role of T_u and S_u in MIMO Feedback Design

Before we summarize our performance and robust stability requirements both for the return ratio L and the closed-loop transfer functions matrices, it would be informative to discuss the role of the closed-loop transfer function matrices T_u and S_u in MIMO control design.

The important question to be asked here is, do we need to consider more than one node when designing MIMO controllers? This means in addition to T_y and S_y do we need to consider T_u and S_u. Referring again to Figure 12.8, we can write

$$y = S_y G_p d_u \tag{12.138}$$

$$\bar{u} = S_u d_u \tag{12.139}$$

Recognize that,

$$G_p S_u = S_y G_p \tag{12.140}$$

$$S_u G_c = G_c S_y \tag{12.141}$$

Using the singular value properties stated above, the following expressions are given for Eq. (12.140):

$$\sigma_{\max}(S_u)\sigma_{\min}(G_p) \le \sigma_{\max}(G_p S_u) = \sigma_{\max}(S_y G_p) \le \sigma_{\max}(S_y)\sigma_{\max}(G_p) \tag{12.142}$$

$$\sigma_{\max}(S_y)\sigma_{\min}(G_p) \le \sigma_{\max}(S_y G_p) = \sigma_{\max}(G_p S_u) \le \sigma_{\max}(S_u)\sigma_{\max}(G_p) \tag{12.143}$$

Now, dividing Eq. (12.142) by $\sigma_{\max}(G_p)$,

$$\frac{\sigma_{\max}(S_u)}{\mathcal{K}(G_p)} \le \sigma_{\max}(S_y) \Rightarrow$$

$$\sigma_{\max}(S_u) \le \sigma_{\max}(S_y)\mathcal{K}(G_p) \tag{12.144}$$

Dividing Eq. (12.143) by $\sigma_{\max}(G_p)$,

$$\frac{\sigma_{\max}(S_y)}{\mathcal{K}(G_p)} \le \sigma_{\max}(S_u) \tag{12.145}$$

Combining these two yields the equation,

$$\frac{\sigma_{\max}(S_y)}{\mathcal{K}(G_p)} \le \sigma_{\max}(S_u) \le \sigma_{\max}(S_y)\mathcal{K}(G_p) \tag{12.146}$$

Similar analysis for Eq. (12.141) yields

$$\frac{\sigma_{\max}(S_y)}{\mathcal{K}(G_c)} \le \sigma_{\max}(S_u) \le \sigma_{\max}(S_y)\mathcal{K}(G_c) \tag{12.147}$$

We can ask, under what conditions will $\sigma_{\max}(S_y) \cong \sigma_{\max}(S_u)$? When $\mathcal{K}(G_p)$ or $\mathcal{K}(G_c) \cong 1$. Therefore, if either $\mathcal{K}(G_p)$ or $\mathcal{K}(G_c) \cong 1$, then feedback design would be sufficient using one node (S_y). If neither of these conditions are true, then one must check S_u when designing MIMO feedback control to make sure the shape of S_u follows S_y!

12.8 Summary of Requirements

This section will summarize the closed-loop requirements for MIMO systems that we have established so far.

12.8.1 Closed-Loop Requirements

- **Disturbance Rejection**: $\sigma_{\max}(S_y)$ as small as possible at low frequencies.
- **Sensor Noise Rejection**: $\sigma_{\max}(T_y)$ as small as possible at high frequencies.
- **Tracking**: $\sigma_{\min}(T_y) \cong \sigma_{\max}(T_y) \cong 1 \rightarrow \mathcal{K}(T_y) \cong 1$ at low frequencies.
- **Minimize Control Effort**: $\sigma_{\max}(Y)$ as small as possible at high frequencies.
- **Robust Stability**: $\sigma_{\max}(T_y)$ as small as possible at high frequencies (model uncertainty increases as $\omega \rightarrow \infty$).

12.8.2 Open-Loop Requirements

- **Disturbance Rejection**: $\sigma_{\min}(L)$ as large as possible at low frequencies.
- **Sensor Noise Rejection**: $\sigma_{\max}(L)$ as small as possible at high frequencies and $\mathscr{K}(L) \approx 1$ at crossover frequency.
- **Tracking**: $\sigma_{\min}(L)$ as large as possible at low frequencies.
- **Minimize Control Effort**: $\sigma_{\max}(G_c)$ as small as possible at high frequencies.
- **Robust Stability**: $\sigma_{\max}(L)$ as small as possible at high frequencies.

12.9 Problems

12.1 Given

$$G_p(I + G_cG_p) = (I + G_pG_c)G_p \tag{12.148}$$

show that

$$G_p(I + G_cG_p)^{-1} = (I + G_pG_c)^{-1}G_p \tag{12.149}$$

12.2 In the field of integrated vehicle dynamics and controls (IVDC), it is well known that control integration of active chassis systems has many benefits, such as an increase in agility and stability (to name a few). The following TFM is computed at low frequency when active differential and active roll control are integrated:

$$G = \begin{bmatrix} -0.0342 & -0.1195 \\ 0.0058 & 0.0214 \end{bmatrix} \tag{12.150}$$

The following singular vectors and singular values are calculated for this TFM:

$$l = \begin{bmatrix} -0.9844 & 0.1757 \\ 0.1757 & 0.9844 \end{bmatrix}, \quad \Sigma = \begin{bmatrix} 0.1262 & 0 \\ 0 & 0.0003 \end{bmatrix}, \quad r^T = \begin{bmatrix} 0.2750 & -0.9614 \\ 0.9614 & 0.2750 \end{bmatrix} \tag{12.151}$$

(a) Compute the condition number K and comment on the difficulty of controlling this plant. Is this plant easy or difficult to control?
(b) What is the gain of the system when the input is in the direction of the first right singular vector?
(c) By looking at the right singular vectors, would you say the gain is higher when the actuators are working in the same direction, or in opposite directions?

12.3 Take

$$G_c = \underset{(2\times2)}{K_c} \tag{12.152}$$

a simple gain matrix. Notice that the return ratio

$$L_y = G_pK_c = [C^T(sI - A)^{-1}B + D]K_c \tag{12.153}$$

Use this simple formula to derive a state variable realization for L_y. (This is an easy task. The answer should be obvious from a mere inspection of the last formula). Repeat this for L_u.

12.4 Consider the following state variable realization for a multivariable vehicle system given below. The parameters b and k are damping and stiffness constraints, while parameters M and m are mass of the cargo and vehicle mass, respectively,

$$A = \begin{bmatrix} 0 & -1/m & 1/M \\ k & -b/m & b/M \\ -k & b/m & -b/M \end{bmatrix}, \quad B = \begin{bmatrix} 0 & 0 \\ 1 & 1 \\ 0 & -1 \end{bmatrix} \tag{12.154}$$

$$C = \begin{bmatrix} 0 & 1/m & 0 \\ -1 & 0 & 0 \end{bmatrix}, \quad D = \begin{bmatrix} 0 & 0 \\ 0 & 0 \end{bmatrix} \tag{12.155}$$

Take

$$G_c = \underset{(2\times2)}{K_c} \tag{12.156}$$

and use the realization derived in Problem 1. Assume normalized values of 1 for all parameters, choose a range of frequencies ω, and obtain

$$\det\left[I + L_y(j\omega)\right] \tag{12.157}$$

Plot this MIMO Nyquist loci for the following compensator matrices, K_c, where I is the identity matrix:
(a) $K_c = 0.1I$
(b) $K_c = 0.5I$
(c) $K_c = I$
(d) $K_c = 2I$
(e) $K_c = 10I$
(f) $K_c = 20I$
Plot each on a separate graph. Which of these choices provide a stable closed loop?

12.5 Suppose that the controlled object, $G_p(s)$, is characterized by:

$$A = \begin{pmatrix} 0 & 1 \\ -2 & -3 \end{pmatrix}, \quad B = \begin{pmatrix} 10 & 0 \\ 0 & 25 \end{pmatrix}, \quad C^T = \begin{pmatrix} 1 & -1 \\ 0 & 2 \end{pmatrix}, \quad D = \begin{pmatrix} 0 & 1 \\ 0 & 0 \end{pmatrix} \tag{12.158}$$

Next, suppose that $G_c = I$ and plot the principal gains of Y, T_y, and S_y. Comment on this design.

12.6 Repeat Problem 12.5, but now suppose that the compensator is the MIMO version of a PI controller:

$$G_c(s) = K_P + K_I/s \tag{12.159}$$

where K_p and K_I are constant matrices with dimensions $m \times r$. The state variable realization of $G_c(s)$ $(m \times r)$ is such that

$$\underset{(n\times n)}{A_c} = \underset{(n\times n)}{0}, \quad \underset{(m\times r)}{K_I} = \underset{(m\times n)}{C_c^T}\underset{(n\times r)}{B_c}, \quad \underset{(m\times r)}{D_c} = \underset{(m\times r)}{K_P} \tag{12.160}$$

where the value of n (the number of controller states) is determined as follows:

$$n = \min\left(\text{rank } B_c,\ \text{rank } C_c^T\right) \tag{12.161}$$

Consider these choices: If $r < m$, take $n = r$, $B_c = I_r$, and select $K_I = C_c^T$ to be full rank. If $m < r$, take $n = m$, $C_c^T = I_m$, and select $B_c = K_I$ to be full rank. Will these always work?
Use any values of B_c, C_c^T, and K_P that provide closed-loop stability. How do you verify that? Comment on the relative merits of the two feedback designs obtained in this problem and in 12.5.

13

Youla Parameterization for Feedback Systems

In the first volume, we established the Youla parameterization method for control system design for single-output, single-input (SISO) systems. In this chapter, we will explore how to apply the Youla parameterization method to MIMO systems as well.

13.1 Neoclassical Control for MIMO Systems

In this section, similar to the SISO feedback loop, we discuss the role of neoclassical control for MIMO feedback control design. We present equivalent MIMO feedback loops which Youla parameterization technique could be utilized either directly (*Internal Model Control*) or utilized to compute a MIMO feedback controller, G_c.

We begin our discussion with a MIMO feedback control loop as illustrated in Figure 13.1.

In neoclassical control, as discussed in the previous chapter, the performance and robustness criteria are achieved by direct frequency loop-shaping of three different closed-loop transfer function matrices, Y, Youla, S, sensitivity, and T, complementary sensitivity. Then, utilizing the Youla transfer function matrix, we compute the corresponding G_c, MIMO feedback controller, as presented below. From Figure 13.1 the following relationships are written as:

$$\underline{u} = Y\underline{r} = G_c\underline{r} - L_u\underline{u} \tag{13.1}$$

where $L_u = G_cG_p$,

$$\Rightarrow\ Y = (I + L_u)^{-1}G_c = S_uG_c \tag{13.2}$$

we can then write,

$$Y = (1 + G_cG_p)^{-1}G_c = G_c(I + G_pG_c)^{-1} = G_cS_y \tag{13.3}$$

Given the realization for Y, the realization for G_c is computed as follows,

$$Y = (I + G_cG_p)^{-1}G_c \tag{13.4}$$

$$G_c = (I + G_cG_p)Y$$

$$G_c = Y + G_cG_pY$$

$$\Rightarrow\ G_c = Y(I - G_pY)^{-1} = (I - YG_p)^{-1}Y \tag{13.5}$$

In terms of a feedback control loop, Youla parameterization can be interpreted as an equivalent open-loop control by re-drawing Figure 13.1 to the system shown in Figure 13.2.

Now knowing that $Y = (1 + G_cG_p)^{-1}G_c$, then the following equivalent control loop can be drawn as the system shown in Figure 13.3.

13.1.1 Internal Model Control

It should be obvious that in terms of feedback control design, a transformation to an open-loop control using Youla parameterization, as illustrated in Figure 13.3, is theoretically insightful and novel and practically straightforward. It should also be clear that the open-loop transformation is only for computation and not implementation purposes. A good question to ask at this stage if Youla transfer function matrix could be directly used in a feedback control loop rather than only utilized

Robust Control: Youla Parameterization Approach, First Edition. Farhad Assadian and Kevin R. Mallon.

Companion Website: www.wiley.com/go/assadian/robust

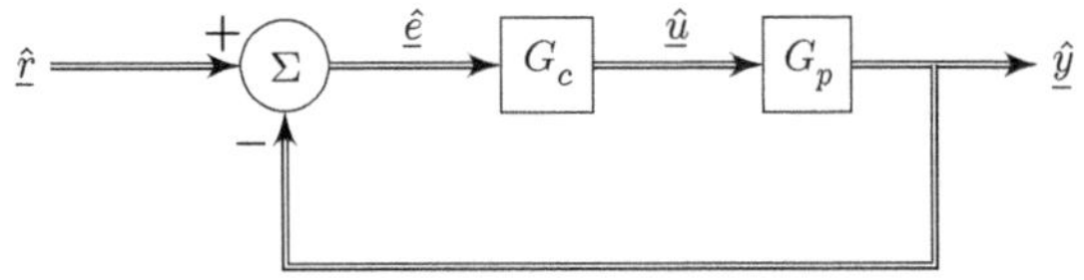

Figure 13.1 Block diagram of a multi-variable control system.

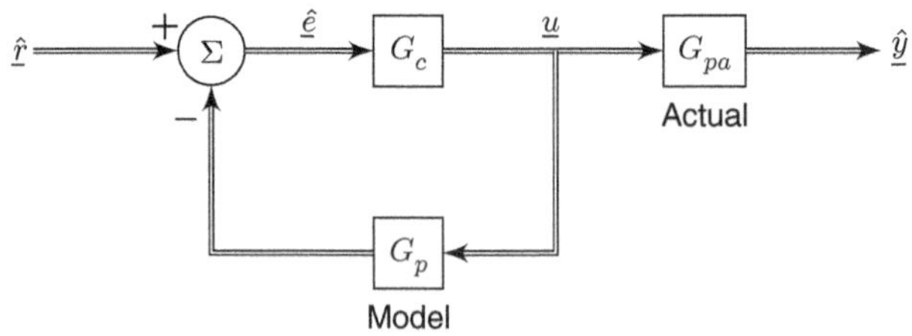

Figure 13.2 Interpretation of a neoclassical control system.

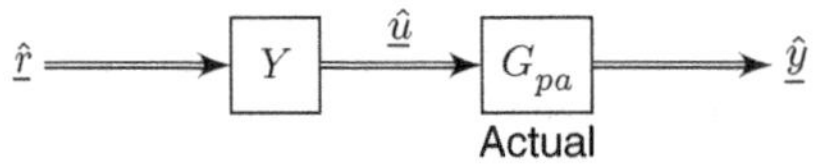

Figure 13.3 Open-loop Youla control.

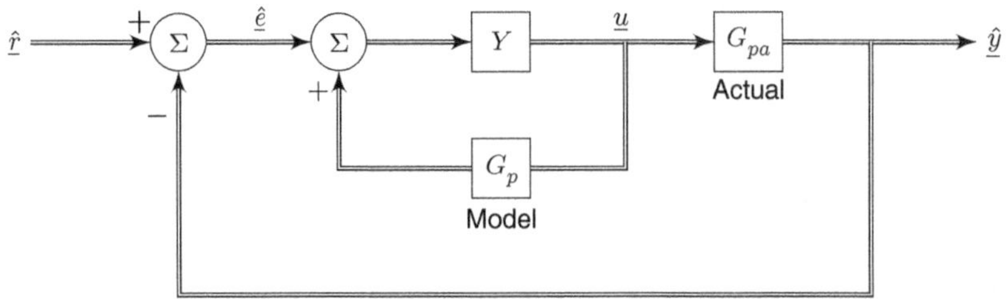

Figure 13.4 Internal model control.

for computation of the feedback controller? This question is previously answered through internal model control (IMC). Let's again find an equivalent feedback control loop to Figure 13.1 by modifying this loop to include Y as illustrated in the Figure 13.4.

An equivalent feedback control loop now could be obtained, known as IMC [13], from Figure 13.4, and is shown in Figure 13.5.

In the above feedback control loop, Y is directly utilized in place of feedback controller G_c. But the benefits of IMC that goes beyond just directly utilize Y, some of these benefits are discussed in the previous chapters when IMC is used in the SISO control design using optimization techniques. Similar optimization techniques could be used for MIMO controller design using this feedback control loop.

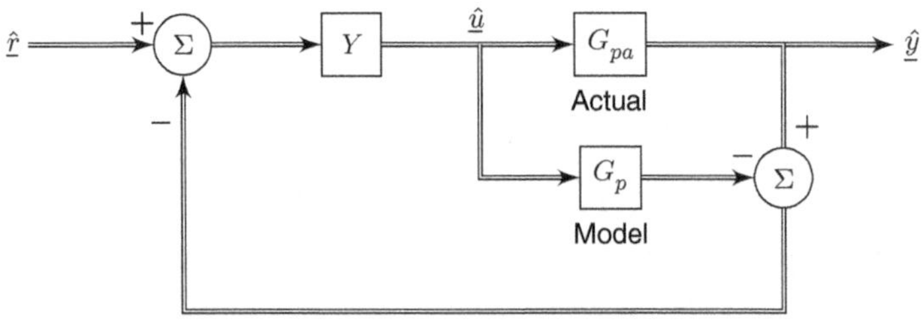

Figure 13.5 Internal model control.

13.2 MIMO Feedback Control Design for Stable Plants

We formulate the MIMO control design for two cases:

(a) Stable plants, G_p.
(b) Unstable plants, G_p.

13.2.1 Procedure to Find the MIMO Controller, G_c

Starting with $G_p(s) \in \mathbb{C}^{r\times m}$,

$$\left[\begin{array}{c|c} I_{r\times r} & G_p \\ \hline * & I_{m\times m} \end{array}\right] \sim \left[\begin{array}{c|c} U_L & M_p \\ \hline * & U_R \end{array}\right] \tag{13.6}$$

where M_p is the Smith-McMillan form of G_p, and U_L and U_R are left and right unimodular matrices:

$$U_L G_p U_R = M_p \Rightarrow G_p = U_L^{-1} M_p U_R^{-1} \tag{13.7}$$

This diagonalizes G_p, and furthermore, aids in decoupling the control design as follows:
We choose,

$$Y = U_R M_Y U_L \tag{13.8}$$

where M_Y is a diagonal matrix of stable proper transfer functions. Then the closed-loop transfer functions matrices at the two nodes can be written as:

$$T_y = G_p Y = U_L^{-1} M_P M_Y U_L \tag{13.9}$$

$$T_u = Y G_p = U_R M_Y M_p U_R^{-1} \tag{13.10}$$

Therefore, we shape the closed-loop transfer function matrices, i.e. assigning the closed-loop poles and zeros, simply by our choices for the diagonal elements of M_Y.

13.2.2 Interpolation Conditions

From the SISO feedback control systems, we know if "z" is any non-minimum phase zeros, in $G_p(s)$, then "z" must be a zero of T_y.

Furthermore, if "p" is a pole of $G_p(s)$ on the $j\omega$-axis or in the right half-plane, then "p" must be a zero of S_y.

But, we also know if G_p is stable, then closed-loop BIBO stability is guaranteed only by selecting a stable and proper Y transfer function matrix. As stated above, this can be done through the proper selection of the diagonal elements of M_Y. Then, we write the following relationship:

$$S_y = I - T_y = I - U_L^{-1} M_P M_Y U_L = U_L^{-1}(I - M_P M_Y) U_L \tag{13.11}$$

Finally, the MIMO controller $G_c = (I - YG_p)^{-1} Y$ which yields

$$\begin{aligned} G_c &= (I - U_R M_Y U_L U_L^{-1} M_P U_R^{-1})^{-1} U_R M_Y U_L \\ &= (U_R U_R^{-1} - U_R M_Y M_P U_R^{-1})^{-1} U_R M_Y U_L \\ &= U_R (I - M_Y M_P)^{-1} M_Y U_L \end{aligned} \tag{13.12}$$

13.3 MIMO Feedback Control Design Examples

Example 13.1 Let G_P to be a stable and proper transfer function matrix:

$$G_p = 20 \begin{bmatrix} \dfrac{1}{(s+2)^2} & \dfrac{s+1}{s+2} \\ -\dfrac{s+1}{s+2} & \dfrac{s+1}{s+2} \end{bmatrix} \tag{13.13}$$

$$= \frac{20}{(s+2)^2} \begin{bmatrix} 1 & (s+1)(s+2) \\ -(s+1)(s+2) & (s+1)(s+2) \end{bmatrix} \tag{13.14}$$

$$= \frac{20}{(s+2)^2} G_{p_n} \tag{13.15}$$

We start with

$$\left[\begin{array}{c|c} I_{r\times r} & G_{p_n} \\ \hline * & I_{m\times m} \end{array}\right] \tag{13.16}$$

where then we can write

$$\left[\begin{array}{cc|cc} 1 & 0 & 1 & (s+1)(s+2) \\ 0 & 1 & -(s+1)(s+2) & (s+1)(s+2) \\ \hline * & * & 1 & 0 \\ * & * & 0 & 1 \end{array}\right] \tag{13.17}$$

Using the standard row/column operations on the numerator matrix,

1. Add C_1 to C_2 into C_2
2. Multiply R_1 by $(s+1)(s+2)$ and add to R_2 into R_2
3. Multiply C_1 by $(s+1)(s+2)$ and subtract from C_2 into C_2
4. Subtract C_1 from C_2 into C_2

These operations result in

$$\left[\begin{array}{cc|cc} 1 & 0 & 1 & 0 \\ (s+1)(s+2) & 1 & 0 & (s+1)(s+2)(s^2+3s+3) \\ \hline * & * & 1 & -(s+1)(s+2) \\ * & * & 0 & 1 \end{array}\right] \tag{13.18}$$

Using this,

$$U_L = \begin{bmatrix} 1 & 0 \\ (s+1)(s+2) & 1 \end{bmatrix}, \quad U_L^{-1} = \begin{bmatrix} 1 & 0 \\ -(s+1)(s+2) & 1 \end{bmatrix} \tag{13.19}$$

$$U_R = \begin{bmatrix} 1 & -(s+1)(s+2) \\ 0 & 1 \end{bmatrix}, \quad U_R^{-1} = \begin{bmatrix} 1 & (s+1)(s+2) \\ 0 & 1 \end{bmatrix} \tag{13.20}$$

$$\tag{13.21}$$

Then we write SM form using $M_p = U_L G_p U_R$,

$$M_p = 20\begin{bmatrix} \dfrac{1}{(s+2)^2} & 0 \\ 0 & \dfrac{(s+1)(s^2+3s+3)}{s+2} \end{bmatrix} = \begin{bmatrix} G_{p_1} & 0 \\ 0 & G_{p_2} \end{bmatrix} \tag{13.22}$$

From above, this system has

- **Poles at:** $-2, -2, -2$
- **Zeros at:** $-1, -\frac{3}{2} \pm j\sqrt{\frac{3}{2}}$

It should be noted that the coefficient of the highest-order polynomial in M_p is not equal to 1, as the gain is equal to 20. This is simply a scaling problem and does not impact the MIMO control design. Now using the SM form, the Youla transfer function for each channel can be selected independently. Let's assume simple form of T_1 and recognizing that the desired $T_1(0) = 1$:

$$T_1 = \frac{4}{(s+2)^2} \tag{13.23}$$

Then Y_1 can computed by:

$$T_1 = Y_1 G_{p_1} \Rightarrow Y_1 = \frac{1}{5} \tag{13.24}$$

This yields the desired first closed-loop control channel. Ideally, we are interested in having $T_2 = T_1$ at all frequencies (based on our previous discussions on the benefits of decoupling), so we will do the following:

$$T_1 = Y_2 G_{p_2} \Rightarrow Y_2 = \frac{1}{5(s+1)(s+2)(s^2+3s+3)} \tag{13.25}$$

Now $T_y = G_p Y$ and $T_y = U_L^{-1} M_P M_Y U_L$, where $M_Y = \begin{bmatrix} Y_1 & 0 \\ 0 & Y_2 \end{bmatrix}$,

$$T_y = \begin{bmatrix} 1 & 0 \\ -(s+1)(s+2) & 1 \end{bmatrix} \underbrace{\begin{bmatrix} \frac{4}{(s+2)^2} & 0 \\ 0 & \frac{4}{(s+2)^2} \end{bmatrix}}_{M_P M_Y} \begin{bmatrix} 1 & 0 \\ (s+1)(s+2) & 1 \end{bmatrix} \tag{13.26}$$

$$= \begin{bmatrix} \frac{4}{(s+2)^2} & 0 \\ 0 & \frac{4}{(s+2)^2} \end{bmatrix} \tag{13.27}$$

$$S_y = I - T_y = \frac{s(s+4)}{(s+2)^2} I \tag{13.28}$$

Now the MIMO controller $G_c = U_R(I - M_Y M_p)^{-1} M_Y U_L$, where

$$U_R = \begin{bmatrix} 1 & -(s+1)(s+2) \\ 0 & 1 \end{bmatrix} \tag{13.29}$$

$$(I - M_Y M_p)^{-1} M_Y = \begin{bmatrix} \frac{(s+2)^2}{5s(s+4)} & 0 \\ 0 & \frac{s+2}{5s(s+1)(s+4)(s^2+3s+3)} \end{bmatrix} \tag{13.30}$$

$$U_L = \begin{bmatrix} 1 & 0 \\ (s+1)(s+2) & 1 \end{bmatrix} \tag{13.31}$$

$$\Rightarrow G_c = \frac{1}{5} \begin{bmatrix} \frac{(s+2)^2}{s(s+4)(s^2+3s+3)} & -\frac{(s+2)^2}{s(s+4)(s^2+3s+3)} \\ \frac{(s+2)^2}{s(s+4)(s^2+3s+3)} & \frac{(s+2)}{s(s+1)(s+4)(s^2+3s+3)} \end{bmatrix} \tag{13.32}$$

In this case, the shape of the final controller is determined based on our selection of M_Y, and, the selection of M_Y is achieved by our choice for the desired complementary sensitivity transfer function matrix.

Now, we are in a position to place this MIMO controller, G_c, in the feedback loop with the plant, G_p as shown in Figure 13.6.

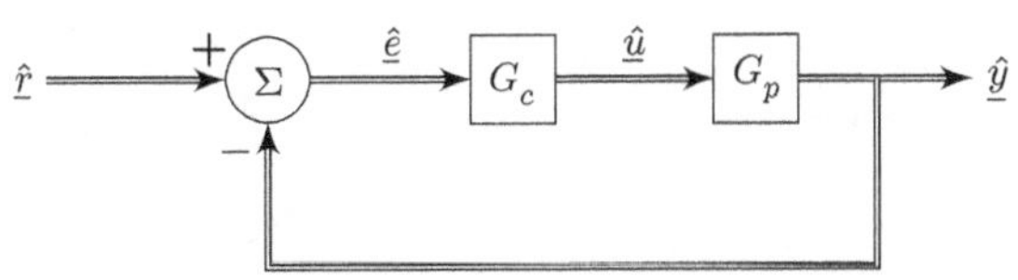

Figure 13.6 Block diagram of a multi-variable control system.

We then investigate the performance and robustness criteria for MIMO feedback controller design given in Chapter 12 and provided below as a reminder:

13.3.1 Summary of Closed-Loop Requirements

- **D**isturbance Rejection: $\sigma_{max}(S_y)$ as small as possible at low frequencies.
- **S**ensor Noise Rejection: $\sigma_{max}(T_y)$ as small as possible at high frequencies.
- **T**racking: $\sigma_{min}(T_y) \cong \sigma_{max}(T_y) \cong 1 \rightarrow \mathscr{K}(T_y) \cong 1$ at low frequencies.
- **M**inimize Control Effort: $\sigma_{max}(Y)$ as small as possible at high frequencies.
- **R**obust Stability: $\sigma_{max}(T_y)$ as small as possible at high frequencies (model uncertainty increases as $\omega \rightarrow \infty$).

13.3.2 Summary of Open-Loop Requirements

- **D**isturbance Rejection: $\sigma_{min}(L)$ as large as possible at low frequencies.
- **S**ensor Noise Rejection: $\sigma_{max}(L)$ as small as possible at high frequencies.
- **T**racking: $\sigma_{min}(L)$ as large as possible at low frequencies.
- **M**inimize Control Effort: $\sigma_{max}(G_c)$ as small as possible at high frequencies.
- **R**obust Stability: $\sigma_{max}(L)$ as small as possible at high frequencies.

The closed-loop complementary, sensitivity, and return ratio singular values are illustrated in Figure 13.7. The maximum and minimum singular values for all these transfer functions matrices are exactly equal. This is the result of our design which enforced a decoupling of the complementary transfer function, T_y. This decoupling is in all frequencies as evident from Figure 13.7. As a result, the return ratio transfer function matrix maximum and minimum singular values have become equal in all frequencies, including the crossover frequency. It is important to point out that this decoupling also results in $S_y = S_u$ and hence, there is no need for considering the second node or S_u in this design. It should be noted that the closed-loop speed of response or the closed-loop bandwidth has been arbitrarily selected in this example. All the closed-loop requirements are met through this design.

The important takeaway here is that with this method, all the tuning knobs for this design are under our control, and unlike many other black box optimization approaches, which we will discuss in the next chapters, this method is very much a white box approach. In addition, the closed-loop transfer function matrix decoupling, as discussed above, has several benefits, and hence, when possible, it is recommended to decouple the closed-loop transfer function matrix when designing MIMO controls.

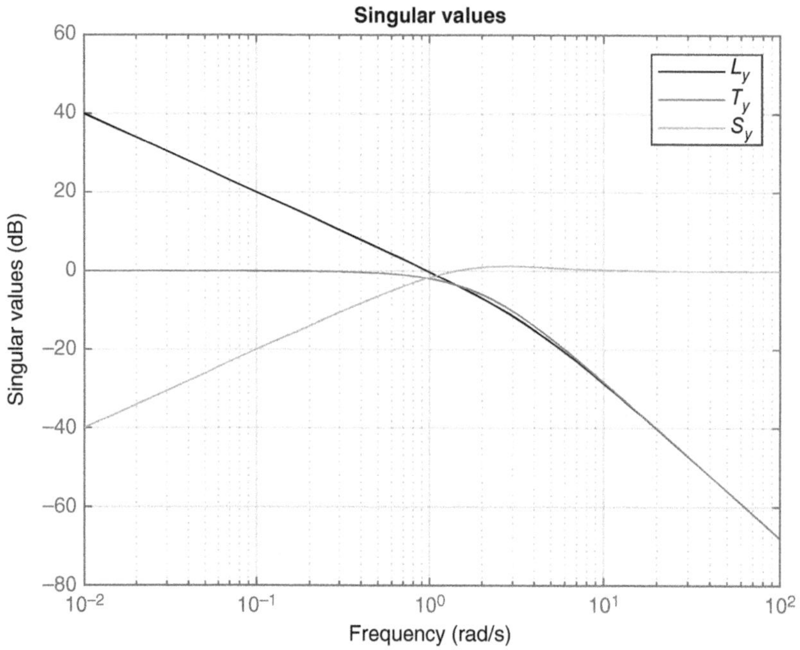

Figure 13.7 Closed-loop and return ratio singular values.

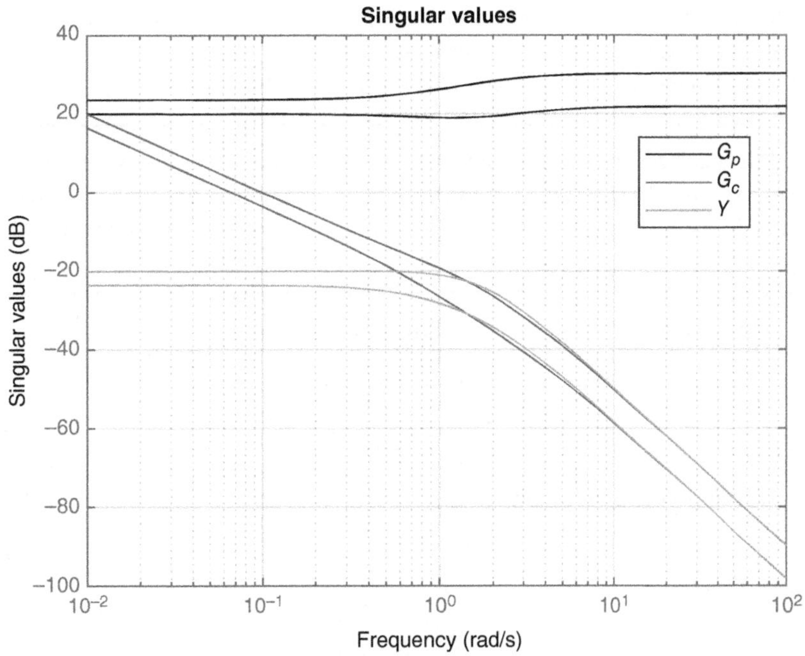

Figure 13.8 MIMO controller, plant and Youla transfer function matrices singular values.

The control effort requirement will be discussed next. In Figure 13.8, the plant, G_p, the Youla, Y, and the MIMO controller, G_c singular values are illustrated. In this example, the plant is high gain and therefore, the Youla transfer function matrix is low gain at low frequency as:

$$\sigma(Y) \to \frac{1}{\sigma(G_p)} \quad when \quad \omega \to 0$$

This is similar to the SISO case as discussed in the previous chapters. In addition, as it can be seen from Figure 13.8:

$$\sigma(Y) \to \sigma(G_c) \quad when \quad \omega \to \infty$$

Another important point, discussed more in Appendix C, is the frequency at which the feedback loses all its benefits. This frequency is specified to be the frequency when the three closed-loop transfer function matrices, T_y, S_y, and Y, are within $\pm 2\%$ to $\pm$ 5% their final values:

$$T_y \to L_y$$
$$S_y \to 1$$
$$Y \to G_c$$

In this case, this frequency is around 2 rad/s.

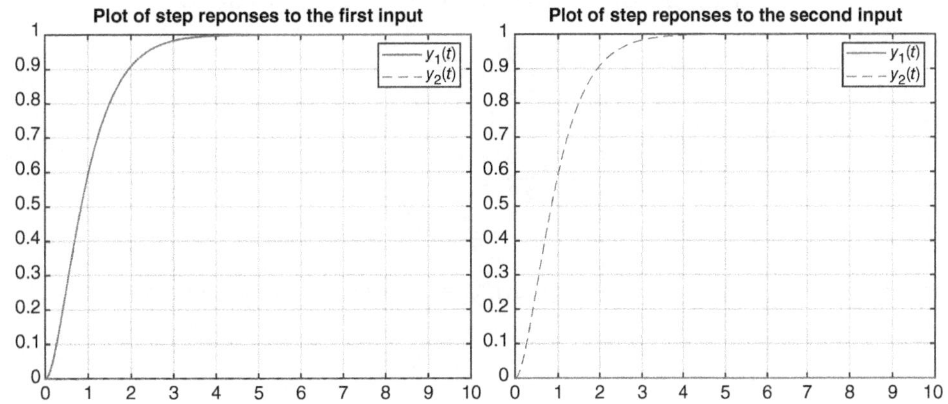

Figure 13.9 Step input responses.

Figure 13.9, step input responses of the closed-loop transfer function, T_y. Again, the decoupling is evident from Figure 13.9. Only the first output responds to the first unit step input, and the second output is exactly equal to 0. While, the response of the first output to the second unit step input is precisely equal to 0.

Computing the unimodular matrices and the SM from using row and column operations are not practical, especially when matrix sizes become relatively large. In the next example, we will use the same plant transfer function matrix as above but this time, we only compute the SM form and the unimodular matrices based on our development in Section 3 of Chapter 10.

Example 13.2 Let G_p to be a stable and proper transfer function matrix:

$$G_p = 20\begin{bmatrix} \frac{1}{(s+2)^2} & \frac{s+1}{s+2} \\ -\frac{s+1}{s+2} & \frac{s+1}{s+2} \end{bmatrix} \tag{13.33}$$

We can compute a right MFD form for G_p as follows:

$$G_p = N_R D_R^{-1} \tag{13.34}$$

where we select N_R and D_R to be

$$D_R = \begin{bmatrix} (s+2)^2 & 0 \\ 0 & s+2 \end{bmatrix} \tag{13.35}$$

$$N_R = 20\begin{bmatrix} 1 & s+1 \\ -(s+1)(s+2) & s+1 \end{bmatrix} \tag{13.36}$$

As we made this transformation arbitrary, we don't know if N_R and D_R are coprime. Next, we compute the SM form using the following steps given in Section 3 of Chapter 10, There are different approaches to compute Smith-McMillan form, the most common approach discussed previously is as follows:

- Define a determinantal divisor D_k = greatest common divisor (GCD) of all $k \times k$ minors and set $D_0 = 1$ with the following steps:
  (a) Factor $G_p(s) = \frac{p(s)}{\delta(s)}$
  (b) Find all minors of $p(s)$
  (c) Compute D_k for each $k \times k$ minor
  (d) Calculate $\varepsilon_k = \frac{D_k}{D_{k-1}}$ for each $i \in \{1, \dots, p\}$
  (e) Find M using $M = \frac{1}{\delta(s)}\text{diag}(\varepsilon_k)$

  We choose $\frac{1}{\delta(s)} = \frac{20}{(s+2)^2}$ and then

$$p(s) = \begin{bmatrix} 1 & (s+1)(s+2) \\ -(s+1)(s+2) & (s+1)(s+2) \end{bmatrix} \tag{13.37}$$

we can then write the determinantal divisors as:

$$D_0 = 1, \quad D_1 = 1, \quad D_2 = (s+1)(s+2)(s^2+3s+3) \tag{13.38}$$

$$\Rightarrow \epsilon_1 = \frac{D_1}{D_0} = 1, \quad \epsilon_2 = \frac{D_2}{D_1} = (s+1)(s+2)(s^2+3s+3) \tag{13.39}$$

Therefore, the SM form can be written as:

$$M_p = 20\begin{bmatrix} \frac{1}{(s+2)^2} & 0 \\ 0 & \frac{(s+1)(s^2+3s+3)}{s+2} \end{bmatrix} \tag{13.40}$$

This is exactly the same SM form that we found in the previous example using row–column operations.

The intent here to compute the unimodular matrices, U_L and U_R. But, before we accomplish this, we need to check if N_R and D_R in the previous MFD are coprime. This is simply accomplished by comparing the zeros of N_R to the zeros of M_p (SM form) and zeros of D_R to poles of the M_p. It is obvious that the poles and zeros of the SM form are exactly the same as zeros of D_R and N_R. So, we conclude that N_R and D_R are coprime.

So, an irreducible realization of $G_p(s)$ is

$$G_p = N_R D_R^{-1} = 20\underbrace{\begin{bmatrix} 1 & s+1 \\ -(s+1)(s+2) & s+1 \end{bmatrix}}_{\text{Zeros}=-1,-\frac{3}{2}\pm j\sqrt{\frac{3}{2}}}, \underbrace{\begin{bmatrix} (s+2)^2 & 0 \\ 0 & s+2 \end{bmatrix}^{-1}}_{\text{Poles}=-2,-2,-2} \tag{13.41}$$

Note that the degree of this denominator polynomial matrix, D_R, is 3, precisely equal to the SM degree.

Next, we must find U_R and U_L, by knowing both the SM form and a coprime MFD:

$$G_p = \begin{bmatrix} 1 & s+1 \\ -(s+1)(s+2) & s+1 \end{bmatrix} \begin{bmatrix} (s+2)^2 & 0 \\ 0 & s+2 \end{bmatrix}^{-1} \sim \underbrace{\begin{bmatrix} \frac{1}{(s+2)^2} & 0 \\ 0 & \frac{(s+1)(s^2+3s+3)}{s+2} \end{bmatrix}}_{\text{SM form}} \tag{13.42}$$

we know,

$$N_R = U_L^{-1}\text{diag}(\eta_1, \eta_2, \dots, \eta_p) \tag{13.43}$$

$$\begin{bmatrix} 1 & s+1 \\ -(s+1)(s+2) & s+1 \end{bmatrix} = U_L^{-1} \begin{bmatrix} 1 & 0 \\ 0 & (s+1)(s^2+3s+3) \end{bmatrix} \tag{13.44}$$

therefore,

$$U_L = \begin{bmatrix} 1 & 0 \\ 0 & (s+1)(s^2+3s+3) \end{bmatrix} \begin{bmatrix} 1 & s+1 \\ -(s+1)(s+2) & s+1 \end{bmatrix}^{-1} = \begin{bmatrix} \frac{1}{s^2+3s+3} & \frac{-1}{s^2+3s+3} \\ (s+1)(s+2) & 1 \end{bmatrix} \tag{13.45}$$

and

$$D_R = U_R\text{diag}(\delta_1, \delta_2, \dots, \delta_p) \tag{13.46}$$

$$\begin{bmatrix} (s+2)^2 & 0 \\ 0 & s+2 \end{bmatrix} = U_R \begin{bmatrix} (s+2)^2 & 0 \\ 0 & s+2 \end{bmatrix} \tag{13.47}$$

so,

$$U_R = \begin{bmatrix} 1 & 0 \\ 0 & 1 \end{bmatrix} \tag{13.48}$$

Now we can verify that,

$$G_p = U_L^{-1} M_p U_R^{-1} \tag{13.49}$$

The computed U_L is not a polynomial matrix, we could use the "smithForm" command in Matlab (this command only works for square transfer function matrices, we will discuss computation of unimodular matrices for non-square transfer function matrices in the next section) to find the two unimodular matrices U_L and U_R, which are polynomials, as follows:

$$U_L = \begin{bmatrix} 1 & 0 \\ -(s^2+3s+2) & -1 \end{bmatrix} \tag{13.50}$$

$$U_R = \begin{bmatrix} 1 & s^2+3s+2 \\ 0 & -1 \end{bmatrix} \tag{13.51}$$

13.4 MIMO Feedback Control Design: Unstable Plants

In this section, we extend our MIMO control design approach to unstable plants (Figure 13.10). As shown previously, there are four independent transfer function matrices that need to be analyzed for internal stability:

$$\begin{bmatrix} \bar{u} \\ y \end{bmatrix} = \begin{bmatrix} Y & I - YG_p \\ G_pY & G_p(I - YG_p) \end{bmatrix} \begin{bmatrix} r \\ d_u \end{bmatrix} \tag{13.52}$$

Necessary conditions for these transfer functions include

- Every uncanceled pole of L must be a zero of S
- Every uncanceled zero of L must be a zero of T

In MIMO systems, the "location" of zeros of T and S is important. In addition, Y, S, T, and G_c must be proper. This is complicated by the fact that equivalence transformations (utilizing SM form for our MIMO control design) do not preserve properness. This must be checked separately!!!.

13.4.1 The Proposed Control Design Method

1. Compute the Smith-McMillan form (M_p) and a MFD ($N_R D_R^{-1}$) of the plant.
2. Find U_L and U_R using either column and row operations or using Smith-McMillan form and coprime factorization ($N_R' D_R'^{-1}$), or other methods, which will be discussed in this section.
3. Choose M_Y to shape $M_T = M_p M_Y$, where M_T is decoupled T_y or select the desired M_T and then compute M_Y. Note that $Y = U_R M_Y U_L$ and check and make sure all entries of Y are proper.
4. Check the equivalent SISO interpolation conditions (discussed below) for each diagonal entry of M_T.
5. Find $T_y = U_L^{-1} M_P M_Y U_L$.
6. Find $S_y = I - T_y = U_L^{-1}(I - M_p M_Y)U_L$.
7. Compute $G_c = U_R(I - M_Y M_p)^{-1} M_Y U_L$.
8. Find $S_y G_p = U_L^{-1}(I - M_p M_Y) M_p U_R^{-1}$ and make sure its shape meets the requirement.
9. Check the $\mathcal{K}$ condition numbers of G_p and G_c to make sure designing for one node does meet the appropriate shapes for T_u, and S_u.

Note that $M_p = \text{diag}\{G_1, G_2, \dots\}$, $M_T = \text{diag}\{T_1, T_2, \dots\}$, and $I - M_T = \text{diag}\{S_1, S_2, \dots\}$, where G_k, T_k, and S_k are transfer functions.

The SISO interpolation conditions that are applied to each transfer function G_k:

- If G_k has a $j\omega$-axis or a RHP pole at p_k, then $S_k(p_k) = 0$ and $T_k(p_k) = 1$.
- If G_k has an non-minimum phase zero at z_k, then $S_k(z_k) = 1$ and $T_k(z_k) = 0$.
- if p_k or z_k is of order m then, in addition to the above conditions, the first to the mth $- 1$ derivatives of S_k and T_k evaluated at p_k or z_k have to be equal to 0, as discussed in the SISO control design part.

The following examples should provide some insight into the proposed MIMO control design method.

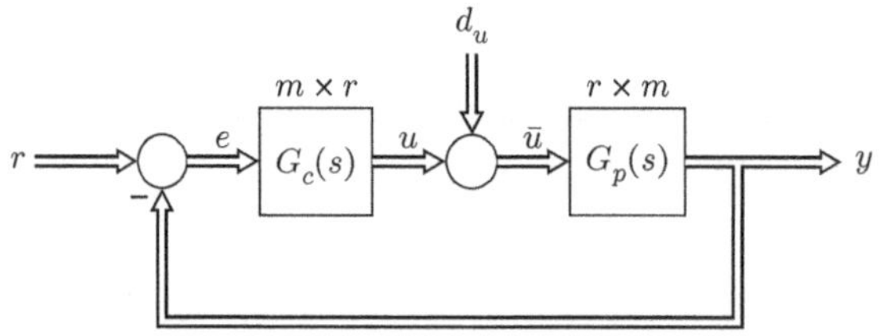

Figure 13.10 MIMO feedback system without noise injection.

Example 13.3

$$G_p = 100\begin{bmatrix} \frac{1}{s+1} & -\frac{1}{s(s+1)} \\ \frac{2}{s+1} & \frac{1}{s+1} \end{bmatrix} \tag{13.53}$$

$$= \underbrace{100\begin{bmatrix} 1 & 0 \\ -s & 1 \end{bmatrix}}_{U_L^{-1}} \underbrace{\begin{bmatrix} \frac{1}{s(s+1)} & 0 \\ 0 & \frac{s+2}{s+1} \end{bmatrix}}_{M_p} \underbrace{\begin{bmatrix} s & -1 \\ 1 & 0 \end{bmatrix}}_{U_R^{-1}} \tag{13.54}$$

Computation of U_L and U_R are not shown and left as an exercise.
Pole are: $-1, -1, 0$
Zeros at: -2
We select the following Y:

$$Y = \frac{1}{100}\underbrace{\begin{bmatrix} 0 & 1 \\ -1 & s \end{bmatrix}}_{U_R}\underbrace{\begin{bmatrix} Y_1 & 0 \\ 0 & Y_2 \end{bmatrix}}_{M_Y}\underbrace{\begin{bmatrix} 1 & 0 \\ s & 1 \end{bmatrix}}_{U_L} \tag{13.55}$$

$$M_pM_Y = \begin{bmatrix} \frac{Y_1}{s(s+1)} & 0 \\ 0 & \frac{Y_2(s+2)}{s+1} \end{bmatrix} \tag{13.56}$$

Therefore,

$$T_1 = \frac{Y_1}{s(s+1)} \quad \text{and} \quad T_2 = \frac{Y_2(s+2)}{s+1} \tag{13.57}$$

Let

$$Y_1 = \frac{s}{\tau_1 s+1} \quad \text{and} \quad Y_2 = \frac{1}{(\tau_1 s+1)(s+2)} \tag{13.58}$$

This yields the following complementary sensitivity transfer functions:

$$T_1 = \frac{1}{(\tau_1 s+1)(s+1)} \quad \text{and} \quad T_2 = \frac{1}{(\tau_1 s+1)(s+1)} \tag{13.59}$$

$T_1(0) = 1$ so it meets the required interpolation condition, as the first entry in M_p is not BIBO stable.
Letting $\tau_1 = 1$ and inserting Y_1 and Y_2 into Eq. (13.55) yields

$$Y = \frac{1}{100}\begin{bmatrix} 0 & 1 \\ -1 & s \end{bmatrix}\begin{bmatrix} \frac{s}{s+1} & 0 \\ 0 & \frac{1}{(s+1)(s+2)} \end{bmatrix}\begin{bmatrix} 1 & 0 \\ s & 1 \end{bmatrix} \tag{13.60}$$

$$= \frac{1}{100}\begin{bmatrix} \frac{s}{(s+1)(s+2)} & \frac{1}{(s+1)(s+2)} \\ \frac{-2s}{(s+1)(s+2)} & \frac{s}{(s+1)(s+2)} \end{bmatrix} \tag{13.61}$$

Then it is straightforward to compute G_c. *Checking the conditions from steps 8 and 9 given above is left as an exercise.*

Example 13.4

$$G_p = \begin{bmatrix} 2 & -\frac{1}{s+1} \\ -\frac{1}{s^2} & \frac{1}{s^2(s+1)} \end{bmatrix} \tag{13.62}$$

$$= \underbrace{\begin{bmatrix} -s^2 & 1 \\ 1 & 0 \end{bmatrix}}_{U_L^{-1}} \underbrace{\begin{bmatrix} \frac{1}{s^2(s+1)} & 0 \\ 0 & 1 \end{bmatrix}}_{M_p} \underbrace{\begin{bmatrix} -(s+1) & 1 \\ 1 & 0 \end{bmatrix}}_{U_R^{-1}} \tag{13.63}$$

Again the computation of U_L and U_R are not shown and left as an exercise. For this example, the poles are $-1, 0$, and 0, and there are no zeros.

So, the two of the poles are at origin and the plant is BIBO unstable. We write the following expression for Y:

$$Y = \underbrace{\begin{bmatrix} 0 & 1 \\ 1 & s+1 \end{bmatrix}}_{U_R} \underbrace{\begin{bmatrix} Y_1 & 0 \\ 0 & Y_2 \end{bmatrix}}_{M_Y} \underbrace{\begin{bmatrix} 0 & 1 \\ 1 & s^2 \end{bmatrix}}_{U_L} \tag{13.64}$$

Knowing that,

$$T_1 = \frac{Y_1}{s^2(s+1)} \quad \text{and} \quad T_2 = Y_2 \tag{13.65}$$

Now, internal stability dictates the shape of Y_2, let

$$Y_1 = \frac{s^2}{\delta_1(s)} \Rightarrow T_1 = \frac{1}{\delta_1(s)(s+1)} \tag{13.66}$$

where $\delta_1(s)$ is some polynomial of s. The interpolation conditions for the double pole at 0 are $T_1(0) = 1$ and $T_1'(0) = 0$, where T_1' is the first derivative of T_1 with respect to s, $T_1'(0) = \frac{dT_1(0)}{ds}$. Taking the derivative of T_1 with respect to s,

$$T_1'(s) = \frac{-(\delta_1'(s)(s+1) + \delta_1(s))}{[\delta_1(s)(s+1)]^2} \tag{13.67}$$

Applying the interpolation conditions yields the following set of conditions that must be met for $\delta_1(s)$. From Eq. (13.66),

$$T_1(0) = 1 \Rightarrow \delta_1(0) = 1 \tag{13.68}$$

From Eq. (13.66),

$$T_1'(0) = 0 \Rightarrow \delta_1(0) = -\delta_1'(0) \tag{13.69}$$

How can we meet these conditions for assuring internal stability?

We will answer this question by selecting a simple polynomial for $\delta_1(s)$. Letting $\delta_1(s) = as + b$, (from the above conditions, $\delta_1(0) = b = 1$, and $\delta_1'(0) = a = -1$), this means that $\delta_1(s) = 1 - s$ which is not stable, obviously not meeting stability requirements of $T_1(s)$. (It should be noted that the higher-order polynomials will have the same stability issue.)

Try a new selection for Y_1 such as,

$$Y_1 = \frac{s^2\eta_1(s)}{\delta_1(s)} \Rightarrow T_1(s) = \frac{\eta_1(s)}{\delta_1(s)(s+1)} \tag{13.70}$$

Then by using the interpolation conditions above, Eqs. (13.66) and (13.67), the following conditions exist,

$$T_1(0) = \frac{\eta_1(0)}{\delta_1(0)} = 1 \Rightarrow \eta_1(0) = \delta_1(0) \tag{13.71}$$

$$T_1'(0) = \frac{\eta_1'(0)\delta_1(0) - \eta_1(0)[\delta_1'(0) + \delta_1(0)]}{\delta_1^2(0)} = 0 \Rightarrow \eta_1'(0) = \delta_1'(0) + \delta_1(0) \tag{13.72}$$

It is not difficult to show that by setting,

$$\eta_1(s) = \cdots + (q+p)s + p \tag{13.73}$$

$$\delta_1(s) = \cdots + qs + p \tag{13.74}$$

will meet the interpolation conditions.

Another concern is the properness of Y, using the transformation,

$$Y = \begin{bmatrix} 0 & 1 \\ 1 & s+1 \end{bmatrix} \begin{bmatrix} \dfrac{s^2\eta_1(s)}{\delta_1(s)} & 0 \\ 0 & \dfrac{1}{\delta_2(s)} \end{bmatrix} \begin{bmatrix} 0 & 1 \\ 1 & s^2 \end{bmatrix} \tag{13.75}$$

$$= \begin{bmatrix} \dfrac{1}{\delta_2(s)} & \dfrac{s^2}{\delta_2(s)} \\ \dfrac{s+1}{\delta_2(s)} & s^2\left(\dfrac{s+1}{\delta_2(s)} + \dfrac{\eta_1(s)}{\delta_1(s)}\right) \end{bmatrix} \tag{13.76}$$

By ensuring that Y is selected to be proper and stable, then all of the other closed-loop transfer function matrices will also be proper.

For the properness of Y, the two requirements that must be met, from Eq. (13.76),

$$\deg \delta_2(s) \geq 2 \tag{13.77}$$

$$\deg \delta_1(s) + \deg \delta_2(s) \geq \max\{2 + \deg \eta_1(s) + \deg \delta_2(s),\ 3 + \deg \delta_1(s)\} \tag{13.78}$$

The degree constraints in Eqs. (13.77) and (13.78) come from $\frac{s^2}{\delta_2(s)}$ and $\frac{s^2(s+1)\delta_1(s)+s^2\eta_1(s)\delta_2(s)}{\delta_1(s)\delta_2(s)}$, the entries of the TFM in Eq. (13.76). If we select $\delta_1(s) = \delta_2(s) = \delta(s)$, then the constraints reduce to

$$\deg \delta(s) \geq \max\{\deg \eta_1(s) + 2,\ 3\} \tag{13.79}$$

Hence, selecting $\deg \eta_1(s) = 1$ makes the minimal degree of $\delta(s)$ to be 3. Setting $\delta(s) = (s+1)^3 = s^3 + 3s^2 + 3s + 1$ with the interpolation conditions of Eqs. (13.73) and (13.74) makes $\eta_1(s) = 4s + 1$.

Therefore,

$$Y_1 = \frac{s^2(4s+1)}{(s+1)^3},\ T_1 = \frac{4s+1}{(s+1)^4} \tag{13.80}$$

$$Y_2 = \frac{1}{(s+1)^3},\ T_2 = \frac{1}{(s+1)^3} \tag{13.81}$$

Now using $S = I - T$

$$S_1 = \frac{s^2(s^2+4s+6)}{(s+1)^4} \tag{13.82}$$

$$S_2 = \frac{s(s^2+3s+3)}{(s+1)^3} \tag{13.83}$$

Finding T_y using the transformation $T_y = U_L^{-1}M_yU_L$

$$T_y = \begin{bmatrix} -s^2 & 1 \\ 1 & 0 \end{bmatrix} \begin{bmatrix} \dfrac{4s+1}{(s+1)^4} & 0 \\ 0 & \dfrac{1}{(s+1)^3} \end{bmatrix} \begin{bmatrix} 0 & 1 \\ 1 & s^2 \end{bmatrix} \tag{13.84}$$

$$= \begin{bmatrix} \dfrac{1}{(s+1)^3} & \dfrac{-3s^3}{(s+1)^4} \\ 0 & \dfrac{4s+1}{(s+1)^4} \end{bmatrix} \tag{13.85}$$

Observe that $T_y(0) = I$ which means decoupling at low frequency. We can then compute the controller G_c. *Checking the conditions of steps 8 and 9 given above is left as an exercise.*

So far, we used our proposed method to design MIMO controllers for square plant matrices. In the next example, we generalize our method to a non-square plant matrix.

Example 13.5

$$G_p = 100\begin{bmatrix} \frac{s}{s+1} & \frac{s}{s+1} & \frac{1}{s+1} \\ \frac{1}{s+1} & \frac{1}{s+1} & \frac{1}{s+1} \end{bmatrix} \tag{13.86}$$

$$= 100\underbrace{\begin{bmatrix} 1 & -1 \\ 1 & 0 \end{bmatrix}}_{U_L^{-1}}\underbrace{\begin{bmatrix} \frac{1}{s+1} & 0 & 0 \\ 0 & \frac{1-s}{s+1} & 0 \end{bmatrix}}_{M_p}\underbrace{\begin{bmatrix} 1 & 1 & 1 \\ 1 & 1 & 0 \\ 1 & 0 & 0 \end{bmatrix}}_{U_R^{-1}} \tag{13.87}$$

Computation of U_L and U_R are done using column–row operations, this is left as an exercise. Below, we will compute U_L and U_R by finding first the SM form using gcd of all $k \times k$ minors, as described in Chapter 10, and coprime MFDs.

First, we find the SM form by writing $G_p = \frac{p(s)}{\delta(s)}$ as:

$$G_p = 100\frac{1}{s+1}\begin{bmatrix} s & s & 1 \\ 1 & 1 & 1 \end{bmatrix} \tag{13.88}$$

Then, the determinantal divisors are written as:

$$D_0 = 1, \tag{13.89}$$

The determinantal divisors D_1 is obtained by computing all 1×1 minors of $p(s)$ as:

$$D_1 = \begin{bmatrix} s & s & 1 & 1 & 1 & 1 \end{bmatrix} = 1 \tag{13.90}$$

The determinantal divisor, D_2, is obtained by computing all 2×2 minors of $p(s)$ as:

$$D_2 = \left[\begin{vmatrix} s & s \\ 1 & 1 \end{vmatrix}, \begin{vmatrix} s & 1 \\ 1 & 1 \end{vmatrix}, \begin{vmatrix} s & 1 \\ 1 & 1 \end{vmatrix}\right] = [0,\ s-1,\ s-1] = s-1 \tag{13.91}$$

$$\Rightarrow \epsilon_1 = \frac{D_1}{D_0} = 1, \quad \epsilon_2 = \frac{D_2}{D_1} = s-1 \tag{13.92}$$

$$M_p = 100\frac{1}{s+1}\begin{bmatrix} 1 & 0 & 0 \\ 0 & s-1 & 0 \end{bmatrix} = 100\begin{bmatrix} \frac{1}{s+1} & 0 & 0 \\ 0 & \frac{s-1}{s+1} & 0 \end{bmatrix} \tag{13.93}$$

We should know that M_p is the same size of G_p but with the last column of zeros. From the SM form, we know

- **Pole** are at: −1, −1
- **Zeros** are at: 1

Now, we write G_p as multiplication of a numerator, N_R and denominator, D_R matrices as follows:

$$G_p = 100\begin{bmatrix} \frac{s}{s+1} & \frac{s}{s+1} & \frac{1}{s+1} \\ \frac{1}{s+1} & \frac{1}{s+1} & \frac{1}{s+1} \end{bmatrix} \tag{13.94}$$

$$= 100\underbrace{\begin{bmatrix} s & s & 1 \\ 1 & 1 & 1 \end{bmatrix}}_{N_R}\underbrace{\begin{bmatrix} s+1 & 0 & 0 \\ 0 & s+1 & 0 \\ 0 & 0 & s+1 \end{bmatrix}^{-1}}_{D_R:\ \text{deg}=3} \tag{13.95}$$

We know this MFD is not a coprime as the degree of D_R is higher than the number of poles in SM form, M_p.

We use our previously presented procedure in Chapter 10 to compute the right common factor, W (this computation is not shown here and is left as an exercise). The right common factor is given below:

$$W = \begin{bmatrix} 1 & 1 & 0 \\ 0 & 0 & 1 \\ s+1 & 0 & 0 \end{bmatrix} \quad \Rightarrow \quad W^{-1} = \begin{bmatrix} 0 & 0 & \frac{1}{s+1} \\ 1 & 0 & -\frac{1}{s+1} \\ 0 & 1 & 0 \end{bmatrix} \tag{13.96}$$

$$N_R' = N_R W^{-1} = \begin{bmatrix} s & s & 1 \\ 1 & 1 & 1 \end{bmatrix} \begin{bmatrix} 0 & 0 & \frac{1}{s+1} \\ 1 & 0 & -\frac{1}{s+1} \\ 0 & 1 & 0 \end{bmatrix} = \begin{bmatrix} s & 1 & 0 \\ 1 & 1 & 0 \end{bmatrix} \tag{13.97}$$

$$D_R' = D_R W^{-1} = \begin{bmatrix} s+1 & 0 & 0 \\ 0 & s+1 & 0 \\ 0 & 0 & s+1 \end{bmatrix} \begin{bmatrix} 0 & 0 & \frac{1}{s+1} \\ 1 & 0 & -\frac{1}{s+1} \\ 0 & 1 & 0 \end{bmatrix} \tag{13.98}$$

$$= \begin{bmatrix} 0 & 0 & 1 \\ s+1 & 0 & -1 \\ 0 & s+1 & 0 \end{bmatrix} \tag{13.99}$$

$$\Rightarrow G_p = \underbrace{100 \begin{bmatrix} s & 1 & 0 \\ 1 & 1 & 0 \end{bmatrix} \begin{bmatrix} 0 & 0 & 1 \\ s+1 & 0 & -1 \\ 0 & s+1 & 0 \end{bmatrix}^{-1}}_{\text{Coprime}} \tag{13.100}$$

Now that we have the coprime factorization and the SM form of G_p, and we can compute U_L and U_R matrices as follows:

$$\begin{aligned} N_R' &= U_L^{-1} \operatorname{diag}(\eta_1, \ldots, \eta_p, 0, \ldots, 0) \\ D_R' &= U_R \operatorname{diag}(\delta_1, \ldots, \delta_p, 1, \ldots, 1) \end{aligned} \tag{13.101}$$

where,

$$\operatorname{diag}(\eta_1, \ldots, \eta_p, 0, \ldots, 0) = \begin{bmatrix} 1 & 0 & 0 \\ 0 & s-1 & 0 \end{bmatrix} \tag{13.102}$$

and

$$\operatorname{diag}(\delta_1, \ldots, \delta_p, 1, \ldots, 1) = \begin{bmatrix} s+1 & 0 & 0 \\ 0 & s+1 & 0 \\ 0 & 0 & 1 \end{bmatrix} \tag{13.103}$$

From Eq. (13.101), we can compute U_L and U_R,

$$U_L = \begin{bmatrix} \frac{1}{s-1} & \frac{-1}{s-1} \\ -1 & s \end{bmatrix} \tag{13.104}$$

$$U_R = \begin{bmatrix} 0 & 0 & 1 \\ 1 & 0 & -1 \\ 0 & 1 & 0 \end{bmatrix} \tag{13.105}$$

these are not the same U_L and U_R as given at the beginning of this problem as the methods of the two computation differ, but one can check that

$$M_p = U_L G_p U_R \tag{13.106}$$

Now the SM form can be computed from the unimodular matrices, U_L and U_R, and G_p, as shown above, to check if we obtain the same SM form as given above. Unfortunately, the left unimodular matrix U_L cannot be utilized to compute a MIMO controller based on our proposed method, as this matrix is not stable (BIBO stability).

We introduce the following robust method to compute the unimodular matrices for both tall and wide matrices. It can be shown that for the tall matrices, where the number of sensors are greater than the actuators (under-actuated plants), using the following method, one can automatically compute the unimodular matrices. This method utilizes the hermite form to compute the left unimodular matrix U_L from $p(s)$, where $G_p = \frac{p(s)}{\delta(s)}$. This could be done using the Matlab following command:

$$[U_L, H] = \textbf{hermiteForm}(p(s)) \tag{13.107}$$

where U_L is the left unimodular matrix and H is the upper triangular matrix. Then knowing U_L, U_R could computed from,

$$U_R = G \backslash M_p \tag{13.108}$$

where $G = U_L G_P$ and "\" stands for left division, that is, the solution of the linear equations $M_p = G U_R$.

When the plant matrix is wide, the number of sensors are less than the actuators (over-actuated plants), such as this case. It can be shown that directly using the above approach results in rank deficiency of the right unimodular matrix, U_R. So, we use a simple trick to get around this problem.

Matrix $p(s)$ is computed as previously but we transpose $p(s)$ before computing the left unimodular matrix, U_L, when using the hermite form given above. Let's call this U_L, U_{L_w}, where w stands for wide matrix:

$$p(s) = p(s)' \tag{13.109}$$

Then, we compute the hermite form as:

$$[U_{L_w}, H] = \textbf{hermiteForm}(p(s)) \tag{13.110}$$

To compute the U_{R_w}, we also need to transpose both the plant matrix G_p and the SM form M_p, given above:

$$G_p = G_p' \tag{13.111}$$

$$M_p = M_p' \tag{13.112}$$

We can then compute the U_{R_w} as given above:

$$U_{R_w} = G \backslash M_p \tag{13.113}$$

where $G = U_{L_w} G_P$, and the final unimodular matrices U_L and U_R to be used in control design are computed as follows:

$$U_R = U_{L_w}' \tag{13.114}$$

and

$$U_L = U_{R_w}' \tag{13.115}$$

Using the above approach, we compute U_L and U_R to be

$$U_L = \begin{bmatrix} 1 & 0 \\ -1 & 1 \end{bmatrix} \tag{13.116}$$

$$U_R = \begin{bmatrix} 0 & 0 & 1 \\ 0 & -1 & -1 \\ 1 & s & 0 \end{bmatrix} \tag{13.117}$$

Now these unimodular matrices could be utilized to compute a MIMO controller based on our proposed method. This is left as an exercise.

13.4.2 Another Approach for MIMO Controller Design

If the plant TFM G_p is minimum phase, stable, and square, we can design the MIMO controller very quickly using the following steps:

- Select $Y = G_p^{-1} T_{yd}$ where T_{yd} is a desired proper stable closed-loop transfer function matrix.
- Then $T_y = G_p Y$.
- Find $S_y = I - T_y$.
- Finally $G_c = Y S_y^{-1}$.

Example 13.6 As above, we use

$$G_p = 20 \begin{bmatrix} \dfrac{1}{(s+2)^2} & \dfrac{s+1}{s+2} \\ -\dfrac{s+1}{s+2} & \dfrac{s+1}{s+2} \end{bmatrix} \tag{13.118}$$

We want

$$Y = G_p^{-1} T_{yd} \tag{13.119}$$

$$= G_p^{-1} \left(\frac{4}{(s+2)^2} \right) I \tag{13.120}$$

$$= \frac{1}{5} \begin{bmatrix} \dfrac{1}{s^2+3s+3} & -\dfrac{1}{s^2+3s+3} \\ \dfrac{1}{s^2+3s+3} & \dfrac{1}{(s+1)(s+2)(s^2+3s+3)} \end{bmatrix} \tag{13.121}$$

Now

$$T_y = G_p Y = \frac{4}{(s+2)^2} I = T_{yd} \tag{13.122}$$

and

$$S_y = I - T_y = \frac{s(s+4)}{(s+2)^2} I \tag{13.123}$$

Finally,

$$G_c = Y S_y^{-1} = \frac{1}{5s(s+4)(s^2+3s+3)} \begin{bmatrix} (s+2)^2 & -(s+2)^2 \\ (s+2)^2 & \dfrac{s+2}{s+1} \end{bmatrix} \tag{13.124}$$

13.5 Problems

13.1 We all reason that it must be impossible to decouple everything, and we reason that way on purely physical grounds. For example, it must be impossible to decouple displacement from velocity. To investigate this,

(a) Compute the transfer function matrix for the following state-space representation of a plant:

$$A = \begin{pmatrix} 0 & 1 \\ -1 & -2 \end{pmatrix}, \quad B = \begin{pmatrix} 0 & 0 \\ 1 & 1 \end{pmatrix}, \quad C^T = \begin{pmatrix} 1 & 0 \\ 0 & 1 \end{pmatrix}, \quad D = \begin{pmatrix} 0 & 0 \\ 0 & 0 \end{pmatrix} \tag{13.125}$$

Notice that the first state equation is

$$\dot{x}_1 = x_2 \tag{13.126}$$

and both displacement x_1 and velocity x_2 are measured.

(b) Briefly comment on your finding.

13.2 Suppose that the controlled object $G_p(s)$ is similar to the controlled object of Problem #3 of the last homework, where

$$A = \begin{pmatrix} 0 & 1 \\ -2 & -3 \end{pmatrix}, \quad B = \begin{pmatrix} 10 & 0 \\ 0 & 25 \end{pmatrix}, \quad C^T = \begin{pmatrix} 1 & -1 \\ 0 & 2 \end{pmatrix}, \quad D = \begin{pmatrix} 0 & 1 \\ 0 & 0 \end{pmatrix} \tag{13.127}$$

Assuming that $G_c = I$ obtain simulations of closed-loop transients, first to a step input to $r_1(t)$ and then to a step input at $r_2(t)$. In each case, plot the response of both sensors, $y_1(t)$ and $y_2(t)$. Also, obtain the response of both actuators, $u_1(t)$ and $u_2(t)$. I think that it is best to obtain the realization for $T_y(s)$ and $Y(s)$ with compositions based on the realizations of $G_c(s)$ and $G_p(s)$.

13.3 Repeat Problem 13.2, but this time with $G_c(s) = K_P + K_I/s$, that is, a MIMO PI controller. (Refer to Problem 12.6, where this controller type was also used.)

13.4 Return to the design of the controlled object using the method based on the Smith-McMillan form of the controlled object. Using the same $G_p(s)$ as the previous problems, attempt to achieve decoupling of the reference signals. Hint: You need to compute the controller $G_c(s)$ by hand, then use Matlab to generate the required plots as in Problem 13.2.

13.5 Given the following controlled object $G_p(s)$,

$$G_p(s) = \begin{bmatrix} 0 & \dfrac{-1}{s+1} \\ 1 & \dfrac{1}{s(s+1)} \end{bmatrix} \tag{13.128}$$

(a) Utilize any method to compute the left and right unimodular matrices and a decoupled plant (hand computation only).

(b) Use Youla parameterization to hand-compute a controller $G_c(s)$.

13.6 Consider a system with plant

$$G_p = 50 \begin{bmatrix} \dfrac{-1}{s} & \dfrac{-(s+1)}{s} \\ \dfrac{1}{s^2+1} & 0 \end{bmatrix} \tag{13.129}$$

(a) Design a controller using Youla parameterization.
(b) Plot the principal gains of the closed-loop, sensitivity and Youla transfer function matrices. On the same graph, plot the inverse of the robust uncertainty and nominal performance filters (use legends or text boxes to clearly label your plots).
(c) On a separate graph, plot the principal gains of the open-loop (G_pG_c) transfer function matrix. On the same graph, plot the nominal performance filter (do not invert this filter) and the inverse of the robust uncertainty filter.
(d) Obtain simulations of the closed-loop transients, first to a unit step input at $r_1(t)$ and then to a unit step input at $r_2(t)$. In each case, plot the response of both sensors, $y_1(t)$ and $y_2(t)$.

Use Matlab to make your plots. Comment on your design.

13.7 For this problem, compute a $Y(j\omega)$, $T(j\omega)$, $S(j\omega)$, and $G_c(j\omega)$ that stabilizes a system with plant

$$G_p = 100\begin{bmatrix} 1 & -1 \\ 1 & 0 \end{bmatrix}\begin{bmatrix} \frac{1}{s+1} & 0 & 0 \\ 0 & \frac{1-s}{s+1} & 0 \end{bmatrix}\begin{bmatrix} 1 & 1 & 0 \\ 1 & 1 & 0 \\ 1 & 0 & 0 \end{bmatrix} \tag{13.130}$$

14

Norms of Feedback Systems

In this chapter, we introduce the norms, which discussed in the SISO section of this book, for vector signals with the applications to MIMO controller design. Similar to SISO systems, we concentrate on two fundamental norms: $\mathscr{H}_2$ and $\mathscr{H}_\infty$.

In addition, we revisit linear fractional transformation (LFT) with application to MIMO controller design optimization. We finish this chapter by introducing the concept of uncertainties, as discussed in the SISO part of this book, for MIMO systems.

These topics have been discussed in many textbooks and we introduce them again so that we could provide comparative results, in the last chapter of this book, between control concepts such as MIMO $\mathscr{H}_\infty$ with previously introduced MIMO Youla control method.

14.1 Norms

A norm measures the "size" of signals or systems but what do we mean by size? let $V_s = \{V_1(\cdot), \ldots, V_k(\cdot)\}$ where $V_i \in \mathbb{R}^n$ and $||V_i(\cdot)||$ is a mapping $\mathbb{R}^n \to \mathbb{R}$.

Definition 14.1 (Norm). $||V_i(\cdot)||$ is a norm if it meets the following requirements:

1. **Nonnegative**: $||V_i(\cdot)|| \geq 0$ for all k
2. **Homogeneity**: $||\alpha V_i(\cdot)|| = |\alpha|\,||V_i(\cdot)||$
3. **Triangle Inequality**: $||V_i(\cdot) + V_j(\cdot)|| \leq ||V_i(\cdot)|| + ||V_j(\cdot)||$

14.1.1 Signal Norms, the Discrete Case

The mean of a signal V is given by:

$$\mu_V \triangleq \frac{1}{N}\sum_{n=0}^{N-1} V_n \tag{14.1}$$

If signal V was a stationary random signal, then the calculation of its mean, as given in Eq. (14.1), still applies but the result would be an estimate of its mean [34]. The total energy of V is defined as:

$$E_V \triangleq \sum_{n=0}^{N-1} |V_n|^2 \tag{14.2}$$

The average power of the signal is defined as:

$$P_V \triangleq \frac{1}{N}\sum_{n=0}^{N-1} |V_n|^2 \tag{14.3}$$

The variance of V is defined as the power of the signal with its mean removed:

$$\sigma_V^2 \triangleq \frac{1}{N}\sum_{n=0}^{N-1} |V_n - \mu_V|^2 \tag{14.4}$$

For real signals, this reduces to $\sigma_V^2 = P_V - \mu_V^2$.

Robust Control: Youla Parameterization Approach, First Edition. Farhad Assadian and Kevin R. Mallon.

Companion Website: www.wiley.com/go/assadian/robust

The 2-norm of a signal is defined as:

$$||V||_2 \triangleq \sqrt{E_V} = \sqrt{\sum_{n=0}^{N-1} |V_n|^2} \tag{14.5}$$

We could choose a normalized 2-norm (rms):

$$||V||_{rms} \triangleq \sqrt{P_V} = \sqrt{\frac{1}{N}\sum_{n=0}^{N-1} |V_n|^2} \tag{14.6}$$

more generally we can define a p-norm as:

$$||V||_p \triangleq \left[\sum_{n=0}^{N-1} |V_n|^p\right]^{1/p} \tag{14.7}$$

The most popular norms are $p = 1,2$, and ∞ with ∞-norm is defined as:

$$||V||_\infty \triangleq \max_{0\le n<N} \{|V_n|\} \tag{14.8}$$

14.1.2 System Norms

For a dynamic system of the form:

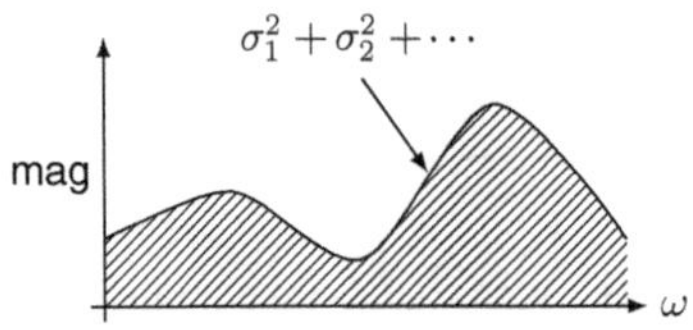

Figure 14.1 $\mathscr{H}_2$ norm of a TFM G in terms of its singular values.

An induced system norm is defined as:

$$\max_{w\neq 0} \frac{\overbrace{||Gw||_p}^{||z||_p}}{||w||_p} = ||G||_{p,i} \tag{14.9}$$

$||G||_{p,i}$ is known as an induced p norm.

Example 14.1 In this example, we illustrate computation of an induced 2-norm for a static matrix A.
Consider a static matrix A, then by definition,

$$\max_{w\neq 0} \frac{||Aw||_2}{||w||_2} = ||A||_{2,i} \tag{14.10}$$

From SVD, $A = U\Sigma V^T$, then, we know that the largest possible gain of A is in the direction of the first right singular vector. Let's assume $w = r_1$, where r_1 is the first right singular vector, we can then write

$$Ar_1 = \sigma_1 l_1 \tag{14.11}$$

plug this equation into (14.10),

$$\frac{||Ar_1||_2}{||r_1||_2} = \frac{||\sigma_1 l_1||_2}{||r_1||_2} = \sigma_1 \tag{14.12}$$

We know that $\sigma_1 = \sigma_{max}$.

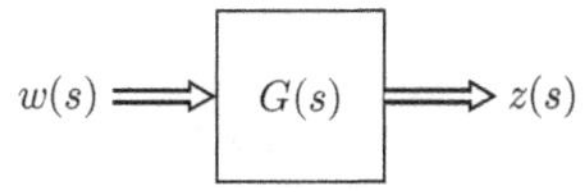

Figure 14.2 Standard I/O relationship.

Consider a transfer function matrix $G(s)$ with each element being strictly proper and stable. Then, the $\mathscr{H}_2$ norm of the TFM is given by:

$$||G||_2^2 = \frac{1}{2\pi}\text{tr}\int_{-\infty}^{\infty} G(j\omega)G^*(j\omega)d\omega \tag{14.13}$$

For a 2-norm,

$$GG^* = \sum_k l_k \sigma_k^2 l_k^*$$

and therefore,

$$||G||_2^2 = \frac{1}{2\pi}\int_{-\infty}^{\infty} \sum_k \sigma_k^2(\omega)d\omega \tag{14.14}$$

So $||G||_2^2$ is the equivalent to the area under the curve of the sum of the squares of the singular values of G, as shown in Figure 14.1.

Again referring to the system of Figure 14.2, we can write

$$z(t) = \int_0^{\infty} G(t-\tau)w(\tau)d\tau \tag{14.15}$$

Let each input be a zero-mean white noise with unit variance, such that, $E[ww^T] = I_m\delta(t-\tau)$ where $\delta(t-\tau)$ is a Dirac delta function and it has a value when $t = \tau$ and equal to 0 when $t \neq \tau$, then it can be show that

$$E[z^T z] = \text{tr}\int_0^{\infty} G^T(t)G(t)dt \tag{14.16}$$

where $E[z^T z] = \lim_{T\to\infty} E[\frac{1}{T}\int_0^T z^T z dt]$.

Theorem 14.1 **(Parseval's Theorem)**

$$tr\int_0^{\infty} G(t)^T G(t)dt = \frac{1}{2\pi}tr\int_{-\infty}^{\infty} G^*(j\omega)G(j\omega)d\omega \tag{14.17}$$

where $G(t)$ is the system impulse response and

$$G(t) = Ce^{At}B \tag{14.18}$$

$e^{At} = \Phi(t)$ is defined as the state transition matrix and $\Phi(t)$ is the inverse Laplace transform $e^{At} = \mathscr{L}^{-1}(sI - A)^{-1}$.

Using Parseval's theorem,

$$E[z^T z] = \frac{1}{2\pi}\text{tr}\int_{-\infty}^{\infty} G^*(j\omega)G(j\omega)d\omega = ||G||_2^2 \tag{14.19}$$

14.1.3 The $\mathscr{H}_2$-Norm

This norm can be defined as:

- Power of the system output when the input is driven by zero-mean white noise (LQG or $\mathscr{H}_2$) (persistent or everlasting input).
- Energy of the system output (LQR or $\mathscr{H}_2$) when the input is transient and not persistent.

Computation of the $\mathscr{H}_2$-norm can be done in both frequency, as presented in the SISO section, and state-space form using the system impulse response. Next, we present this computation in the state-space form. Let $G(t)$ be the system impulse response or

$$G(t) = Ce^{At}B \tag{14.20}$$

Matrices A, B, and C are as defined previously, and

$$\|G\|_2^2 = \text{tr} \int_0^\infty G(t)G^T(t)dt \tag{14.21}$$

$$= \text{tr} \int_0^\infty Ce^{At}BB^Te^{A^t}C^T dt \tag{14.22}$$

$$= \text{tr}\, C \underbrace{\left[\int_0^\infty e^{At}BB^Te^{A^t}dt \right]}_{P} C^T \tag{14.23}$$

P is defined as the controllability grammian, and it satisfies the Lyapunov equation:

$$AP + PA^T + BB^T = 0 \tag{14.24}$$

For the proof see [35].

We can utilize the MATLAB command `P=lyap(A,Q)` *with $Q = BB^T$ to compute the solution for the controllability grammian P.*

Similar to the controllability grammian, the observability grammian could be used for this computation, where the $\mathcal{H}_2$-norm is written as:

$$\|G\|_2^2 = \text{tr} \int_0^\infty G(t)^T G(t)dt \tag{14.25}$$

This is left as an exercise.

14.1.4 The $\mathcal{H}_\infty$-Norm

Recall that $\mathcal{H}_\infty$-Norm for MIMO systems provides the maximum singular value of a system across all frequencies, or $\|G\|_\infty = \max_{\omega} \bar{\sigma}[G(j\omega)]$, such as shown in Figure 14.3.

Also recall that,

$$\sigma_k = \frac{\|z\|_2}{\|w\|_2} \tag{14.26}$$

Theorem 14.2 *The ∞-norm is an induced norm, i.e.*

$$\|G\|_\infty = \max_{w \neq 0} \frac{\|z\|_2}{\|w\|_2} \tag{14.27}$$

Proof:

$$\|z\|_2^2 = \|z(j\omega)\|_2^2 \tag{14.28}$$

$$= \frac{1}{2\pi}\text{tr} \int_{-\infty}^{\infty} |G(j\omega)w(j\omega)|^2 d\omega \tag{14.29}$$

$$= \frac{1}{2\pi}\text{tr} \int_{-\infty}^{\infty} |G(j\omega)|^2 |w(j\omega)|^2 d\omega \tag{14.30}$$

$$\leq \|G\|_\infty^2 \frac{1}{2\pi}\text{tr} \int_{-\infty}^{\infty} |w(j\omega)|^2 d\omega \tag{14.31}$$

$$\leq \|G\|_\infty^2 \|w\|_2^2 \tag{14.32}$$

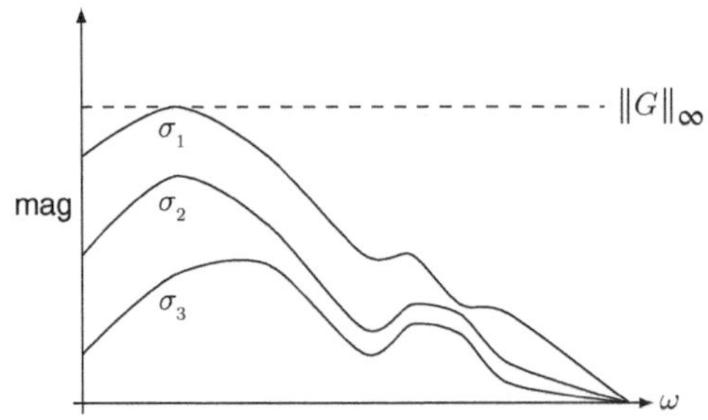

Figure 14.3 $\mathcal{H}_\infty$-Norm of MIMO systems.

Now take the maximum of the left hand side to obtain,

$$||G||_{\infty} = \max_{w \neq 0} \frac{||z||_2}{||w||_2} \tag{14.33}$$

■

In other words if

$$||G||_{\infty} \leq \gamma \tag{14.34}$$

where $\gamma \in \mathbb{R}_+$ then,

$$\frac{||z||_2}{||w||_2} \leq \gamma \tag{14.35}$$

This relation can be written as:

$$||z||_2^2 - \gamma^2 ||w||_2^2 \leq 0 \tag{14.36}$$

Writing the system ∞-norm as proposed in Eq. (14.36) helps the derivation of the state-space solution for H_∞ problem as discussed in the later chapters.

A physical interpretation of the H_∞-norm is appropriate here. Recall that the square of the 2-norm is equivalent to signal energy. Therefore, by specifying a bound or limit γ on the ratio of the output signal energy z to the disturbance energy w, we will therefore minimize the impact of worst possible disturbance on the system output when minimizing the H_∞-norm.

14.2 Linear Fractional Transformations (LFT)

In this section, we generalize the LFT concept, which is discussed in the SISO part of this book, to MIMO optimization and control design.

We generalize linear control design by the following representation in Figure 14.4: where $A(s)$ represents the actuator dynamics, $G_p(s)$ is the physical plant, $S(s)$ represents the sensors, $K(s)$ is the controller, with one or more degrees of freedom as discussed below, and d represents the vector of system disturbances. Here, $G = SG_pA$.

The vector disturbance signals d can be decomposed as follows: where d_u are the input vector process disturbances, d_y are the output vector disturbances (low frequency), and n is the sensor noise vector (high frequency), all of which are illustrated in Figure 14.5.

In general, control design will follow one of typologies showing in Figure 14.6. It is important to note the following points about the higher degrees of freedom controller:

1. The prefilter ($K_1(s)$) in two degrees of freedom controller could aid in relaxing the target following constraint ($T(0) = 1$) for the feedback control design. For example, the feedback controller ($K_2(s)$) could be only designed to stabilize the loop, and the prefilter ($K_1(s)$) could then be designed to meet the target following constraint.

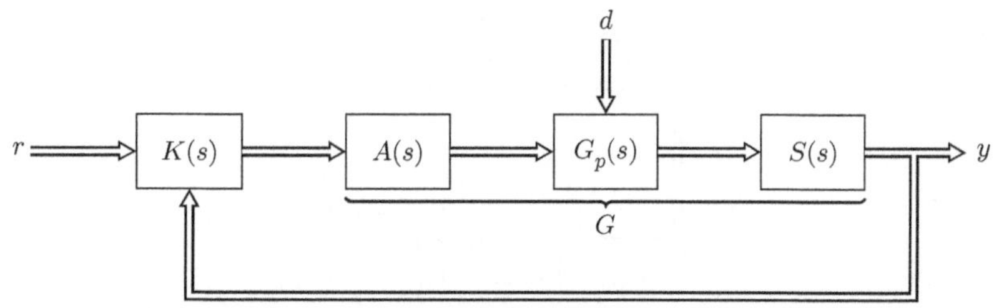

Figure 14.4 General linear control feedback design.

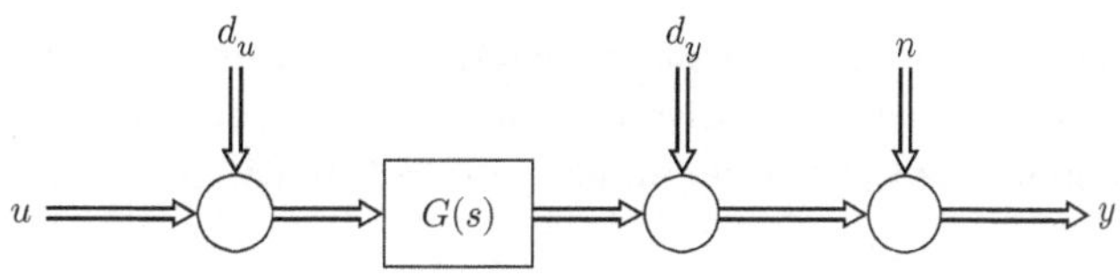

Figure 14.5 Disturbance decomposition.

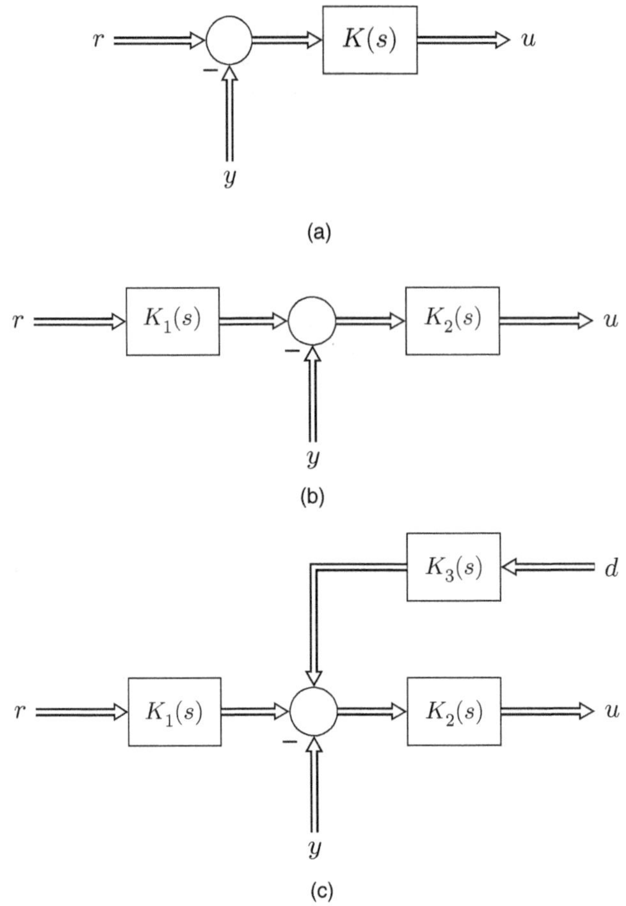

Figure 14.6 Degrees of freedom for control design: (a) 1 degree of freedom controller; (b) 2 degrees of freedom controller; (c) 3 degrees of freedom controller.

2. In three degrees of freedom controller, $K_3(s)$ controller/estimator could help in increasing the feedback loop performance and decreasing the actuator efforts.

We can design various degrees of freedom controllers with the use of a framework called the LFT of the system, more specifically this is called the lower LFT, as opposed to the upper LFT, which will be discussed later in this section. All types of controllers with various degrees of freedom, as discussed in the previous section, can be transformed into this framework for the application of norm optimization and control design. This framework is illustrated in Figure 14.7:

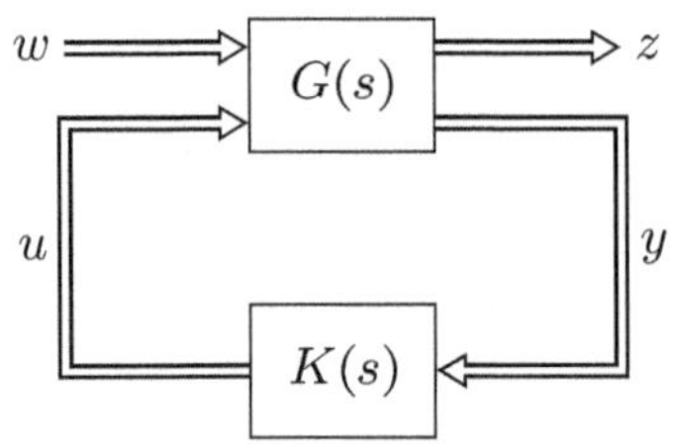

Figure 14.7 Lower LFT on systems.

where:

- The vector w is composed of the exogenous inputs including references, disturbances, and sensor noises.
- The vector z is composed of the signals that we wish to regulate tracking errors, actuator effort signals that must be kept small and within certain limits, and system states that must be kept within certain limits. These are known as the regulated outputs.

- The vector y is the sensor output measurements that are available to the controller.
- The vector u is the controlled inputs of the system.

14.3 Linear Fractional Transformation Explained

In this section, we will provide some insight into the LFT concept. Assuming a rational function of the form,

$$F(s) = \frac{as+b}{cs+d} \tag{14.37}$$

after manipulation, it can be shown that

$$F(s) = e + fs(1-gs)^{-1}h \tag{14.38}$$

where e, f, g, and h are constants and

$$e = bd^{-1} \tag{14.39}$$

$$f = a - bd^{-1}c \tag{14.40}$$

$$g = -d^{-1}c \tag{14.41}$$

$$h = d^{-1} \tag{14.42}$$

This results in the lower LFT shown in Figure 14.8. In general form, the lower LFT looks like the diagram in Figure 14.9.

Where the matrix M can be written as:

$$\begin{bmatrix} z \\ y \end{bmatrix} = \begin{bmatrix} M_{11} & M_{12} \\ M_{21} & M_{22} \end{bmatrix} \begin{bmatrix} w \\ u \end{bmatrix} \tag{14.43}$$

Then, the closed-loop map from w to z, using Eq. (14.38), can be written as:

$$F_l(M, \Delta_l) = M_{11} + M_{12}\Delta_l(I + M_{22}\Delta_l)^{-1}M_{21} \tag{14.44}$$

We derive Eq. (14.44) as follows:

$$z = M_{11}w + M_{12}u \tag{14.45}$$

$$y = M_{21}w + M_{22}u \tag{14.46}$$

$$u = \Delta_l y \tag{14.47}$$

Combining Eqs. (14.46) and (14.47) yields

$$y = (I - M_{22}\Delta_l)^{-1}M_{21}w \tag{14.48}$$

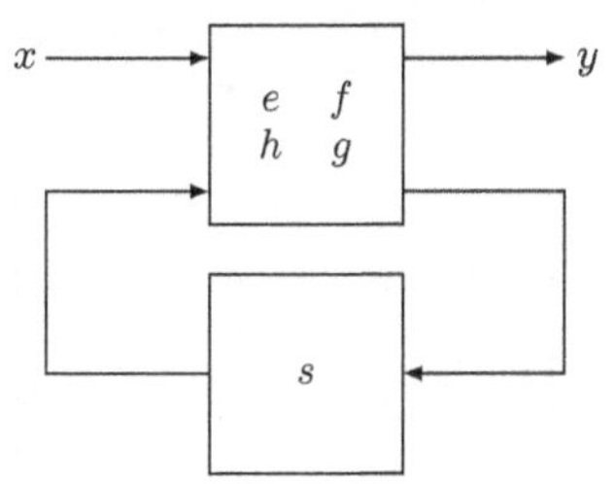

Figure 14.8 Lower LFT.

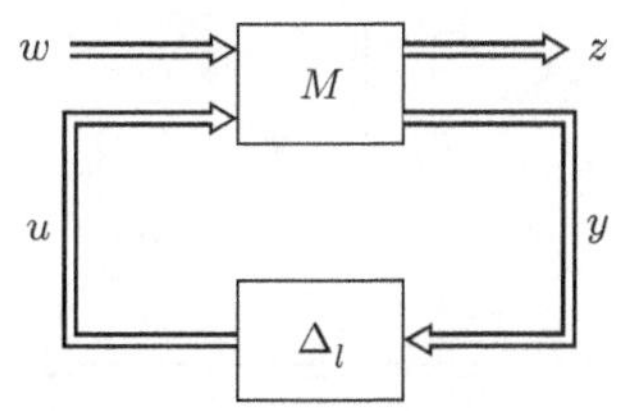

Figure 14.9 Lower MIMO LFT.

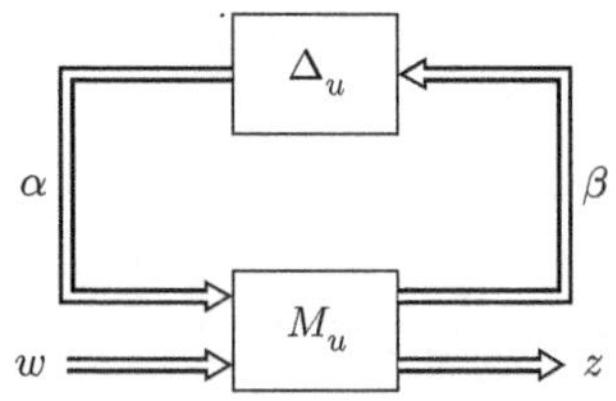

Figure 14.10 Upper system LFT.

Substituting this into Eq. (14.47) yields,

$$u = \Delta_l(I - M_{22}\Delta_l)^{-1}M_{21}w \tag{14.49}$$

Finally, substituting this expression for u into Eq. (14.45) yields

$$z = (M_{11} + M_{12}\Delta_l(I - M_{22}\Delta_l)^{-1}M_{21})w \tag{14.50}$$

As you might expect, if there is a lower LFT, then there is also an upper LFT. This takes the form shown in Figure 14.10.

$$\begin{bmatrix} \beta \\ z \end{bmatrix} = \begin{bmatrix} M_{u11} & M_{u12} \\ M_{u21} & M_{u22} \end{bmatrix} \begin{bmatrix} \alpha \\ w \end{bmatrix} \tag{14.51}$$

The closed-loop map from w to z is

$$F_u(M_u, \Delta_u) = M_{u22} + M_{u21}\Delta_u(I - M_{u11}\Delta_u)^{-1}M_{u12} \tag{14.52}$$

The upper LFT is used for modeling uncertainties, while the lower LFT is used for control design, as will be discussed in this and the following sections.

14.3.1 LFTs in Control Design

With the system in Figure 14.11,

- The exogenous inputs are $w = [w_1 \;\; w_2 \;\; w_3]^T = [r \;\; d \;\; n]^T$, where $d = [d_u \;\; d_y]$
- The regulated outputs are selected to be $z_1 = e = r - y'$, $z_2 = \bar{u}$,
- The measured output is $y' = y + n$
- Controlled input is u

This is equivalent to the block diagram in Figure 14.12. The relationships become

$$\begin{bmatrix} z \\ e \end{bmatrix} = \begin{bmatrix} r - n - y' \\ \bar{u} \\ r - n - y' \end{bmatrix} = \begin{bmatrix} r - n - d_y - G_p d_u - G_p u \\ u + d_u \\ r - n - d_y - G_p d_u - G_p u \end{bmatrix} \tag{14.53}$$

or

$$\begin{bmatrix} z_1 \\ z_2 \\ e \end{bmatrix} = \begin{bmatrix} -I & -I & I & -G_p & -G_p \\ 0 & 0 & 0 & I & I \\ -I & -I & I & -G_p & -G_p \end{bmatrix} \begin{bmatrix} d_y \\ n \\ r \\ d_u \\ u \end{bmatrix} \tag{14.54}$$

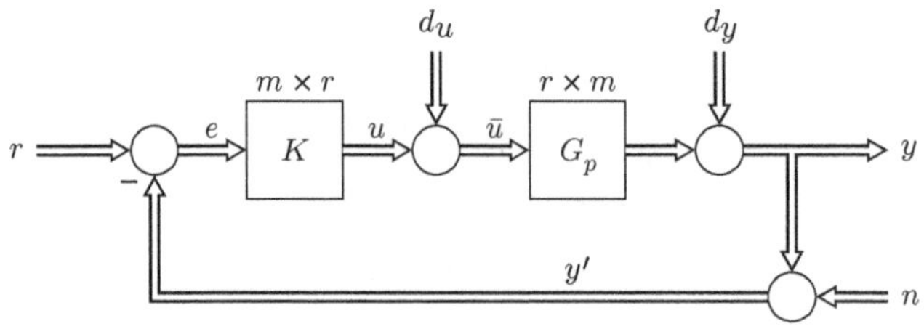

Figure 14.11 MIMO feedback system.

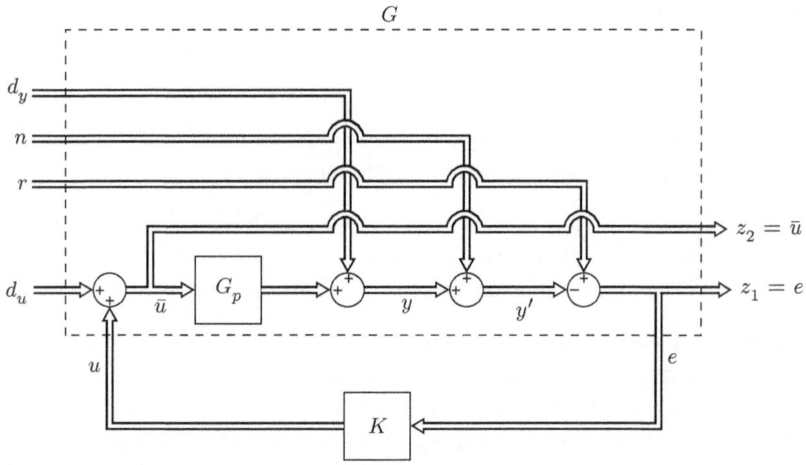

Figure 14.12 LFT block diagram.

However, as discussed in the SISO section, for meeting performance and robustness measures, we must augment this plant with the required filters. This is done so that the optimization techniques, which will be discussed later, could be utilized to compute a controller so that the closed-loop singular values meet the specified requirements. This is known as *MIMO Mixed Sensitivity Method.*

The simplified version of the MIMO feedback loop, where $w = n$, and the other exogenous inputs are ignored, is shown in Figure 14.13, where

- **Exogenous Input**: $w = n$
- **Regulated Outputs**: $z = \begin{bmatrix} z_1 \\ z_2 \\ z_3 \end{bmatrix} = \begin{bmatrix} -W_p(y+w) \\ W_Y u \\ W_\Delta y \end{bmatrix} = \begin{bmatrix} -W_p(G_p u + n) \\ W_Y u \\ W_\Delta G_p u \end{bmatrix}$
- **Measured Outputs**: $e = -y - w = -G_p u - n$
- **Controlled Input**: u

The input–output relationships for the generalized plant are

$$\begin{bmatrix} z \\ e \end{bmatrix} = G \begin{bmatrix} w \\ u \end{bmatrix} = \begin{bmatrix} -W_p & -W_p G_p \\ 0 & W_Y \\ 0 & W_\Delta G_p \\ -I & -G_p \end{bmatrix} \begin{bmatrix} n \\ u \end{bmatrix} \tag{14.55}$$

We will utilize the LFT framework for norm optimization and control design in the later sections.

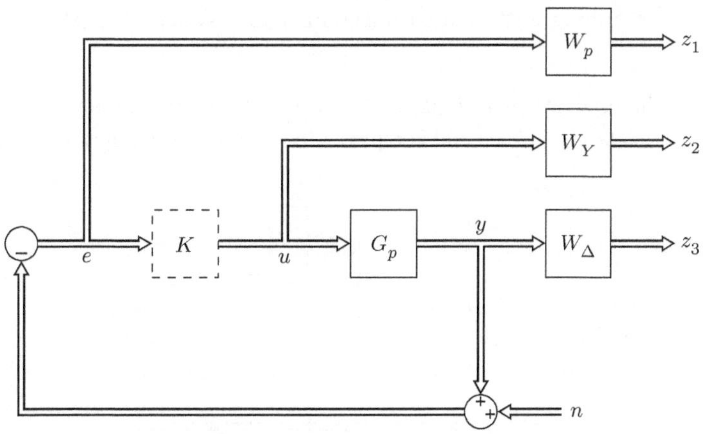

Figure 14.13 Augmented closed-loop feedback system.

14.4 Modeling Uncertainties

14.4.1 Uncertainties

Every dynamic model includes a certain amount of uncertainty. The reason for not being able to derive a precise dynamic model could be one of the followings or a combination of the followings:

(a) It could be time prohibitive or even not possible to derive a precise model for a highly complex dynamic system.
(b) Precise or high-fidelity models are computationally intensive.
(c) High-fidelity models are not necessarily useful for the control design applications.

These statements are true across all areas of research and design.

Control systems are designed using models which are not exact; they merely approximate the real system. It is then necessary to account for modeling errors in the design process. Modeling requires quantification of the model errors, even if this cannot be done precisely. A reasonable design allows a "margin" for model errors; a design with this feature is called robust.

One approach to deal with impreciseness of our dynamic models or modeling errors is to introduce uncertainties to these models. In general, uncertainties of dynamic systems in engineering fall into two categories:

1. Unstructured uncertainty (this could be due to ignoring high-frequency dynamics in our models (complex-valued uncertainty)):
   - Uses a specific single perturbation of the entire system.
   - The conditions for robust stability can be expressed as bounds on the TFM, i.e. $\bar{\sigma}(T)$, $\bar{\sigma}(S)$, $\bar{\sigma}(Y)$ or a combination of these principal gains.
   - Results in conservative closed-loop robustness, i.e. sacrificing closed-loop performance.

   Introducing this type of uncertainty and compensating for it guarantees robust stability and nominal performance, as discussed in the SISO section.
2. Structured uncertainty (this could be due to the combination of low-frequency parameter variations of our models (real-valued uncertainty) and high-frequency uncertainties similar to unstructured uncertainty):
   - The individual sources of uncertainty are identified.
   - Does not result in conservative closed-loop robustness.
   - The disadvantage of this procedure results in non-analytical solution, i.e. solved by numerical optimization.

Compensation for structured uncertainty guarantees both robust performance and stability (see [36]).

14.4.2 Descriptions of Unstructured Uncertainty

In this section, we will only cover unstructured uncertainty and deal indirectly with structured uncertainty to obtain robust performance for the MIMO feedback design to avoid using numerical optimization.

There are different ways of describing unstructured uncertainty in a system. We present three of the most popular approaches below:

- Additive uncertainty where the modeled plant $\tilde{G}_p(s)$ is the sum of the nominal plant $G_{p_0}(s)$ and function $L_A(s)$ where $\tilde{G}_p(s) = G_{p_0}(s) + L_A(s) \Rightarrow L_A(s) = \tilde{G}_p(s) - G_{p_0}(s)$ and is represented by the block diagram in Figure 14.14.

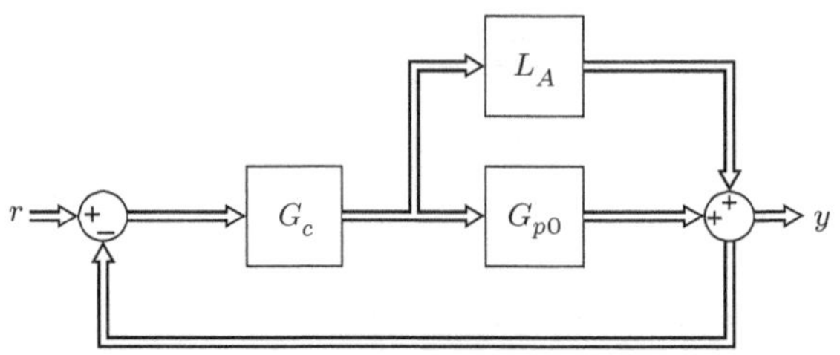

Figure 14.14 Additive uncertainty topology.

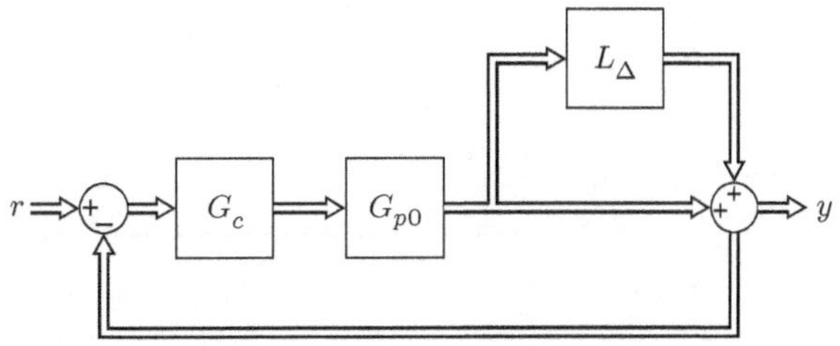

Figure 14.15 Multiplicative uncertainty topology.

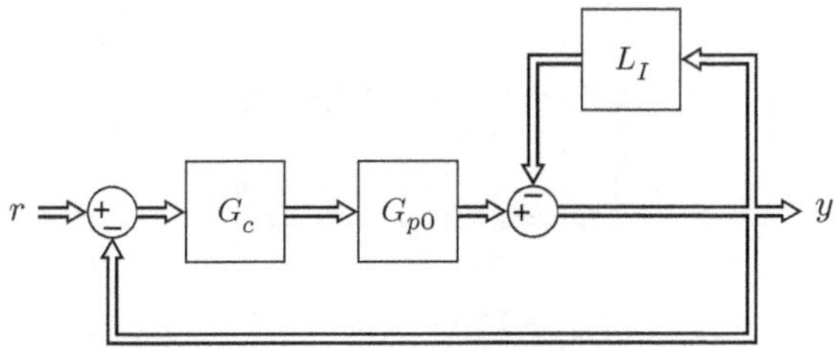

Figure 14.16 Inverse multiplicative uncertainty topology.

- Multiplicative output uncertainty where the modeled plant $\tilde{G}_p(s)$ is the product of the nominal plant $G_{p_0}(s)$ and the sum $I + L_\Delta(s)$ where $\tilde{G}_p(s) = (I + L_\Delta(s))G_{p_0}(s) \Rightarrow L_\Delta(s) = (\tilde{G}_p(s) - G_{p_0}(s))G_{p_0}^{-1}(s)$ and is represented by the block diagram in Figure 14.15.
- Inverse multiplicative output uncertainty where the modeled plant $\tilde{G}_p(s)$ is the ratio of the nominal plant $G_{p_0}(s)$ and sum $I + L_I(s)$ where $\tilde{G}_p(s) = (I + L_I(s))^{-1}G_{p_0}(s) \Rightarrow L_I(s) = (G_{p_0}(s) - \tilde{G}_p(s))\tilde{G}_p^{-1}(s)$ and is represented by the block diagram in Figure 14.16.

The conditions for robust stability depend on which perturbation is chosen to describe the uncertainty. Only the case of multiplicative uncertainty is discussed in the following section.

This type of modeling for uncertainty is abstract and counterintuitive, after all if there is uncertainty, then what is the meaning of modeling for uncertainty? If through all this development, we keep in mind that our perturbed plant, $\tilde{G}_p$, is the actual plant and the nominal plant or the model that we use for control design is G_{p0}, then perhaps the above analysis would make more sense.

14.5 General Robust Stability Theorem

Due to its practical importance, in this section, we investigate the impact of multiplicative uncertainties on the closed-loop robust stability. Consider the system shown in Figure 14.17. Now if $L_\Delta = W_\Delta \Delta$ with $\bar{\sigma}(\Delta) \leq 1$, where the bound W_Δ can be interpreted as a weight on a normalized perturbation Δ, the topology becomes as shown in Figure 14.18. Now we are interested in how the uncertainty Δ affects the closed-loop behavior of the system. First, find the closed-loop transfer function matrix from i to v':

$$v' = -G_{p_0}G_c(i + v') \Rightarrow (I + G_{p_0}G_c)v' = -G_{p_0}G_c i \tag{14.56}$$

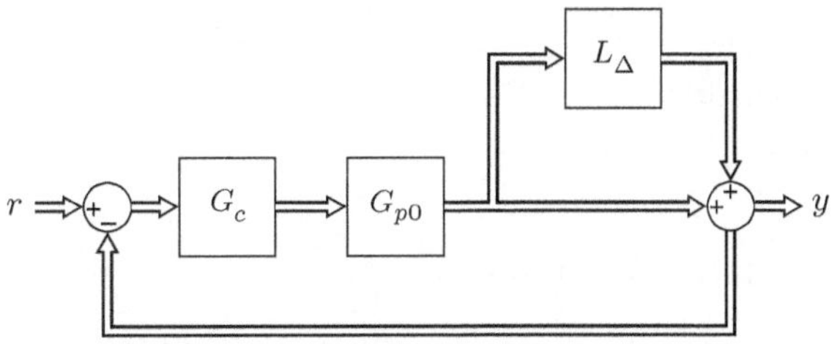

Figure 14.17 Multiplicative uncertainty topology.

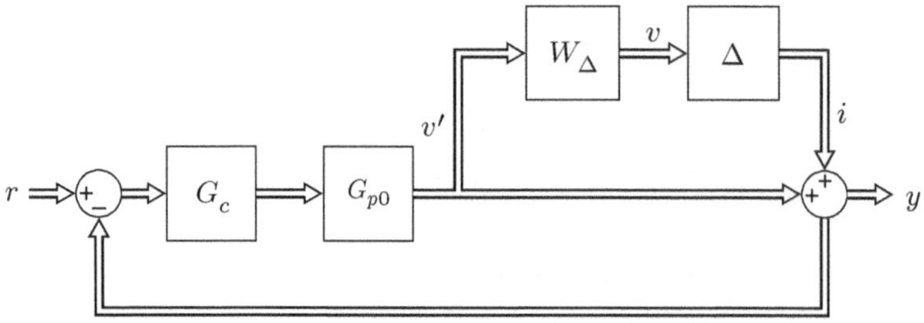

Figure 14.18 Multiplicative uncertainty topology with factored L_o.

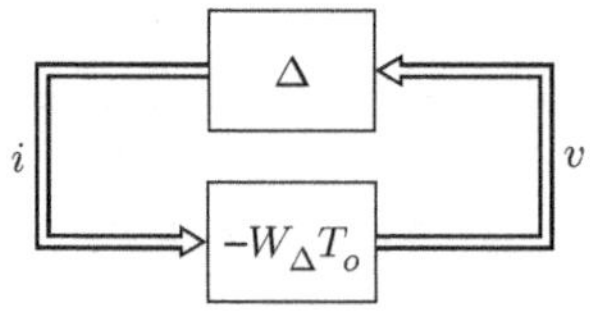

Figure 14.19 Upper LFT.

letting $L = G_{p_0} G_c$ shows that,

$$v' = -(I + L)^{-1} Li \tag{14.57}$$

$$= -T_o i \tag{14.58}$$

$$v = -W_\Delta T_o i \tag{14.59}$$

which conveniently fits into the framework, and where T_0 is the nominal closed-loop transfer function matrix shown in Figure 14.19. If we concern ourselves only with the gains of this transfer function matrix, then the negative sign can be ignored, without the loss of generality.

Theorem 14.3 (Small-Gain or Small-Signal Theorem) *Assume $W_\Delta T_o$ is stable, then the actual closed-loop system is robustly stable for all Δ with $\bar{\sigma}(\Delta) \leq 1$ if and only if $\bar{\sigma}(W_\Delta T_o \Delta) < 1 \Leftrightarrow \det(I - W_\Delta T_o \Delta) \neq 0$.*

For the proof of this theorem, see [13].

Other descriptions of uncertainties result in bounds on different closed-loop transfer matrices singular values.

14.5.1 SVD Properties Applied

To find some of the conditions for robust stability, we use properties of the SVD:

$$\bar{\sigma}(W_\Delta T_o \Delta) \leq \bar{\sigma}(W_\Delta)\bar{\sigma}(T_o) \underbrace{\bar{\sigma}(\Delta)}_{\leq 1} \leq \bar{\sigma}(W_\Delta)\bar{\sigma}(T_o) \tag{14.60}$$

So the condition for robust stability becomes

$$\bar{\sigma}(W_\Delta)\bar{\sigma}(T_o) < 1 \text{ or } \bar{\sigma}(T_o) < \frac{1}{\bar{\sigma}(W_\Delta)} \tag{14.61}$$

Equivalently for nominal performance, the condition is

$$\bar{\sigma}(W_p)\bar{\sigma}(S_o) < 1 \text{ or } \bar{\sigma}(S_o) < \frac{1}{\bar{\sigma}(W_p)} \tag{14.62}$$

where W_p is a performance weighting matrix.

Then, the control design constraints can be written as:

- **Nominal Performance**: $\bar{\sigma}(S_o) < \frac{1}{\bar{\sigma}(W_p)}$
- **Robust Stability (Multiplicative Uncertainty)**: $\bar{\sigma}(T_o) < \frac{1}{\bar{\sigma}(W_\Delta)}$

The inverse of shaping filters can be interpreted as upper bounds for $\bar{\sigma}(T_o)$ and $\bar{\sigma}(S_o)$. Interpret these with graphical representations results in Figure 14.20.

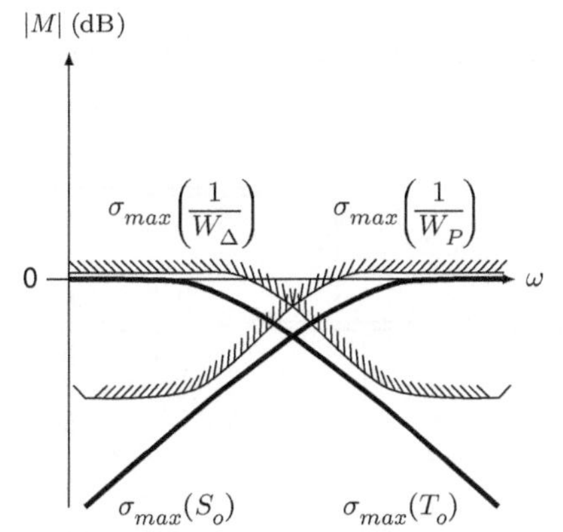

Figure 14.20 Frequency plot of MIMO mixed sensitivity loop-shaping technique.

14.5.2 Robust Performance

Robust performance means that the performance does not deteriorate for not only the nominal plant but also for a set of plants as defined by the uncertainty bound. The robust performance is dealt with by giving a structure to the unstructured uncertainty (such as the multiplicative uncertainty for robust stability) with an addition of a performance uncertainty, such that the uncertainty block is written as:

$$\Delta_S = \begin{bmatrix} \Delta & 0 \\ 0 & \Delta_P \end{bmatrix} \tag{14.63}$$

then the technique such as μ synthesis [3] could be utilized to guarantee robust performance and stability. But, as mentioned, techniques such as μ synthesis do not have analytical solutions. Hence, in this section, we will present a conservative solution to the mixed sensitivity technique to guarantee robust performance.

We start with the constraints on both S_o and T_o which can be written as $\bar{\sigma}(W_\Delta T_o) < 1$ and $\bar{\sigma}(W_p S_o) < 1$, and the condition for simultaneously achieving nominal performance and robust stability is

$$\max\{\bar{\sigma}(W_\Delta T_o),\ \bar{\sigma}(W_p S_o)\} < 1 \tag{14.64}$$

This condition is a modified version of the SISO constraint shown in Chapter 6 where it was shown that

$$\bar{\sigma}(W_\Delta T_o) + \bar{\sigma}(W_p S_o) < 1 \tag{14.65}$$

is a necessary and sufficient condition for robust performance. Since

$$\max\{\bar{\sigma}(W_\Delta T_o),\ \bar{\sigma}(W_p S_o)\} \le \bar{\sigma}(W_\Delta T_o) + \bar{\sigma}(W_p S_o) \le 2max\{\bar{\sigma}(W_\Delta T_o),\ \bar{\sigma}(W_p S_o)\} \tag{14.66}$$

and in addition,

$$\frac{2}{\sqrt{2}}\bar{\sigma}\begin{bmatrix} W_\Delta T_o \\ W_p S_o \end{bmatrix} \le 2\max\{\bar{\sigma}(W_\Delta T_o),\ \bar{\sigma}(W_p S_o)\} \le 2\bar{\sigma}\begin{bmatrix} W_\Delta T_o \\ W_p S_o \end{bmatrix} \tag{14.67}$$

the condition for simultaneously achieving *nominal performance and robust stability* can be written as:

$$\bar{\sigma}\begin{bmatrix} W_p S_o \\ W_\Delta T_o \end{bmatrix} < 1 \tag{14.68}$$

Additionally, the *robust performance* condition can be written as:

$$\bar{\sigma}\begin{bmatrix} W_p S_o \\ W_\Delta T_o \end{bmatrix} < \frac{1}{2} \tag{14.69}$$

It is usual to incorporate the weighting filters in the generalized plant so that the regulated variable z satisfies

$$z = \underbrace{\begin{bmatrix} W_p S_o \\ W_\Delta T_o \end{bmatrix}}_{T_{zw}} w \tag{14.70}$$

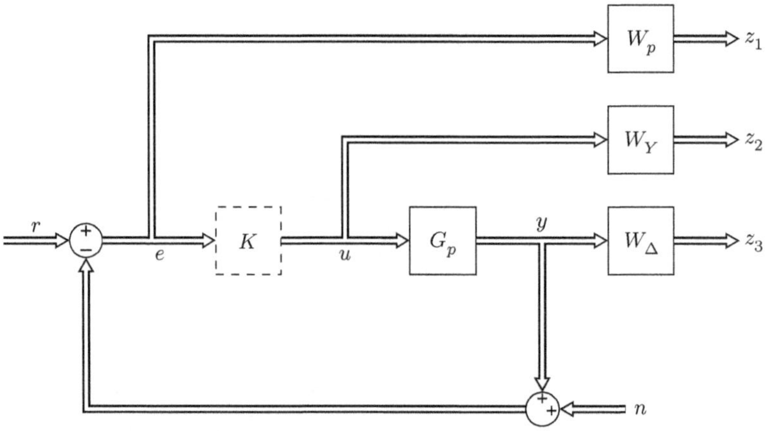

Figure 14.21 Augmented plant block diagram.

The augmented plant, G, can be obtained by including the weighting filters to the general feedback loop shown in Figure 14.21 and finding the closed-loop TFM:

$$z = \underbrace{\begin{bmatrix} W_p S_o \\ W_Y Y_o \\ W_\Delta T_o \end{bmatrix}}_{T_{zw}} w \tag{14.71}$$

where $w = \begin{bmatrix} r \\ n \end{bmatrix}$ and $z = \begin{bmatrix} z_1 \\ z_2 \\ z_3 \end{bmatrix}$. The optimization objective can be written as:

$$\bar{\sigma} \begin{bmatrix} W_p S_o \\ W_Y Y_o \\ W_\Delta T_o \end{bmatrix} < 1 \tag{14.72}$$

14.6 Problems

14.1 A feedback control system can be rearranged as an LFT, as shown in Figure 14.22. such that

$$\begin{bmatrix} z \\ y \end{bmatrix} = \underbrace{\begin{bmatrix} G_{11} & G_{12} \\ G_{21} & G_{22} \end{bmatrix}}_{G(s)} \begin{bmatrix} w \\ u \end{bmatrix}, \quad u = Ky \tag{14.73}$$

where w is all exogenous inputs, u is the controlled inputs, z is regulated outputs, and y is the measured outputs. Then,

$$T_{zw} = F_l(G, K) = G_{11} + G_{12}K(I - G_{22}K)^{-1}G_{21} \tag{14.74}$$

where $F_l(G, K)$ is the closed-loop transfer function matrix between the regulated signals and exogenous inputs. In H_∞ control problems, the objective is to minimize $\| F_l(G, K)\|_\infty$.

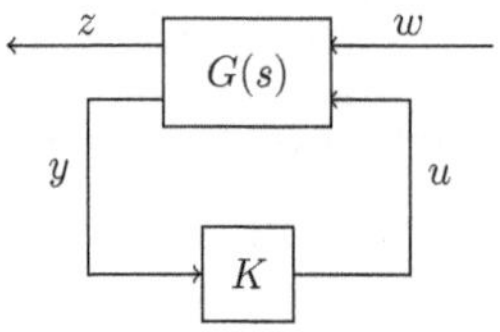

Figure 14.22 Linear Fractional Transformation (LFT).

(a) Knowing what you have learned, show by adding the weighting filters to the classical control loop as shown in Figure 14.23, the augmented closed-loop transfer function will be equal to

$$\begin{bmatrix} z_1 \\ z_2 \\ z_3 \end{bmatrix} = \begin{bmatrix} W_1 S \\ W_2 Y \\ W_3 T \end{bmatrix} w \tag{14.75}$$

(b) Could you transform the classical control loop to an equivalent LFT as shown in Figure 14.23

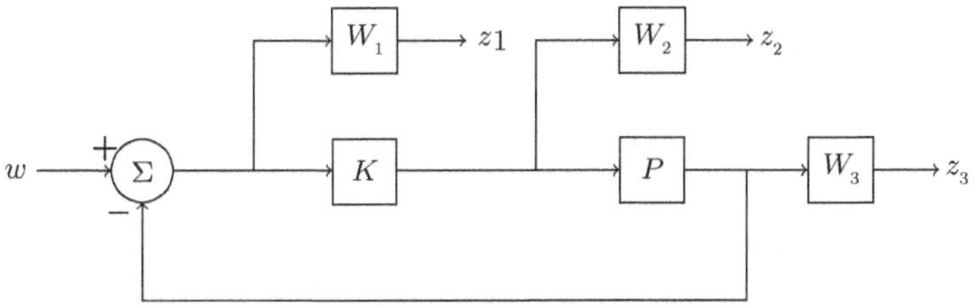

Figure 14.23 Augmented classical control feedback loop.

14.2 Consider a system with G_p given as:

$$G_p(s) = \begin{bmatrix} 0 & \frac{-1}{s+1} \\ 1 & \frac{1}{s(s+1)} \end{bmatrix} \tag{14.76}$$

Use the LFT command in Matlab (`Tzw=lft(P,K)`) to compare the $\| T_{zw} \|_\infty$ found in Problem 1 part (a) with the result from the LFT command. Hint: Don't forget to scale T_{zw} with γ such that $T_{zw,new} = \gamma \cdot T_{zw,ord}$.

14.3 Compute the H_2 norm of

$$G_p(s) = \begin{bmatrix} \frac{1}{(s+2)^2} & \frac{s+1}{s+2} \\ \frac{-(s+1)}{s+2} & \frac{s+1}{s+2} \end{bmatrix} \tag{14.77}$$

Comment on your result. Is this a correct H_2 norm of this plant?

14.4 The transfer function for the (SISO) control system shown in Figure 14.24 below is

$$\frac{y(s)}{u(s)} = \frac{bs + k}{ms^2(ms^2 + 2bs + 2k)} \tag{14.78}$$

A controller will be designed on the basis of the rigid body mode only, and the vibratory mode will be considered as the residual mode (uncertain mode).

(a) You are asked to write the transfer function in the form of a multiplicative uncertainty as:

$$\frac{y(s)}{u(s)} = G(s)(1 + \Delta(s)), \text{ where } G(s) = \frac{1}{2ms^2} \Leftarrow \text{Rigid Body Mode} \tag{14.79}$$

(b) Provide a Bode plot of $\Delta(s)$ if $m = 1, k = 1$, and $b = 0.1$.

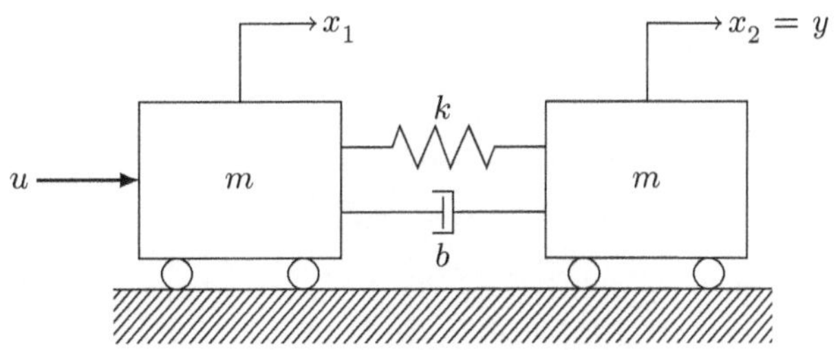

Figure 14.24 A two-mass system.

15

Optimal Control in MIMO Systems

In this chapter, we utilize the norms discussed in the previous chapter to derive MIMO controllers. As the discussion of $\mathscr{H}_2$ and $\mathscr{H}_\infty$ MIMO control design are topics of robust optimal control domain, it would be beneficial to start with having a brief discussion on LQG (Linear Quadratic Gaussian) control, which is a well-established MIMO technique. Our intent is to discuss what the main drawbacks of LQG method were and how the techniques such as $\mathscr{H}_\infty$ control method address these drawbacks. It should be noted that all the optimal controllers are derived based on the linearity and infinite horizon assumptions, as $t \to \infty$.

Again, these topics have been discussed in many textbooks and publications and we introduce them for the sake of completeness and so that we would be able to provide a comprehensiveness comparative results between these control concepts such as MIMO $\mathscr{H}_\infty$ with previously introduced MIMO Youla control method.

15.1 Output Feedback Control

The first linear optimal controller, LQR (Linear Quadratic Regulator), is derived given the following cost function:

$$J_{LQR} = \int_0^\infty x^T Qx + u^T Ru\,\mathrm{d}t \tag{15.1}$$

where u is the control input and x are the system states. It has been shown [29] that by solving this cost function, utilizing algebraic Riccati equation, results in a control input to be a linear function of the states, $u = -K_c x$. In addition, it has shown that this optimal control method has very good robustness characteristics, such as 60° phase margin and from 6 dB to infinite gain margin. However, this method requires having access to the entire dynamic system states. It should be obvious that this last requirement is expensive or not very practical at times.

The general dynamics system in a differential equation form is written as:

$$\begin{bmatrix} \dot{x} \\ z \\ y \end{bmatrix} = \begin{bmatrix} A & B_1 & B_2 \\ C_1 & 0 & D_{12} \\ I & D_{21} & 0 \end{bmatrix} \begin{bmatrix} x \\ w \\ u \end{bmatrix} \tag{15.2}$$

where I is the identity matrix and D_{11} and D_{22} are assumed to be equal to zero. As discussed before, the H_2 norm of the closed loop will not be finite if D_{11} is not precisely equal to zero. The second assumption, $D_{22} = 0$, is done without loss of generality and make the derivation of the equations easier. However, we will remove this second assumption, when discussing minimization of H_2 norm of the closed-loop system.

Now assume that all of the states are not available. Then, the dynamic system above is rewritten as:

$$\begin{bmatrix} \dot{x} \\ z \\ y \end{bmatrix} = \begin{bmatrix} A & B_1 & B_2 \\ C_1 & 0 & D_{12} \\ C_2 & D_{21} & 0 \end{bmatrix} \begin{bmatrix} x \\ w \\ u \end{bmatrix} \tag{15.3}$$

or

$$G = \begin{bmatrix} A & B_1 & B_2 \\ C_1 & 0 & D_{12} \\ C_2 & D_{21} & 0 \end{bmatrix} \tag{15.4}$$

where the identity matrix I is replaced by matrix C_2 or the read out/output matrix.

Robust Control: Youla Parameterization Approach, First Edition. Farhad Assadian and Kevin R. Mallon.

Companion Website: www.wiley.com/go/assadian/robust

In order to get around this constraint, the researchers in 1960s derived two approaches for solving this problem. The first approach provides a general solution to the static output feedback problem using a nonlinear optimization method. This approach results in a controller with static gains, similar to the LQR problem [34].

In the second approach, a new optimal control method based on estimation of the states with use of Kalman filtering technique is derived (assuming the dynamic system is observable/stabilizable) and then the estimated states are used in place of the measured states for devising an optimal feedback controller similar to the LQR method. This technique is known as the LQG (as the noise signals introduced in the system assumed to be normal/Gaussian distribution). It is interesting that although the differential equations written to solve LQG are stochastic (see next section), the problem can still be solved deterministically [34]. It is shown that the optimal solution for the computation of control gains is given by the set of equations:

$$u = -K_c\hat{x} \tag{15.5}$$

$$K_c = R^{-1}B_2^T P_c \tag{15.6}$$

$$0 = P_c A + A^T P_c - P_c B_2 R^{-1} B_2^T P_c + Q \tag{15.7}$$

With the only difference here being that $\hat{x}$ is the set of states estimated from the measured outputs y (this assumes the separation between state estimation and state feedback control design). The matrix P is a no-negative definite matrix and a solution to the algebraic Riccati equation above. The general LQG is discussed next.

15.1.1 LQG Control

For a general stochastic systems with process and sensor noise, which are assumed to be Gaussian distributed, the dynamic system differential equation, using the system differential Eq. (15.2), is written as:

$$\dot{x} = Ax + B_2 u + w_d \tag{15.8}$$

$$y = C_2 x + w_n \tag{15.9}$$

$$z = C_1 x + D_{12} u \tag{15.10}$$

where w_d and w_n are stochastic process and sensor noises modeled by assuming zero-mean Gaussian white noise, where $w_d = B_1 w$ and $w_n = D_{21} w$.

The LQG cost is similar to the LQR cost and is given by:

$$J_{LQG} = E\{z^T z\} \tag{15.11}$$

where this time, due to the stochastic nature of the problem, as expectation operator has been used and $z^T z = x^T C_1^T C_1 x + u^T D_{12}^T D_{12} u$ with cross terms equal to zero without loss of generality. For the infinite horizon case and assuming a stabilizing feedback controller, this cost function can be written as $E\{z^T z\} = \lim_{T\to\infty} E\left[\frac{1}{T}\int_0^T z^T(t)z(t)dt\right]$, and, furthermore,

$$E\left\{\begin{bmatrix} w_d(t) \\ w_n(t) \end{bmatrix}\begin{bmatrix} w_d^T(t) & w_n^T(t) \end{bmatrix}\right\} = \begin{bmatrix} W & 0 \\ 0 & V \end{bmatrix}\delta(t-\tau) \tag{15.12}$$

where W and V are variances (strength of the impulse) and $\delta(t-\tau) = 0$ when $t \neq \tau$ and, obviously $\delta(t-\tau) = \delta(0)$ when $t = \tau$, hence, these noise signals are assumed to be uncorrelated. In addition, it could be shown that these variances are equal to the power spectral densities (PSDs) of these signals, e.g. $S_{w_d}(\omega) = \sigma_{w_d}^2 = W$.

For a signal $w(s)$ (zero-mean white noise) with the I/O relationship shown in Figure 15.1.

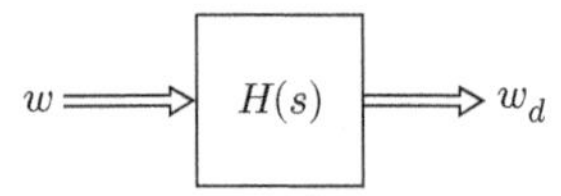

Figure 15.1 PSD I/O relationship.

The PSD of the output signal w_d can be written as:

$$S_{w_d}(j\omega) = |H(j\omega)|^2 S_w(j\omega) \tag{15.13}$$

Therefore, with $S_w = I$, this implies that $S_{w_d} = W$, if $|H(j\omega)| = W^{1/2} = B_1$ (same applies to w_n, where $|H(j\omega)| = V^{1/2} = D_{21}$). In addition, if one selects $C_1 = Q^{1/2}$ where Q is a weighting matrix, and $D_{12} = R^{1/2}$ where R is another weighting matrix, then the following relationships hold for an LQG type control system:

$$\begin{bmatrix} w_d \\ w_n \end{bmatrix} = \begin{bmatrix} W^{1/2} & 0 \\ 0 & V^{1/2} \end{bmatrix} w \tag{15.14}$$

$$z = \begin{bmatrix} z_1 \\ z_2 \end{bmatrix} = \begin{bmatrix} Q^{1/2} & 0 \\ 0 & R^{1/2} \end{bmatrix} \begin{bmatrix} x \\ u \end{bmatrix} \tag{15.15}$$

Then the LFT framework is given by the system diagram shown in Figure 15.2.

This gives the relationship:

$$\begin{bmatrix} \dot{x} \\ z_1 \\ z_2 \\ y \end{bmatrix} = \begin{bmatrix} A & W^{\frac{1}{2}} & B \\ Q^{\frac{1}{2}} & 0 & 0 \\ 0 & 0 & R^{\frac{1}{2}} \\ C & V^{\frac{1}{2}} & 0 \end{bmatrix} \begin{bmatrix} x \\ w \\ u \end{bmatrix} \tag{15.16}$$

Now the square of the 2-norm of the closed-loop transfer function, $T_{zw} = F_l(G, K)$, where the input, w is white noise with the PSD of $S_w = I$, can be written as,

$$J_{LQG} = E\{z^T z\} = \|T_{zw}\|_2^2 \tag{15.17}$$

Now the LQG cost is written as:

$$J_{LQG} = \lim_{T\to\infty} E\left[\frac{1}{T}\int_0^T z^T(t)z(t)\mathrm{d}t\right] \tag{15.18}$$

$$= \lim_{T\to\infty} E\left[\frac{1}{T}\int_0^T x^T Qx + u^T Ru\,\mathrm{d}t\right] \tag{15.19}$$

The linear controller K_{LQG} includes two separable parts. The first part is the feedback control portion similar to LQR. It can be shown [34] that the optimal cost V°, in this case, can be written as,

$$V^0 = x^T P_c x + C \tag{15.20}$$

where P_c is a solution to the algebraic Riccati equation and $x = \hat{x} + \tilde{x}$, where, $\hat{x}$ are the estimated states and $\tilde{x}$ are the error states. Note the presence of the additional constant term C which is independent of x; this is the additional cost caused by the white noise disturbance. But, because it is independent of x, the optimal controller is the same as in the deterministic case, LQR. The second part of the controller K_{LQG} is the Kalman filter, which is discussed next.

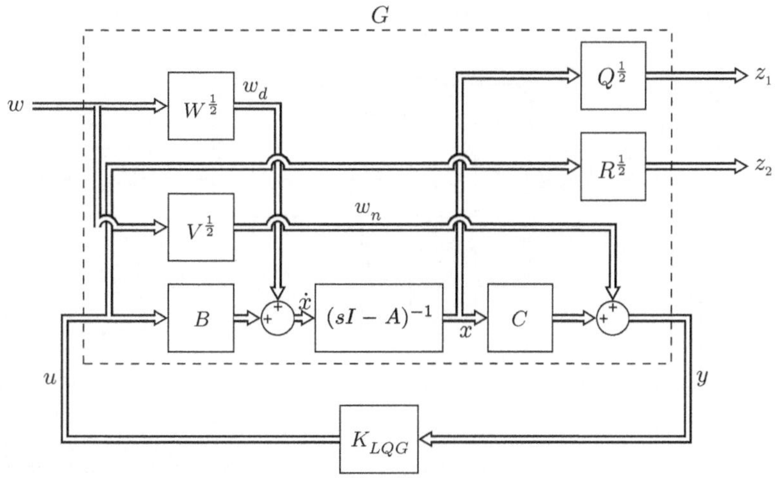

Figure 15.2 LQG using LFT framework.

15.1.2 Kalman Filter

To construct the estimator, we use the following system differential equation:

$$\dot{\hat{x}} = A\hat{x} + Bu + K_f(y - \hat{y}) \tag{15.21}$$

where, as stated earlier, $x = \hat{x} + \tilde{x}$, and this system differential equation corresponds to the topology shown in Figure 15.3. The filter feedback gains K_f are computed to minimize the variance of the estimation error, $\tilde{x} = x - \hat{x}$. Now let, P_f denote the variance $P_f = E\{\tilde{x}^T(t)\tilde{x}(t)\}$. The Ricatti equation from the LQR design can be used to find P_f with the following substitutions $A \rightarrow A^T, B \rightarrow C^T, Q \rightarrow W$, and $R \rightarrow V$. This results in the following equations:

$$K_f = P_f C^T V^{-1} \tag{15.22}$$

$$0 = P_f A^T + AP_f - P_f C^T V^{-1} C P_f + W \tag{15.23}$$

Adding this to the full-state feedback controller above results in the feedback diagram in Figure 15.4.

The control K_{LQG} can be redrawn as shown in Figure 15.5.

LQR is an optimal controller while Kalman filter is an optimal estimator. Both LQR and Kalman filter have very good robustness characteristics. It is then natural for one to assume that by combining these two optimal control and estimator, for generating the LQG controller K_{LQG}, the outcome is robust as well. But, interestingly, it has been shown [14] that this is not strictly true. The work for addressing the robustness of LQG controller has been referred to as the loop transfer recovery (LTR). In the LTR, one tries to recover the robustness characteristics of the LQR controller when designing the LQG control.

One approach is to modify the Kalman filter gains, K_f so that the return ratio for the LQG control approaches the return ratio of the LQR control [14]. This approach works well for stable minimum phase plants (no right half-plane zeros). Newer procedures have been derived for the case when $G(s)$ is not minimum phase. It is important to keep in mind that by changing

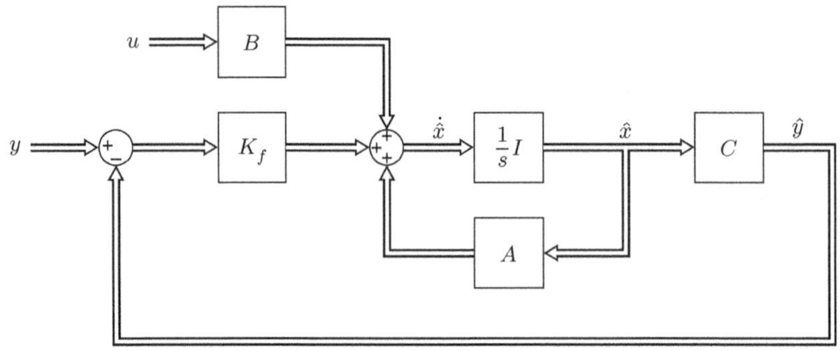

Figure 15.3 Kalman filter topology.

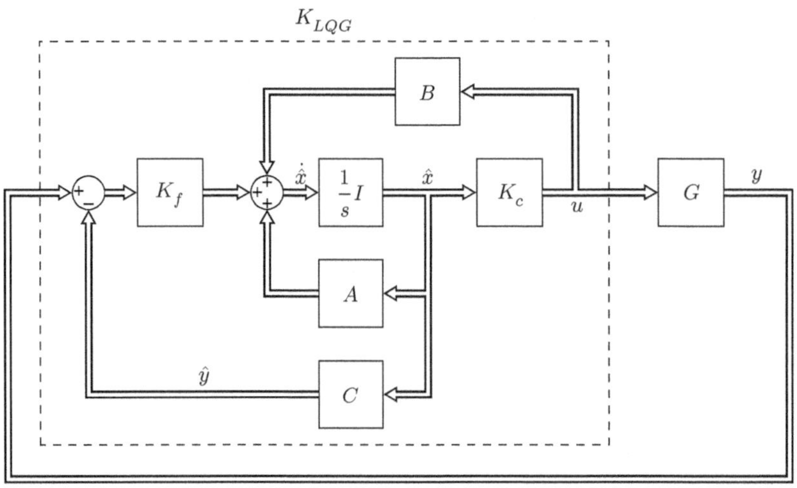

Figure 15.4 Full LQG feedback diagram.

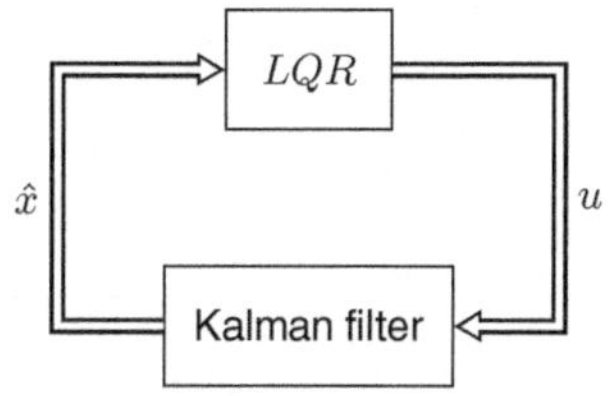

Figure 15.5 K_{LQG}.

the Kalman filter gains, we will trade off robustness with optimality or performance. This is unfortunate but it is a price that we have to pay for this recovery.

It should be pointed out that $\mathscr{H}_2$ control method could be looked at as the deterministic version of the *LQG* control as will be discussed next. Furthermore, LTR using $\mathscr{H}_2$ control is much more straightforward when comparing to *LQG* control approach.

15.1.3 $\mathscr{H}_2$ Control

We now have all the necessary ingredients for exploiting LQG optimal control theory to solve multivariable control design problems where the objectives are robust stability and performance.

The basic design objectives have been formulated as constraints on $||S||_2$, $||Y||_2$ and $||T||_2$. As discussed earlier, we may transform this, at least approximately, into a constraint on $||T_{zw}||$, the norm of the transfer function from w to z, where w are the exogenous (disturbance/reference) signals and z the regulated (cost/performance) outputs; this transfer function is

$$T_{zw} = \begin{bmatrix} W_p S_0 \\ W_Y Y_0 \\ W_\Delta T_0 \end{bmatrix} \tag{15.24}$$

The weights are chosen to reflect the constraints. For example, in the scalar case, W_p is chosen so that $|W_p^{-1}(j\omega)|$ is the upper bound for $|S_0(j\omega)|$ so that the constraint on S_0 is equivalent to the constraint:

$$|W_p(j\omega)S_0(j\omega)| < 1 \tag{15.25}$$

for all (relevant) ω. This means that W_p is chosen so that $W_p(j\omega)$ is large at low frequencies and small at high frequencies. The weighting functions W_Y and W_Δ are similarly chosen to reflect the constraints on Y and T.

We next employ LQG optimal control theory to minimize the H_2 norm of T_{zw}. As discussed in the LQG control part of this section, this problem is equivalent to a suitably formulated LQG problem; we give the detailed formulation below. Now, LQG control does not enforce strict satisfaction of the design constraints at each ω, but it does tend to make $||T_{zw}||$ small and constant in the pass band of the optimal system, and this, in turn, enforces approximate satisfaction of the constraints.

The weighting functions are dynamic systems, so that the system G_p must be suitably augmented. How this is done is shown in Figure 15.6. The augmented plant G_A whose regulated outputs z looks at $e = Sw$, $u = KSw = Yw$, and $y_p = Tw$. The output $y = e$ is fed back to the controller K; this minor change does not effect our conclusions.

The LQG/H_2 problem corresponding to Figure 15.6 is,

$$\min \lim_{T\to\infty} \int_0^T z^T z dt \tag{15.26}$$

subject to

$$\dot{x} = Ax + B_1 w + B_2 u \tag{15.27}$$

$$z = C_1 x + D_{11} w + D_{12} u \tag{15.28}$$

$$y = C_2 x + D_{21} w + D_{22} u \tag{15.29}$$

where x is the state of the augmented plant, $B_1 w$ is the process noise, $D_{11} = 0$ (as H_2 norm from w to z will be infinite when this term is included), and $D_{21} w$ is the sensor noise.

It should be noted that the cost is no longer stochastic, and it is deterministic as the disturbance inputs are considered impulsive now.

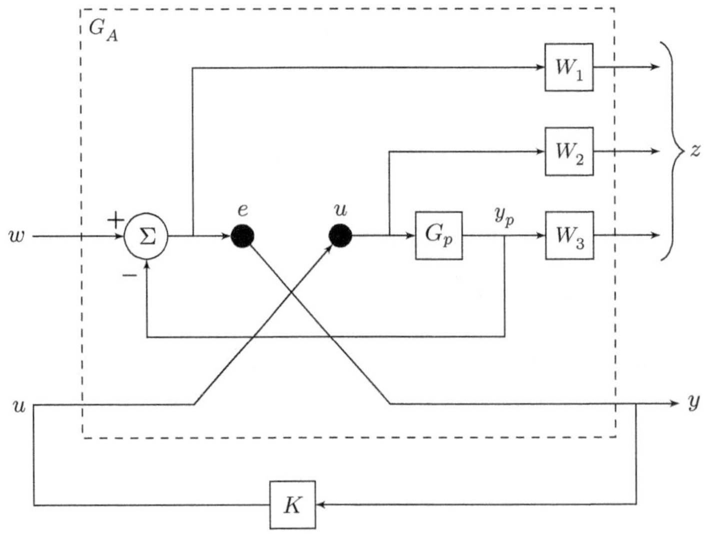

Figure 15.6 Augmented plant.

Making use of the definition of z, the LQG/H_2 problem may be written in the form, and assuming $w(t)$ is impulsive. Then,

$$J_{LQR} = \|T_{zw}\|_2^2 = \lim_{T\to\infty} \int_0^T z^T(t)z(t)dt \tag{15.30}$$

where $T_{zw} = F_l(G, K)$, the closed-loop map from w to z. Using the equations above,

$$z^T z = x^T C_1^T C_1 x + u^T D_{12}^T D_{12} u + 2x^T C_1^T D_{12} u \tag{15.31}$$

Without the loss of generality, assume that $C_1^T D_{12} = 0$, and this makes the cross-product of u and x to be zero: where

$$z = \begin{bmatrix} C_1 x \\ D_{12} u \end{bmatrix} \tag{15.32}$$

$$z^T = \begin{bmatrix} x^T C_1^T & u^T D_{12}^T \end{bmatrix} \tag{15.33}$$

then,

$$Q = C_1^T C_1 \tag{15.34}$$

$$R = D_{12}^T D_{12} \tag{15.35}$$

The variance matrix $\sum_{w_d w_n}$ of the process noise, $w_d = B_1 w$ and the sensor noise, $w_n = D_{21} w$, assuming $E(w(t)w^T(\tau)) = I\delta(t - \tau)$, is

$$\sum_{w_d w_n} = \sum = \begin{bmatrix} B_1 B_1^T & B_1 D_{21}^T \\ D_{21} B_1^T & D_{21} D_{21}^T \end{bmatrix} \tag{15.36}$$

where the process noise variance, W, as defined before is equal to $B_1 B_1^T$ and the sensor noise variance, V is equal to $D_{21} D_{21}^T$. If we assume the cross-correlation terms are zero, then $B_1 D_{21}^T = D_{21} B_1^T = 0$.

15.1.3.1 Kalman Filter Dynamic Model

The dynamic model for the estimation is given as:

$$\dot{\hat{x}} = A\hat{x} + B_2 u + K_f(y - C_2\hat{x} - D_{22}u) \tag{15.37}$$

The solution to the Kalman filter gains for the general case of nonzero cross-correlation noise terms are written as:

$$K_f = \left(P_f C_2^T + \sum_{12} \right) \sum_{22}^{-1} = \left(P_f C_2^T + B_1 D_{21}^T\right) \left(D_{21} D_{21}^T\right)^{-1} \tag{15.38}$$

where P_f satisfies algebraic Ricatti equation (ARE),

$$P_f A^T + AP_f + \sum_{11} - \left(P_f C_2^T + \sum_{12} \right) \sum_{22}^{-1} \left(C_2 P_f + \sum_{21} \right) = 0 \tag{15.39}$$

If we assume zero for the cross-correlation terms and knowing $\sum_{22} = (D_{21} D_{21}^T) = V$ then,

$$K_f = P_f C_2^T V^{-1} \tag{15.40}$$

This is similar to Equation (15.22), which is derived previously.

15.1.3.2 State Feedback

Then the H_2 control cost is exactly equal to LQR control cost. Therefore, the optimal control is given by $u = -K_c \hat{x}$ where, for the general case with the cross-product term $N = C_1^T D_{12}$, we can write

$$K_c = R^{-1}(B_2^T P_c + N^T) = (D_{12}^T D_{12})^{-1}(B_2^T P_c + D_{12}^T C_1) \tag{15.41}$$

where P_c is computed from the following ARE:

$$P_c A + A^T P_c - (P_c B_2 + N)(D_{12}^T D_{12})^{-1}(B_2^T P_c + N^T) + C_1^T C_1 = 0 \tag{15.42}$$

The complete controller is given as:

$$K \sim \left[\begin{array}{cc} A - K_f C_2 - B_2 K_c + K_f D_{22} K_c & K_f \\ K_c & 0 \end{array} \right] \tag{15.43}$$

which maps $\begin{bmatrix} \hat{x} \\ y \end{bmatrix} \rightarrow \begin{bmatrix} \dot{\hat{x}} \\ u \end{bmatrix}$. It can be shown that by letting $D_{21} \rightarrow 0$, which is letting the sensor noise decrease in comparison with the process noise (or the process noise increase in comparison with the sensor noise) full state feedback LTR is obtained.

For this design procedure to work, the following conditions must be satisfied:

- (A, B_2) is stabilizable.
- (A, C_2) is detectable.
- D_{11} must be zero.
- D_{12} and D_{21} must have full rank.

15.2 $\mathcal{H}_\infty$ Control Design

In this section, we generalize the linear fractional transformation concept which is discussed in the SISO part of this book to MIMO optimization and control design. We only discuss the infinite horizon case (as $T \rightarrow \infty$) as the controller gains in this case are constants and hence, time-independent.

We start the section by assuming access to the full states of a dynamic system (this is also known as the full information case), then we introduce H_∞ filter for state estimation where access to full states is not possible (this is also known as the partial information case). Finally, we combine H_∞ control and filter design (similar to the LQG problem) and present the solution for this combined case.

Similar to LQR and LQG optimal control topics, these material are fully covered in other texts and technical papers. We present these topics again for completeness and comparative analysis with the proposed Youla MIMO control design method.

Understanding that the closed-loop system of using lower LFT, as shown in Figure 15.7 is equivalent to Figure 15.8, where $z(s) = T_{zw}(s)w(s)$.

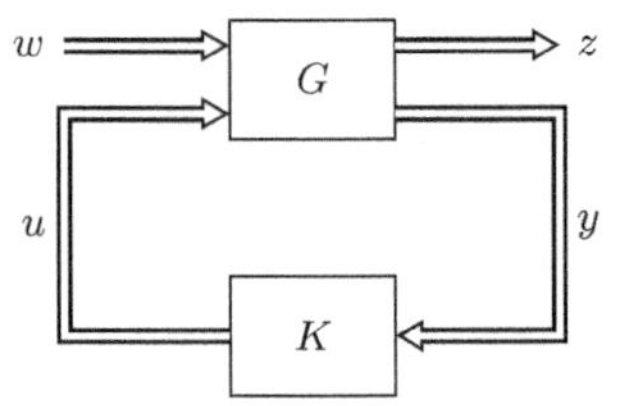

Figure 15.7 Lower LFT.

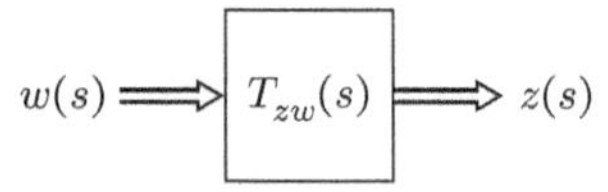

Figure 15.8 Standard I/O relationship.

For H_∞ optimal robust control design, the objective can be written as:

$$\min_K \|T_{zw}\|_\infty \tag{15.44}$$

where the H_∞ norm of the closed-loop transfer function, T_{zw}, is the objective function. The minimization of this objective function turns out to be nontrivial and hence, a difficult problem. One approach, for solving this problem, is to find a lower bound for this norm optimization, and therefore, a suboptimal solution as discussed next.

Recall that

$$\|T_{zw}\|_\infty = \max_w \frac{\|z\|_2}{\|w\|_2} \tag{15.45}$$

In words, this is the largest possible gain over all possible disturbances or the gain for the largest disturbance input. If we assume that $\|T_{zw}\|_\infty \leq \gamma$ for $\gamma \in \mathbb{R}_+$, this implies that for any arbitrary disturbances:

$$\frac{\|z\|_2}{\|w\|_2} \leq \gamma \tag{15.46}$$

or

$$\|z\|_2^2 - \gamma^2 \|w\|_2^2 \leq 0 \tag{15.47}$$

This bound reduces (instead of minimizing) the impact of disturbance energy on the output energy. (Note that these are actually signal energy not a true disturbance or output energy.)

Now we can write the previous objective function as:

$$J(x_0, w) = \int_0^\infty (z^T z - \gamma^2 w^T w)\mathrm{d}t \tag{15.48}$$

where x_0 is the initial condition. If we ignore the disturbance input in the above objective function, then this cost or objective function would be precisely equal to the H_2 or LQR cost function. The bound γ acts a weighting to the disturbance input, at very large γ, the effect of disturbance input becomes insignificant and the solution to this optimal control problem approaches LQR (H_2) solution. It is important to note that disturbance input appears with a negative sign in this cost/objective function. Therefore, at small γ where the disturbance input becomes significant, it tries to maximize this cost function. Hence, the H_∞ problem is a zero-sum game. We can write the optimal control problem as:

$$\min_u \max_w J(x, u, w) \tag{15.49}$$

In the simplest setting, a zero-sum game with two players u and w involves a cost $J(u, w)$; player u seeks to minimize J and player w to maximize J (or minimize $-J$). One player's gain is the other player's loss, the sum of the two objectives (J and $-J$) being zero; hence its name.

Every optimization problem can be cast either in the primal minimization form or the dual maximization, the solution of which provides a lower bound for the solution of the primal problem. If the two solutions are equal, then this is called strong duality, where a saddle point exists. If a duality gap exists, then this is known as weak duality. Graphically, this is represented by Figure 15.9. For this application in optimal control problems, the following is always true:

$$J(u^*, w) \leq J(u^*, w^*) \leq J(u, w^*) \tag{15.50}$$

for all $u \in U$, for all $w \in W$, the game admits a saddle point. It should be noted that not every game has a saddle point. However,

Theorem 15.1 *if U and W are convex, J is continuous, convex in u for all $w \in W$, and concave in w for all $u \in U$, then the game admits a saddle point. The saddle point is unique if J is strictly convex-concave.*

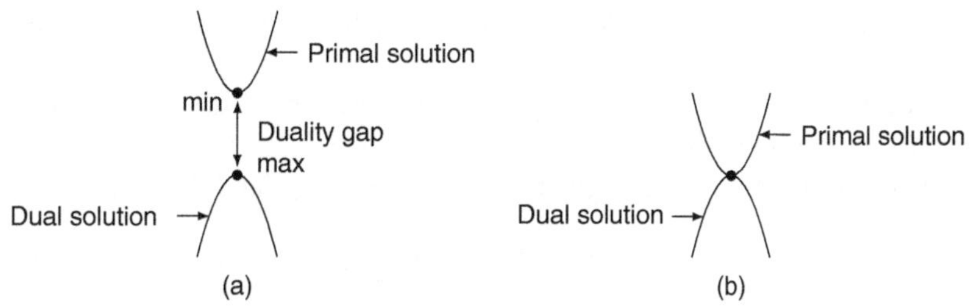

Figure 15.9 Duality in optimization. (a) Weak duality. (b) Strong duality.

For the proof, see [37].

In other words, we have transformed the original minimization problem of the H_∞ norm of the closed-loop transfer function matrix to a problem which can be analytically solved, by assuming a bound on the infinity norm of the closed-loop transfer function matrix. The state-space solution is discussed next.

15.2.1 State Feedback (Full Information) $\mathscr{H}_\infty$ Control Design

Consider the previous dynamic game defined by:

$$\min_u \max_w J(x_0, 0; u, w) \tag{15.51}$$

where:

$$J(x_0, 0; u, w) = \int_0^\infty (z^T z - \gamma^2 w^T w)\mathrm{d}t \tag{15.52}$$

where $x(.)$ is the solution of,

$$\dot{x} = Ax + B_2 u + B_1 w \tag{15.53}$$

$$z = C_1 x + D_{12} u \tag{15.54}$$

$$y = x \tag{15.55}$$

The following assumptions are made about the dynamic system to solve this optimal control problem:

- (A, B_2) are stabilizable (necessary)
- (C_1, A) is detectable (necessary)
- $C_1^T D_{12} = 0$ (orthogonal) and $D_{12}^T D_{12} = I$ (neither are necessary)

From the LQR/LQG solutions, we assume that the worst-case disturbance w and the best-case control u have the following structure:

$$w(t) = K_{d\infty} x(t) \tag{15.56}$$

$$u(t) = K_{c\infty} x(t) \tag{15.57}$$

therefore,

$$z = C_1 x + D_{12} u \tag{15.58}$$

$$= (C_1 + D_{12} K_{c\infty}) x \tag{15.59}$$

substituting this into Eq. (15.52),

$$J = \int_0^\infty x^T \underbrace{(C_1^T C_1 + K_{c\infty}^T K_{c\infty} - \gamma^2 K_{d\infty}^T K_{d\infty})}_{Q_c} x\mathrm{d}t \tag{15.60}$$

$$= \int_0^\infty x^T Q_c x\mathrm{d}t \tag{15.61}$$

where $Q_c \geq 0$, a positive semi-definite matrix, and

$$\dot{x} = Ax + B_1 K_{d\infty} x + B_2 K_{c\infty} x \tag{15.62}$$

$$= \underbrace{(A + B_1 K_{d\infty} + B_2 K_{c\infty})}_{A_c} x \tag{15.63}$$

Now, the derivations of the two Eqs. (15.60) and (15.63) transform the min-max problem to the general LQR-like optimal control problem.

Next, we discuss how to compute the gains $K_{c\infty}$ and $K_{d\infty}$ such that the conditions for minimization of the worst disturbance of the cost function J is met and the solution is stable.

We know from Lyapunov analysis that for an energy function $V(x) = x^T P_\infty x$ and $P_\infty \geq 0$,

$$\dot{V}(x) = \dot{x}^T P_\infty x + x^T P_\infty \dot{x} \tag{15.64}$$

$$= x^T A_c^T P_\infty x + x^T P_\infty A_c x \tag{15.65}$$

$$= x^T (A_c^T P_\infty + P_\infty A_c) x \tag{15.66}$$

Now for a stable solution, we have to enforce that rate of change of the Lyapunov function is negative, therefore, we select

$$A_c^T P_\infty + P_\infty A_c = -Q_c \tag{15.67}$$

Substituting the values from Eqs. (15.60) and (15.63) into Eq. (15.67) results in

$$\begin{aligned}(A + B_1 K_{d\infty} + B_2 K_{c\infty})^T P_\infty + P_\infty (A + B_1 K_{d\infty} + B_2 K_{c\infty})\\ + C_1^T C_1 + K_{c\infty}^T K_{c\infty} - \gamma^2 K_{d\infty}^T K_{d\infty} = 0\end{aligned} \tag{15.68}$$

because this is an infinite horizon problem, we know that P_∞ is a constant matrix, and the following holds

$$\nabla P_\infty = 0 \tag{15.69}$$

Taking the gradient of Eq. (15.68) with respect to $K_{d\infty}$ yields,

$$A_c^T \underbrace{\nabla_{K_{d\infty}} P_\infty}_{=0} + B_1^T P_\infty + \underbrace{\nabla_{K_{d\infty}} P_\infty}_{=0} A_c + P_\infty B_1 - 2\gamma^2 K_{d\infty} = 0 \tag{15.70}$$

Solving for $K_{d\infty}$,

$$K_{d\infty} = \frac{1}{\gamma^2} B_1^T P_\infty \tag{15.71}$$

Following similar development for $K_{c\infty}$ results in

$$K_{c\infty} = -B_2^T P_\infty \tag{15.72}$$

These two gains are the min-max and stabilizing solution to J.

For computing P_∞, we substitute these two gains into Equation (15.68), which results in the following ARE:

$$P_\infty A + A^T P_\infty + C_1^T C_1 - P_\infty \left(B_2 B_2^T - \frac{1}{\gamma^2} B_1 B_1^T \right) P_\infty = 0 \tag{15.73}$$

It can be shown that the ARE will always have a stabilizing solution, P_∞, when $\gamma \gg 1$ and in this case, the optimal $u = K_c x$ where $K_c = -B_2^T P_\infty$, i.e. this is an equivalent LQR solution!

The min-max optimization routine is solved iteratively, through two principal stages. The condition for the first stage is $\|T_{zw}\|_\infty < \gamma$, for any arbitrary disturbances, where γ is selected to be a large value, and this implies

1. The ARE has a solution such that $P_\infty \geq 0$
2. $A_c = A + B_1 K_{d\infty} + B_2 K_{c\infty}$ is Hurwitz (stable).

The second stage iterates until γ_{min} is found which still implies that $\|T_{zw}\|_\infty < \gamma_{min}$. This is done by iterating until there is no solution to the ARE and/or A_c is not Hurwitz.

15.2.2 $\mathscr{H}_\infty$ Filtering

In order to compensate for missing states, $\mathscr{H}_\infty$ filter is utilized to estimate the states. We assume that the governing equations for generating signal y is

$$\dot{x} = Ax + \overbrace{B_1 w}^{\text{Process Noise},w_d} + B_2 u \tag{15.74}$$

$$y = C_2 x + \underbrace{D_{21} w}_{\text{Sensor Noise},w_n} \tag{15.75}$$

$$z = C_1 x \tag{15.76}$$

where $D_{21}D_{21}^T = I$ and $d = (w_d, w_n)$ is assumed to be energy-bounded (two-norm is finite) with no assumption on the statistics of these noise signals.

The objective is to find an estimate $\hat{z}$ of $z = C_1 x$ which minimizes $||T_{ed}||_\infty$ where $e = z - \hat{z}$. As with the $\mathscr{H}_\infty$ control, this filter is designed in two stages. In the first stage, the estimate $\hat{z} = C_1\hat{x}$ is chosen to ensure

$$\|e\|_2^2 = \|z - \hat{z}\|_2^2 < \gamma^2 \|d\|_2^2 \tag{15.77}$$

such that,

$$\|T_{ed}\|_\infty < \gamma \tag{15.78}$$

In the second stage, γ^* (the smallest possible γ) is found such that a solution to the first stage exists. The filter $K_{f\infty}$ corresponds to γ^* is the solution to the H_∞ filtering problem. Assume the following estimation model,

$$\dot{\hat{x}} = A\hat{x} + B_2 u + K_{f\infty}(y - \hat{y}) \tag{15.79}$$

$$\hat{y} = C_2\hat{x} \tag{15.80}$$

$$\hat{z} = C_1\hat{x} \tag{15.81}$$

Then the error dynamics for $\tilde{x} = x - \hat{x}$ are

$$\dot{\tilde{x}} = (A - K_{f\infty}C_2)\tilde{x} + (B_1 - K_{f\infty}D_{21})w \tag{15.82}$$

$$e = C_1\tilde{x} \tag{15.83}$$

We can rewrite these equations in term of noise vector input, d:

$$\dot{\tilde{x}} = (A - K_{f\infty}C_2)\tilde{x} + [I - K_{f\infty}]d \tag{15.84}$$

$$e = C_1\tilde{x} \tag{15.85}$$

where d is a column vector of w_d and w_n. The problem is to choose $K_{f\infty}$ to minimize the induced two-norm (∞ norm) of the system mapping d to e. Similar to LQG problem, it can easily be shown that this is a dual problem to the H_∞ control problem where,

- $A \to A^T$
- $B_2 \to C_2^T$
- $B_1 \to C_1^T$

This results in the following ARE for the estimation filter:

$$P_{f\infty}A^T + AP_{f\infty} + B_1B_1^T - P_{f\infty}\left(C_2^TC_2 - \frac{1}{\gamma^2}C_1^TC_1\right)P_{f\infty} = 0 \tag{15.86}$$

and the filter gains are

$$K_{f\infty} = P_{f\infty}C_2^T \tag{15.87}$$

More precisely, the above solution ensures, for given γ,

$$\|T_{ed}\|_\infty < \gamma \tag{15.88}$$

provided that a solution to the ARE Eq. (15.86) exists; the solution to the H_∞ estimation problem is then obtained by determining the minimum γ for which this solution exists.

An important fact to observe is that the solution of the H_∞ filtering problem, unlike that of the corresponding LQG/H_2 filtering problem, does not depend on the input noise.

15.3 $\mathcal{H}_\infty$- Robust Optimal Control

In this section, we combine the results of H_∞ state feedback and filtering to develop an output feedback (partial state feedback) H_∞ control. Similar to the *LQG* control, the partial state/output feedback H_∞ utilizes an H_∞ filter for state estimation. However, unlike Kalman filter, the H_∞ filter is a deterministic filter, which no assumptions regarding the statistics of the noise inputs are made, except their energy boundedness. Furthermore, similar to the *LQG*, two AREs needs to be solved for finding an appropriate solution to infinite horizon case. However, unlike the *LQG*, the two AREs are coupled as the worst disturbance input, as discussed earlier, play a role both in control and filtering problems. Therefore, in the most basic form of the H_∞ output feedback control, the separation of control and filtering, as was the case for LQG, cannot be assumed. In addition and most importantly, the LTR, as discussed earlier for LQG controllers, for H_∞ output feedback control design becomes unnecessary.

As stated earlier, finding all possible controllers $K(s)$ which minimizes $\|T_{zw}\|_\infty$ is a difficult problem to solve.

In general, this is solved using a suboptimal solution, as discussed earlier, such that:

(a) Choose a $\gamma > 0$,
(b) Find a controller such that $\|T_{zw}\|_\infty < \gamma$,
(c) If such a controller exists, then decrease γ, otherwise, increase γ.

The following general state-space representation of the dynamic system is utilized to solve this problem:

$$\dot{x} = Ax + B_2u + B_1w \tag{15.89}$$

$$z = C_1x + D_{11}w + D_{12}u \tag{15.90}$$

$$y = C_2x + D_{21}w + D_{22}u \tag{15.91}$$

The following assumptions are made:

(A1) Pair $(A, \; B_2 \; C_2)$ is stabilizable and detectable, (necessary)
(A2) $D_{11} = 0$ and $D_{22} = 0$, (not necessary for H_∞ but necessary for H_2)
(A3) D_{12} and D_{21} have full rank, (necessary)
(A4) $D_{12}^T \begin{bmatrix} C_1 & D_{12} \end{bmatrix} = \begin{bmatrix} 0 & I \end{bmatrix}$, (not necessary)
(A5) $\begin{bmatrix} B_1 \\ D_{21} \end{bmatrix} D_{21}^T = \begin{bmatrix} 0 \\ I \end{bmatrix}$, (not necessary)
(A6) Pair $(A, \; B_1 \; C_1)$ is stabilizable and detectable. (Necessary)

These assumptions have the following purposes:

- (A1) is for the existence of stabilizing controllers,
- (A2) is the conventionality in H_2 control. $D_{11} = 0$ makes G_{11}, the open-loop transfer function from w to z, strictly proper. $D_{22} = 0$ makes G_{22}, the open-loop transfer function from u to y, strictly proper and simplifies the formulas in the algorithms:
- (A3) is to ensure the controllers are proper and hence realizable,
- (A4) means that C_1 and D_{12} are orthogonal so that the penalty on $z = C_1x + D_{12}u$ includes a nonsingular normalized penalty on the control u,

- (A4) and (A5) drastically simplify the mathematical equations but they are not necessary,
- (A6) is to ensure that the optimal controller does not try to cancel poles or zeros on the imaginary axis which would result in closed-loop instability.

Remark: If there is a control input with no weighting filter, then D_{12} does not have full rank. If this happens, then we need to add a small weight on u to make D_{12} full rank. In addition, no $j\omega$ poles for H_∞ computation are allowed.

These assumptions will result in the following general state-space representation of the dynamic system:

$$\dot{x} = Ax + B_2u + B_1w \tag{15.92}$$

$$z = C_1x + D_{12}u \tag{15.93}$$

$$y = C_2x + D_{21}w \tag{15.94}$$

The synthesis of the output feedback H_∞ control is based on the following two AREs, where the first is for the synthesis of the full state feedback and the second is for the synthesis of H_∞ state estimation:

$$P_{c\infty}A + A^TP_{c\infty} + P_{c\infty}(\gamma^{-2}B_1B_1^T - B_2B_2^T)P_{c\infty} + C_1^TC_1 = 0 \tag{15.95}$$

$$AP_{f\infty} + P_{f\infty}A^T + P_{f\infty}(\gamma^{-2}C_1^TC_1 - C_2^TC_2)P_{f\infty} + B_1B_1^T = 0 \tag{15.96}$$

These two AREs are quadratic equations in terms of the two unknowns matrices $P_{c\infty}$ and $P_{f\infty}$. These matrices are symmetric with dimension $n \times n$, where n is dimension of the state space. Both equations depend on the value of γ. We are interested in the stabilizing solutions of these two quadratic equations. As the worst disturbance plays a role in both AREs, to find at least one stabilizing solution to these equations, it can be shown that the solution to H_∞ state estimation, $P_{f\infty}$, has to be scaled such that

$$P_{f\infty_s} = P_{f\infty}Z_\infty \tag{15.97}$$

where $Z_\infty = (I - \gamma^{-2}P_{c\infty}P_{f\infty})^{-1}$. Again, similar to control solution, we see that the state estimation solutions scaled versus non-scaled, become equal when γ is very large.

We conclude that a solution to the output feedback H_∞ control exists if and only if there exists a stabilizing solution to the control ARE, $P_{c\infty}$, and the filtering ARE, $P_{f\infty_s}$. From (15.97), this is equivalent to the existence of solutions for $P_{c\infty}$ and $P_{f\infty}$ and the coupling condition that $(I - \gamma^{-2}P_{c\infty}P_{f\infty})^{-1}$ is non-singular.

The conditions for the general output feedback H_∞ control is summarized in the following theorem:

Theorem 15.2 *There exists a linear stabilizing feedback controller for the system defined by (15.92) to (15.94) that satisfies*

$$\|T_{zw}\|_\infty < \gamma \tag{15.98}$$

if and only if

1. *The Algebraic Riccati Equation (15.95) has a stabilizing solution* $P_{c\infty} = P_{c\infty}^T \geq 0$.
2. *The Algebraic Riccati Equation (15.96) has a stabilizing solution* $P_{f\infty} = P_{f\infty}^T \geq 0$.
3. $\overline{\sigma}(P_{c\infty}P_{f\infty}) < \gamma^2$.

Items 2 and 3 are equivalent to the existence of a solution $P_{f\infty_s}$. Moreover, in this case, the dynamics of one such controller is written as:

$$\dot{\hat{x}} = (A + B_1K_{d\infty} + B_2K_{c\infty} + Z_\infty K_{f\infty}C_2)\hat{x} - Z_\infty K_{f\infty}y \tag{15.99}$$

$$u = K_{c\infty}\hat{x} \tag{15.100}$$

Next, using LFT framework, we compare the structure of H_2 and H_∞ controllers shown in Figures 15.10 and 15.11.

Given the dynamic equations for the H_∞ output feedback in (15.100) and (15.99), the control block diagram is in Figure 15.12.

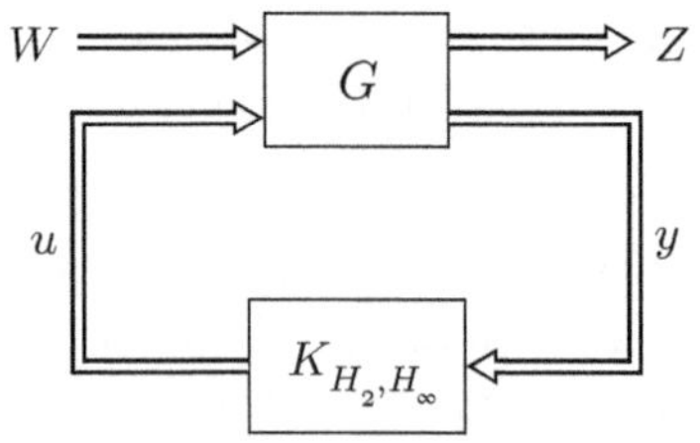

Figure 15.10 H_2 and H_∞ LFT.

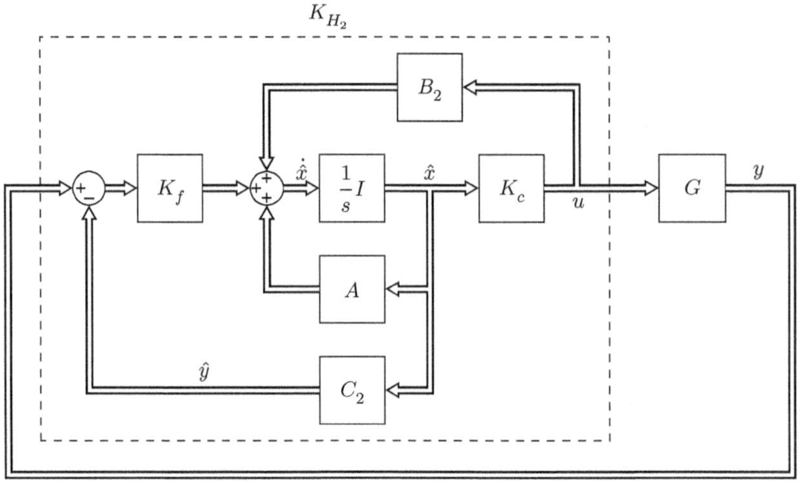

Figure 15.11 LQG/H_2 block diagram.

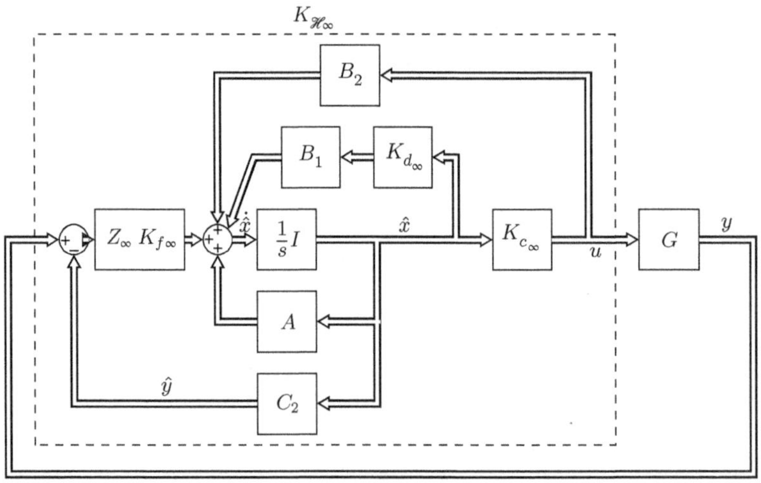

Figure 15.12 $\mathscr{H}_\infty$ block diagram.

15.4 Problems

15.1 Repeat Problem 13.2, where the plant dynamics are given by:

$$A = \begin{pmatrix} 0 & 1 \\ -2 & -3 \end{pmatrix}, \quad B = \begin{pmatrix} 10 & 0 \\ 0 & 25 \end{pmatrix} \quad C^T = \begin{pmatrix} 1 & -1 \\ 0 & 2 \end{pmatrix}, \quad D = \begin{pmatrix} 0 & 1 \\ 0 & 0 \end{pmatrix}, \tag{15.101}$$

15.2 Repeat Problem 13.5, where G_p is given as:

$$G_p(s) = \begin{bmatrix} 0 & \frac{-1}{s+1} \\ 1 & \frac{1}{s(s+1)} \end{bmatrix} \tag{15.102}$$

(a) This time, choose a filter for W_Δ (the robustness uncertainty filter) and W_P (performance filter) and use Matlab to compute a controller $K(s)$ such that

$$||T_{zw}||_\infty < 1 \tag{15.103}$$

(b) Plot the closed-loop step responses and closed-loop transfer function Bode responses. Briefly discuss your results.

15.3 Repeat Problem 13.6, where G_p is given as

$$G_p = 50 \begin{bmatrix} \frac{-1}{s} & \frac{-(s+1)}{s} \\ \frac{1}{s^2+1} & 0 \end{bmatrix} \tag{15.104}$$

This time choose a W_Δ (the robustness uncertainty filter) and W_P (performance filter) and compute an H_∞ controller.

15.4 Repeat Problem 13.7, where G_p is given as:

$$G_p = 100 \begin{bmatrix} 1 & -1 \\ 1 & 0 \end{bmatrix} \begin{bmatrix} \frac{1}{s+1} & 0 & 0 \\ 0 & \frac{1-s}{s+1} & 0 \end{bmatrix} \begin{bmatrix} 1 & 1 & 0 \\ 1 & 1 & 0 \\ 1 & 0 & 0 \end{bmatrix} \tag{15.105}$$

This time choose a W_Δ (the robustness uncertainty filter) and W_P (performance filter) and compute an H_∞ controller.

16

Estimation Design for MIMO Using Parameterization Approach

In Chapter 7 of Volume 1, we demonstrated how the Youla Parameterization approach could be adapted for state feedback for a single-input, single-output system, a method called the Youla Controller Output Observer.

In this chapter, we expand this method for multi-input, multi-output systems and provide step-by-step discussion of the estimation method for the MIMO case. The final section presents a MIMO Youla COO example for an automotive suspension with both input and state estimation by utilizing the measured outputs of a validated six chassis degrees-of-freedom nonlinear vehicle model. The estimation results are then compared to the nonlinear model inputs and states. Additionally, the estimation is shown to work in conjunction with a fully active Skyhook controller and as a road roughness coefficient estimator.

The reader should recall that the generalized concept for the YCOO was developed in Sections 7.1 and 7.2. The work in these sections applies to both SISO and MIMO YCOO, and so **this derivation will not be repeated here**. The general structure for a MIMO YCOO is shown in Figure 16.1. Here, controllable inputs, $\hat{u}$, are used to minimize the error between the estimated signals, $\hat{y}$, and the measured signals, Y, with the use of the feedforward controller, $Y(s)$. The estimated states, $\hat{x}$, are also shown in this structure. The matrices A_e, B_e, C_e, and D_e were discussed previously. The left partition represents signals coming from an actual vehicle (containing sensor noise), while the right is contained entirely within the computer for estimation purposes.

16.1 YCOO Concept for MIMO

The general concept of the YCOO method was introduced in Sections 7.1 and 7.2, and the controller derivation was provided for SISO systems in Section 7.2. This section will briefly cover the controller derivation for MIMO systems.

Like in the SISO estimator design, the classical feedback loop (as illustrated in Figure 16.2a) is transformed to an equivalent feedforward loop, Figure 16.2b, to directly shape complementary sensitivity, or closed-loop transfer function, by selecting a stable and proper Youla parameter, where

$$G_{p,e}(s) = C_e(sI - A_e)^{-1}B_e \tag{16.1}$$

assuming that the transfer function for the observer is strictly proper without loss of generality. This model is assumed to be always linear. Equivalently,

$$G_p(s) = C(sI - A)^{-1}B \tag{16.2}$$

represents the actual linear plant model, however, this model could also be nonlinear. Therefore, we have transformed an estimator design to a neoclassical feedback control design utilizing Youla parameterization. It is important to note that the fundamental difference between designing an estimator using this technique and its dual control design is that in the case of an estimator design, the observer model (estimated model of the plant), $G_{p,e}(s)$, is an integral part of the feedback loop. In the case of the control design, on the other hand, this model is discarded and the controller is implemented on the actual plant.

To develop the MIMO YCOO, the estimator's plant model must be decoupled

$$G_{p,e}(s) = U_L^{-1}G_{p,e}^{SM}(s)U_R^{-1} \tag{16.3}$$

where the unimodular matrices are computed as

$$U_L = (N_P N_{SM}^{-1})^{-1} \tag{16.4}$$

Robust Control: Youla Parameterization Approach, First Edition. Farhad Assadian and Kevin R. Mallon.

Companion Website: www.wiley.com/go/assadian/robust

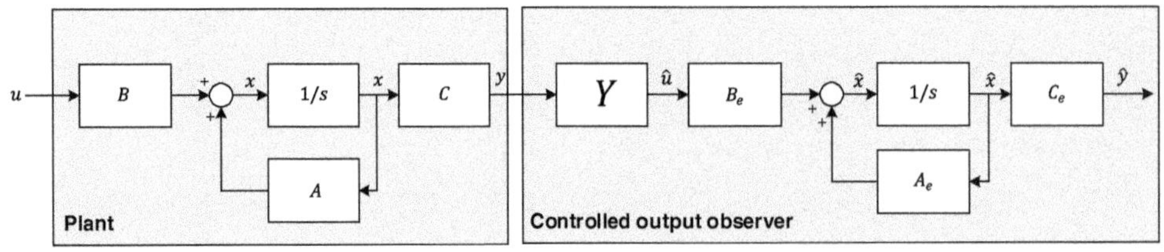

Figure 16.1 Youla Controller Output Observer.

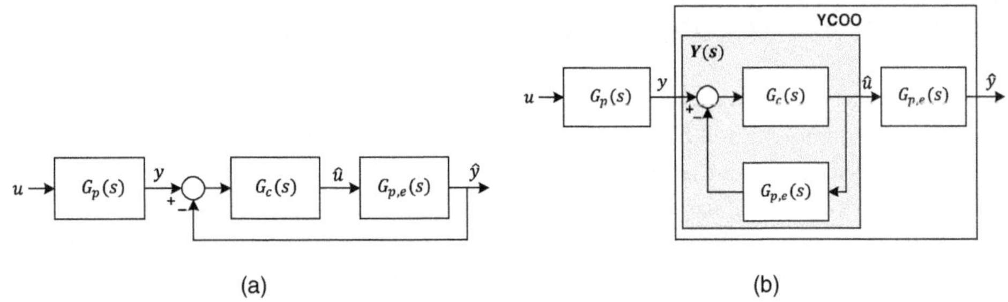

Figure 16.2 Comparison of classical feedback loop and Youla feedforward transformation. (a) Classical feedback loop. (b) Youla feedforward transformation.

$$U_R = D_P D_{SM}^{-1} \tag{16.5}$$

which in turn come from a matrix fractional description of the original and decoupled plant.

$$G_{p,e} = N_P {D_P}^{-1} \tag{16.6}$$

$$G_{p,e}^{SM} = N_{SM} D_{SM}^{-1} \tag{16.7}$$

The procedures laid out in Sections 13.2 and 13.4 can then be used to a Youla transfer function matrix $Y(s)$.

$$Y(s) = U_R M_Y U_L \tag{16.8}$$

Given this realization for $Y(s)$, the controller $G_c(s)$ is obtained from

$$Y(s) = (I + G_c G_{p,e})^{-1} G_c = G_c (I + G_{p,e} G_c)^{-1} \tag{16.9}$$

or

$$G_c(s) = Y(I - G_{p,e} Y)^{-1} = (I - Y G_{p,e})^{-1} Y \tag{16.10}$$

Then, the closed-loop transfer functions are given by

$$T_y = G_{p,e} Y \tag{16.11}$$

$$S_y = I - T_y \tag{16.12}$$

$$T_u = Y G_{p,e} \tag{16.13}$$

$$S_u = I - T_u \tag{16.14}$$

The MIMO YCOO design procedure are given as follows

1. Given an observer plant transfer function, $G_{p,e}(s)$, find a decoupled plant $G_{p,e}^{SM}(s) = U_L G_{p,e}(s) U_R$ with unimodular matrices U_L and U_R.
2. Select a desired complementary sensitivity transfer function for each element of the decoupled plant $G_{p,e}^{SM}(s)$
3. Compute a Youla transfer function for each element of the decoupled system that meets the aforementioned interpolation conditions and assemble them into a diagonal matrix M_Y.

4. Using the unimodular matrices and M_Y, obtain the Youla transfer function matrix for the original system as $Y(s) = U_R M_Y U_L$.
5. From the expression given in Eq. (16.10), compute the controller $G_c(s)$ which satisfies all the performance, robustness, and stability criterion.
6. Finally, from $G_{p,e}$ and Y, compute the closed-loop transfer functions T_y, S_y, T_u, S_u.

16.2 MIMO Estimator Design

We will use the suspension system from the last SISO example to derive a MIMO estimator. In this example, both the road disturbance and the force input and are treated as two inputs and the sprung and unsprung mass accelerations are the measured outputs. Therefore, the system includes two inputs and two outputs. The block diagram in Figure 16.1 will be utilized to design this estimator using the previously developed quarter car suspension model. However, this time, the results of the state estimation are compared to a validated nonlinear six chassis degrees-of-freedom vehicle model, G_p. Due to the space constraints, the development of this model is not discussed in this book [38]. It is important to note that in this example, the plant model, G_p, and the observer model, $G_{p,e}$ are not the same as it was the case for the SISO design. In fact, G_p is a nonlinear model as mentioned above and the observer model $G_{p,e}$ is, as described in the following paragraphs, a linear transfer function matrix.

In Chapter 7, it was stated that the force input, F_a, is a controlled input and that the road input velocity, v_i, is a disturbance input that can be treated as a controlled input. The sprung mass, $\dot{v}_s$, and unsprung mass, $\dot{v}_{us}$, accelerations are the signals we wish to track. This is represented by $G_{p,e}$ in Eq. (16.15),

$$\begin{bmatrix} \dot{v}_s \\ \dot{v}_{us} \end{bmatrix} = G_{p,e} \begin{bmatrix} F_a \\ v_i \end{bmatrix} \tag{16.15}$$

where

$$G_{p,e} = C_e(sI - A_e)^{-1}B_e + D_e \tag{16.16}$$

using the state-space model given in Eqs. (7.13)–(7.24).

Using the methods described in Chapters 9 and 10, the decoupled transfer functions are found for the diagonals of the Smith-McMillan form.

$$G_{p,e}^{SM}(s) = \begin{bmatrix} G_{sm1} & 0 \\ 0 & G_{sm2} \end{bmatrix} \tag{16.17}$$

These diagonal entries are defined below. Asterisks (*) represent the portion of the plant with no unstable zeros.

$$G_{sm1} = \frac{\beta_1 s}{(s^2 + \alpha_{21}s + \alpha_{22})(s^2 + \alpha_{12}s + \alpha_{11})} \tag{16.18}$$

$$= sG_{sm1}^* \tag{16.19}$$

$$G_{sm2} = \beta_2 s^2 \tag{16.20}$$

$$= s^2 G_{sm2}^* \tag{16.21}$$

where the α and β terms correspond to system natural frequencies, damping ratios, and gains. Then,

$$G_{p,e}^{SM} = U_L G_{p,e} U_R \tag{16.22}$$

where the unimodular matrices U_L and U_R are found from the matrix fractional description of $G_{p,e}^{SM}$ and $G_{p,e}$, such that

$$U_L = (N_R N_{SM}^{-1})^{-1} \tag{16.23}$$

$$U_R = D_R D_{SM}^{-1} \tag{16.24}$$

as shown in Chapter 10.

16.3 State Estimation

16.3.1 First Decoupled System (G_{sm1})

Based on the procedures outlined for SISO design in Chapter 7, the Youla transfer function, Y_1, is shaped to give a desirable complementary sensitivity transfer function, T_1. Normally, T_1 is shaped as a low-pass Butterworth filter where the magnitude is one at low frequencies and decreases at high frequencies for good tracking and noise rejection, respectively. However, in this case, it is impossible to achieve a DC magnitude of one because there exists a zero at the origin. Therefore, T_1 is designed similar to a bandpass filter where the magnitude is one except at very low frequencies. This process was applied in the earlier SISO control development. A pole, p_1, is placed near the origin to approximately cancel the effect of the unstable zero. A first-order lag and a second-order filter are designed to achieve the required performance and robustness characteristics. To ensure the properness of Y_1, a high frequency pole, p_2, is included. A gain was chosen so that the magnitude of T_1 is equal to one in the low-frequency region up to the closed-loop bandwidth. The results are summarized in Eqs. (16.25)–(16.26) and Figure 16.3.

$$Y_1 = \frac{k_1}{(s+p_1)(s^2+2\zeta\omega_n s+\omega_n^2)(s+p_2)}\frac{1}{G_{sm1}^*} \tag{16.25}$$

$$\begin{aligned} T_1 &= Y_1 G_{sm1} \\ &= \frac{k_1 s}{(s+p_1)(s^2+2\zeta\omega_n s+\omega_n^2)(s+p_2)} \end{aligned} \tag{16.26}$$

16.3.2 Second Decoupled System (G_{sm2})

A similar process is performed for the second decoupled system, G_{sm2}. With this new plant, there are now two zeros at the origin. The two zeros are dealt with as before, this time using a complex set of poles (defined by ζ_1 and ω_{n1}) at a very low frequency. Then, a complex set of poles (defined by ζ_2 and ω_{n2}) are used to shape T_2 after the closed-loop bandwidth and at a higher frequency. A gain was chosen so that the magnitude of T_2 is equal to one in the low-frequency region up to the closed-loop bandwidth. The results are summarized in Equations (16.27) and (16.28) and Figure 16.3.

$$Y_2 = \frac{k_2}{(s^2+2\zeta_1\omega_{n1}s+\omega_{n1}^2)(s^2+2\zeta_2\omega_{n2}s+\omega_{n2}^2)}\frac{1}{G_{sm2}^*} \tag{16.27}$$

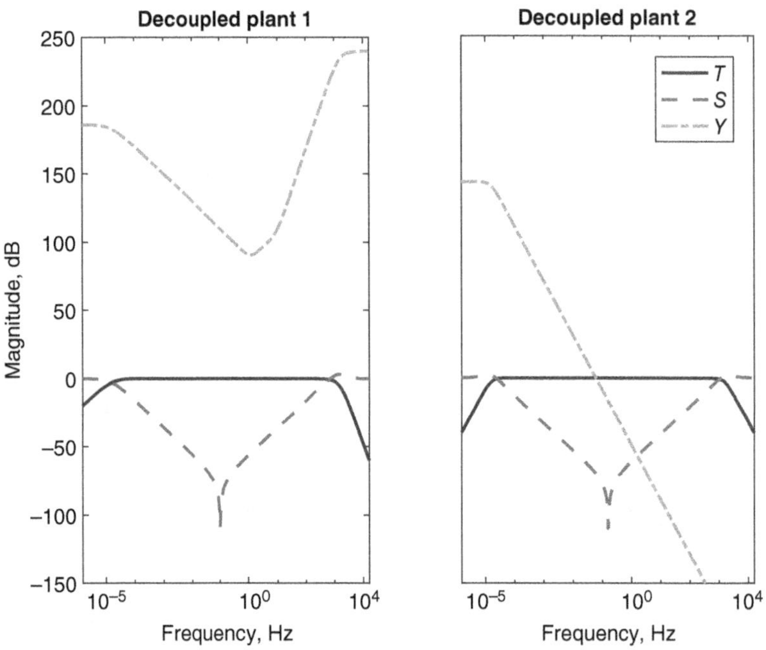

Figure 16.3 Bode for decoupled system.

$$T_2 = Y_2 G_{sm2} = \frac{k_2 s^2}{(s^2 + 2\zeta_1 \omega_{n1} s + \omega_{n1}^2)(s^2 + 2\zeta_2 \omega_{n2} s + \omega_{n2}^2)} \tag{16.28}$$

16.3.3 Coupled System

The decoupled Youla transfer functions are coupled by utilizing the previously computed unimodular matrices. Complementary sensitivity, sensitivity and the closed-loop controller transfer functions for the coupled system are derived in Eqs. (16.29)–(16.32).

$$Y = U_R M_Y U_L \tag{16.29}$$

$$T = G_p Y \tag{16.30}$$

$$S = I_2 - T \tag{16.31}$$

$$G_c = YS^{-1} \tag{16.32}$$

where

$$M_Y = \begin{bmatrix} Y_1 & 0 \\ 0 & Y_2 \end{bmatrix} \tag{16.33}$$

and

$$I_2 = \begin{bmatrix} 1 & 0 \\ 0 & 1 \end{bmatrix} \tag{16.34}$$

The coupled system Bode plots are shown in Figure 16.4.

16.4 Applications

The utility of this estimator is made apparent by two example applications: Skyhook-based damping force estimation for closed-loop control in active/semi-active suspensions and road roughness coefficient estimation.

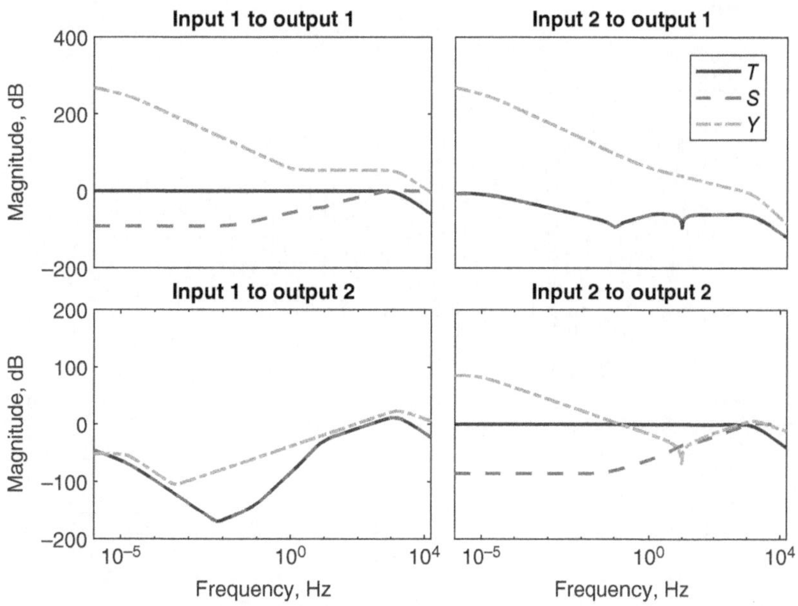

Figure 16.4 Bode for the coupled system.

16.4.1 States Estimation: Four States

As with the SISO case, state estimation is performed over a typical D-road (Farm road). The states include the sprung and unsprung mass velocity (Figures 16.5 and 16.6), wheel stroke (Figure 16.7), and tire deflection (Figure 16.8). The estimates are comparable to the actual signals generated from the nonlinear vehicle model.

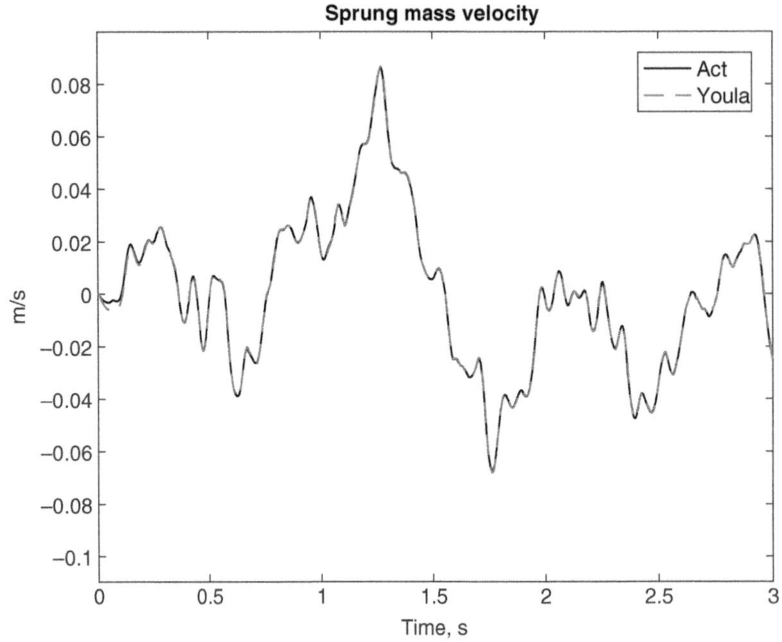

Figure 16.5 MIMO state estimation with D-Road PSD input – sprung mass velocity.

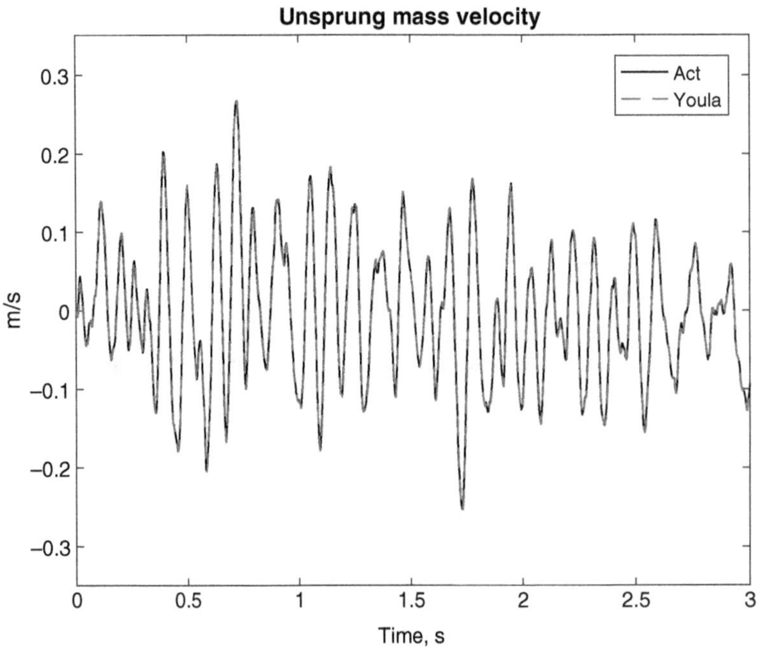

Figure 16.6 MIMO state estimation with D-Road PSD input – unsprung mass velocity.

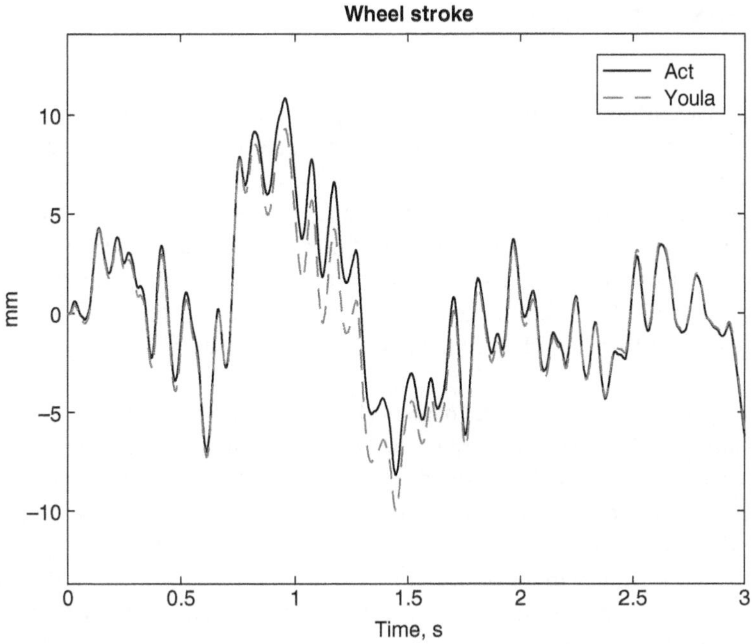

Figure 16.7 MIMO state estimation with D-Road PSD input – wheel stroke.

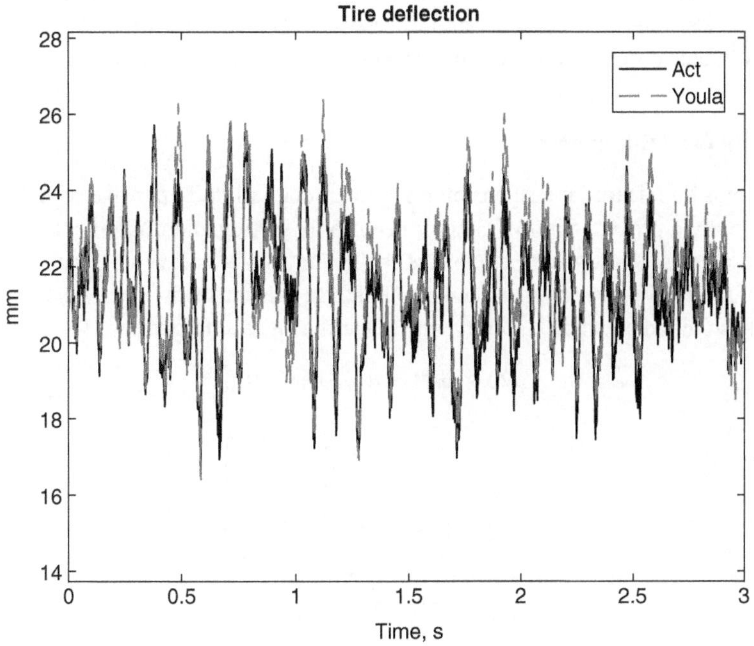

Figure 16.8 MIMO state estimation with D-Road PSD input – tire deflection.

16.4.2 Input Estimation: Skyhook Based Control

The estimator is applied to a Skyhook controller, a common strategy with variants deployed on many vehicle brands. It was proposed that an optimal suspension is one in which the sprung mass is connected through a linear damper to some inertial reference [22]. Due to the inability to build such a contraption, this ideal damping force is instead applied within the suspension between sprung and unsprung masses either by using a force actuator or through control of the damper. Although suboptimal, this strategy has proven highly effective. The ideal damping force is given by Eq. (16.35), where b_{SH}

is a tuning parameter and v_s is the sprung mass vertical velocity.

$$F_{d,SH} = b_{SH} v_s \tag{16.35}$$

An idealized fully active suspension using the Skyhook strategy is applied to a high order vehicle model developed in [39, 40]. Each corner suspension is modeled as a nonlinear spring and damper (lookup tables) with an active element capable of applying ideal forces.

For optimizing the active or semi-active suspension performance, it is recommended to utilize the damper force in the feedback loop. Therefore, either a measurement or estimation of this signal is required. In this section, comparative performance results are obtained for scenarios where ideal access to the damping force measurement is available versus when an estimation of this force using the YCOO concept is made. It will be shown that the influence of the estimator does very little to inhibit the performance of the controller when compared to the ideal case of having access to the measured damping force.

The control structure in Figure 16.9 is proposed and described in this paragraph. A simple controller, $G_{c,SH}$, different from the YCOO discussed previously, is developed that drives the effective damping force toward the ideal Skyhook damping force, $F_{d,SH}$, using the controlled input force, F_i. This effective damping force is the combination of both passive damping and the controllable damping force. The estimate of the damping force, $\hat{F}_d$, will inherently contain both these forces. Note that d represents additional system inputs and the *YCOO* block represents the estimator.

Success of the estimator is measured by the level of degradation in root-mean-square (RMS) heave acceleration as this value is strongly correlated with ride comfort. A comparison is made between uncontrolled, controlled with no estimator, and controlled with estimator. This is done in two scenarios: (i) driving over a motorway at 100 kph and (ii) driving over a large bump (0.1 m tall) at 40 kph. The results are summarized in Table 16.1 where RMS heave acceleration is shown for the uncontrolled case and percent change for cases with control. Both control scenarios are shown: with the ideal case of direct damping force feedback, and with the estimated damping force feedback. It is apparent that little degradation in RMS heave acceleration exists due to the inclusion of the estimator versus direct measurement feedback. Additionally, Figure 16.10 shows the estimated damping force versus that generated from the nonlinear model for motorway conditions.

16.4.3 Input Estimation: Road Roughness

Road roughness estimation is important for suspension parameter adaptation. Expensive sensors may be used to measure the road directly, but the method proposed in this chapter needs only sprung and unsprung mass accelerations, which are readily available on production vehicles. A road roughness coefficient, G_0, is a curve-fit parameter for Eq. (16.36), which is used to find the best fit of the road power-spectral-density (PSD). In this curve fit, v is the spatial frequency and v_0 is the cutoff spatial frequency, both in rad/m, and G_z is the road PSD at that spatial frequency [41].

$$G_z(v) = G_0[1 + (v_0/v)^2]/(2\pi v)^2 \tag{16.36}$$

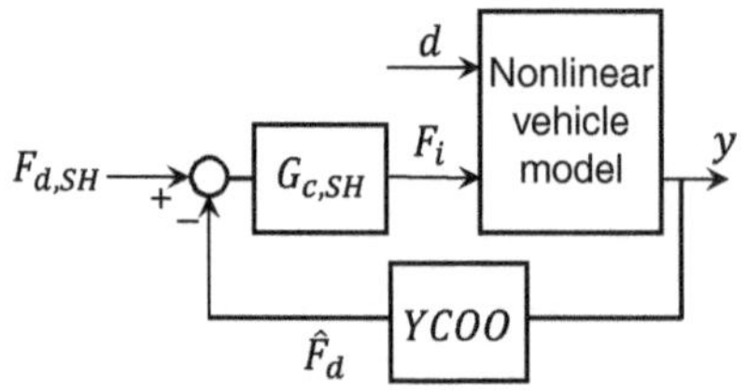

Figure 16.9 Damping force feedback control structure.

Table 16.1 Heave acceleration results for both cases, with and without the estimator.

Test	Motorway	Bump
No control (RMS, m/s^2)	0.0279	0.1192
Control with ideal measurement (% change)	−52.3176	−30.2342
Control with estimator (% change)	−46.5230	−22.0213

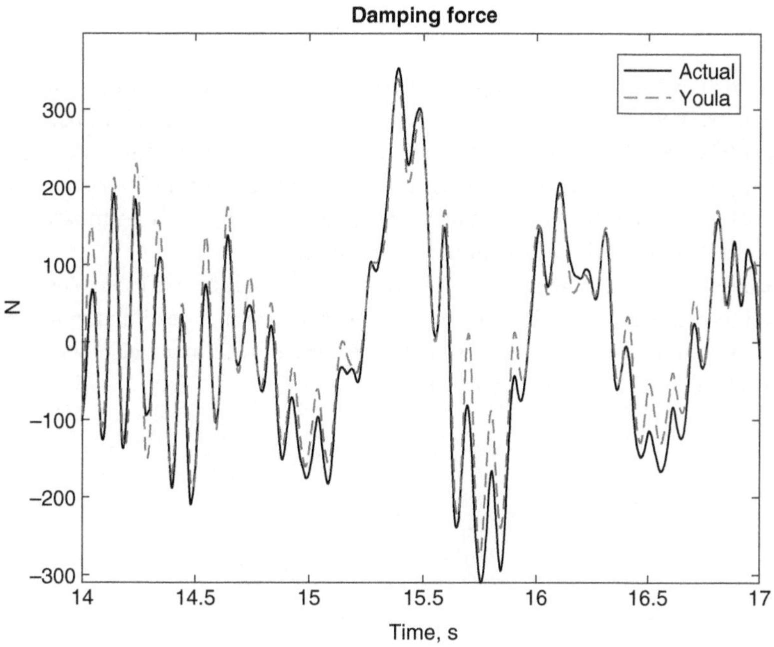

Figure 16.10 Actual versus estimated damping force over a motorway at 100 kph.

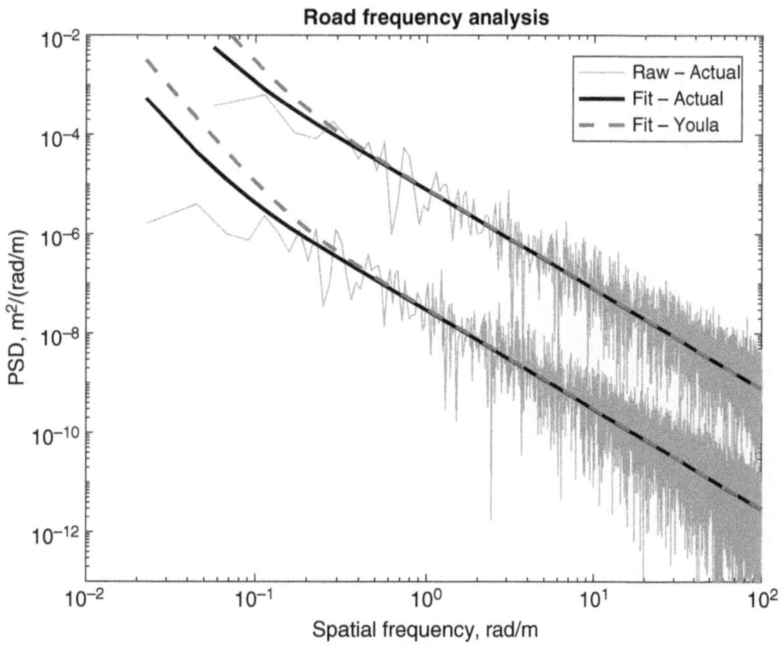

Figure 16.11 Power spectral density of road profile for a motorway (bottom) and farm road (top).

Such a plot is shown in Figure 16.11 where results are shown for a motorway (bottom) and farm road (top). Shown in solid light gray, solid black, and dashed gray are the PSD from the actual road data, the best fit of the actual road PSD, and a best fit of the estimated road PSD, respectively. The general shape of this plot is a result of decreasing road variation over smaller distances (larger spatial frequency). It is apparent from Figure 16.11 that the best fit of both the actual and estimated road data are very similar over a large frequency range.

16.5 Final Remarks

In Chapter 7, it was demonstrated how the Youla parameterization approach to controller design could be equivalently utilized for state and input estimation. The comparative results between YCOO and Kalman filter were presented in the linear domain including robustness characteristics of YCOO versus Kalman filtering method. In this chapter, this method was expanded for multi-input, multi-output state estimation and was demonstrated for state and input estimation on a nonlinear vehicle model. The resulting YCOO proved a capable and robust method of estimating the four suspension state variables as well as the controller and disturbance inputs to the plant.

There are many open questions that remain to be answered. The performance and robustness characteristics of YCOO in the nonlinear domain and its comparative analysis against established nonlinear estimation techniques, such as Extended Kalman and Unscented Kalman filters, will be explored in an example in Chapter 17. Other techniques such as Sliding Mode Observers and Particle filters require further investigation.

17

Practical Applications

In this chapter, we present four practical applications for which we investigate a comparison between Multiple Input Multiple Output (MIMO) Youla control design and MIMO H_∞, mixed sensitivity, control design. These applications include active suspension control (ASC) development, fuel injection control for engine start-up for hybrid diesel powertrain with a combination of an electric motor and a diesel engine, a free-flying space robot, and estimation with Youla compared to Kalman filtering for vehicle lateral dynamics.

17.1 Active Suspension

Active vibration control has been extensively studied and its applications stretch across a wide range of fields. Precision industrial process including semiconductor development relies on lasers to develop nanoscale devices. Vibration control is essential to these precision devices due to their nanoscales. Precision motion control systems for such applications in the semiconductor manufacturing have been developed utilizing conventional PID controls [42]. Application of vibration control in the automotive industry utilizes active dampers (known as semi-active suspension control) to improve occupant comfort and vehicle handling.

Simple quarter-car mathematical models for vehicle suspension systems have been developed and utilized for the purpose control design [43]. Furthermore, quarter-car models of active suspension systems utilizing a multitude of control systems have been studied. Control methodologies such as velocity feedback, LQR and H_∞ are among some of the popular control methods which have been used for active control of suspension system. The minimization of disturbances in automotive suspension systems utilizing robust control methods has proven to be an affective approach in increasing vehicle comfort and handling. Furthermore, active vehicle suspension design is among some of the most commonly studied MIMO systems.

In this part, we will focus on a two active force control actuators using a simple quarter-car model as shown in Figure 17.1 and with parameters given in Table 17.1. The first actuator, which is used in many previous studies, is located between the unsprung and sprung masses (wheel and vehicle body). The second actuator, which is unique in this development and has only been a topic of one previous study [44], is positioned between the road disturbance input and the unsprung mass, and it represents active tire pressure control.

The objectives for the control design are to minimize the acceleration of the sprung mass, therefore improving the comfort of passengers, by not violating the suspension deflection constraint (the available space between the sprung and unsprung masses). This constraint not only put a limit on the suspension deflection, but it would also indirectly help in road holding and hence, improving vehicle handling.

17.1.1 Model and Control Design

In this section, we provide the state equations for the quarter-car model, which is utilized for the control development, our simplified assumption for the actuator model, and our control development strategies including their comparisons.

Robust Control: Youla Parameterization Approach, First Edition. Farhad Assadian and Kevin R. Mallon.

Companion Website: www.wiley.com/go/assadian/robust

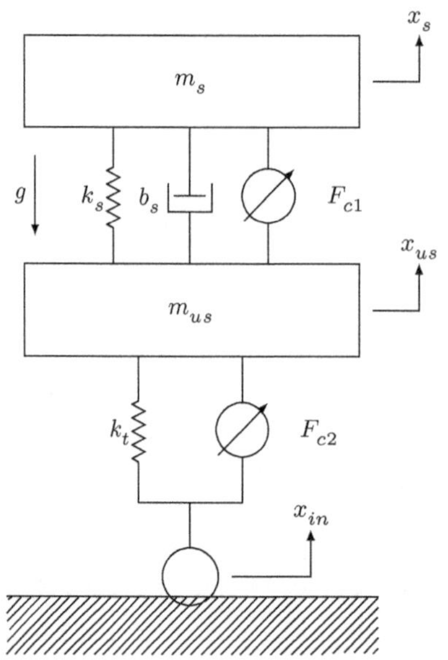

Figure 17.1 Quarter-car model with two active control forces.

Table 17.1 Nomenclature.

Symbol	Description	Units
m_s	Quarter of vehicle body mass	kg
m_u	Wheel mass	kg
x_s	Vehicle body displacement	m
x_u	Wheel displacement	m
x_in	Road disturbance input displacement	m
k_s	Suspension spring stiffness	N/m
k_t	Tire spring stiffness	N/m
b_s	Suspension damping coefficient	N/(m/s)
F_{c1}	Suspension actuator active force	kN
F_{c2}	Tire actuator active force	kN

The equations of motion for the quarter-car model are written as follows:

$$m_s\ddot{x}_s = k_s(x_{us} - x_s) + b_s(\dot{x}_{us} - \dot{x}_s) + 10^3F_{c1} \tag{17.1}$$

$$m_{us}\ddot{x}_{us} = k_s(x_s - x_{us}) + b_s(\dot{x}_s - \dot{x}_{us}) - 10^3F_{c1} + 10^3F_{c2} + k_t(x_{in} - x_{us}) \tag{17.2}$$

where the constant, 10^3, is a conversion factor from "kN" unit to "N".
The state variables are defined as follows,

$$x_1 = x_s, \quad x_2 = \dot{x}_s, \quad x_3 = x_{us}, \quad x_4 = \dot{x}_{us} \tag{17.3}$$

The measured outputs used for control design are the suspension deflection, $s_d = x_1 - x_3$ and the sprung mass acceleration, $a_s = \dot{x}_2$.

The state-space representation of this system is as follows:

$$\dot{x} = Ax + B_w w + B_u u \tag{17.4}$$

$$y = Cx + D_w w + D_u u \tag{17.5}$$

where $w = x_{in}$ is the road disturbance input, and u is the vector of active control forces, $\begin{bmatrix} F_{c1} \\ F_{c2} \end{bmatrix}$.

The state-space representation is rewritten as:

$$\begin{bmatrix} \dot{x}_1 \\ \dot{x}_2 \\ \dot{x}_3 \\ \dot{x}_4 \end{bmatrix} = \begin{bmatrix} 0 & 1 & 0 & 0 \\ \frac{-k_s}{m_s} & \frac{-b_s}{m_s} & \frac{k_s}{m_s} & \frac{b_s}{m_s} \\ 0 & 0 & 0 & 1 \\ \frac{k_s}{m_{us}} & \frac{b_s}{m_{us}} & \frac{-k_s - k_t}{m_{us}} & \frac{-b_s}{m_{us}} \end{bmatrix} \begin{bmatrix} x_1 \\ x_2 \\ x_3 \\ x_4 \end{bmatrix} + \begin{bmatrix} 0 & 0 \\ \frac{10^3}{m_s} & 0 \\ 0 & 0 \\ \frac{-10^3}{m_{us}} & \frac{10^3}{m_{us}} \end{bmatrix} \begin{bmatrix} F_{c1} \\ F_{c2} \end{bmatrix} + \begin{bmatrix} 0 \\ 0 \\ 0 \\ \frac{k_t}{m_{us}} \end{bmatrix} [x_{in}] \tag{17.6}$$

$$y = \begin{bmatrix} 1 & 0 & -1 & 0 \\ \frac{-k_s}{m_s} & \frac{-b_s}{m_s} & \frac{k_s}{m_s} & \frac{b_s}{m_s} \end{bmatrix} \begin{bmatrix} x_1 \\ x_2 \\ x_3 \\ x_4 \end{bmatrix} + \begin{bmatrix} 0 & 0 \\ \frac{10^3}{m_s} & 0 \end{bmatrix} \begin{bmatrix} F_{c1} \\ F_{c2} \end{bmatrix} + \begin{bmatrix} 0 \\ 0 \end{bmatrix} [x_{in}] \tag{17.7}$$

The suspension parameter values are considered for our proposed model are given in Table 17.2.

The general block diagram for the design of the feedback controller is shown in Figure 17.2.

Youla parameterization, as discussed in the previous chapters, is a frequency domain control method. Therefore, the plant (quarter-car model) as given in Eqs. (17.4 and 17.5) has to be Laplace transformed to obtain its transfer function matrix (TFM). Before making this transformation, we will rewrite the state-space representations as:

$$\dot{x} = Ax + B_2u \tag{17.8}$$

$$y = C_2x + D_2u \tag{17.9}$$

where $C_2 = \begin{bmatrix} 1 & 0 & -1 & 0 \\ \frac{-k_s}{m_s} & \frac{-b_s}{m_s} & \frac{k_s}{m_s} & \frac{b_s}{m_s} \end{bmatrix}$ and $D_2 = \begin{bmatrix} 0 & 0 \\ \frac{10^3}{m_s} & 0 \end{bmatrix}$. We can then compute the TFM, G_p, using the following:

$$G_p = C_2(sI - A)^{-1}B_2 + D_2 \tag{17.10}$$

Table 17.2 Nomenclature.

Parameter	Value
m_s	300 kg
m_u	60 kg
k_s	16 000 N/m
k_t	190 000 N/m
b_s	1000 N/(m/s)

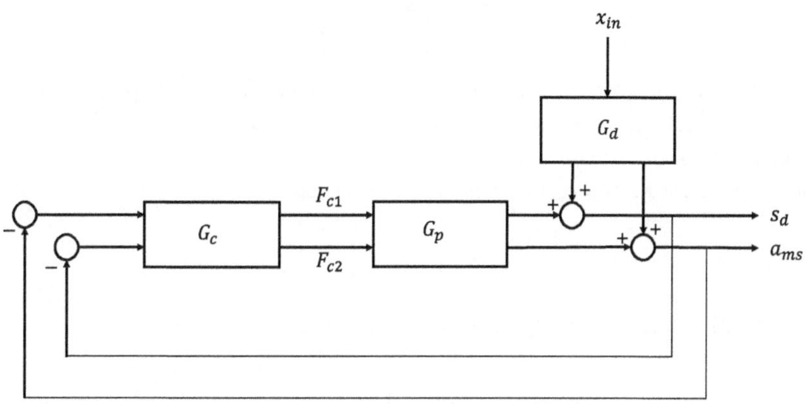

Figure 17.2 General MIMO feedback control loop.

The transfer function G_p is written as:

$$G_p = \begin{bmatrix} G_{11} & G_{12} \\ G_{21} & G_{22} \end{bmatrix} \tag{17.11}$$

where the inputs are $\begin{bmatrix} F_{c1} \\ F_{c2} \end{bmatrix}$ and the outputs are $\begin{bmatrix} s_d \\ a_{ms} \end{bmatrix}$. This TFM is used in Section 17.1.2 for designing a MIMO Youla controller.

17.1.2 MIMO Youla Control Design

The MIMO Youla control design approach, introduced in this book, starts by first computing the Smith-McMillan (SM) form of the plant TFM.

For computing the SM form, we first compute a polynomial matrix numerator, $P(s)$, and a denominator matrix of the plant, $\delta(s)$, such as:

$$G_p(s) = P(s)\delta(s)^{-1} \tag{17.12}$$

The plant TFM is

$$G_p(s) = \begin{bmatrix} \dfrac{20(s^2+527.8)}{(s^2+2.864s+49.85)*(s^2+17.14s+3388)} & \dfrac{-16.667(s^2)}{(s^2+2.864s+49.85)*(s^2+17.14s+3388)} \\ \dfrac{3.3333s^2(s^2+3167)}{(s^2+2.864s+49.85)*(s^2+17.14s+3388)} & \dfrac{55.556s^2(s+16)}{(s^2+2.864s+49.85)*(s^2+17.14s+3388)} \end{bmatrix}$$

where $P(s) = \begin{bmatrix} 20(s^2+527.8) & -16.667(s^2) \\ 3.3333s^2(s^2+3167) & 55.556s^2(s+16) \end{bmatrix}$ and $\delta(s) = (s^2+2.864s+49.85)*(s^2+17.14s+3388)$.

Knowing polynomial matrix $P(s)$ and the denominator $\delta(s)$, we can compute the SM form, M_p, using the greatest common denominator of minors of $P(s)$ as follows:

- GCD of the 1×1 minor of $P(s)$, D_1: 1
- GCD of the 2×2 minor of $P(s)$, D_2: $s^2\delta(s)$

SM form for this plant is computed as follows:

$$M_p = \frac{1}{\delta(s)} \begin{bmatrix} \epsilon_1 & 0 \\ 0 & \epsilon_2 \end{bmatrix} \tag{17.13}$$

where $\epsilon_i = \frac{D_i}{D_{i-1}}$ for $i = 1,2$, and knowing that $D_0 = 1$.

Therefore, we can write M_p as:

$$M_p = \begin{bmatrix} \dfrac{1}{\delta(s)} & 0 \\ 0 & s^2 \end{bmatrix} \tag{17.14}$$

The roots of $\delta(s)$ are all in the LHP and hence are stable. The SM form has four stable poles and two zeros at $s = 0$. These zeros are on the $j\omega$-axis and cannot be canceled. The right and left unimodular matrices are found using a method described in the last example in Section 13.3.

The decoupled Youla TFM M_Y for MIMO control design is written as:

$$M_Y = \begin{bmatrix} Y_1 & 0 \\ 0 & Y_2 \end{bmatrix} \tag{17.15}$$

We also know that the decoupled complementary sensitivity TFM M_T is

$$M_T = M_p M_Y = \begin{bmatrix} \dfrac{Y_1}{\delta(s)} & 0 \\ 0 & Y_2 s^2 \end{bmatrix} \tag{17.16}$$

We select Y_1 and Y_2 so that the diagonal entries and exactly the same, this makes designing this MIMO controller easier. The TFM, M_T is written as:

$$M_T = \begin{bmatrix} \dfrac{\alpha\tau^2 s^2}{\delta(s)(\tau s+1)^2} & 0 \\ 0 & \dfrac{\alpha\tau^2 s^2}{\delta(s)(\tau s+1)^2} \end{bmatrix} \tag{17.17}$$

As pointed out earlier, the two zeros of the SM form cannot be canceled with the selection of Y_2; otherwise, the internal stability will be compromised. The interpolation conditions in this case for meeting internal stability enforces that

$$T(0) = 0 \quad \text{and} \quad \frac{\mathrm{d}T(0)}{\mathrm{d}s} = 0 \tag{17.18}$$

which are both met if the zeros of M_p are not canceled.

The constant α is used as a tuning parameter. If the tuning parameter is selected to be 1, then in the frequency range set by τ and $\delta(s)$, the closed-loop transfer function, M_T, will be equal to 1. But, as our objective here to reject disturbances and not target following, the tuning parameter is selected to be 0.8 in this range. In addition, the two additional poles are provided to counter the effect of the two zeros, in the numerator of the plant, for correctly shaping the complementary sensitivity TFM at low frequency.

The shapes of T_y and S_y are illustrated in Figure 17.3. As shown, the sensitivity TFM, S_y, is shaped to be a bandstop filter around body resonant frequency. By shaping S_y in this manner, we will also reduce the actuator effort, and it can be shown as we increase the frequency range of the bandstop filter, and this will result in an increase in the actuator effort.

By selecting and computing, M_Y, M_T, the unimodular matrices, U_L and U_R, we can then compute the final coupled closed-loop TFMs and the controller as follows:

$$Y = U_R M_Y U_L \tag{17.19}$$

$$T_y = U_L^{-1} M_p M_Y U_L \tag{17.20}$$

$$S_y = I - T_y \tag{17.21}$$

$$G_c = U_R(I - M_Y M_p)^{-1} M_Y U_L \tag{17.22}$$

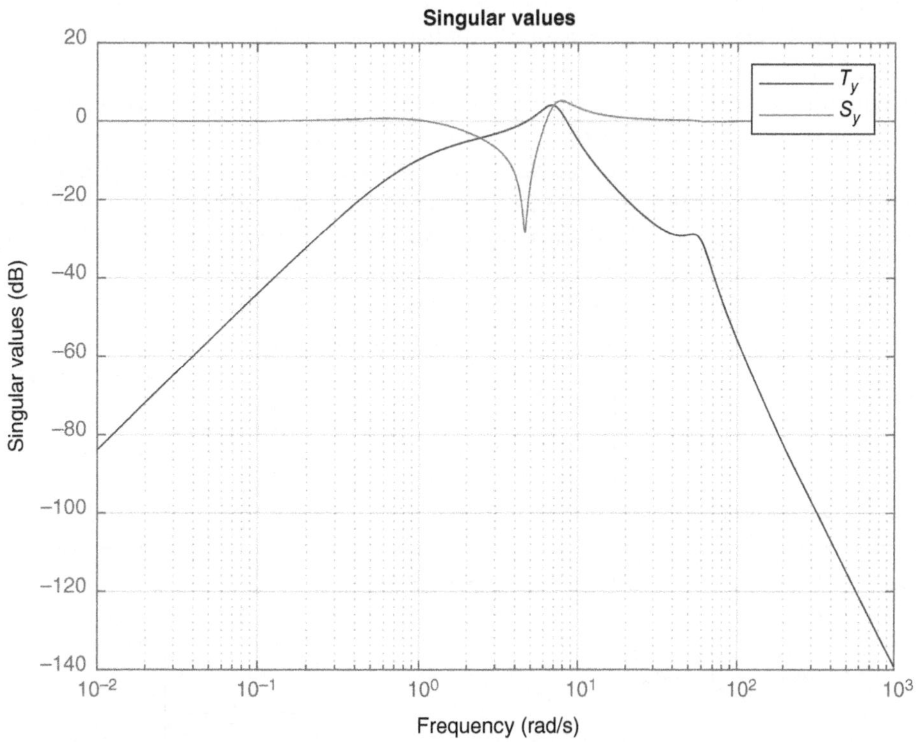

Figure 17.3 Sensitivity and complementary sensitivity closed-loop transfer function matrices.

We will compare the results of the MIMO Youla control development after introducing the MIMO H_∞ control design in Section 17.1.3. The *m* file for MIMO Youla control design is included in the Appendix D.

17.1.3 H_∞ Control Design Technique

The H_∞ controller, in this section, is previously designed under in the "Robust Control of an Active Suspension" example for Matlab's Robust Control Toolbox [45]. This example includes one actuator only, and in this section, we augment this plant with the second actuator as discussed in Section 17.1.2. The simulation results for the H_∞ controller is then compared with the Youla controller developed in Section 17.1.2.

The mixed sensitivity loop-shaping approach as discussed in the previous chapters requires augmentation of the nominal plant by shaping filters, as illustrated in Figure 6.7. However, in this example, a general loop-shaping approach is utilized, which in a sense more flexible than mixed sensitivity approach, as illustrated in Figure 17.4. The only difference between this model and the model in Section 17.1.2 is the addition of actuator dynamics ("Act" block) as shown in Figure 17.4. When comparing the results in Section 17.1.4, we will illustrate that we need to also augment the model in Section 17.1.2 with actuator dynamics to produce comparable results. We have also changed the names of the variables to be consistent with the Matlab example.

As pointed out in the "Design Setup" section of the Matlab example, the main control objectives for H_∞ control design are formulated in terms of passenger comfort and road handling, similar to Section 17.1.2. These objectives relate to body acceleration, a_b, same as the a_{ms} measurement output of Section 17.1.2, and suspension travel s_d. It is stated that other factors that influence control design include the characteristics of the road disturbance, the quality of sensor measurements, and the limits on the available control forces. To use H_∞ synthesis algorithms, these objectives are expressed as a single-cost function to be minimized by augmenting the base model with appropriate shaping filters as shown in Figure 17.4.

The quarter-car model used in this control design is exactly the same as the model introduced in the Youla control design. The disturbance input to this model, as shown in Figure 17.4, is the road disturbance input d_1, which is filtered to produce input r, or x_{in} as used in the model for Youla control design. The controlled inputs are the two actuator inputs, f_{s1} and f_{s2}, which are previously labeled as F_{c1} and F_{c2}, and the only difference here is the inclusion of the actuator dynamics, Act block, which filters the controller outputs, u_1 and u_2. For comparison basis, we will also augment the model for Youla control design with actuator dynamics.

Figure 17.5 illustrates the frequency responses, a_b and s_d, of the passive model to the road disturbance input r and the two actuator inputs, f_{s1} and f_{s2}.

The transfer function from the first actuator to body acceleration and suspension travel has two zeros on $j\omega$-axis at 56.3 and 23 rad/s. These frequencies are called wheel-hop and rattle space frequencies. Because of these zeros, the feedback control cannot improve the responses from the road disturbance to body acceleration and suspension deflection at these frequencies. However, it seems that at these frequencies, the second actuator would be able to improve these responses. But, Figure 17.5 assumes no scaling of the second actuator relative to the first. In this work, we will assume that the second

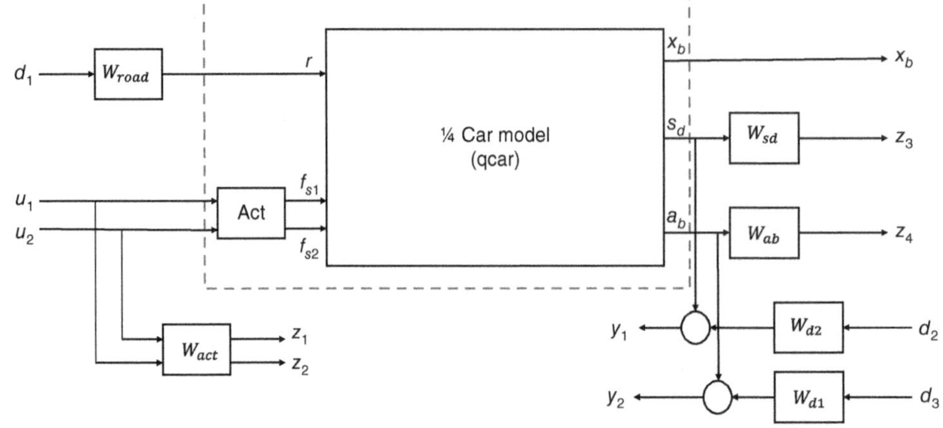

Figure 17.4 Disturbance rejection formulation.

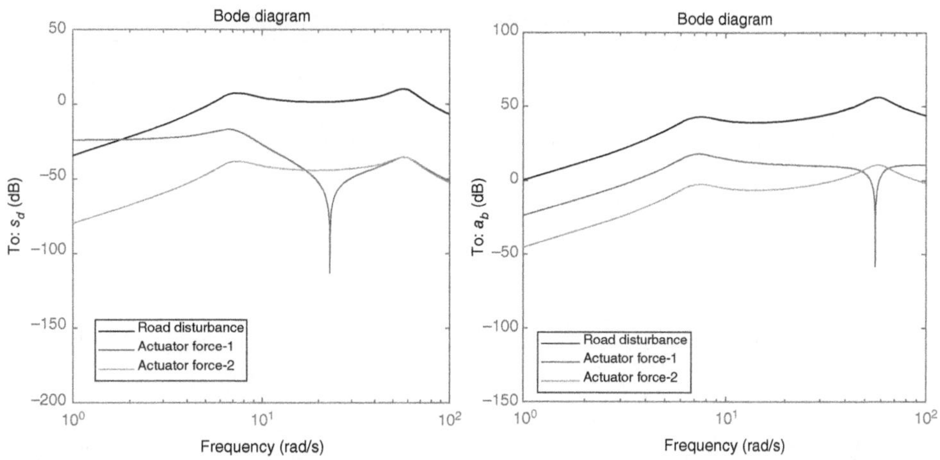

Figure 17.5 Frequency gain from road disturbance and actuator forces to suspension travel and body acceleration.

actuator can only produce a force in the range of 1/5 to 1/50 of the first actuator. This seems to be a wide range, but this is due to the lack of experimental work in this area.

17.1.4 Uncertain Actuator Model

For H_∞ control development, a hydraulic actuator is assumed to be used for ASC which is connected between the body mass, m_s, and the wheel assembly mass, m_{us}. The second actuator is used to modulate tire pressure by solenoid valves and an air compressor and is simply modeled by an actuator which generates a force and is connected between the road input and the wheel assembly.

The nominal dynamics of actuators are represented by the first-order and the second-order transfer functions. It is assumed that the first actuator is four times faster than the second actuator which uses tire pressure modulation for force generation.

The first-order transfer function which characterizes the dynamics of the first actuator is

$$ActNom1 = \frac{1}{1/60s + 1} \tag{17.23}$$

The second-order transfer function which characterizes the dynamics of the second actuator is

$$ActNom2 = \frac{1}{0.0025s^2 + 0.1s + 1} \tag{17.24}$$

Figure 17.6 illustrates the step responses of the nominal actuator dynamics.

These nominal models only approximate the physical actuator dynamics. A family of actuator models to account for modeling errors and variability in the actuators and quarter-car models can be utilized. For the first actuator, at low frequency, below 3 rad/s, the model can vary up to 40% from its nominal value. At 15 rad/s, the uncertainty increases to 100% and at around 1000 rad/s, the uncertainty reaches 300%. For the second actuator, at low frequency, below 1 rad/s, the model can vary up to 40% from its nominal value. At 4 rad/s, the uncertainty increases to 100% and at around 250 rad/s, the uncertainty reaches 300%. The weighting functions W_{unc1} and W_{unc2} are selected to set the level of uncertainties with frequencies. Figure 17.7 shows the Bode magnitude plot of the actuator transfer functions including their frequency weighted uncertainties. In this figure, the nominal actuator models are represented with plus-marks, and a family of uncertain models is shown as solid lines.

17.1.5 Design Setup

The H_∞ control is designed using the augmented plant for disturbance rejection formulation in Figure 17.4. In this section, the weighting filters are selected for disturbance inputs, which includes road and sensor noise, d_i, and the controlled inputs, u_i, and regulated outputs, z_i.

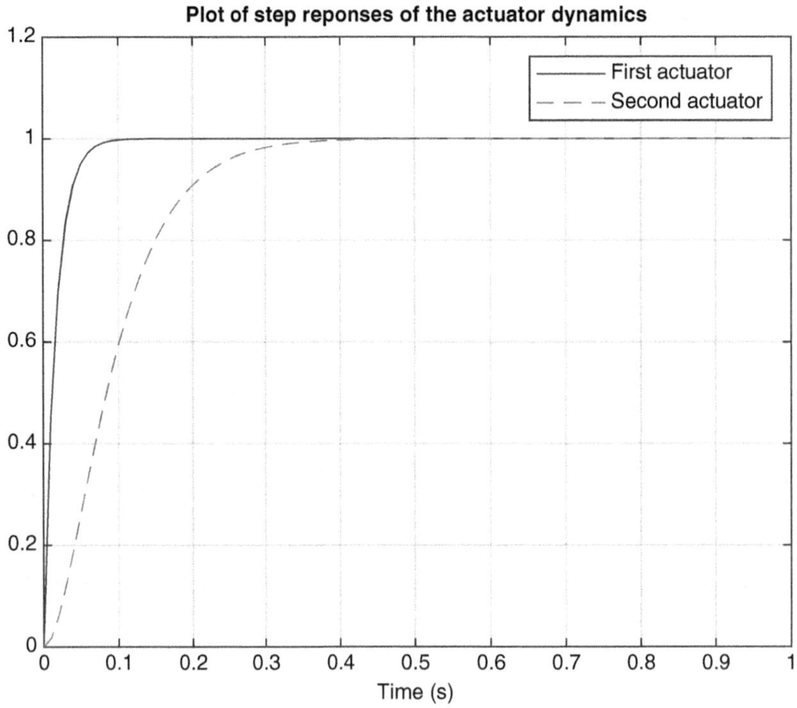

Figure 17.6 Step responses of the nominal actuator dynamics.

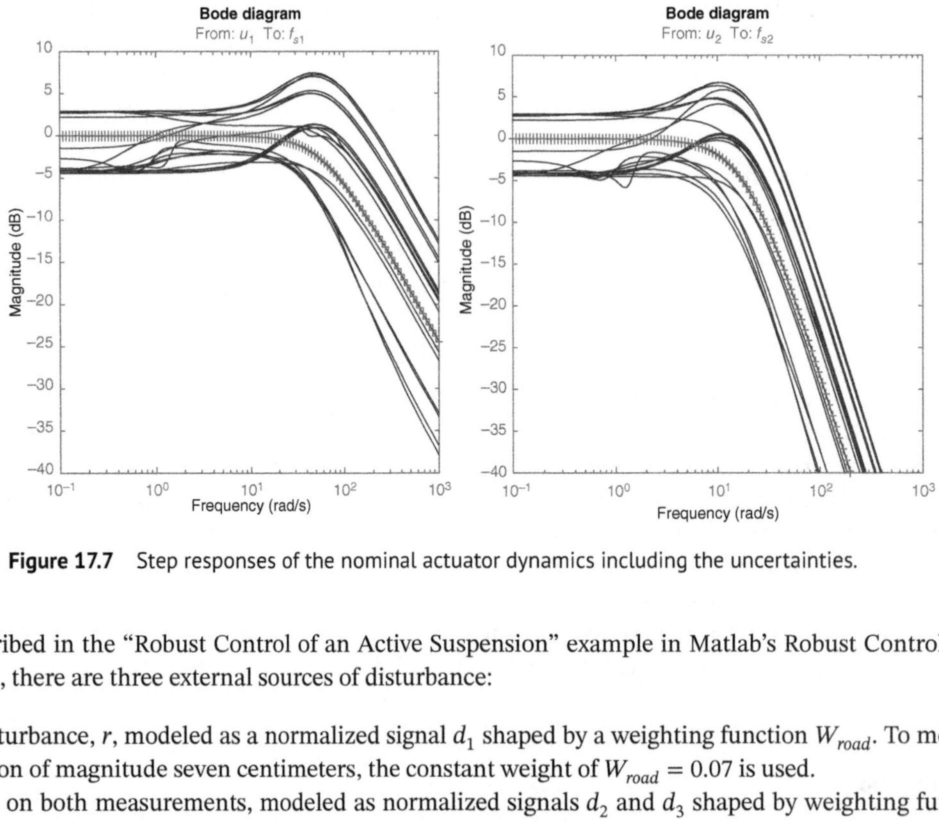

Figure 17.7 Step responses of the nominal actuator dynamics including the uncertainties.

As it is described in the "Robust Control of an Active Suspension" example in Matlab's Robust Control Toolbox documentation [45], there are three external sources of disturbance:

- The road disturbance, r, modeled as a normalized signal d_1 shaped by a weighting function W_{road}. To model broadband road deflection of magnitude seven centimeters, the constant weight of $W_{road} = 0.07$ is used.
- Sensor noise on both measurements, modeled as normalized signals d_2 and d_3 shaped by weighting function W_{d2} and W_{d3}. It is proposed that $W_{d2} = 0.01$ and $W_{d3} = 0.5$ to model broadband sensor noise of intensity 0.01 and 0.5, respectively. It is stated that in a more realistic design, these weights would be frequency dependent to model the noise spectrum of the displacement and acceleration sensors.

- Closed-loop targets from road disturbance r to suspension deflection s_d for handling improvement and to body acceleration a_b for comfort improvement are specified as high pass filters up to only 10 rad/s due to the plant zeros on $j\omega$-axis and the actuator uncertainties. These closed-loop transfer functions are selected as:

$$HandlingTarget = 0.04\frac{(1/8)s + 1}{(1/80)s + 1} \tag{17.25}$$

$$ComfortTarget = 0.04\frac{(1/0.45)s + 1}{(1/150)s + 1} \tag{17.26}$$

The open-loop responses and the closed-loop targets are shown in the Bode magnitude plots of Figure 17.8.
The performance weights W_{sd} and W_{ab} are the reciprocal of these handling and comfort targets. It is well known that in ASC design, there is a trade-off between handling and passenger comfort. To account for this trade-off, it is suggested to construct three sets of weights $(\beta W_{sd}, (1-\beta)W_{ab})$ corresponding to three different trade-offs: comfort ($\beta = 0.01$), balanced ($\beta = 0.5$), and handling ($\beta = 0.99$). Here, we only consider the balanced trade-off where $\beta = 0.5$. Therefore, our W_{sd} and W_{ab} weighting functions are written as a function of β, as follows:

$$W_{sd} = \frac{\beta}{HandlingTarget} \tag{17.27}$$

$$W_{ab} = \frac{1-\beta}{ComfortTarget} \tag{17.28}$$

The control objective is reinterpreted as disturbance rejection goal: minimize the impact of the disturbances d_i on a weighted combination of control effort u_i, suspension travel s_d, and body acceleration a_b. Then, the H_∞ norm optimization is used to design a controller that minimizes the H_∞ norm from the disturbance inputs d_i to the regulated outputs z_i, where

$$d_i = \begin{bmatrix} d_1 \\ d_2 \\ d_3 \end{bmatrix} \tag{17.29}$$

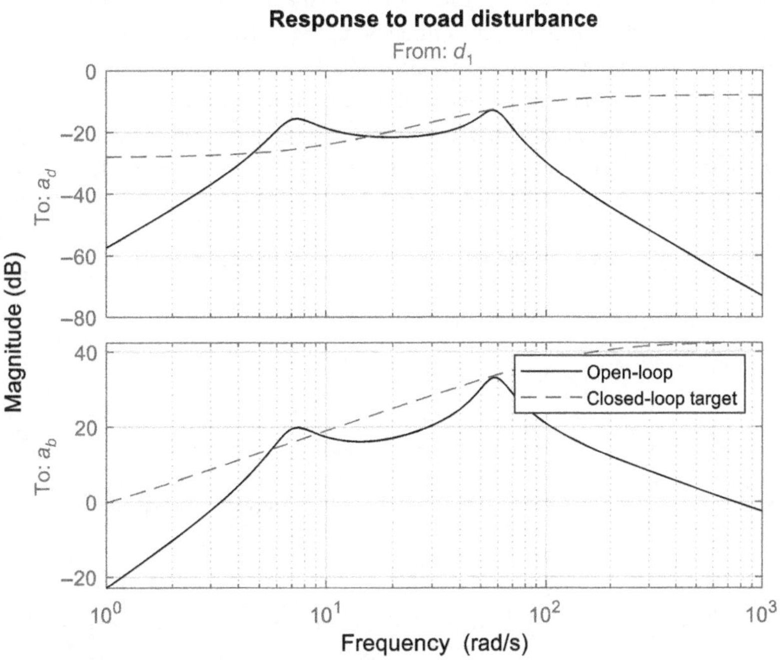

Figure 17.8 Suspension deflection and body acceleration frequency responses to road disturbance input.

and

$$z_i = \begin{bmatrix} z_1 \\ z_2 \\ z_3 \\ z_4 \end{bmatrix} \tag{17.30}$$

17.1.6 Simulation Results

In this section, we provide the comparative analysis between the H_∞ and the Youla MIMO control design using the quarter-car model with two active force actuators.

We first simulate both active systems in time domain by a smooth bump with a height of 0.05 m as a road disturbance input, where $r = 0.025 * (1 - cos(8 * pi * t)$.

The body deflection, x_b, body acceleration a_b, suspension deflection, sd, and actuator forces, f_{s1} and f_{s2} for both H_∞ and MIMO Youla are shown below.

Figure 17.9 illustrates the simulation resulted for a time repose to a bump road input for MIMO Youla and H_∞ control design. It is interesting to observe that the results for body acceleration and suspension deflection are similar. However, the body travel results differ and this is fundamentally due to the difference in active control forces. Note that "**Open-Loop**" means passive system.

The MIMO Youla is designed without any consideration to the actuator dynamics and saturation (in the linear case, saturation is referred to the steady-state values) constraints, as described in Section 17.1.5. However, in the MIMO H_∞ design, these constraints are addressed through the combination of weighting functions and specifying transfer functions for modeling actuator dynamics. But, the fundamental difference between the MIMO Youla and the H_∞ controllers is that the Youla controller is internally stable, while the H_∞ controller is not!

Next, we show by including actuator dynamics and scaling the second actuator force, we can produce very similar results to H_∞ controller. We have modified our Youla TFMs, described above, so that the closed-loop TFM, M_T, is equal to,

$$M_T = \begin{bmatrix} \dfrac{\alpha\tau^2 s^2}{\delta(s)(\tau s+1)^2(0.02s+1)^2} * ActNom1 & 0 \\ 0 & \beta * \dfrac{\alpha\tau^2 s^2}{\delta(s)(\tau s+1)^2(0.02s+1)^2} * ActNom2 \end{bmatrix} \tag{17.31}$$

We have included the actuator dynamics and scaling factor β in Youla design. The additional transfer function, $\frac{1}{(0.02s+1)^2}$, is used to shape/lower the actuator effort in the high-frequency region (5 times greater than 10 rad/s). The Matlab m-file for this design and the H_∞ control design are included in Appendix D.

By scaling (the scaling factor β is 1/50 in these results) and including actuator dynamics in Youla design, we will obtain very similar results to H_∞ control at lower actuator efforts, as shown in Figure 17.10. There are couple of important points to mention here. Firstly, after adding all the dynamics to the Youla design, we have increased its number of states to 43, when the H_∞ controller includes 13 states. It should be pointed out that using controller order reduction technique (included

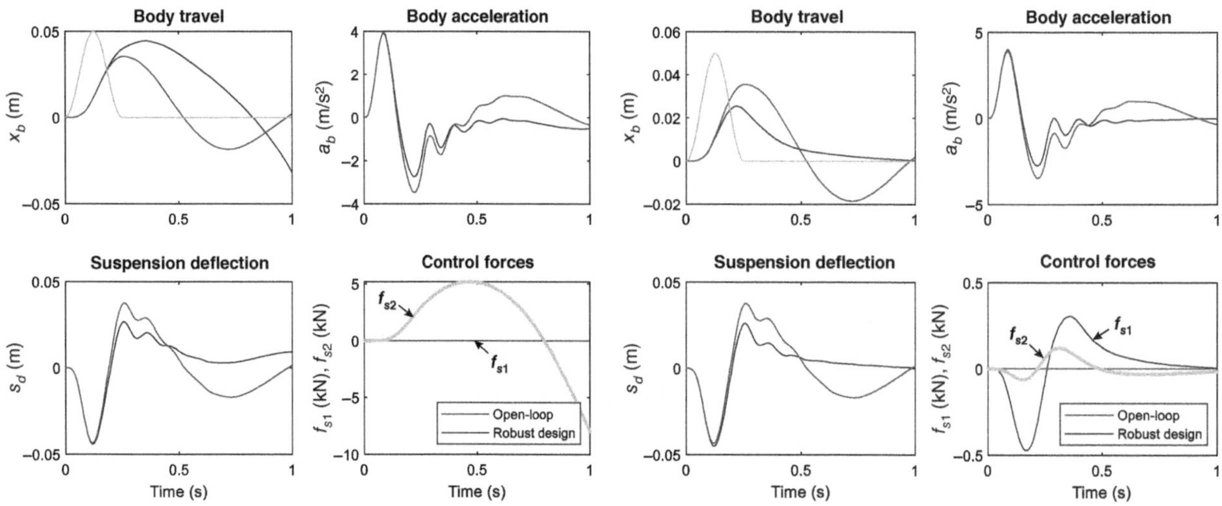

Figure 17.9 H_∞, right, and MIMO Youla, left, comparison for a smooth bump road input.

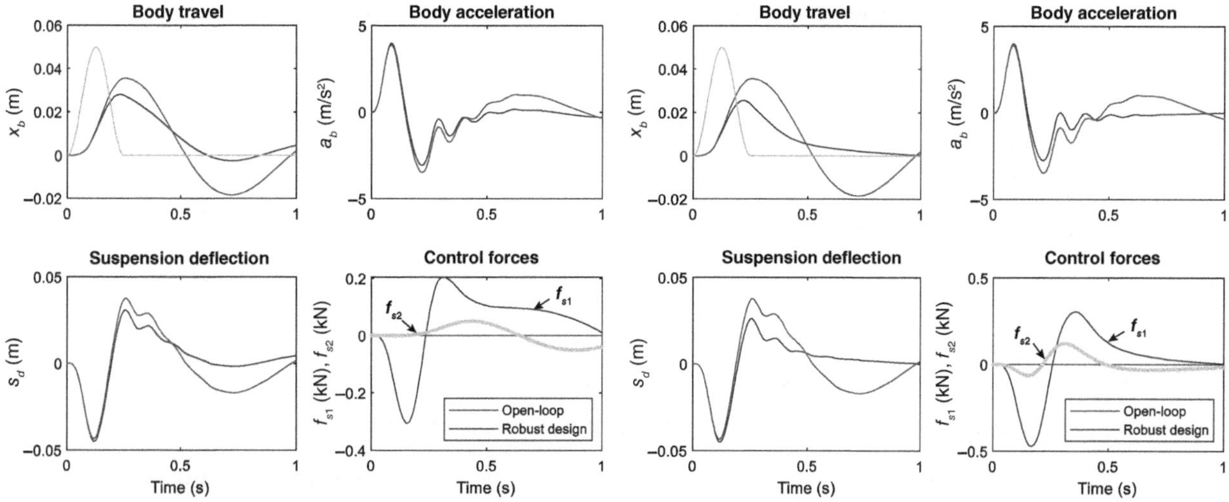

Figure 17.10 H_∞, right, and scaled MIMO Youla with actuator dynamics, left, comparison for a smooth bump road input.

in the m-file), the order of the Youla controller could be safely reduced to 18 without sacrificing performance or internal stability. The H_∞ controller on the other hand is not an internally stabilizing controller!

Before ending this section, we will present the frequency domain simulation results, where the Bode magnitude of road input to body acceleration, suspension deflection, and body travel, for active and passive systems. In addition, we provide the frequency domain results for actuator effort comparison of the MIMO Youla and H_∞ control.

The simulation results in Figure 17.11 illustrates that MIMO Youla design do not isolate the body around its resonant frequency as well as H_∞ control; however, to accomplish this, H_∞ control compromises body responses in low-frequency region.

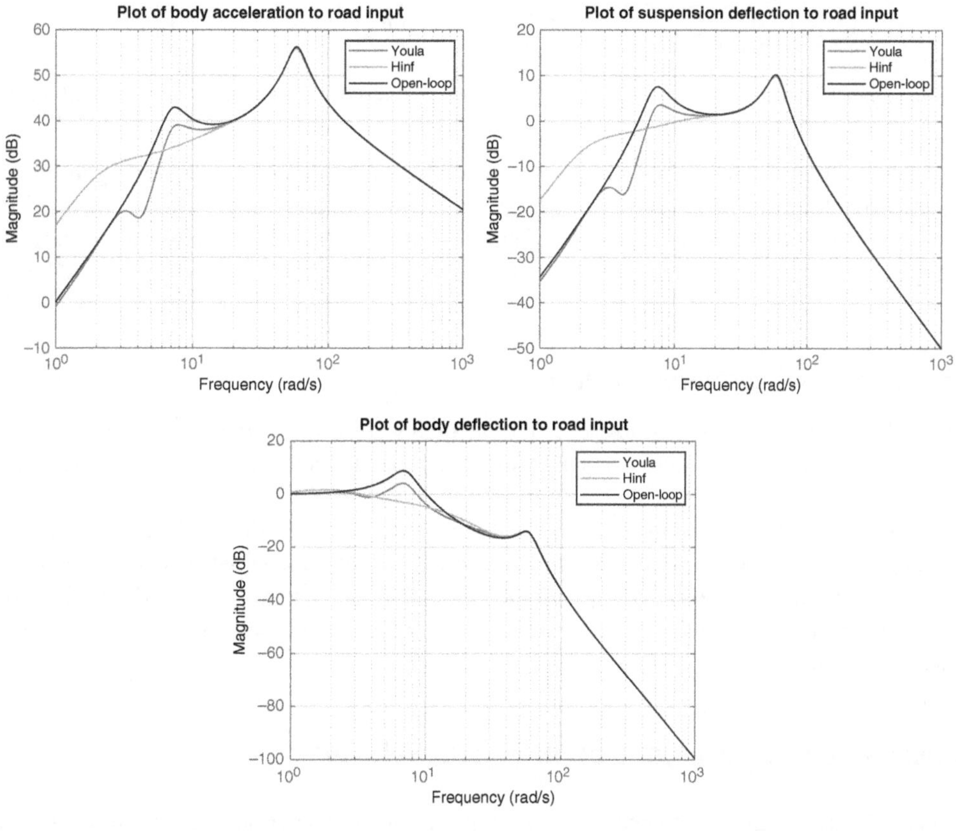

Figure 17.11 Body acceleration, suspension deflection, and body travel frequency responses to road disturbance input.

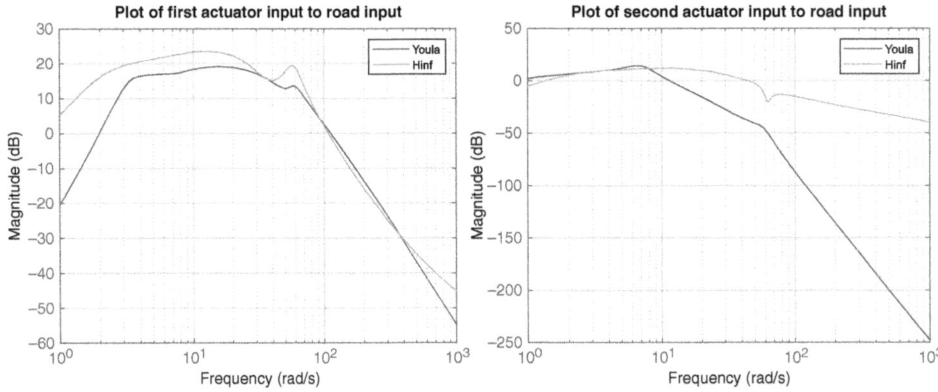

Figure 17.12 Actuator effort frequency responses to road disturbance input.

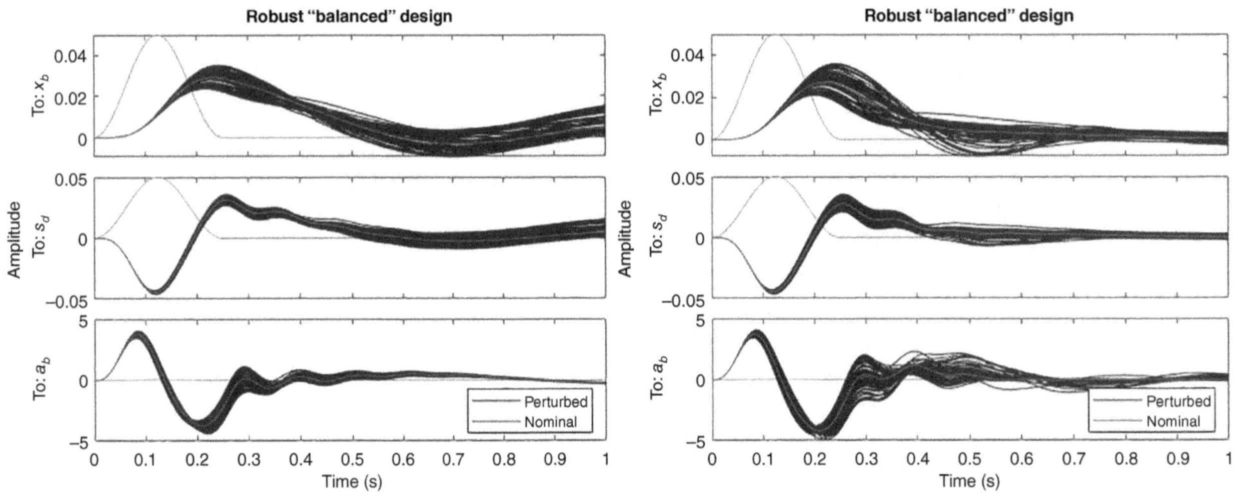

Figure 17.13 Time domain robustness test to a road bump for Youla, left, and H_∞, right, control using 100 different actuator models.

The frequency response magnitude of actuator efforts to road disturbance input is shown in Figure 17.12. It is apparent that overall the actuator efforts of both control strategies are comparable except with MIMO Youla that has slight edge in the low-frequency region of the first actuator. The H_∞ control requires high second actuator effort in the high-frequency region when comparing to Youla controller. This is due to the way H_∞ controller utilizes weighting functions to weight actuator efforts.

17.1.7 Robustness Test: Actuator Model Variations

Figure 17.7 shows a family of actuator models to account for modeling errors and variability in the actuator and quarter-car models. As described previously, this family consists of a nominal model with a frequency-dependent amount of uncertainty.

The simulation results presented in Figure 17.13 illustrate the performance of the Youla and H_∞, controllers to a road bump for 100 actuator models randomly selected from the uncertain actuator model set. The robustness of body travel and suspension deflection for both controllers are fairly similar. However, when comparing the body acceleration robustness to the actuator variations, interestingly Youla control design has much tighter response with respect to its nominal response in comparison with H_∞ control.

17.2 Advanced Engine Speed Control for Hybrid Vehicles

The requirements for the design and development of engine speed control of a hybrid electric vehicle (HEV) are drastically different than those of a conventional vehicle. In a hybrid propulsion-based system, the vehicle could be moving in electric

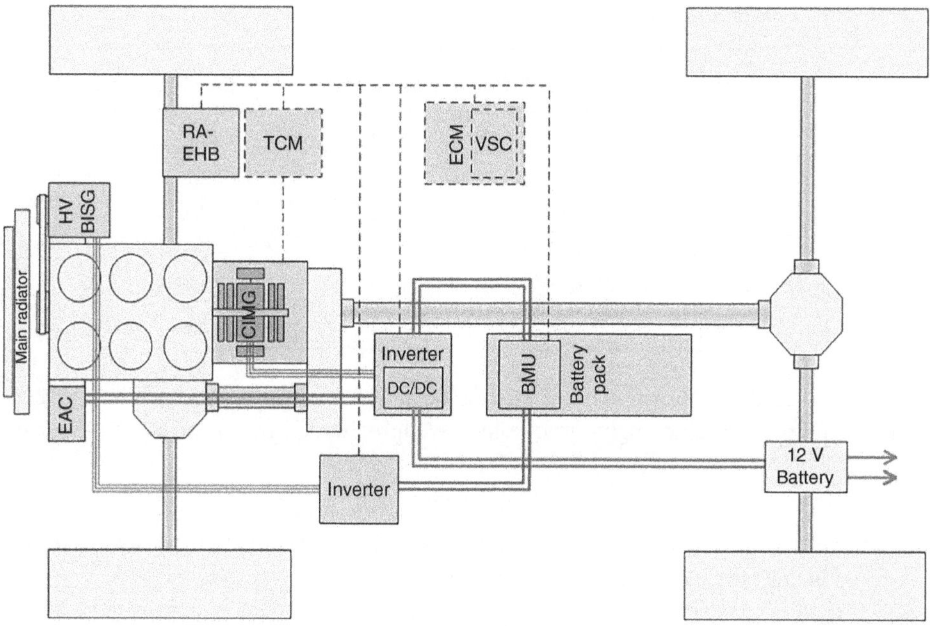

Figure 17.14 Schematic of the HEV powertrain with a BISG and a CISG.

mode only and when required, the internal combustion engine (ICE) has to be turned on, and possibly operated at higher speeds, in a prescribed time, and in a smooth manner in order to maintain driveability. In addition, in a hybrid vehicle, several actuators are to be utilized for the purpose of ICE speed control – these actuators should be driven by appropriate control efforts. Thus, both hybrid topology and control algorithm play a key role in the challenging ICE speed control development [46].

In this section, we study the impact of two new MIMO Youla control strategies and compare these results with the standard H_∞ control method given only one hybrid topology. The interested readers could examine the reference above for the impact of different hybrid topologies on ICE speed control design. An illustration of two different hybrid topologies (Belt-Integrated Starter Generator (BISG) and Crank-Integrated Starter Generator (CISG)) is given in Figure 17.14.

We will only consider the hybrid topology with ICE and CISG in this section. The rest of this section is organized as follows: in the first part, model for the hybrid electric vehicle is presented using an empirical diesel ICE model and CISG. We will then present the design of advanced control system for the application of hybrid powertrain engine speed control. The effectiveness of the proposed control methods, based on given performance requirements, is illustrated through simulation results. In addition, we present experimental results from on-board vehicle testing.

17.2.1 Diesel Hybrid Electric Vehicle Model

The diesel ICE model includes a mapping from the delayed fueling input, u_{fuel}, to the indicated torque, T_{ind}, therefore has limited bandwidth due to internal combustion dynamic effects, arising mainly from combustion lag and turbo delay. In a transfer function form, we can write this model as:

$$T_{ind} = \frac{e^{-\tau_c(\omega)s}}{\tau_r(\omega)s + 1} u_{fuel} \tag{17.32}$$

where τ_r is the speed-dependent time constant due to combustion lag, τ_c is the speed-dependent time delay due to fueling, and u_{fuel} is the fueling input.

This empirical simplified diesel engine model is derived from experimental tests of a 2.0L I4 turbo diesel engine. The speed-dependent fueling delay (τ_c) and combustion lag (τ_r) are given in Table 17.3.

These results are captured by measuring the response of the engine brake torque to a step change in the equivalent engine fueling torque.

The HEV diesel model also includes a speed-dependent torque loss T_{Loss} due to friction torque, ancillary torque, and pumping losses:

$$T_B = T_{ind} - T_{Loss} \tag{17.33}$$

Table 17.3 Nomenclature.

ω (rad/s)	ΔT_B (Nm)	τ_c (ms)	τ_r (ms)
1200	50	100	144
1200	100	140	142
1600	50	84	140
1600	100	96	137
2000	50	80	140
2000	100	72	134

This torque loss is typically a nonlinear function of the engine speed and could be represented by a lookup table. However, for the purpose of this example, the total engine torque loss is approximated by an uncertain linear function of ω and modeled as:

$$T_{Loss} = m\omega \tag{17.34}$$

where ω (rad/s) is engine speed and $m = [0.05 \quad 0.15]$ is a real parametric uncertainty that captures nonlinearities associated with engine torque losses in any possible engine operating condition.

As the engine speed is required to either follow the idle speed (say, e.g. 800 rpm) or synchronized with the CISG when the vehicle is moving at higher speeds, the rotational dynamics of the hybrid powertrain (including the crankshaft and motor inertia) is given as follows:

$$J\dot{\omega} + b\omega = T_B + T_M \tag{17.35}$$

where J is the effective combined moment of inertia, b is the effective joint damping coefficient, and T_M is the CISG motor torque.

The four equations given above form the simplified but realistic model of HEV powertrain. A schematic representation of the simplified parallel diesel hybrid electric vehicle (DHEV) is shown in Figure 17.15.

The diesel ICE model includes one actuator in the form of complex-valued (complex exponential) dynamics due to the uncertain value in the fueling control time delay, τ_c. This complex-valued dynamics is models by the first order Padè approximation and τ_c is assumed to be a real uncertain parameter as given below. Also, the diesel HEV model consists of four uncertain real parameters associated with the diesel engine combustion lag, τ_r, engine torque loss coefficient, m, the hybrid effective combined moment of inertia, J, and the effective damping coefficient, b. These parameters are modeled as their nominal values and their uncertain parameter values are used to test the robustness of both Youla and H_∞ controllers. The electric CISG is modeled as the following transfer function:

$$T_M = \frac{1}{\tau_e(\omega)s + 1} u_{volt} \tag{17.36}$$

where τ_e is the speed-dependent first-order time constant to account for the electrical model of the electric machine.

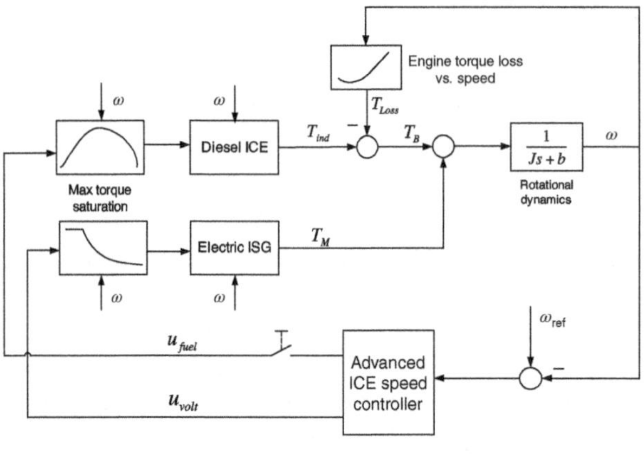

Figure 17.15 Block diagram of the simplified parallel diesel hybrid electric vehicle (DHEV) model.

The real uncertain parameter values for the DHEV case study are

1. $\tau_r \in [0.13, 0.15]\,\text{s} \rightarrow \overline{\tau}_r = 0.14, \tilde{\tau}_r = 0.1$
2. $m \in [0.05, 0.15]\,\text{N/(rad/s)} \rightarrow \overline{m} = 0.1, \tilde{m} = 0.05$
3. $J \in [0.4, 1.2]\,\text{Nm/(rad/s}^2) \rightarrow \overline{J} = 0.8, \tilde{J} = 0.4$
4. $b \in [0.05, 0.25]\,\text{Nm/(rad/s)} \rightarrow \overline{b} = 0.15, \tilde{b} = 0.1$
5. $\tau_c \in [0.075, 0.14]\,\text{s} \rightarrow \overline{\tau}_c = 0.105, \tilde{\tau}_c = 0.035$
6. $\tau_e \in [0.015, 0.075]\,\text{s} \rightarrow \overline{\tau}_c = 0.045, \tilde{\tau}_c = 0.03$

These uncertain parameters will be indirectly modeled and used for robustness tests for Youla and H_∞ controllers.

17.2.2 MISO Youla Control Design

The following TFM is derived from the DHEV model of Section 17.2.1,

$$G_p = [G_{pe} \quad G_{pm}] \tag{17.37}$$

where $G_{pe} = \frac{1}{\tau_r s+1} G_d \frac{1}{Js+b+m}$ is the Diesel ICE model including the rotational dynamics and torque loss, and $G_{pm} = \frac{1}{\tau_e s+1}\frac{1}{Js+b}$ is the electric motor model including the rotational dynamics, and G_d is the first-order Padè approximation of $e^{-\tau_c(\omega)s}$ time delay.

Figure 17.16 illustrates the feedback loop for both MISO (multiple inputs single output) Youla and H_∞ control design.

We propose two methods for deriving two different Youla controllers. Both methods could be used to address the shortcomings of over-actuated systems, such as the case here, where at least one actuator is not utilized effectively. Neither of the proposed methods uses unimodular matrices.

17.2.3 First Youla Method

In the first method, we select the Youla controller such as to transform the closed loop to a SISO transfer function. Knowing,

$$T_y = G_p Y \tag{17.38}$$

Then, we select $Y = \begin{bmatrix} Y_1 \\ \alpha Y_1 \end{bmatrix}$, and given that $G_p = [G_{pe} \quad G_{pm}]$, we can write the closed-loop transfer function as:

$$T_y = (G_{pe} + \alpha G_{pm})Y_1 \tag{17.39}$$

which is a SISO transfer function. The interesting fact about this structure for Youla parameterization is the variable α could be selected to be an adaptation parameter and this structure could be easily transformed to an adaptive robust controller or a Youla controller with dyanmic allocation. However, in this example, we selected the parameter α to be equal to a constant, similar to static allocation. We then select Y_1 to be,

$$Y_1 = \frac{1}{G_p} \frac{\omega_1^2 \omega_2^2}{(s^2 + 2\zeta\omega_1 s + \omega_1^2)(s^2 + 2\zeta\omega_2 s + \omega_2^2)} \tag{17.40}$$

By tuning ω_1, ω_2, ζ, and α, we can meet our design requirements for both our actuator efforts and time response measures, such as rise time, settling time, and overshoot as will be discussed in the simulation results.

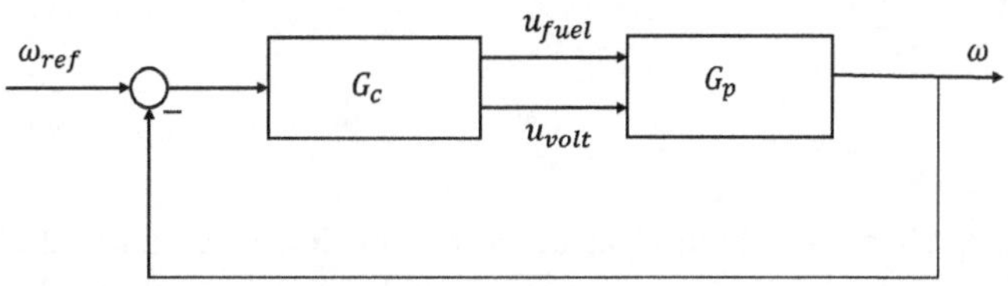

Figure 17.16 MISO feedback control loop.

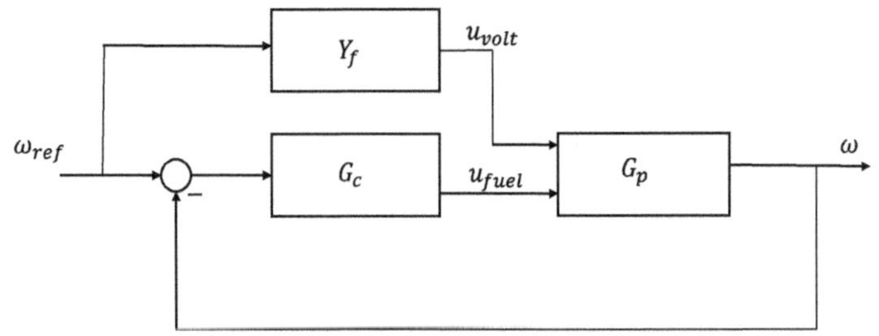

Figure 17.17 MISO feedforward/feedback control loop.

17.2.4 Second Youla Method

In order to better utilize actuator effectiveness in the over-actuated systems, such as the case, we propose the following control structure as depicted in Figure 17.17.

In this method, we design both a feedforward and feedback controller as shown in Figure 17.17. Feedforward control is designed to increase the effectiveness of the electric motor actuator such that it would aid in increasing the system response speed. The feedback controller is a SISO controller derived using SISO Youla control design technique. The feedback control adds robustness to the fast feedforward control. In this way, we could speed up the response without compromising robustness.

The feedforward controller is designed in a straightforward manner simply by the choice of Y_f as follows:

$$Y_f = \frac{1}{G_{pm}} \frac{\omega_1^2}{(s^2 + 2\zeta\omega_1 s + \omega_1^2)} \tag{17.41}$$

By tuning ω_1 and ζ, we can obtain a desired feedforward control response.

The feedback controller is also designed in a straightforward manner by the choice of Y as follows:

$$Y = \frac{1}{G_{pe1}} \frac{\omega_2^2}{(s^2 + 2\zeta\omega_2 s + \omega_2^2)} \tag{17.42}$$

where G_{pe1} is invertible part of G_{pe}. We selected ω_2 to be four times slower than ω_1 so that the feedback loop helps to increase the robustness and the feedforward control increases the speed of response. The closed-loop transfer function, T_y, and the feedback controller, G_c, are computed as discussed in the SISO section of this book. Furthermore, a Matlab m-file for these developments are included in Appendix D.

In order to simulate the results that are shown in the simulation results section, the plant is augmented as shown in Figure 17.18.

The augmented plant inputs and outputs are written as:

$$\begin{bmatrix} \omega \\ u_{volt} \\ u_{fuel} \\ r \\ e \end{bmatrix} = \begin{bmatrix} 0 & G_{pm} & G_{pe} \\ 0 & 1 & 0 \\ 0 & 0 & 1 \\ 1 & 0 & 0 \\ 1 & -G_{pm} & -G_{pe} \end{bmatrix} \begin{bmatrix} r \\ u_{volt} \\ u_{fuel} \end{bmatrix} \tag{17.43}$$

It should be noted that u_{volt} and u_{fuel} have units of Nm (units of torques such as engine and electric motor torques).

We then utilize the `lft` command to compute the closed-loop TFM from w (r or ω_{ref}) (reference input) to z. The reference input is selected to be 3000 rpm (300 rad/s) (this is the engine speed when the vehicle is moving) and the simulation results are presented.

17.2.5 H_∞ Control Design

We utilize mixed sensitivity loop-shaping technique, as described previously, to derive a H_∞ controller. The same plant augmentation as in Figure 17.18 is used but this time, we augment the plant with three filters, W_P, W_Δ, and W_Y as shown in Figure 17.19.

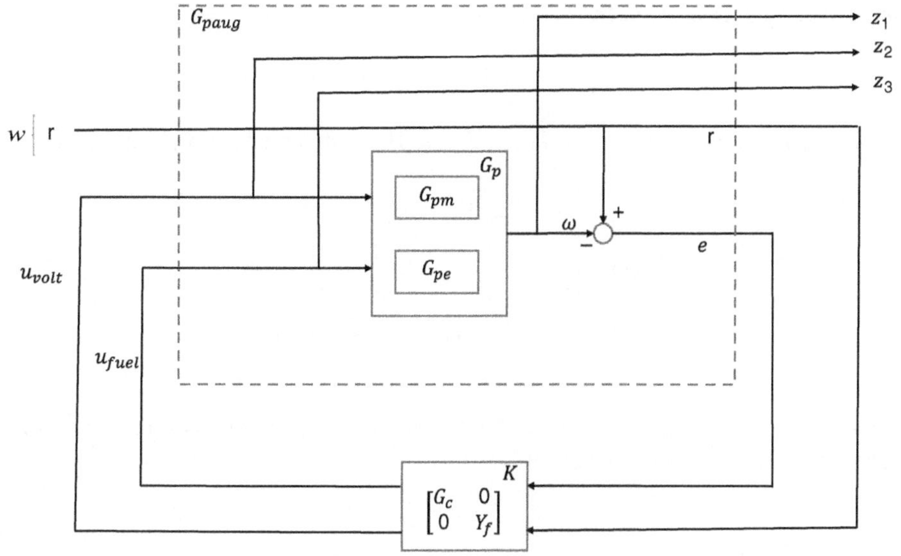

Figure 17.18 Plant augmentation.

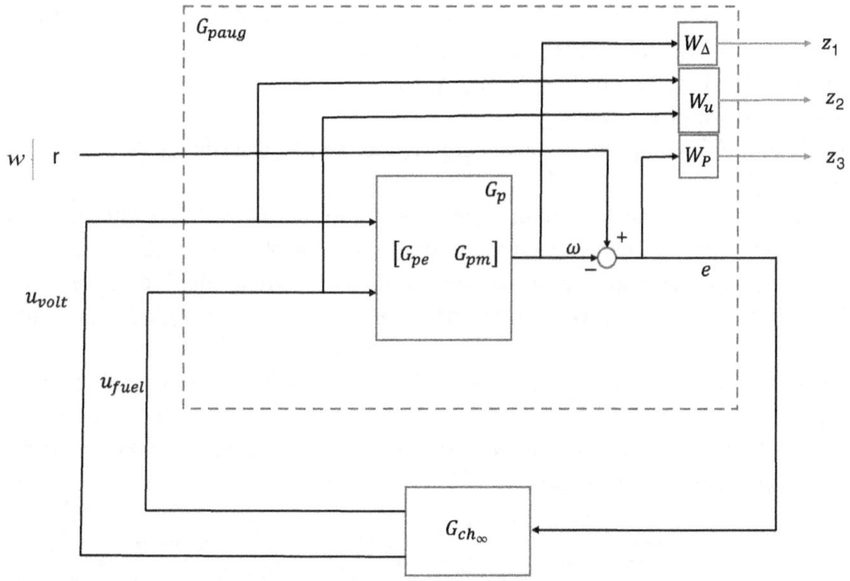

Figure 17.19 H_∞ plant augmentation.

In this general mixed sensitivity configuration, the controller has only access to the error signal and not the target speed. The plant is two inputs and one output (1×2), and the shaping filters W_Δ for shaping the closed-loop transfer function, T_y, W_P for shaping the sensitivity transfer function, S_y, and finally, W_u for shaping Youla TFM Y are selected as follows.

First, we define an ideal T_y such as,

$$T_y = \frac{\omega_c^2}{s^2 + 2\zeta\omega_c s + \omega_c^2} \tag{17.44}$$

We have tuned ω_c to be 5 rad/s and $\zeta = 0.707$, and we then compute the S_y (where $S_y = 1 - T_y$). The filter are selected to be,

$$W_\Delta = \frac{0.5}{T_y}$$

$$W_P = \frac{1}{S_y}$$

Finally, the W_u filter is selected to be a 2×2 constant matrix:

$$W_u = \begin{bmatrix} 0.25 & 0 \\ 0 & 0.50 \end{bmatrix} \tag{17.45}$$

Filter W_u is tuned to account for the actuator saturation limits. The augmented plant inputs and outputs for H_∞ control design is written as:

$$\begin{bmatrix} z_1 \\ z_{2a} \\ z_{2b} \\ z_3 \\ e \end{bmatrix} = \begin{bmatrix} 0 & G_{pm}W_\Delta & G_{pe}W_\Delta \\ 0 & W_u & 0 \\ 0 & 0 & W_u \\ 1 & -G_{pm}W_P & -G_{pe}W_P \\ 1 & -G_{pm} & -G_{pe} \end{bmatrix} \begin{bmatrix} r \\ u_{volt} \\ u_{fuel} \end{bmatrix} \tag{17.46}$$

With the selected filters, the `hinfopt` control command in Matlab is used to derive a controller and simulation results for all the control design methods are presented in Section 17.2.6.

17.2.6 Simulation Results

In this section, we present the simulation results for the Youla and H_∞ controllers. In addition to the MISO control results, for comparison, we have designed a SISO controller for the diesel engine only, and we present the results for this controller. The time domain responses are simulated with a reference engine speed, (ω_{ref}), of 300 rad/s (or 3000 rpm). The diesel engine speed responses for the Youla first method and H_∞ are illustrated in Figure 17.20.

Interestingly, the responses for both H_∞ and Youla first method are identical. Both reach the target speed in less than 1.0 second with the error signals shown in black.

It is informative to look at the actuator efforts for the engine fueling and electric motor torques (u_{fuel} and u_{volt}). These results are shown in Figure 17.21.

The first Youla method requires around 400 Nm of engine fueling torque and around 250 Nm of electric motor torque, while H_∞ control, for similar output response, requires around 500 Nm of engine fueling torque and over 300 Nm of electric motor torque during transient response. In addition, it is interesting to observe the H_∞ at steady-state requires negative engine fueling torque, and this forces the electric motor torque to be higher than Youla method at steady state.

To understand the benefits of MISO vs. SISO control, we designed a SISO H_∞ controller only with the electric motor and no fueling. In addition, we ensured that the saturation torque limit of CISG is not violated, which is around 250 Nm, while designing this controller. It should be noted that similar responses can be obtained with SISO Youla design. We present the simulation and on-vehicle testing results in Figure 17.22.

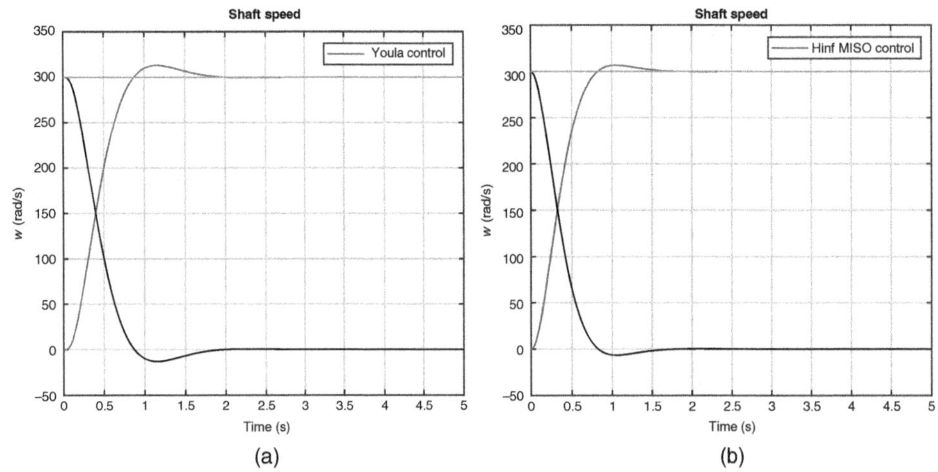

Figure 17.20 H_∞, (b), and Youla first method, (a), time response comparison for an engine reference speed of 300 rad/s.

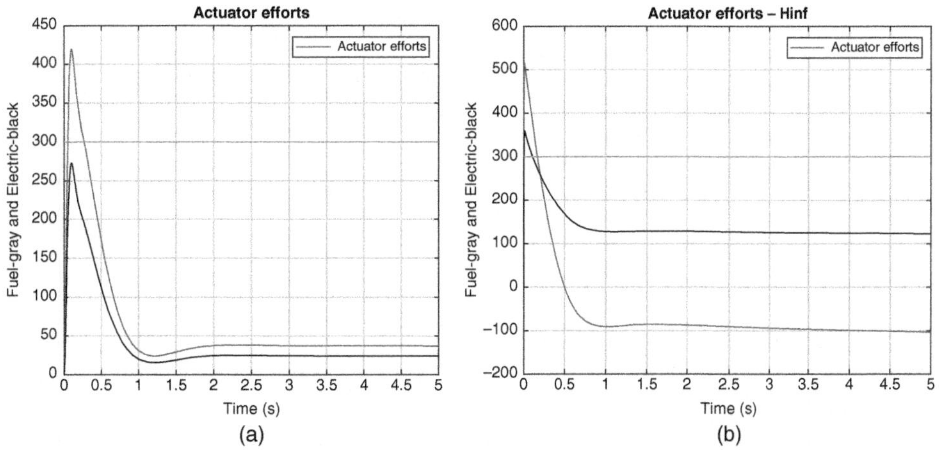

Figure 17.21 H_∞, (b), and Youla first method, (a), actuator effort comparisons for an engine reference speed of 300 rad/s.

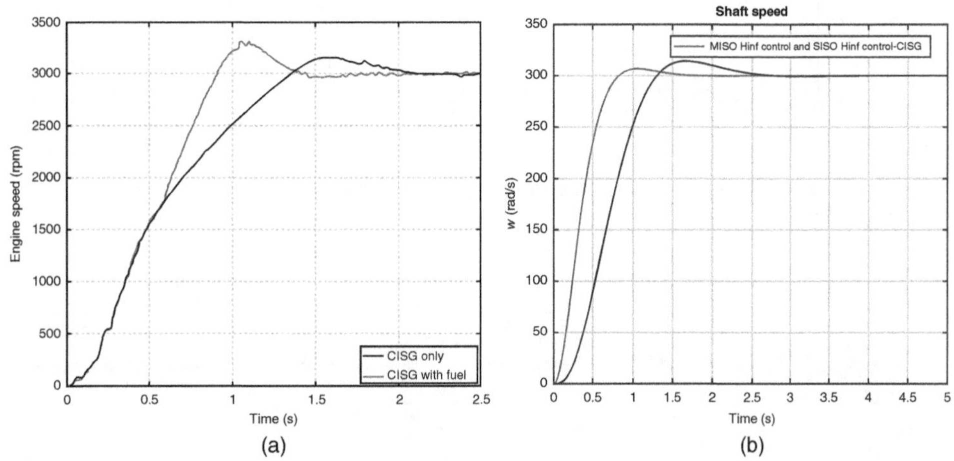

Figure 17.22 H_∞, MISO and SISO control on-vehicle results, (a), simulation results, (b).

The nonlinear (on-vehicle) and linear (simulation) responses are evident during the transient. However, the settling times for on-vehicle and simulation results are almost identical. MISO on-vehicle result has a higher overshoot than the MISO simulation result (in gray). This is due to parameter differences between the model and the actual system.

In Figure 17.23, we present the results for the second Youla method where the engine fueling is controlled by a feedforward control and the electric motor torque is controlled by a feedback control.

The benefit of this strategy is more effective use of actuators for the over-actuated systems and easier controller tuning. As can be observed in Figure 17.23, the rise time is less than 0.5 second, which is a drastic improvement over exclusively feedback controllers. However, there is compromise between robustness and fast response with this method, at least in this example. This method could better lend itself for higher dimension over-actuated systems where the extra actuators can be controlled by feedforward controllers and the remaining system can be transformed to a square matrix for feedback control design.

17.2.7 Robustness Test

Based on the real uncertain parameter values defined previously for DHEV, we simulated the model with 20 randomly selected samples of these uncertain parameters, as shown in Figure 17.24.

It is evident that both controller types have excellent robustness characteristics given the uncertain ranges for these real parameters. The Youla control method performs better in steady-state; however, the H_∞ control outperforms Youla during transient.

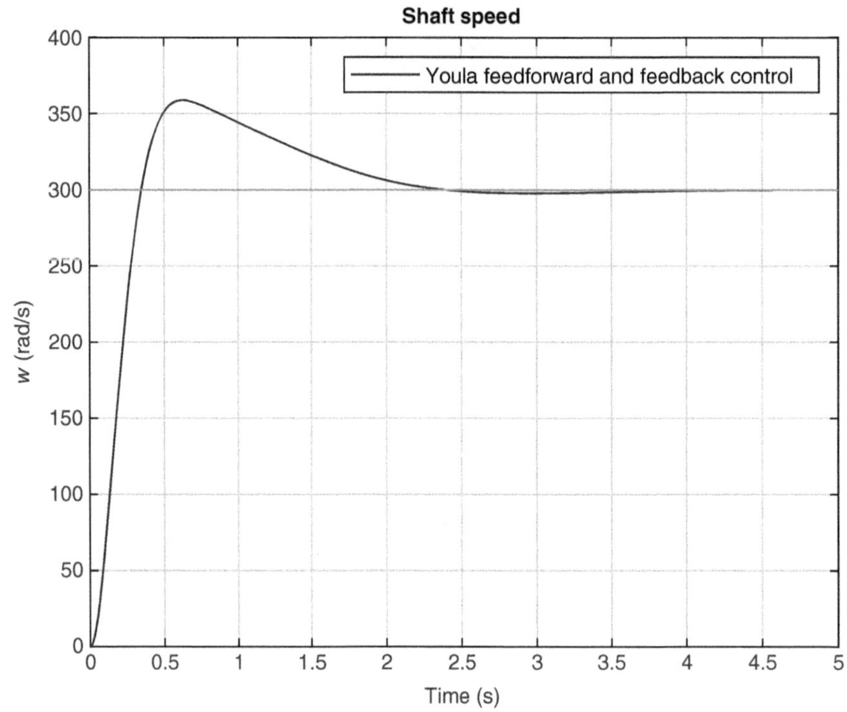

Figure 17.23 Youla second method, time response for an engine reference speed of 3000 rpm.

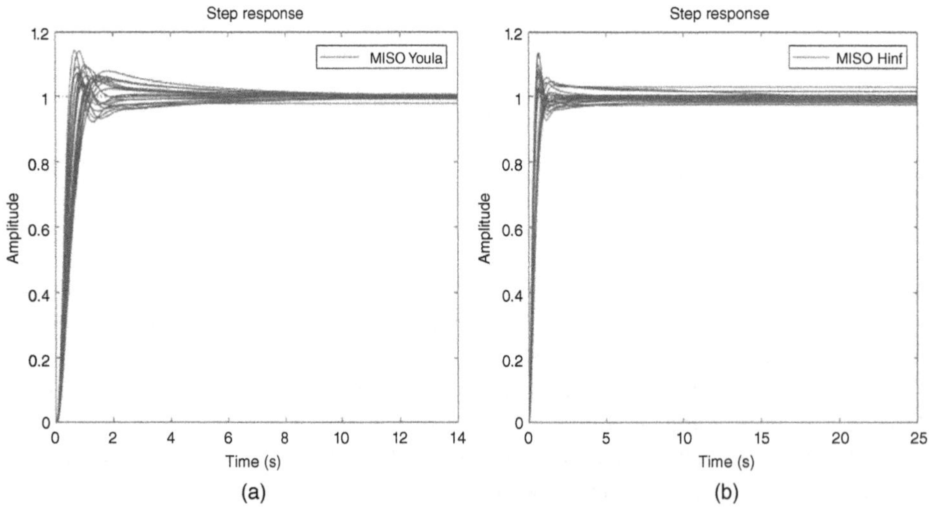

Figure 17.24 Time domain robustness to a step reference input for Youla first method, (a), and H_∞, (b), control using 20 random samples of uncertain parameters.

17.3 Robust Control for the Powered Descent of a Multibody Lunar Landing System

With human missions to the Moon on the horizon with the NASA Artemis lunar exploration program [47], there has been a renewed interest in lunar lander design, especially in the domain of guidance, navigation, and control. One of the most critical aspects of a human lunar landing mission is the powered-descent phase, which terminates in safe-landing on the surface of the Moon. The Apollo missions employed polynomial guidance for powered-descent and manual control for terminal-descent [48]. Although all the attempted landings were successful, the descent trajectories were not propellant-optimal, and the dimensions of the landing dispersion ellipses were in the order of kilometers [49]. Recent advances in technology have made way for a more robust, autonomous approach to precision lunar landing. Precision landing reduces the size of the landing uncertainty ellipse considerably, bringing it down to a precision of meters.

Recent breakthroughs in the formulation of real-time interior point method algorithms have led to the invention of powered-descent guidance (PDG) algorithms that are computationally efficient, making them desirable for real-time implementation [50, 51]. These algorithms rely on simplified dynamical models; however, implementing them in a real scenario in an open-loop could lead to large trajectory tracking errors due to model inaccuracies, such as nonlinearities in the actual system that are not accounted for by the dynamical model. Hence, it becomes necessary to adopt a feedback control design strategies that ensures the closed-loop systems are insensitive to these nonlinearities, while also being robust to system parameter variations and external disturbances. Propellant-optimal guidance and augmented inertial navigation in conjunction with robust trajectory tracking control would enable a full-stack closed-loop guidance, navigation, and feedback control, as depicted in Figure 17.25, for autonomous precision landing for future human and robotic missions to the Moon and beyond.

In this section, we provide a summary of both the nonlinear and linear dynamic models. The nonlinear version is used as an actual plant and the linearized version is utilized for MIMO control development. The trajectory optimization for PDG is not provided and the full description can be found in [52, 53].

17.3.1 Multibody Dynamics Model

The lunar lander is modeled as a holonomic system using a multibody dynamics modeling framework known as Kane's method and the extended Kane's equation for variable mass systems [54].

The body of the lander is considered to be a rigid cube with length, ℓ, and uniform mass, M, as shown in Figure 17.26. The main engine is located at point P, at which the generated thrust is applied. The inertial effects of the main engine on the overall lander dynamics are assumed to be negligible. The onboard propellant is assumed to be a particle with variable mass, m, midway between the body mass center and P. The mass center of the lander body-propellant system is assumed to be fixed.

N is the Moon-fixed inertial reference frame, L is the reference frame attached to the body of the lander, and E is the reference frame attached to the gimbaled main engine. Consider six generalized coordinates $[d_x, d_y, d_z, q_1, q_2, q_3]$, where $[d_x, d_y, d_z]$ represent the position of the mass center of the lander with respect to a fixed point, O, in N, and $[q_1, q_2, q_3]$ represent the Euler rotation angles of L relative to N, their time derivatives as the generalized velocities $[\dot{d}_x : v_x, \dot{d}_y : v_y, \dot{d}_z : v_z, \dot{q}_1 : v_1, \dot{q}_2 : v_2, \dot{q}_3 : v_3]$, and the propellant mass, $[m]$, the modeled system is an under-actuated system, 6 degrees of freedom (DOF), 13 states, and 3 control inputs – the mass depletion rate, $\dot{m}$, and the two main engine gimbal angles, e_1 and e_2. The thrust vector always acts along $\hat{e}_x$ and has a magnitude $C\dot{m}$, where C, the exit velocity of the ejected propellant, is a positive constant.

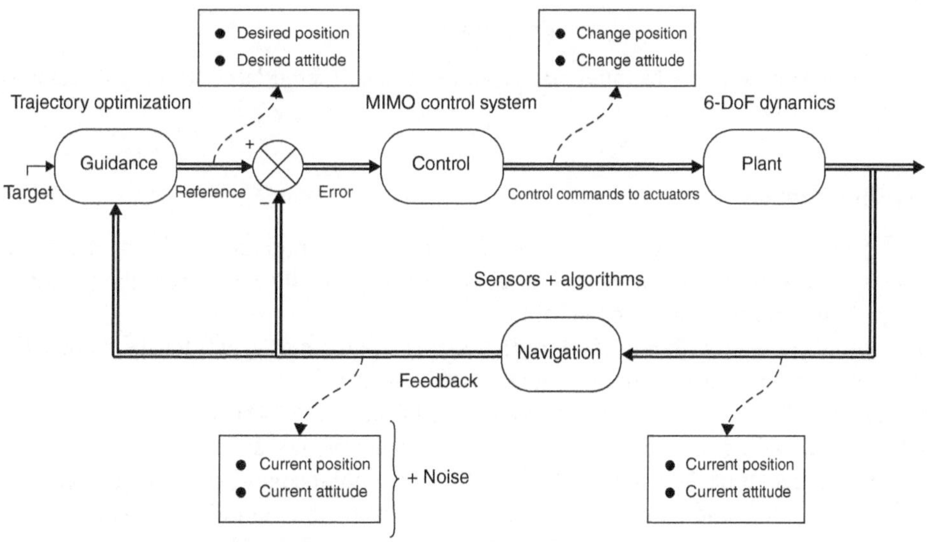

Figure 17.25 A high-level guidance, navigation, and control block diagram.

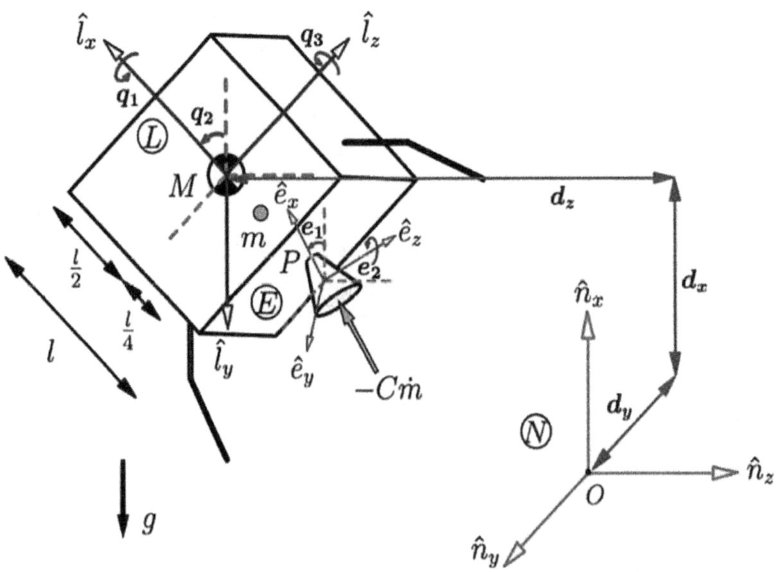

Figure 17.26 A multibody dynamics model of a lunar landing system.

The extended Kane's equations for holonomic variable mass systems take the following form:

$$F_r + F_r^* + F_r^{**} = 0 \quad (r = 1, \ldots, 6) \tag{17.47}$$

where $F_r (r = 1, \ldots, 6)$ are the generalized nonconservative forces, which account for the resultant of all body contact forces and torques, $F_r^* (r = 1, \ldots, 6)$ are the generalized inertial forces, which account for the body inertia forces and torques, and $F_r^{**} (r = 1, \ldots, 6)$ are the generalized thrusts, which account for the forces due to the time derivative of mass.

The resulting equations of motion can be expressed in the implicit form with the kinematics and dynamics combined:

$$\mathbb{M}\dot{x} = \mathbb{F} \tag{17.48}$$

where $\mathbb{M}$ is the mass matrix of the combined equations, $\mathbb{F}$ is the corresponding forcing vector, and $\dot{x}$ is the derivative of the state vector including 13 states: the generalized coordinates, $[d_x, d_y, d_z, q_1, q_2, q_3]$, the generalized velocities $[v_x, v_y, v_z, v_1, v_2, v_3]$, and the propellant mass, $[m]$. This system of equation can be linearized about a chosen operating point using Jacobian linearization to obtain an explicit first-order state-space representation of the system, as given in Eq. (17.49), which will serve as the plant model for MIMO control system design:

$$\delta\dot{x} = \mathbf{A}\delta x + \mathbf{B}\delta u \tag{17.49}$$

where $\mathbf{A}_{13\times13}$ is the state matrix, $\mathbf{B}_{13\times13}$ is the input matrix, and δu is the differential control input vector, where u includes the inputs, $[\dot{m}, e_1, e_2]$.

17.3.2 Trajectory Optimization

We consider an x–z planar lunar landing scenario for the purposes of simulation and demonstration of the control design methodology, which will be provided next. The chosen parameter values and nominal boundary conditions for propellant-optimal guidance trajectory are listed in Table 17.4.

The details of the optimization procedure is given in [52, 53]. The parameter values for the modeled lunar landing, as provide in Table 17.5, are chosen to closely approximate the Altair Lander.

The first-order ordinary differential equations (ODEs) obtained from Eq. (17.47) are integrated to obtain the evolution of the states. For the planar landing scenario, the control inputs from guidance are applied to the nonlinear model in an "**open-loop**", and the mass center of the lander body is tracked. The initial conditions of the lander are set to the values listed in Table 17.4. The resulting landing translation trajectory traced by the nonlinear model is shown in Figure 17.27, and the corresponding pitch trajectory is shown in Figure 17.28. As shown, the pitch of the lander body is unstable due to the fact that no explicit attitude control torque commands are provided for the lander.

Table 17.4 Guidance parameter values and boundary conditions.

Parameter	Value	Description
r_0	$[-500,0.0,400]$ m	Initial position
v_0	$[40.0,0.0,-13.0]$ m/s	Initial velocity
r_f	$[0.0,0.0,0.0]$ m	Target landing position
v_f	$[0.0,0.0,0.0]$ m/s	Touchdown velocity
T_{max}	83000 N	Full-thrust magnitude
ρ_1	$30\% T_{max}$	Lower thrust bound (minimum throttle)
ρ_2	$80\% T_{max}$	Upper thrust bound (maximum throttle)
g	$-1.625\,\text{m/s}^2$	Lunar gravitational acceleration
m_{wet}	25 000 kg	Wet-mass of the lander (at r_0)
m_{dry}	10 000 kg	Dry-mass of the lander
α	0.00023 s/m	Thrust-specific fuel consumption (TSFC)

Table 17.5 Parameter values for Lander model.

Parameter	Value	Description
ℓ	9.35 m	Length of the lander
$M = m_{dry}$	10 000 kg	Mass of the lander body
$m = m_{wet} - m_{dry}$	15 000 kg	Initial propellant-mass
$C = \frac{1}{\alpha}$	4413.97 m/s	Exit-velocity of the propellant

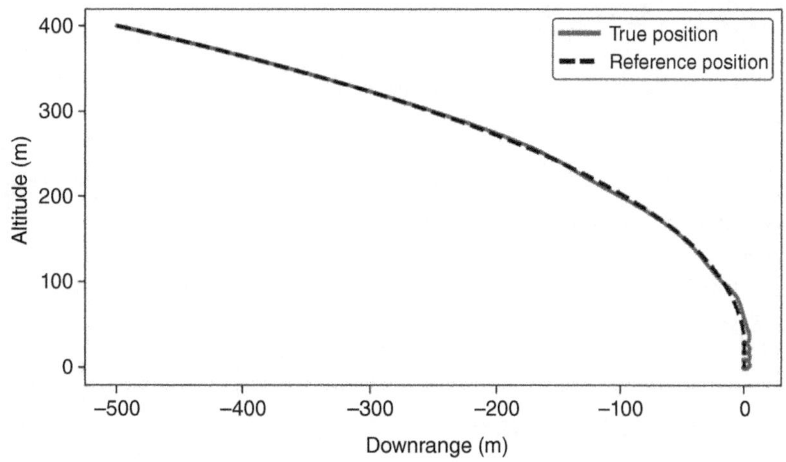

Figure 17.27 Translation trajectory traced by the mass center of the nonlinear lander model with open-loop control.

17.3.3 MIMO Youla Control Design

The following algorithm summarizes the Youla MIMO control design from Chapter 13.

Our nonlinear model of Eq. (17.48) with 13 states with 6 DOF is reduced to 3-DOF (downrange, altitude, and pitch) planar landing scenario for control design. This reduced-order model includes seven states: $[v_x, v_z, v_2, d_x, d_z, q_2, m]$, two control inputs: $[\dot{m}, e_1]$, and three measured outputs: $[d_x, d_z, q_2]$. The resulting combined mass matrix, state derivative vector, and

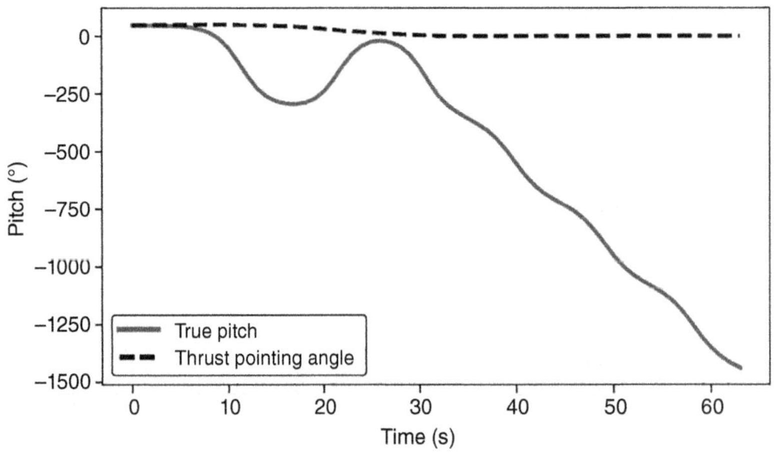

Figure 17.28 Pitch trajectory traced by the body of the nonlinear lander model with open-loop control.

Algorithm Multivariable Feedback Control Design via Youla Parameterization.

procedure MIMO(G_p) ▷ Plant TFM
  compute the SM-form of $G_{p_{(m\times n)}}$, $M_{p_{(m\times n)}}$ ▷ Smith-McMillan Form
  find unimodular $U_L, U_R \mid M_p = U_L G_p U_R$
  choose M_Y to shape $M_T \mid M_T = M_p M_Y$
  check: $Y = U_R M_Y U_L \rightarrow$ all entries of Y are proper TFs ▷ Youla TFM
  for $k = 1 : \min\{m, n\} \mid M_p(k,k) = G_k$ **do**
    SISO(G_k) ▷ Internal Stability
  end for
  compute $T_y = U_L^{-1} M_p M_Y U_L$ ▷ Complementary-Sensitivity TFM
  compute $S_y = I - T_y = U_L^{-1}(I - M_p M_Y)U_L$ ▷ Sensitivity TFM
  compute $G_c = U_R(I - M_Y M_p)^{-1} M_Y U_L$ ▷ Controller TFM
  check: $S_y G_p = U_L^{-1}(I - M_p M_Y) M_p U_R^{-1} \rightarrow$ meets requirement ▷ Loop-Shaping Check
  return G_c
end procedure
function SISO(G) ▷ SISO Interpolation Conditions
  if G has a $j\omega$-axis or RHP pole p of multiplicity a_p **then** ▷ Unstable Poles
    $S(p) = 0; T(p) = 1$
    for $j = 1 : a_p - 1$ **do**
      $\frac{d^j S(p)}{ds^j} = 0; \frac{d^j T(p)}{ds^j} = 0$
    end for
  end if
  if G has an NMP zero z of multiplicity a_z **then** ▷ Non-Minimum Phase Zeros
    $S(z) = 1; T(z) = 0$
    for $j = 1 : a_z - 1$ **do**
      $\frac{d^j S(z)}{ds^j} = 0; \frac{d^j T(z)}{ds^j} = 0$
    end for
  end if
end function

forcing vector are given in equations below:

$$\mathbb{M} = \begin{bmatrix} -M-m & 0 & -\frac{\ell m \sin(q_2)}{2} & 0 & 0 & 0 & -C\cos(e_1) \\ 0 & -M-m & -\frac{\ell m \cos(q_2)}{2} & 0 & 0 & 0 & C\sin(e_1) \\ -\frac{\ell m \sin(q_2)}{2} & -\frac{\ell m \cos(q_2)}{2} & -\frac{\ell^2(8M+3m)}{48} - \frac{\ell^2 m}{4} & 0 & 0 & 0 & \frac{C\ell \sin(e_1-q_2)}{2} \\ 0 & 0 & 0 & 1 & 0 & 0 & 0 \\ 0 & 0 & 0 & 0 & 1 & 0 & 0 \\ 0 & 0 & 0 & 0 & 0 & 1 & 0 \\ 0 & 0 & 0 & 0 & 0 & 0 & 1 \end{bmatrix} \tag{17.50}$$

$$\dot{x} = \begin{bmatrix} \dot{v}_x \\ \dot{v}_z \\ \dot{v}_2 \\ \dot{d}_x \\ \dot{d}_z \\ \dot{q}_2 \\ \dot{m} \end{bmatrix} \tag{17.51}$$

$$\mathbb{F} = \begin{bmatrix} g(M+m) + \frac{\ell m v_2^2 \cos(q_2)}{2} \\ -\frac{\ell m v_2^2 \sin(q_2)}{2} \\ \frac{g\ell m \sin(q_2)}{2} \\ v_x \\ v_z \\ v_2 \\ \dot{m} \end{bmatrix} \tag{17.52}$$

The nonlinear equations of motion are linearized about the ideal touchdown state (the final guidance state) and with use of Laplace transform a MIMO plant TFM is obtained. The resulting TFM, G_p, of the linearized MIMO plant is provided below. The linear plant is a 3×2 TFM, with two inputs, $[\dot{m}, e_1]$, and three outputs, $[d_x, d_z, q_2]$. The plant is therefore under-actuated but controllable:

$$G_p = \begin{bmatrix} -\frac{0.182}{s^2} & 0 \\ 0 & -\frac{1.467(s-1.082)(s+1.082)}{s^2(s-0.863)(s+0.863)} \\ 0 & -\frac{0.307}{(s-0.863)(s+0.863)} \end{bmatrix} \tag{17.53}$$

Following the procedure described in the control design algorithm, we find the SM form, M_p of the linearized plant, G_p, as given below. The instability of the plant is also manifested by SM form with five BIBO unstable poles. Interestingly, the SM form does not include any non-minimum phase zeros. In addition, the corresponding left and right unimodular matrices are provided below, and it can be verified that $M_p = U_L G_p U_R$, and the determinants of the unimodular matrices are scalars:

$$M_p = \begin{bmatrix} \frac{1}{s^2(s-0.863)(s+0.863)} & 0 \\ 0 & \frac{1}{s^2} \\ 0 & 0 \end{bmatrix} = \begin{bmatrix} G_1(s) & 0 \\ 0 & G_2(s) \\ 0 & 0 \end{bmatrix} \tag{17.54}$$

$$U_L = \begin{bmatrix} 0 & 0.582 & -2.786 \\ -5.500 & 0 & 0 \\ 0 & 0.854s^2 & -4.087(s-1.082)(s+1.082) \end{bmatrix} \tag{17.55}$$

$$U_R = \begin{bmatrix} 0 & 1 \\ 1 & 0 \end{bmatrix} \tag{17.56}$$

The feedback controller is designed based on the entries of M_P, such as two individual decoupled SISO controllers by designing the appropriate Youla transfer functions. The closed-loop complementary sensitivity transfer functions are given as follows: $T_1(s) = Y_1(s)G_1(s)$ and $T_2(s) = Y_2(s)G_2(s)$, where $Y_1(s)$ and $Y_2(s)$ are the Youla transfer functions that are to be designed so that to stabilize the entries of M_p and ensure that $T_1(s)$ and $T_2(s)$ meet the required performance (trajectory-tracking) and the interpolation conditions for internal stability. The interpolation conditions are provided in line 15 of the algorithm.

The interpolation conditions for the unstable poles of G_1 are

$$T_1(s)\Big|_{s=0.863} = 1 \quad T_1(s)\Big|_{s=0} = 1 \quad \frac{dT_1(s)}{ds}\Big|_{s=0} = 0 \tag{17.57}$$

The designed Youla transfer function that satisfies these conditions is given below:

$$Y_1(s) = \frac{Ks^2(s-0.863)(s+0.863)(\tau_{z_1}s+1)(\tau_{z_2}s+1)}{(s^2+2\zeta\omega_n s+\omega_n^2)(\tau_{p_1}s+1)(\tau_{p_2}s+1)(\tau_{p_3}s+1)^4} \tag{17.58}$$

The selected parameter values for $Y_1(s)$ are listed in table (17.6), where τ_{z_1}, τ_{z_2}, τ_{p_1}, and τ_{p_2} are the solution variables selected to satisfy the system of equations given by the interpolation conditions above. The fourth-order pole with time constant τ_{p_3} is included so that $Y_1(s)$ and $T_1(s)$ are strictly proper transfer functions as strictly proper $Y(s)$ results in lowering actuator effort at high frequency and strictly proper $T(s)$ better rejects high- frequency disturbances such as sensor noise.

The resulting stable strictly proper closed-loop (complementary sensitivity) transfer function, $T_1(s)$, is given below:

$$T_1(s) = \frac{K(\tau_{z_1}s+1)(\tau_{z_2}s+1)}{(s^2+2\zeta\omega_n s+\omega_n^2)(\tau_{p_1}s+1)(\tau_{p_2}s+1)(\tau_{p_3}s+1)^4} \tag{17.59}$$

In order to ensure good reference tracking for the final coupled MIMO closed-loop system, it is desirable to make $T_1(s) = T_2(s)$. To accomplish this, we will select $Y_2(s)$ transfer function to be

$$Y_2(s) = \frac{Ks^2(\tau_{z_1}s+1)(\tau_{z_2}s+1)}{(s^2+2\zeta\omega_n s+\omega_n^2)(\tau_{p_1}s+1)(\tau_{p_2}s+1)(\tau_{p_3}s+1)^4} \tag{17.60}$$

Using the Youla transfer functions $Y_1(s)$ and $Y_2(s)$ given above, we can then formulate the TFM M_Y as follows:

$$M_Y = \begin{bmatrix} Y_1(s) & 0 & 0 \\ 0 & Y_2(s) & 0 \end{bmatrix} \tag{17.61}$$

The decoupled closed-loop (complementary sensitivity) TFM, M_T, can then be written as follows:

$$M_T = M_P M_Y = \begin{bmatrix} Y_1(s)G_1(s) & 0 & 0 \\ 0 & Y_2(s)G_2(s) & 0 \\ 0 & 0 & 0 \end{bmatrix} = \begin{bmatrix} T_1(s) & 0 & 0 \\ 0 & T_2(s) & 0 \\ 0 & 0 & 0 \end{bmatrix} \tag{17.62}$$

We then compute the decoupled sensitivity transfer function M_S, where $M_S = I - M_T$, $I \in \mathbb{R}^{3\times3}$.

The designed decoupled TFMs, M_Y, M_T, and M_S, are pre- and post-multiplied by the unimodular matrices, as described in the algorithm and Chapter 13, to obtain the Youla TFM, Y, the closed-loop (complementary sensitivity) TFM, T_y, sensitivity

Table 17.6 Parameter values for Lander model.

Parameter	Value	Description
ω_n	3.75 rad/s	Natural frequency of the second-order pole
K	14.062	Gain = ω_n^2
ζ	0.7071	Damping ratio
τ_{z_1}	2.12 s	Time-constant for the first zero
τ_{z_2}	1.83 s	Time-constant for the second zero
τ_{p_1}	3.14 s	Time-constant for the first pole
τ_{p_2}	0.027 s	Time-constant for the second pole
τ_{p_3}	0.1 s	Time constant for the third pole

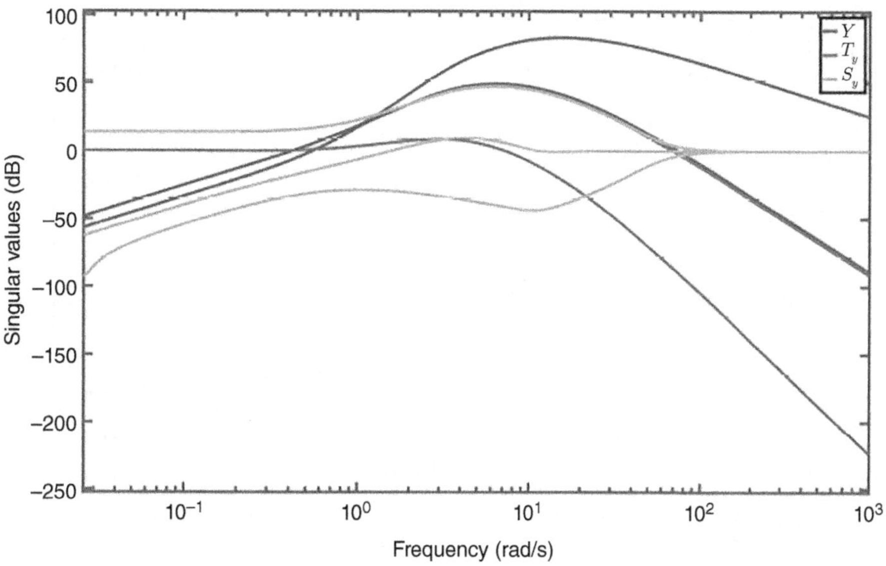

Figure 17.29 Singular values of Y, T_y, and S_y.

Table 17.7 Closed-loop requirements for Y, T_y, and Sy.

Objective	Requirement
$\sigma_{max}(T_y) = \sigma_{min}(T_y) = 1$ at low frequency	Trajectory tracking
$\sigma_{max}(T_y) \ll 0$ at high frequency	Sensor noise rejection and robust stability
$\sigma_{max}(S_y) \ll 0$ at low frequency	Distrubance rejection
$\sigma_{max}(Y) \ll 0$ at high frequency	Small actuator effort

TFM, S_y, and finally, the internally stabilizing robust MIMO controller, G_c. The singular values of Y, T_y, and S_y are shown in Figure 17.29.

The closed-loop requirements listed in Table 17.7.

It is evident from Figure 17.29 that $\sigma_{max}(S_y) > 0$ (this is the third singular value) at low frequency and in addition, $\sigma_{min}(T_y) \neq 1$ at low frequency (this is the third singular value and not shown). Violating these two requirements correspond to not being able to robustly track the pitch angle trajectory of the rocket. This is inherently due to the under-actuated nature of the control problem.

The linear model of the lander with the designed feedback controller tracks the reference altitude and down range trajectories, as shown in Figures 17.30 and 17.31, respectively. This closed-loop system, however, displays oscillatory behavior for pitch angle reference tracking, as illustrated in Figure 17.32. This is principally due to the under-actuated nature of the plant which results in the third output not being properly controlled.

It should be noted that we have designed and simulated a MIMO H_∞ controller with the same plant and the results, not presented here, are very similar to the Youla method presented above.

17.3.4 Youla Method for Under-Actuated Systems

In this section, we propose a modified MIMO Youla control design for under-actuated systems.

The control method proposed in this section is intended for addressing under-actuated systems with highly unstable plants, such as this case. This method combines Youla with an allocation strategy and root locus approach, as shown in Figure 17.33.

In Figure 17.33, G_{cAL} is a 2×3 allocation strategy given as follows:

$$G_{cAL} = \begin{bmatrix} 1 & 0 & 0 \\ 0 & \alpha & 1-\alpha \end{bmatrix} \tag{17.63}$$

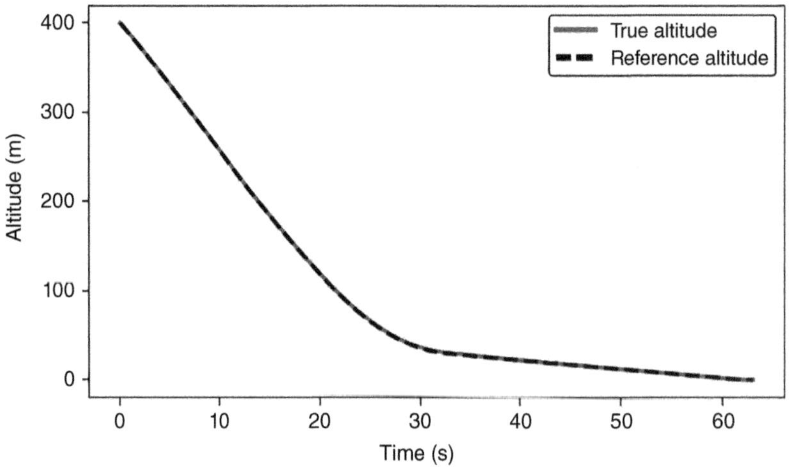

Figure 17.30 Tracking trajectory of the reference altitude (dash-black), true or model output (solid-gray).

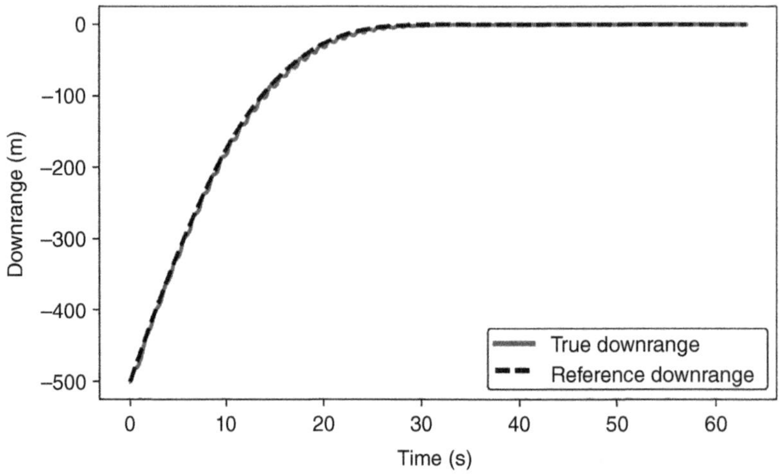

Figure 17.31 Tracking trajectory of the reference downrange (dash-black), true or model output (solid-gray).

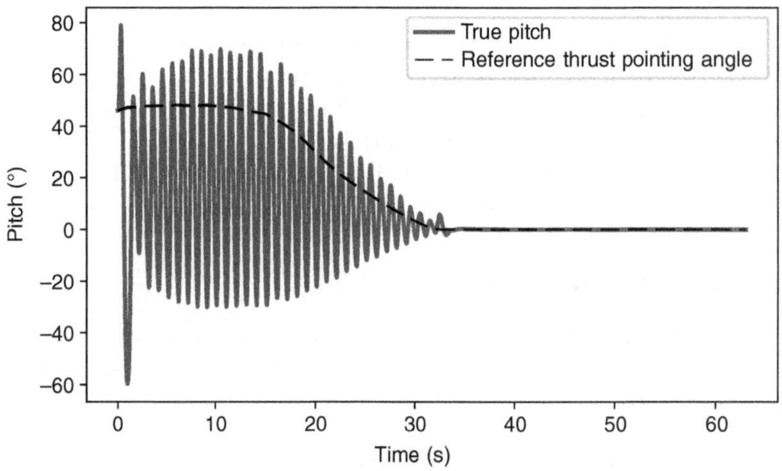

Figure 17.32 Tracking trajectory of the reference pitch angle (dash-black), true or model output (solid-gray).

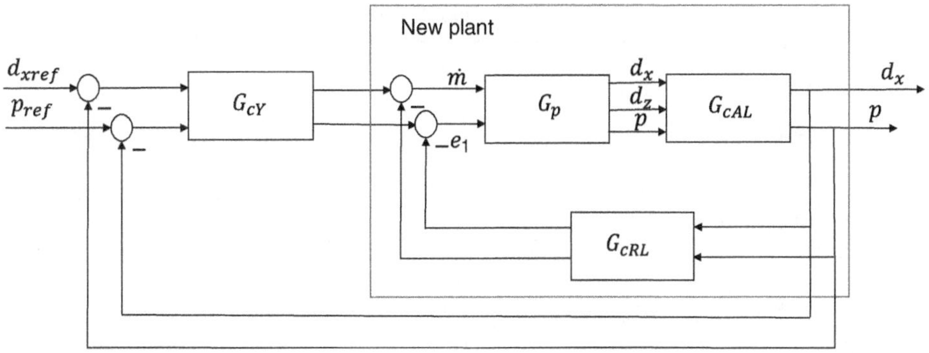

Figure 17.33 The block diagram of the proposed control method for unstable under-actuated lunar landing system.

In this way, by selecting α, we can prioritize downrange or pitch angle control. The α is a constant here; however, this constant can be selected dynamically based on the requirements.

The root locus method is used to stabilize this unstable plant, although, it should be pointed out that this is not necessary as both performance (trajectory following) and stability could be addressed by the Youla method. However, for highly unstable plants, it makes the control design a more straightforward task when the tasks of stabilization and trajectory following are divided between different controllers.

For a plant with double pole at zero (the first entry $G_p(1,1)$), it is known from root locus that by adding a stable zero, this plant could be stabilized as shown in Figure 17.34.

In order to make the controller realizable, we selected the first entry of the $G_{cRL}(1,1)$ to be $\frac{-3.5*(s+1)}{(0.01s+1)}$. The negative value is chosen because the first entry of G_p has a negative sign and a fast pole is added to the controller to make it realizable.

For a plant with one unstable pole ($G_p(3,2)$), we select $G_{cRL}(2,2)$ entry to be equal to $G_{cRL}(1,1)$. This plant could be stabilized as shown in Figure 17.35.

Stabilization of $G_p(2,2)$ using this method is left as an exercise.

The root locus controller, G_{cRL} is written as:

$$G_{cRL} = \begin{bmatrix} G_{cRL}(1,1) & 0 \\ 0 & G_{cRL}(2,2) \end{bmatrix} \tag{17.64}$$

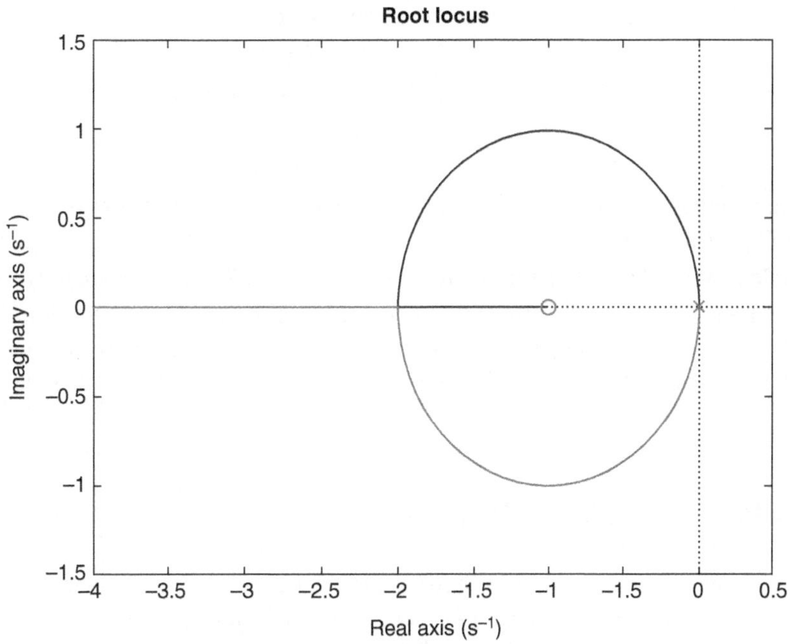

Figure 17.34 Stabilizing a plant with double pole at zero with root locus.

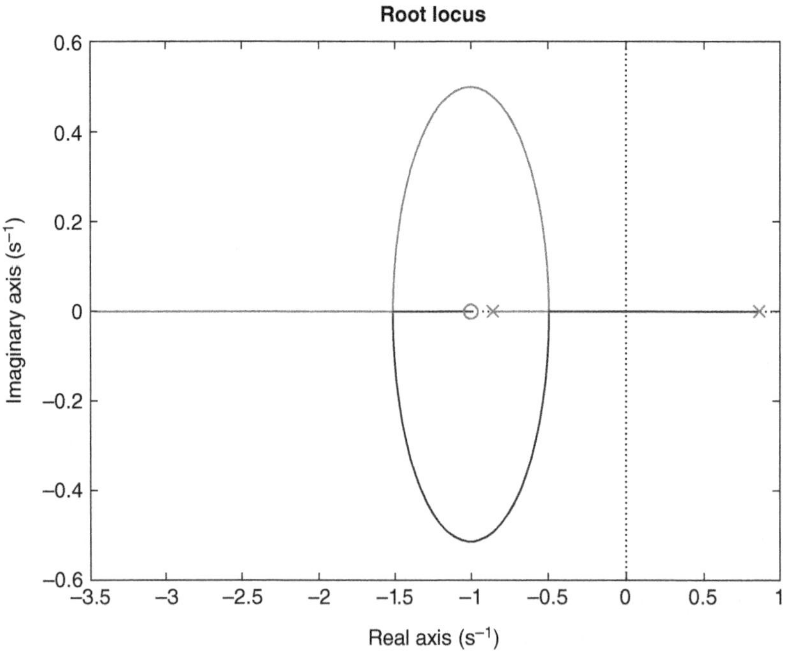

Figure 17.35 Stabilizing a plant with double pole at zero and a RHP pole with root locus.

Then the TFM for the new plant, as illustrated in Figure 17.33, to be used for the design of controller with Youla method is given by:

$$G_{pnew} = (I + (G_{cAL} G_p G_{cRL}))^{-1}(G_{cAL} G_p) \tag{17.65}$$

This will results in the following TFM of the new plant:

$$G_{pnew} = \begin{bmatrix} G_{pnew}(1,1) & 0 \\ 0 & G_{pnew}(2,2) \end{bmatrix} \tag{17.66}$$

The new plant is a 2×2 with diagonal stable entries for altitude and pitch angle control, when $\alpha = 0$. This is in contrast with the original 3×2 with unstable entries. These transfer functions are

$$G_{pnew}(1,1) = \frac{-0.182(s + 100)}{(s + 99.37)(s^2 + 0.6346s + 0.6411)} \tag{17.67}$$

$$G_{pnew}(2,2) = \frac{-0.307(s + 100)}{(s + 98.92)(s^2 + 1.075s + 0.3336)} \tag{17.68}$$

This plant is used for the outer loop controller design using Youla method. As both entries of G_{pnew} are stable, for designing the outer controller G_{cY}, using Youla method, they were simply inverted. Furthermore, Youla transfer functions were selected so that the complementary closed-loop transfer functions $T(1,1)$ and $T(2,2)$ will have the form of $\frac{\omega_n^2}{s^2+2\zeta\omega_n s+\omega_n^2}$, where $\omega_n = 100$ and $\zeta = 0.707$. Then an outer loop controller is computed using Youla parameterization.

The linear model of the lander with the designed feedback controller tracks the reference altitude similar to Figure 17.30 and not shown here. However, this time, the closed-loop system, tracks the reference pitch angle perfectly, as shown in Figure 17.36.

17.4 Vehicle Yaw Rate and Sideslip Estimation

To reveal the benefits of the Youla Controller Output Observer (YCOO), developed in Chapters 7 and 16, this example provides a comparative study of MIMO YCOO and various types of Kalman filters, such as (Linear) Kalman Filter (LKF), Extended Kalman Filter (EKF) and Unscented Kalman Filter (UKF), in estimating yaw rate and sideslip of a vehicle.

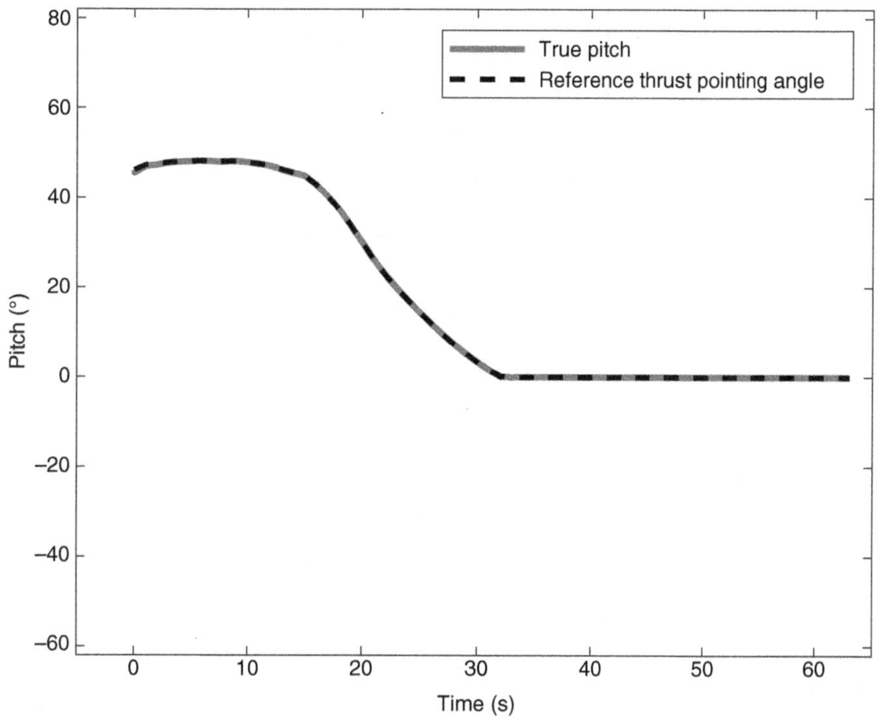

Figure 17.36 Tracking trajectory of the reference pitch angle (dash-black), true or model output (solid-gray) using modified Youla method.

(A similar comparison for SISO systems can be found in Section 8.2.) We compare how each estimation technique performs and test the robustness of these estimators under plant parameters and environmental uncertainties, such as vehicle mass and tire–road coefficient of friction variations. We consider realistic situations such as sensor bias and provide analysis and potential solutions for YCOO when dealing with these practical problems.

17.4.1 Background

Over the past years, there has been a tremendous amount of work in the vast area of dynamic system state and parameter estimation and virtual sensing has become very popular. The strategies used to devise state and parameter estimations include well-developed estimation techniques such as Kalman filtering [23] to nonlinear observers which present a robust estimator by a BMIs (bilinear matrix inequality) and finding a convex solution to this estimator by posing the problem as a LMI (linear matrix inequality) [55]. In almost all these studies, practical implementations of these estimators such as computation and calibration overheads are ignored. Most importantly, there has not been any investigation to illustrate the benefits, such as performance and robustness, of one technique over the other.

In Chapters 7 and 16, we presented the YCOO, a deterministic estimation technique. This example intends to provide further insights into the YCOO by performing a comparative analysis of different types of Kalman filtering techniques with YCOO in a popular vehicle dynamics application. We discuss the practical implication of these estimation strategies where YCOO would require no to low calibration and computation overhead relative to other estimation approaches.

Two important points are noteworthy. The first is that the YCOO, similar to stochastic filtering techniques, is applicable to both stable and unstable plants, even though, the plant under consideration in this example is stable. The second point is all traditional estimation concepts require access to the system input u. However, a Controller Output Observer (COO) (and by extension, a YCOO) does not have this constraint, as shown in Figures 17.37 and 17.38. In fact, if the input is not known, an estimation of the unknown input can be obtained from the YCOO.

This example will build upon Section 8.2; we will begin by briefly reintroducing the vehicle model, then expanding it for MIMO control by introducing a rear wheel steering element. We will then simulate the system compared to the various Kalman filter models.

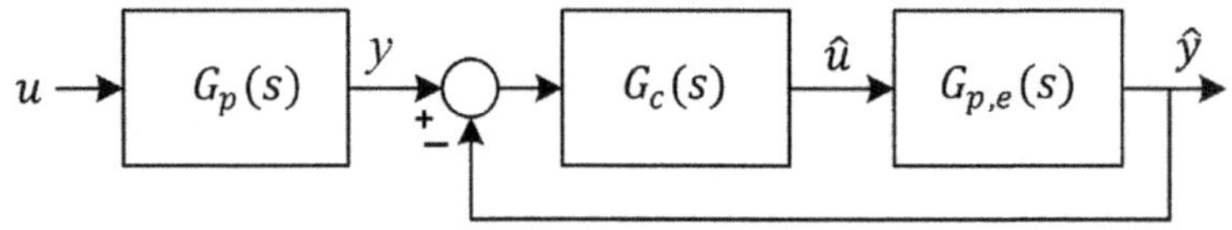

Figure 17.37 General controller output observer.

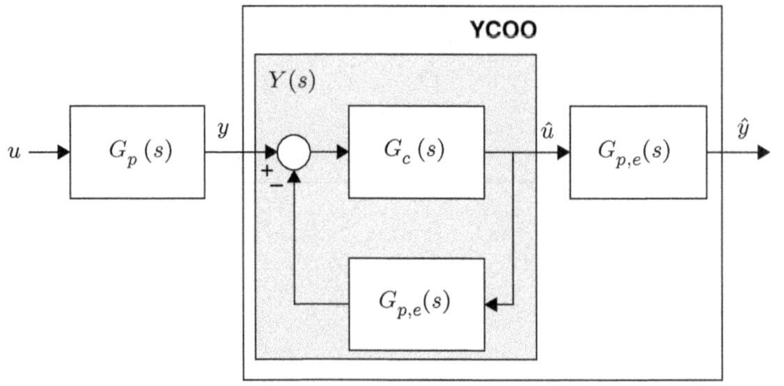

Figure 17.38 Youla controller output observer.

17.4.2 Vehicle Modeling

17.4.2.1 Nonlinear Bicycle Model With Pacejka Tire Model

The equations of motion for this model were developed in Section 8.2 and are repeated here without elaboration.

With body-fixed coordinates, the equations of motion of the bicycle model are

$$\dot{V} = \frac{F_f \cos\delta + F_r}{m} - U\omega \tag{17.69}$$

$$\dot{\omega} = \frac{F_f \cos\delta \cdot a - F_r \cdot b}{J} \tag{17.70}$$

The lateral forces exerted on the front and rear tires, F_f, and F_r, are functions of the slip angles of the front and rear tires. The slip angles of the front and rear are related to the system input and the states as given below:

$$\alpha_f = \delta - \tan^{-1}\left(\frac{V + a\omega}{U}\right) \tag{17.71}$$

$$\alpha_r = \tan^{-1}\left(\frac{b\omega - V}{U}\right) \tag{17.72}$$

The tire forces are functions of the tire slip angle and are computed using the Pacejka "Magic Formula" fire model [56] as given in Section 8.2.

The lateral acceleration at c.g. is related to the front and rear tire forces and is given by:

$$a_y = \frac{F_f \cos\delta + F_r}{m} \tag{17.73}$$

17.4.2.2 Kinematic Relationship

Before moving on to the MIMO model, we must consider a kinematic relationship between the lateral acceleration at the front, rear, and center of gravity. The lateral acceleration is kinematically related to the states V and ω: given the distances between the c.g. and the axles, by measuring the lateral accelerations at the front and rear axles, we can kinematically estimate the yaw rate. As shown in Figure 17.39, the lateral accelerations a_f and a_r, with distances of S_1 and S_2 to the c.g., can be related to the states V and ω as:

$$a_f = \dot{V} + U\omega + S_1 \cdot \dot{\omega} \tag{17.74}$$

$$a_r = \dot{V} + U\omega - S_2 \cdot \dot{\omega} \tag{17.75}$$

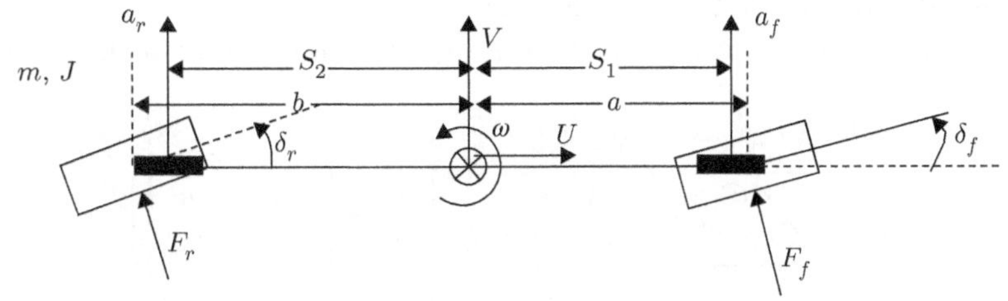

Figure 17.39 Front and rear steering bicycle car model system diagram with two accelerometers.

17.4.2.3 Multi-Input Model

By adding another input, such as controlling the rear road wheel angle, and by measuring both front and rear axle lateral accelerations, we can transform the vehicle model from the SISO system from Section 8.2 to a MIMO system as shown in Figure 17.39. With one more input, the state equations – Eqs. (8.20) and (8.21) from Section 8.2 – can be rewritten as:

$$\dot{V} = \frac{F_f \cos\delta_f + F_r \cos\delta_r}{m} - U\omega \tag{17.76}$$

$$\dot{\omega} = \frac{aF_f \cos\delta_f - bF_r \cos\delta_r}{J} \tag{17.77}$$

Then, Eqs. (17.71) and (17.72) are re-written as:

$$\alpha_f = \delta_f - \tan^{-1}\left(\frac{V + a\omega}{U}\right) \tag{17.78}$$

$$\alpha_r = \delta_r - \tan^{-1}\left(\frac{V - b\omega}{U}\right) \tag{17.79}$$

The vehicle sideslip (β) is the angle between the actual direction of vehicle and the direction of longitudinal velocity. We can use small angle approximation and it can be expressed as the ratio of the lateral velocity V and the longitudinal velocity U:

$$\beta = \frac{V}{U} \tag{17.80}$$

Since we assume the longitudinal velocity, U, to be constant, instead of using the lateral velocity, V, we can use the vehicle sideslip β as one of the states. In this way, our state estimation will consist of estimating the vehicle sideslip and the yaw rate. Then using Eqs. (17.80) and (17.76), we can write

$$\dot{\beta} = \frac{F_f \cos\delta_f + F_r \cos\delta_r}{mU} - \omega \tag{17.81}$$

Then by substituting β in place of V into Eqs. (17.78) and (17.79), the front and rear tire slip angles can be expressed as:

$$\alpha_f = \delta_f - \tan^{-1}\left(\beta + \frac{a\omega}{U}\right) \tag{17.82}$$

$$\alpha_r = \delta_r - \tan^{-1}\left(\beta - \frac{b\omega}{U}\right) \tag{17.83}$$

Then the lateral forces will be computed exactly the same as the SISO case.

The front and rear lateral accelerations can be related to the new state β by modifying Eqs. (17.74) and (17.75) as follows:

$$a_f = U\dot{\beta} + U\omega + S_1\dot{\omega} \tag{17.84}$$

$$a_r = U\dot{\beta} + U\omega - S_2\dot{\omega} \tag{17.85}$$

This MIMO model then consists of two inputs, the front and rear road wheel angles δ_f and δ_r, and two outputs, the vehicle lateral accelerations at the front and rear axles a_f and a_r.

17.4.2.4 Linearizing the Bicycle Model for SISO and MIMO Cases

Knowing that the states for the MIMO case are the vehicle sideslip β and yaw rate ω, the state-space representation for this case can be linearized as:

$$\begin{bmatrix} \dot{\beta} \\ \dot{\omega} \end{bmatrix} = \begin{bmatrix} -\dfrac{(C_f + Cr)}{mu} & -\dfrac{aC_f - bC_r}{mu^2} - 1 \\ -\dfrac{aC_f - bC_r}{J} & -\dfrac{a^2C_f + b^2C_r}{Ju} \end{bmatrix} \begin{bmatrix} \beta \\ \omega \end{bmatrix} + \begin{bmatrix} \dfrac{C_f}{mu} & \dfrac{C_r}{mu} \\ \dfrac{aC_f}{J} & -\dfrac{bC_r}{J} \end{bmatrix} \begin{bmatrix} \delta_f \\ \delta_r \end{bmatrix} \tag{17.86}$$

$$\begin{bmatrix} a_f \\ a_r \end{bmatrix} = \begin{bmatrix} -\dfrac{C_f + C_r}{m} - S_1\dfrac{aC_f - bC_r}{J} & -\dfrac{aC_f - bC_r}{mu} - S_1\dfrac{a^2C_f + b^2C_r}{Ju} \\ -\dfrac{C_f + C_r}{m} + S_2\dfrac{aC_f - bC_r}{J} & -\dfrac{aC_f - bC_r}{mu} + S_2\dfrac{a^2C_f + b^2C_r}{Ju} \end{bmatrix} \begin{bmatrix} \beta \\ \omega \end{bmatrix} + \begin{bmatrix} \dfrac{C_f}{m} + S_1\dfrac{aC_f}{J} & \dfrac{C_r}{m} - S_1\dfrac{bC_r}{J} \\ \dfrac{C_f}{m} - S_2\dfrac{aC_f}{J} & \dfrac{C_r}{m} + S_2\dfrac{bC_r}{J} \end{bmatrix} \begin{bmatrix} \delta_f \\ \delta_r \end{bmatrix} \tag{17.87}$$

This model is utilized to estimate the vehicle yaw rate and sideslip angle.

17.4.3 State Estimation

For the MIMO case, we have increased the number of vehicle model inputs to two by adding rear road wheel angle and making the vehicle model four-wheel steering. We expect a positive impact on both the performance and robustness of these estimators due to the addition of one more input to the system, w. The vehicle model includes two inputs: the front and rear road wheel angles (δ_f and δ_r), and two outputs: the front and rear axle lateral accelerations (a_f and a_r). The model states are the vehicle sideslip angle β and the yaw rate ω, as shown in Figure 17.39.

17.4.3.1 Youla Parameterization Control Design

Given the state-space representation of the vehicle model, with known matrices A_e, B_e, C_e, and D_e, we can derive the TFM of the plant using the following equation:

$$G_p = B_e(sI - A_e)^{-1}C_e + D_e \tag{17.88}$$

Based on the YCOO MIMO control and estimation design in [27], with the resulting TFM G_p, we can write the SM form of this TFM. First, the plant TFM is written as:

$$G_p = N_r D_r^{-1} = \begin{bmatrix} N_{11} & N_{12} \\ N_{21} & N_{22} \end{bmatrix} \begin{bmatrix} DEN & 0 \\ 0 & DEN \end{bmatrix}^{-1} \tag{17.89}$$

N_r is the numerator polynomial matrix and D_r is the decoupled denominator TFM. Then,

$$D_0 = 1, \quad D_1 = 1, \quad D_2 = \det N_r$$

$$G_p^{SM} = \begin{bmatrix} \dfrac{D_1}{D_0} & 0 \\ 0 & \dfrac{D_2}{D_1} \end{bmatrix} D_r^{-1} = \begin{bmatrix} \dfrac{1}{DEN} & 0 \\ 0 & \dfrac{D_2}{DEN} \end{bmatrix} = N_{SM} D_{SM}^{-1} \tag{17.90}$$

where D_0, D_1, and D_2 are the greatest common divisors of the polynomial matrix N_r, $N_{SM} = \begin{bmatrix} 1 & 0 \\ 0 & D_2 \end{bmatrix}$, and $D_{SM} = D_r$.

Given the following relationship between G_p^{SM} and G_p,

$$G_p(s) = U_L^{-1} G_p^{SM}(s) U_R^{-1} \tag{17.91}$$

then the unimodular matrices U_L and U_R can be computed by

$$U_L = (N_r N_{SM}^{-1})^{-1} \tag{17.92}$$

$$U_R = D_r D_{SM}^{-1} \tag{17.93}$$

The S–M form or the decoupled plant TFM has the following form:

$$G_p^{SM} = \begin{bmatrix} G_{p_1} & 0 \\ 0 & G_{p_2} \end{bmatrix} = \begin{bmatrix} \dfrac{1}{(s+p_1)(s+p_2)} & 0 \\ 0 & gs^2 \end{bmatrix} \tag{17.94}$$

The plant has two LHP poles and two zeros at the origin. The Youla transfer functions designed for the decoupled plant have the following structures:

$$Y_1 = \frac{1}{G_{p_1}} \frac{\omega_n^2}{(s^2 + 2\zeta\omega_n s + \omega_n^2)(\tau_1 s + 1)^2} \tag{17.95}$$

$$Y_2 = \frac{K}{(s^2 + 2\zeta\omega_n s + \omega_n^2)(\tau_1 s + 1)^2(\tau_2 s + 1)^2} \tag{17.96}$$

where the gain K is

$$K = \frac{\tau_2^2 + 2\zeta\omega_n\tau_2 + \omega_n^2}{g\tau_2^2} \tag{17.97}$$

The gain K is selected so that the closed-loop TFM T is at 0 dB in the low-frequency region for good target or reference following. The extra poles at $s = 1/\tau_1$ are for reducing the gain of Y at high frequency. The extra poles at $s = 1/\tau_2$ in Y_2 are for canceling the effect of zeros at the origin so that the closed-loop TFM target following at low frequency is not drastically compromised as discussed below in detail.

The coupled Youla TFM can be computed from the decoupled Youla TFM using the previously calculated unimodular matrices:

$$Y = U_R M_y U_L \tag{17.98}$$

where

$$M_y = \begin{bmatrix} Y_1 & 0 \\ 0 & Y_2 \end{bmatrix} \tag{17.99}$$

where M_y is a diagonal matrix with Youla transfer functions, Y_1 and Y_2, designed for the decoupled plant. Then we can derive the close-loop TFM T_y and sensitivity TFM S_y from the decoupled system using the following relationship:

$$T_y = U_L^{-1} G_p^{SM} M_y U_L \tag{17.100}$$

$$S_y = U_L^{-1}(I - G_p^{SM} M_y) U_L \tag{17.101}$$

Then the TFM of the resulting controller, G_c, can be computed by:

$$G_c = U_R(I - M_y G_p^{SM})^{-1} M_y U_L \tag{17.102}$$

The singular values plot of T_y, S_y, and Y are shown in Figure 17.40. Both maximum and minimum value of T_y are at 0 dB in the desired region of frequency, which enforces a good target following for the closed-loop system. Since the second decoupled transfer function, G_{p2}, of the plant includes two zeros at the origin, we need to modify the closed-loop transfer function by adding two poles at $s = 1/\tau_2$ so that the closed-loop transfer function behave as a bandpass filter for ensuring its target following capability in the low frequency is not significantly compromised, as shown in Figure 17.40. The singular values of Y is low at high frequency which results in good sensor noise rejection at the actuator input.

17.4.4 Simulation and Estimation Result

The yaw rate estimation for 2° road wheel angle step inputs is shown in Figure 17.41. With an additional input, the rear road wheel angle, the YCOO's transient response performance is significantly improved. The response of YCOO is very similar to both the EKF and UKF. The LKF has a small steady-state error due to the small difference between the linear model used for estimation and the nonlinear vehicle and tire models of the reference generator. The vehicle sideslip estimations for 2° road wheel angle step inputs are shown Figure 17.42. Except for the LKF, all other strategies have excellent transient and steady-state responses.

The yaw rate estimation that results from increasing the wheel angle step input to 4° is illustrated in Figure 17.43. The steady-state error of LKF has clearly increased as the difference between the linear estimation model and the reference generator model is becoming more significant. Again the YCOO, EKF, and UKF have excellent transient and steady-state

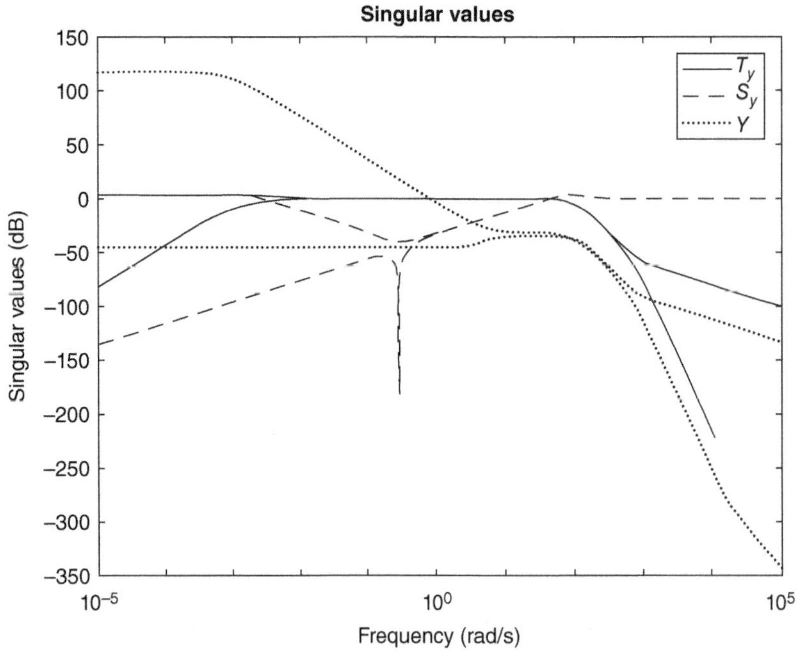

Figure 17.40 Singular values of T_y, S_y, and Y.

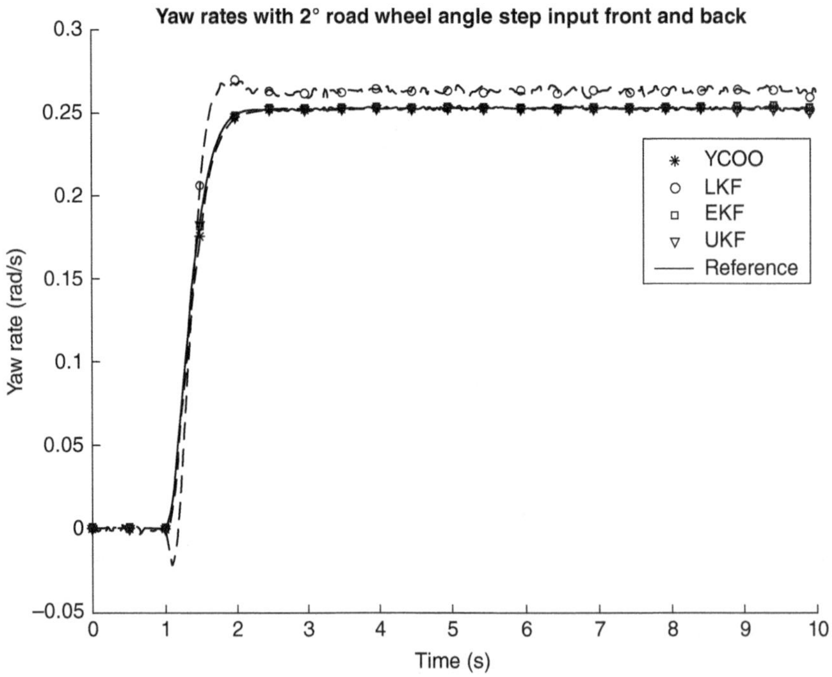

Figure 17.41 The yaw rate estimation results for different strategies.

responses. The excellent performance of the YCOO, in this case, is unexpected as YCOO, similar to LKF, uses a linear model for the state estimation. The vehicle sideslip estimation for 4° road wheel angle step inputs is shown in Figure 17.44. In this case, it seems that the steady-state error of the LKF has reduced when compared to the smaller wheel angle input results. However, this is misleading because, as this is the case with the LKF, large steady-state error in the yaw rate estimation could result in a reduced steady-state error in the vehicle sideslip estimation. All other strategies again have excellent performances in both transient and steady-state responses.

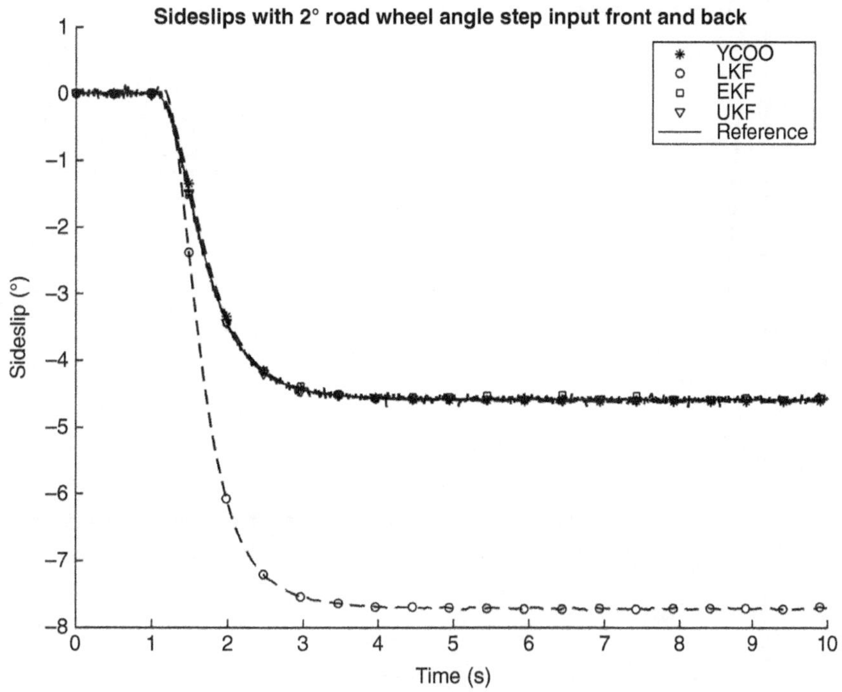

Figure 17.42 The vehicle sideslip estimation for different strategies.

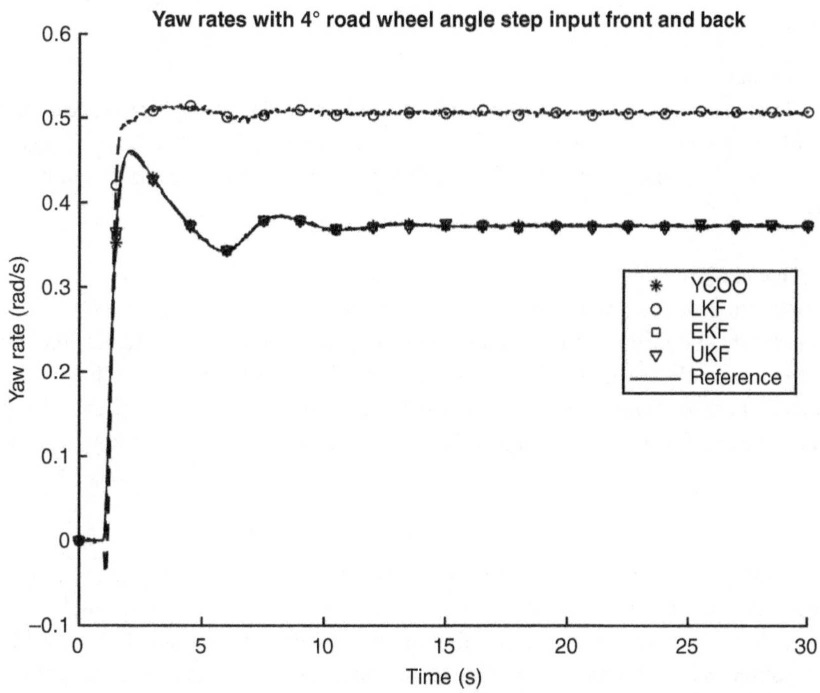

Figure 17.43 The yaw rate estimation results for different strategies with 4° road wheel angle step input.

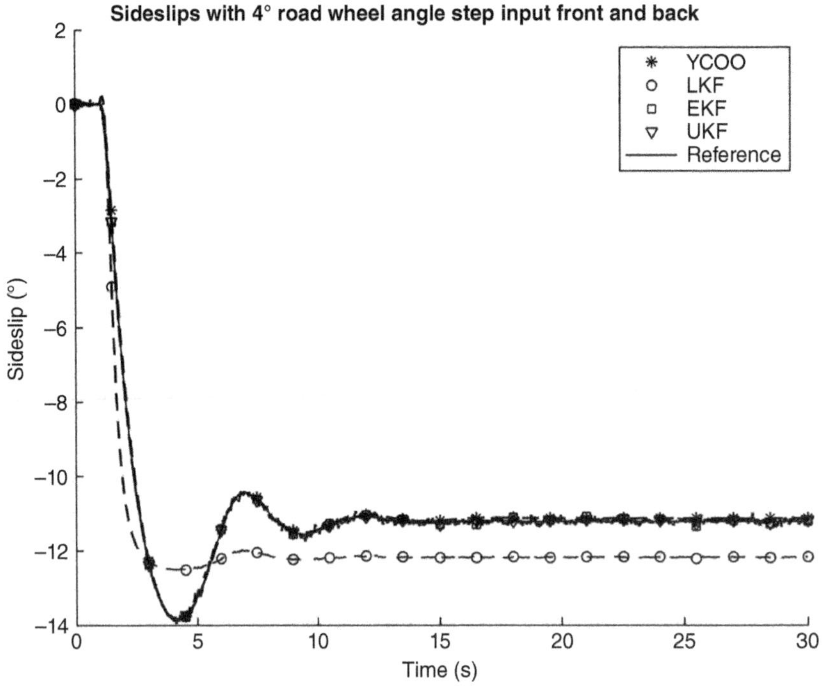

Figure 17.44 The vehicle sideslip estimation results for different strategies with 4° road wheel angle step input.

17.4.5 Robustness Test

17.4.5.1 Vehicle mass variation

The yaw rate and vehicle sideslip estimation results with the addition of extra 30% mass are shown in Figures 17.45 and 17.46. All Kalman filtering types do not have good robustness with respect to the vehicle mass variation, the results in this case are similar to the SISO case. However, the YCOO has excellent robustness characteristics with respect to the vehicle mass variation.

17.4.5.2 Tire–road coefficient of friction

The yaw rate estimations for the reduced tire–road coefficient of friction, $\mu = 0.5$, are shown in Figure 17.47. Unlike the SISO case, by the addition of rear wheel steering to the vehicle model, all Kalman filtering strategies have become more sensitive to the scaling of lateral tire forces. However, the YCOO has capitalized on this second input, based on its MIMO estimation design strategy, to drastically improve both its transient and steady-state responses with respect to the changes of this parameter. For the vehicle sideslip estimation results of Figure 17.48, similar arguments as for the yaw rate estimations apply.

17.4.6 Sensor Bias

Besides the sensor noise, the sensor bias is another source of uncertainties, as sensor measurement normally includes an offset. This offset includes the calibration error and drift over life time due to temperature and supply voltage variations. The sensor bias of the accelerometer type proposed in this example is within the range of $\pm 1m/s^2$ [26]. If we assume both the front and rear accelerometers have the same value for sensor bias, then in the case of YCOO, the yaw rate estimation is not affected by this offset, as illustrated in Figure 17.49. Although we are using a dynamic model in YCOO, the observation matrices (C_e and D_e) are derived from the kinematic relationship between lateral accelerations and states. Therefore, the yaw rate estimation using YCOO is the same as yaw rate estimation using the dynamic relationship,

$$\omega = \int \frac{a_f - a_r}{S_1 + S_2} \, dt \tag{17.103}$$

where a_f and a_r are the estimated lateral accelerations. As YCOO estimates the yaw rate, the sensor bias of the front and rear accelerometers cancel out.

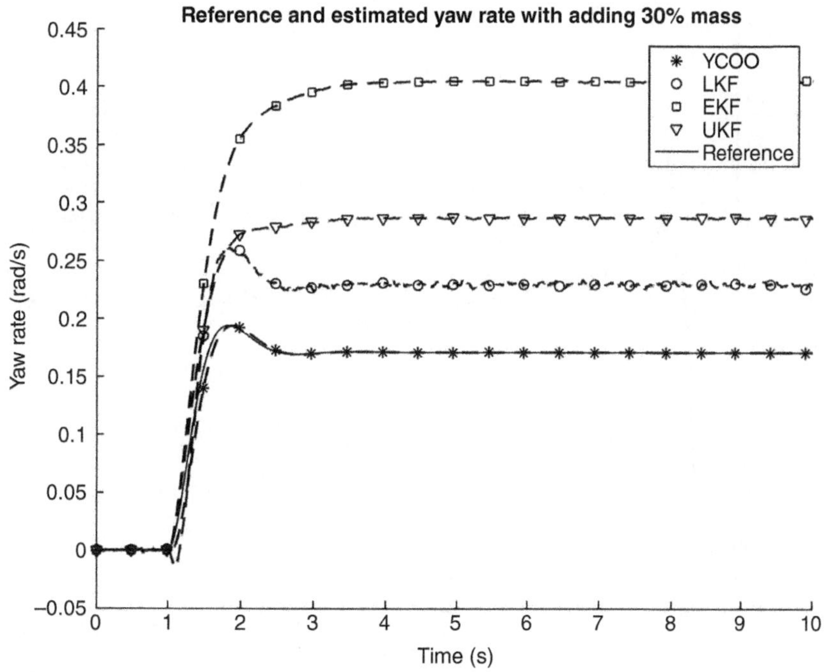

Figure 17.45 The yaw rate estimations for different strategies with 30% additional mass.

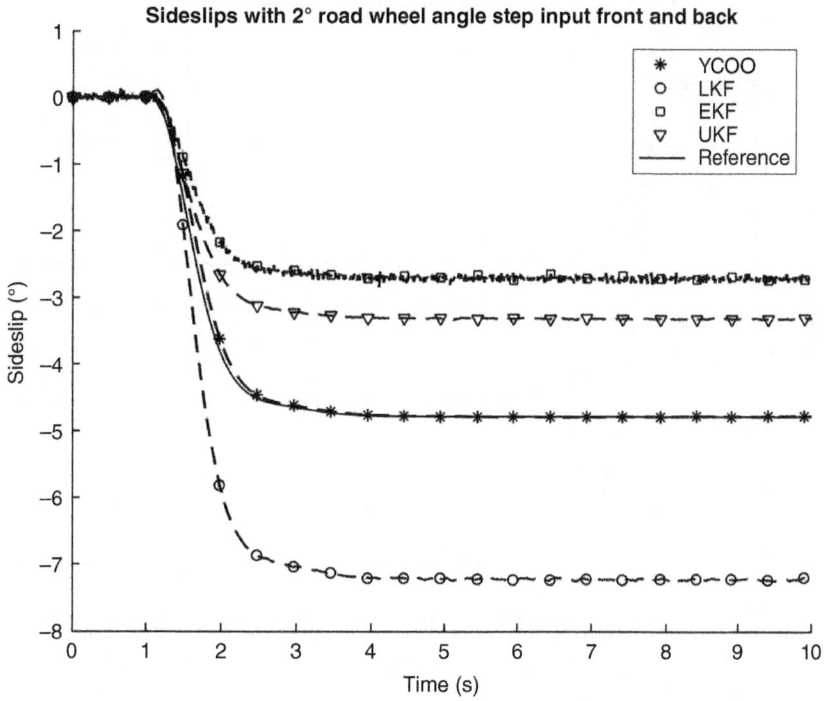

Figure 17.46 The vehicle sideslip estimations for different strategies with 30% additional mass.

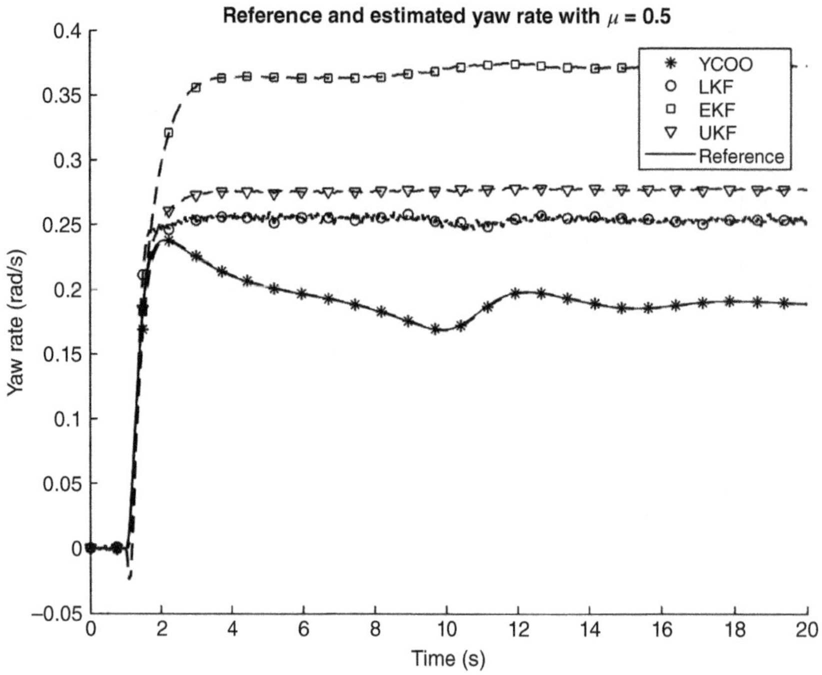

Figure 17.47 The yaw rate estimations for different strategies with $\mu = 0.5$.

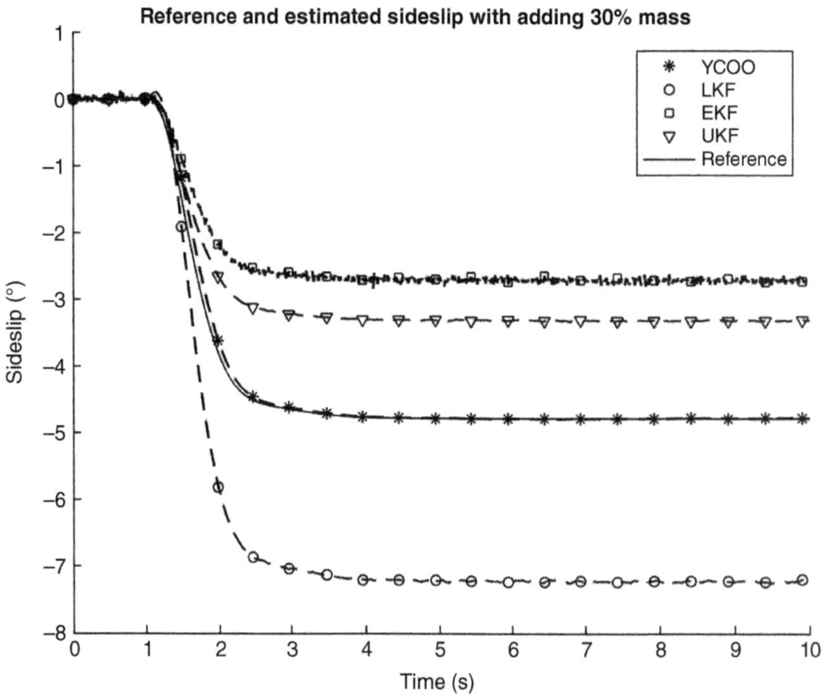

Figure 17.48 The vehicle sideslip estimations for different strategies with $\mu = 0.5$.

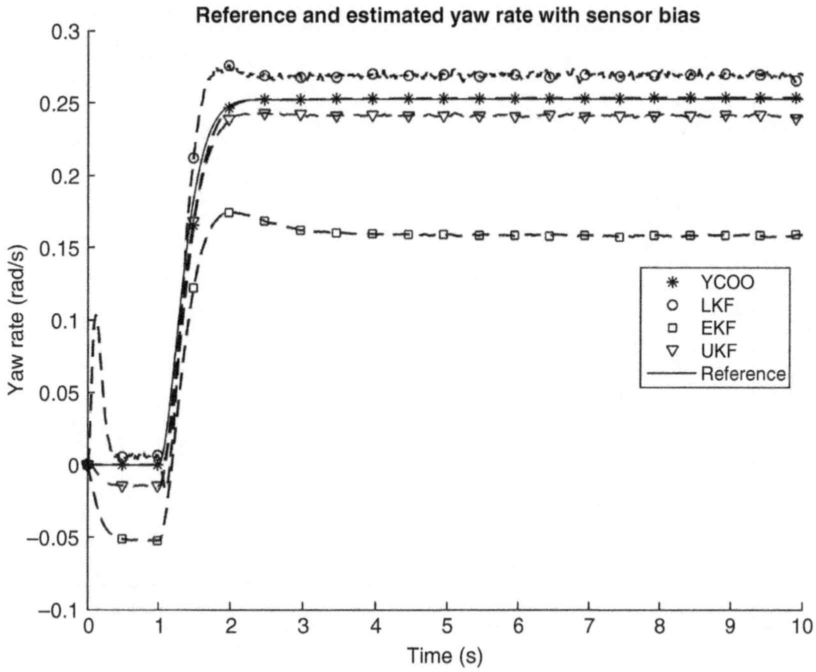

Figure 17.49 The yaw rate estimation of different strategies with a sensor bias of 1 m/s^2.

However, the story is different for the case of vehicle sideslip estimation, as the bias has a large influence on the estimation result. Since the YCOO tries to track the reference generator output signal (measured signal), the error due to the sensor bias will accumulate over time and estimation will have a significant offset with respect to the reference/measured signal as shown in Figure 17.50. In the case of Kalman filters, the LKF and EKF have relatively large steady-state error due to the

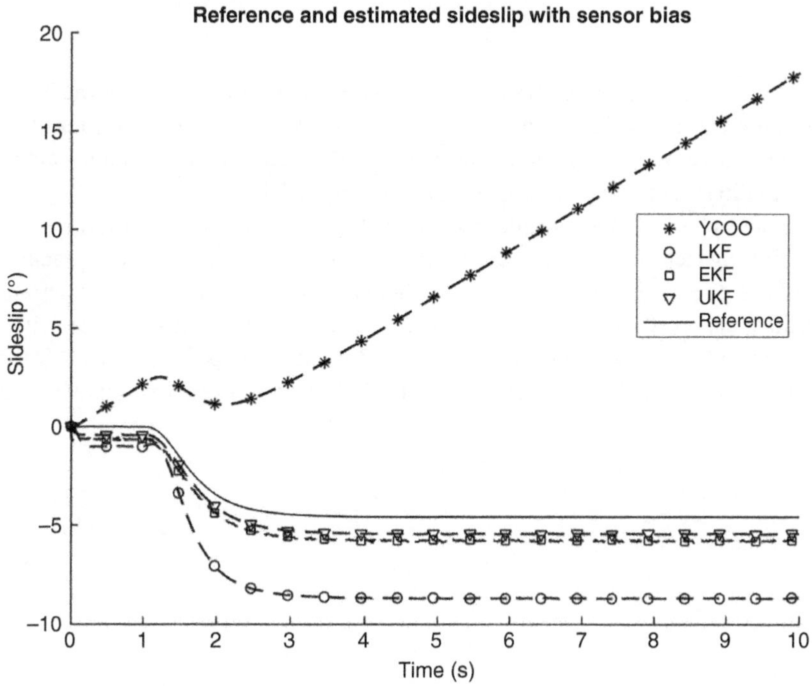

Figure 17.50 The vehicle sideslip estimations for different strategies with a sensor bias of 1 m/s^2.

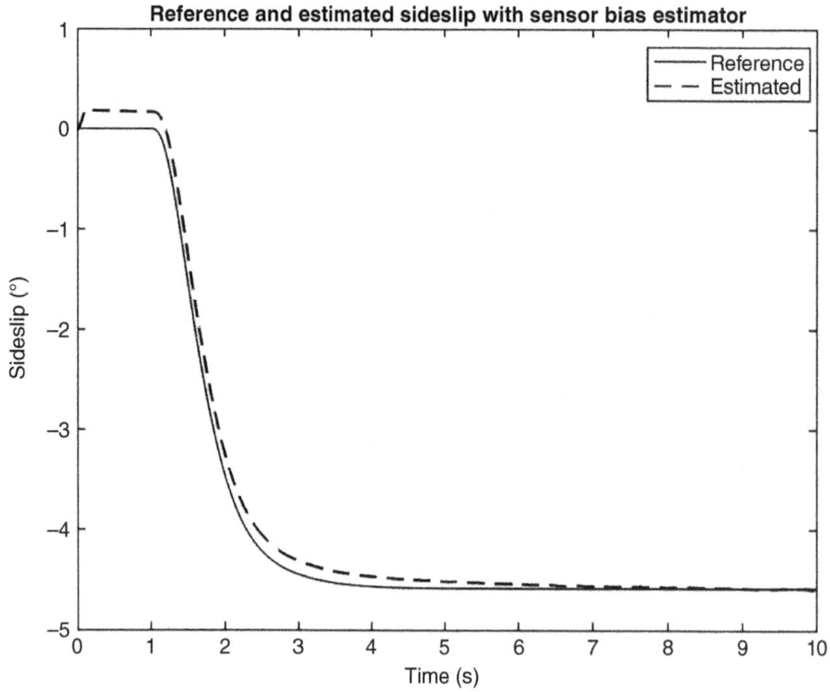

Figure 17.51 The vechile sideslip estimation of YCOO including sensor bias estimator.

sensor bias. However, the UKF, due to its statistical sampling technique for selecting appropriate sample points, is relatively unaffected by the sensor bias.

We have designed a sensor bias estimation strategy to eliminate the negative impact of the sensor bias on the YCOO estimation strategy, as illustrated in Figure 17.51. The sensor bias estimation strategy is a subject of a current invention disclosure.

17.4.7 Final Remarks

We investigated the performance and robustness of various estimation concepts using a four-wheel steering vehicle with two lateral acceleration measurements at the front and rear axles. Various road wheel angle step inputs were used to illustrate the performance and robustness of these estimators. The YCOO MIMO estimation technique showed excellent potential when compared to Kalman filtering techniques such as LKF, EKF, and UKF.

Kalman Filters did not show relatively strong results in the robustness tests using this platform. However, one can imagine using adaptive Kalman filtering techniques to increase their robustness characteristics when facing parameter variations such as vehicle mass and tire–road coefficient of friction. Although the MIMO YCOO does illustrate excellent robustness results when varying vehicle mass and tire–road coefficient of friction, it turns out that it is sensitive to the forward speed, which is assumed constant in all the presented simulations. Therefore, it is also possible to modify the YCOO, for example, by using linear parameter varying (LPV) techniques, to reduce its sensitivity to all parameter variations, including the forward speed.

A

Cauchy Integral

This appendix explores the Cauchy integral and its relation to control systems. It will begin with some necessary definitions and conclude with discussion of the Nyquist contour and development of the Nyquist stability criterion.

A.1 Contour Definitions

Functions $f(x)$ are unique mappings of x to $f(x)$. We only know the calculus of **functions**, but we are often interested in **nonfunctions** in the complex plane. Figure A.1 shows some example functions and nonfunctions.

We can introduce a parameter, ζ for instance, for arc length, as shown in Figure A.2. Then, the real and imaginary parts of the original mapping can be treated as functions of ζ.

Definition A.1 An arc is **continuous** if $\omega(\zeta)$ and $\sigma(\zeta)$ are continuous.

Definition A.2 An arc is a **Jordan arc** if no two values of ζ correspond to the same (σ, ω) pair. A Jordan and non-Jordan curve are demonstrated in Figure A.3.

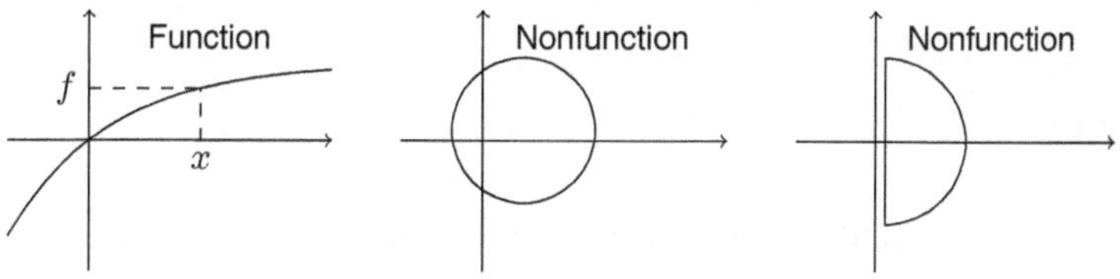

Figure A.1 Functions and nonfunctions.

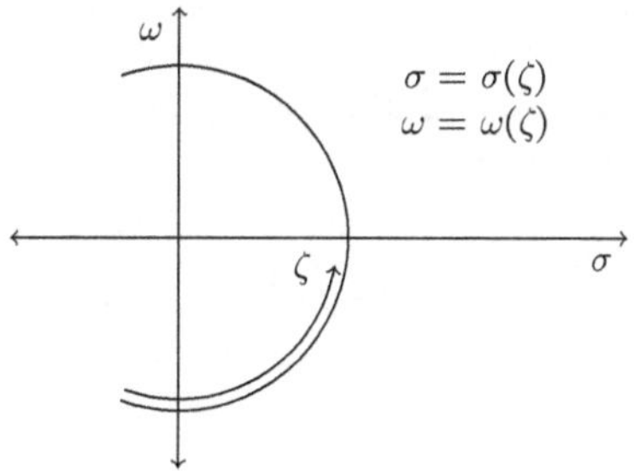

Figure A.2 Parameterization of arc length.

Figure A.3 Jordan and non-Jordan arcs.

Robust Control: Youla Parameterization Approach, First Edition. Farhad Assadian and Kevin R. Mallon.

Companion Website: www.wiley.com/go/assadian/robust

Figure A.4 Jordan curve.

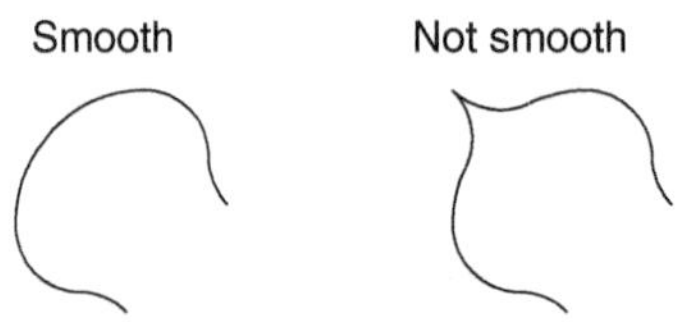

Figure A.5 Smooth and non-smooth curves.

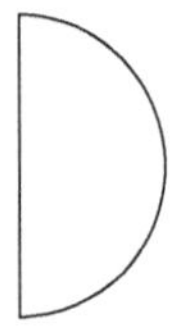

Figure A.6 Basic Jordan contour.

Definition A.3 A curve is a **Jordan curve** if $\sigma(\alpha) = \sigma(\beta)$ and $\omega(\alpha) = \omega(\beta)$ but has equality at no other values of ζ. An example is shown in Figure A.4.

Definition A.4 An arc is **smooth** if $\sigma'(\zeta)$ and $\omega'(\beta)$ are continuous, as shown in Figure A.5.

Definition A.5 A **Jordan contour**, as demonstrated in Figure A.6, is a curve made up of a finite set of smooth Jordan arcs.

A.2 Contour Integrals

This section will expand on the Jordan contours. A general such contour is shown in Figure A.7.

Suppose that $p(\sigma, \omega)$ and $q(\sigma, \omega)$ are continuous, real-valued functions. Then,

$$\int_{\Gamma} (\mathrm{p}\mathrm{d}\sigma + \mathrm{q}\mathrm{d}\omega) = \int_{\alpha}^{\beta} \left(p\frac{\mathrm{d}\sigma}{\mathrm{d}\zeta} + q\frac{\mathrm{d}\omega}{\mathrm{d}\zeta} \right) \mathrm{d}\zeta \tag{A.1}$$

Theorem A.1 **(Greene's Theorem)** *If $p(\sigma, \omega)$ and $q(\sigma, \omega)$, together with their first partial derivatives, are continuous throughout a closed region Ω and its boundary Γ, then*

$$\int_{\Gamma} (\mathrm{p}\mathrm{d}\sigma + \mathrm{q}\mathrm{d}\omega) = \iint_{\Omega} \left(\frac{\partial q}{\partial \sigma} - \frac{\partial p}{\partial \omega} \right) \mathrm{d}\sigma\, \mathrm{d}\omega \tag{A.2}$$

where Ω and Γ are indicated in Figure A.8.

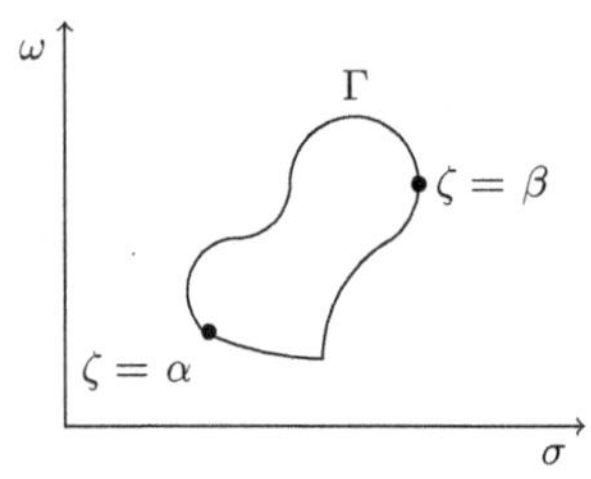

Figure A.7 Generic Jordan contour.

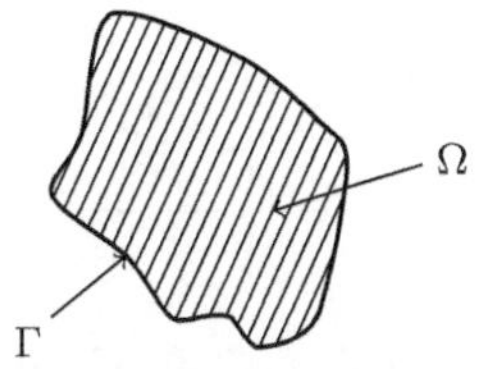

Figure A.8 Contour Γ and enclosed region Ω.

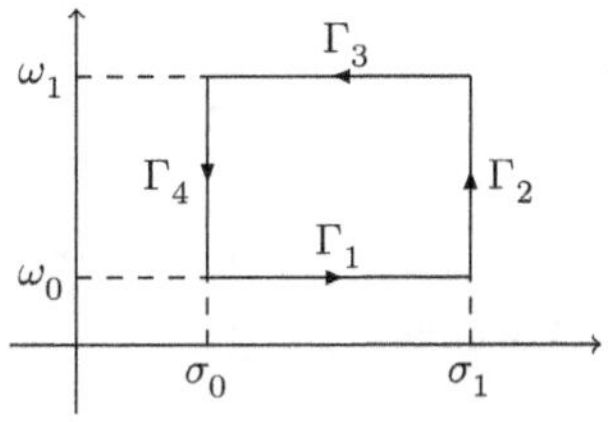

Figure A.9 A simple contour to prove Greene's theorem.

Proof: It suffices to show that

$$\int_{\Gamma} p\mathrm{d}\sigma = -\iint_{\Omega} \frac{\partial p}{\partial \omega} \mathrm{d}\sigma\, \mathrm{d}\omega \tag{A.3}$$

and

$$\int_{\Gamma} q\mathrm{d}\omega = \iint_{\Omega} \frac{\partial q}{\partial \sigma} \mathrm{d}\sigma\, \mathrm{d}\omega \tag{A.4}$$

We will prove the first only. The second can be obtained in a similar manner. Consider the contour in Figure A.9.

$$\iint_{\Omega} \frac{\partial q}{\partial \sigma} \mathrm{d}\sigma\, \mathrm{d}\omega = -\int_{\sigma_0}^{\sigma_1} \int_{\omega_0}^{\omega_1} \frac{\partial p}{\partial \omega} \mathrm{d}\sigma\, \mathrm{d}\omega \tag{A.5}$$

$$= -\int_{\sigma_0}^{\sigma_1} \left(p(\sigma, \omega_1) - p(\sigma, \omega_0)\right) \mathrm{d}\sigma$$

$$= \int_{\sigma_0}^{\sigma_1} p(\sigma, \omega_0)\mathrm{d}\sigma + \int_{\sigma_1}^{\sigma_0} p(\sigma, \omega_1)\mathrm{d}\sigma$$

$$= \int_{\Gamma_1} p(\sigma, \omega) + \int_{\Gamma_3} p(\sigma, \omega)$$

$$= \int_{\Gamma} p(\sigma, \omega) \tag{A.6}$$

■

A.3 Complex Analysis Definitions

In this section we briefly touch on some necessary definitions for complex analysis.

First, recall from previous chapters that the 2-norm is a measure of size. For a point λ shown in (Figure A.10).

$$|\lambda|^2 = \sigma^2 + \omega^2 \tag{A.7}$$

Multiplication cannot be done in the real plane, but can be done in the complex plane:

$$f(\lambda) = w + jv \tag{A.8}$$

$$\lambda = \sigma + j\omega \tag{A.9}$$

Definition A.6 f_0 is the **limit** of $f(\lambda)$ at $\lambda = \lambda_0$ if for any $\epsilon > 0$, there exists $\delta > 0$ such that

$$|f(\lambda) - f_0| < \epsilon \tag{A.10}$$

$$\lambda - \lambda_0 < \delta \tag{A.11}$$

as shown in Figure A.11.

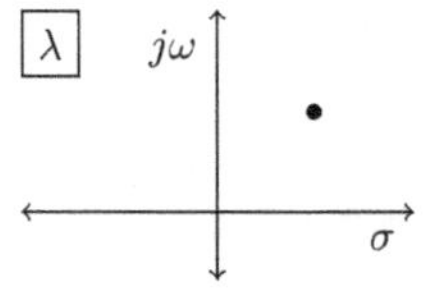

Figure A.10 A point in the complex plane.

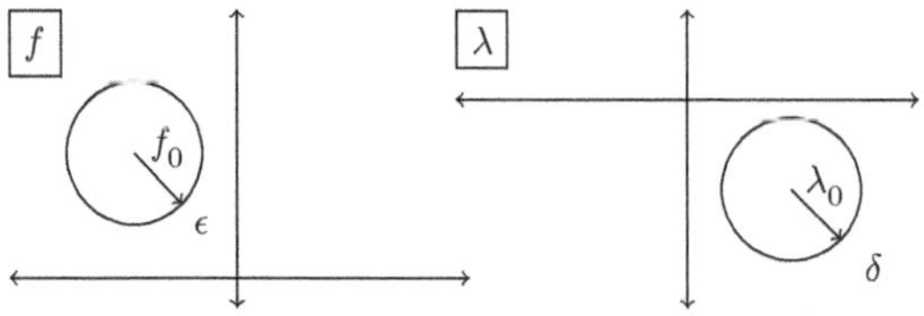

Figure A.11 The limit f_0.

A.4 Cauchy–Riemann Conditions

We are interested in the transformation of the line integral to the surface integral:

$$\int_{\Gamma} (p\mathrm{d}\sigma + q\mathrm{d}\omega) = \iint_{\Omega} \left(\frac{\partial q}{\partial \sigma} - \frac{\partial p}{\partial \omega} \right) \mathrm{d}\sigma\, \mathrm{d}\omega \tag{A.12}$$

for the closed region Ω and its boundary Γ indicated in Figure A.12.

Keep in mind the complex analysis variable, $\lambda = \sigma + j\omega$. Then, a function of λ can be written as:

$$f(\lambda) = w(\sigma, \omega) + jv(\sigma, \omega) \tag{A.13}$$

(Note that the reader should be careful not to confuse w and ω in this section.) f_0 is the limit of $f(\lambda)$ at $\lambda = \lambda_0$ if, for any $\epsilon > 0$, there exists $\delta > 0$ such that

$$|f(\lambda) - f_0| < \epsilon \tag{A.14}$$

wherever

$$|\lambda - \lambda_0| < \delta \tag{A.15}$$

This mapping is shown in Figure A.13.

We say that $f(\lambda)$ is continuous at $\lambda = \lambda_0$ if $w(\sigma, \omega)$ and $v(\sigma, \omega)$ are continuous.

By definition, the derivative of $f(\lambda)$ is

$$\frac{\mathrm{d}f}{\mathrm{d}\lambda} = \lim_{\Delta\lambda \to 0} \frac{f(\lambda - \Delta\lambda) - f\lambda}{\Delta\lambda} \tag{A.16}$$

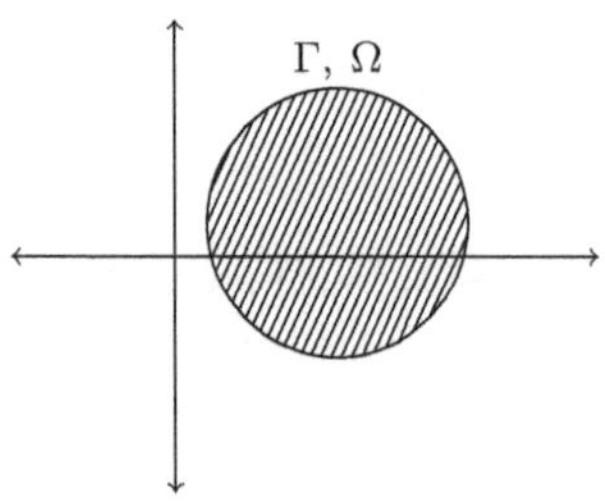

Figure A.12 Line and surface integral in the complex plane.

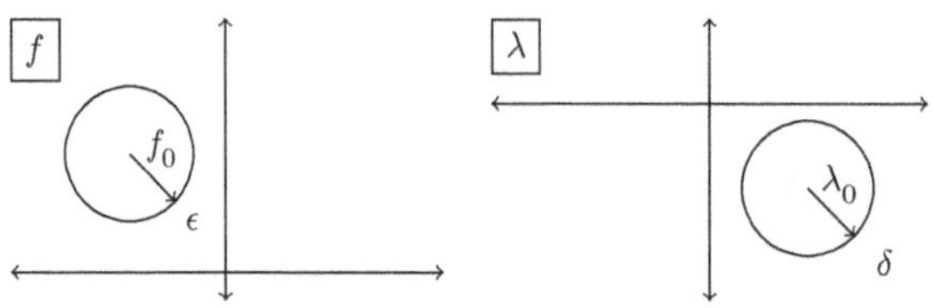

Figure A.13 Mapping the f and λ planes.

λ is complex, therefore two-dimensional. $\Delta\lambda \to 0$ can be considered from any direction of approach. In particular, consider $\Delta\lambda = \Delta\sigma$ (approaching on the positive real axis). Then,

$$\frac{\Delta f}{\Delta \lambda} = \frac{\Delta w + j\Delta v}{\Delta \sigma} = \frac{\Delta w}{\Delta \sigma} + j\frac{\Delta v}{\Delta \sigma} \tag{A.17}$$

As $\Delta\lambda \to 0$,

$$\frac{df}{d\lambda} = \frac{dw}{d\sigma} + j\frac{dv}{d\sigma} \tag{A.18}$$

Now, suppose that $\Delta\lambda = j\Delta\omega$ (approaching on the positive imaginary axis). Then,

$$\frac{df}{d\lambda} = \frac{dw}{jd\omega} + j\frac{dv}{jd\omega} = -j\frac{dw}{d\omega} + \frac{dv}{d\omega} \tag{A.19}$$

A necessary condition of the derivative of $f(\lambda)$ is that it must not depend on the direction of approach:

$$\frac{\partial w}{\partial \sigma} = \frac{\partial v}{\partial \omega}, \quad \frac{\partial w}{\partial \omega} = -\frac{\partial v}{\partial \sigma} \tag{A.20}$$

Cauchy–Riemann conditions are sufficient if w, v are continuous at λ.

Example A.1 Consider a complex function $f(\lambda)$ such that

$$f(\lambda) = \overline{\lambda} = \sigma - j\omega \tag{A.21}$$

$$w = \sigma, \quad v = -\omega \tag{A.22}$$

and $f(\lambda)$ is continuous everywhere.

$$\frac{\partial w}{\partial \sigma} = 1 \neq -1 = \frac{\partial v}{\partial \omega} \tag{A.23}$$

Therefore, the derivative does not exist anywhere!

Example A.2 Consider a complex function $f(\lambda)$ with enclosed region Ω and contour Γ as in Figure A.14, such that

$$f(\lambda) = \lambda^2 = \sigma^2 - \omega^2 + 2j\sigma\omega \tag{A.24}$$

and $f(\lambda)$ is continuous everywhere. For $f(\lambda)$ defined as in Eq. (A.13),

$$\frac{\partial w}{\partial \sigma} = 2\sigma = \frac{\partial v}{\partial \omega}, \quad \frac{\partial w}{\partial \omega} = -2\omega = -\frac{\partial v}{\partial \sigma} \tag{A.25}$$

Therefore, the derivative exists everywhere.

Consider a complex function $f(\lambda)$. The line integral can be expressed as follows:

$$\int_\Gamma f(\lambda)d\lambda = \int_\Gamma (w(\lambda) + jv(\lambda))(d\sigma + jd\omega) \tag{A.26}$$

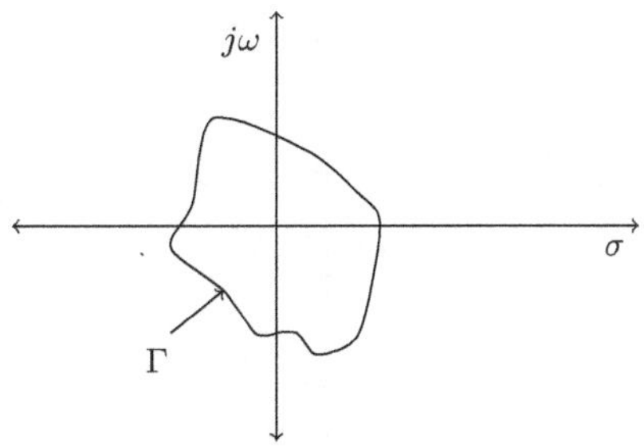

Figure A.14 Contour Γ and enclosed region Ω for Example A.2.

$$= \int_\Gamma (w\mathrm{d}\sigma - v\mathrm{d}\omega) + j\int_\Gamma (v\mathrm{d}\sigma + w\mathrm{d}\omega)$$
$$= \iint_\Omega \left(-\frac{\partial v}{\partial \sigma} - \frac{\partial w}{\partial \omega}\right)\mathrm{d}\sigma\,\mathrm{d}\omega + j\iint_\Omega \left(\frac{\partial w}{\partial \sigma} - \frac{\partial v}{\partial \omega}\right)\mathrm{d}\sigma\,\mathrm{d}\omega \tag{A.27}$$

Theorem A.2 (Cauchy–Goursat) *If the derivative of $f(\lambda)$ exists throughout Ω and on Γ, then*

$$\int_\Gamma f(\lambda)\mathrm{d}\lambda = 0 \tag{A.28}$$

A.5 Cauchy Integral Theorem

We can now approach the Cauchy integral theorem. First, consider the open disc shown in Figure A.15.

Definition A.7 We say that $f(\lambda)$ is **analytic** at $\lambda = \lambda_0$ if $\frac{\mathrm{d}f}{\mathrm{d}\lambda}$ exists at $\lambda = \lambda_0$ and at all points in some open disc about λ_0.

Theorem A.3 *If f is analytic throughout Ω ad its boundary Γ, then*

$$\int_\Gamma f(\lambda)\mathrm{d}\lambda = 0 \tag{A.29}$$

Suppose that $f(\lambda)$ is analytic throughout Ω and Γ. Then,

$$\frac{f(\lambda)}{\lambda - \lambda_0} \tag{A.30}$$

is analytic throughout Ω and Γ *except* at $\lambda = \lambda_0$. From Cauchy–Goursat,

$$\int_\Gamma \frac{f(\lambda)}{\lambda - \lambda_0}\mathrm{d}\lambda = \int_{C_0} \frac{f(\lambda)}{\lambda - \lambda_0}\mathrm{d}\lambda \tag{A.31}$$

$$\int_\Gamma \frac{f(\lambda)}{\lambda - \lambda_0}\mathrm{d}\lambda = \int_\Gamma \frac{f(\lambda) - f(\lambda_0)}{\lambda - \lambda_0}\mathrm{d}\lambda + \int_\Gamma \frac{f(\lambda_0)}{\lambda - \lambda_0}\mathrm{d}\lambda \tag{A.32}$$

The contour C_0 can be expressed in polar coordinates as indicated in Figure A.16.

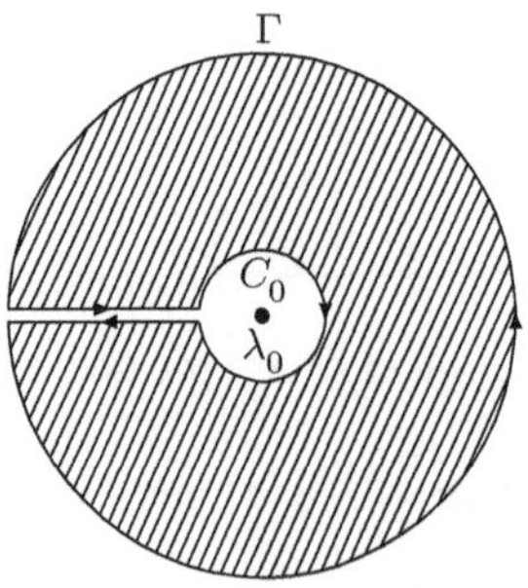

Figure A.15 Open disc about λ_0 with inner contour C_0.

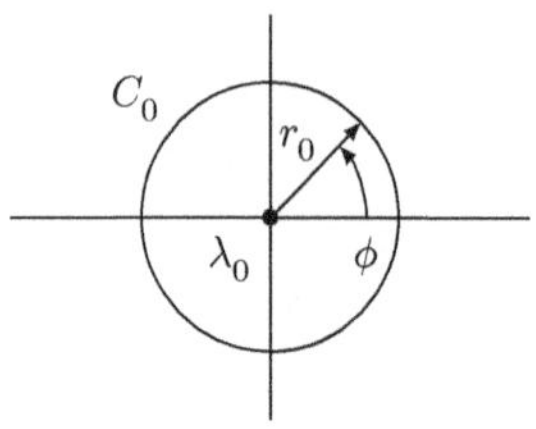

Figure A.16 The modulus $|f|$ and and argument ϕ.

Then,

$$\lambda - \lambda_0 = r_0 e^{j\phi}, \quad d\lambda = jr_0 e^{j\phi} d\phi \tag{A.33}$$

$$\begin{aligned} f(\lambda_0)\int_\Gamma \frac{d\lambda}{\lambda - \lambda_0} &= jf(\lambda_0)\int_{C_0} d\phi \\ &= 2\pi j f(\lambda_0) \end{aligned}$$

$$\int_\Gamma \frac{f(\lambda)}{\lambda - \lambda_0} d\lambda = \int_\Gamma \frac{f(\lambda) - f(\lambda_0)}{\lambda - \lambda_0} d\lambda + 2\pi j f(\lambda_0) \tag{A.34}$$

$f(\lambda)$ is continuous at $\lambda = \lambda_0$, so for any ϵ there exists a δ such that

$$|f(\lambda) - f(\lambda_0)| < \epsilon \tag{A.35}$$

if

$$|\lambda - \lambda_0| < \delta \tag{A.36}$$

Let $r_0 = \delta$. Then,

$$\left|\int_\Gamma \frac{f(\lambda) - f(\lambda_0)}{\lambda - \lambda_0} d\lambda\right| \leq \int_\Gamma \frac{|f(\lambda) - f(\lambda_0)|}{|\lambda - \lambda_0|} |d\lambda| \leq \frac{\epsilon}{\delta}\int_{C_0} |d\lambda| \tag{A.37}$$

If ϵ is small,

$$\frac{\epsilon}{\delta}\int_{C_0} |d\lambda| = \frac{\epsilon}{\not\delta} 2\pi \not\delta = 2\pi\epsilon \tag{A.38}$$

Theorem A.4 (Cauchy Integral Theorem) *Consider an open disc with contour* Γ *and centered about a point* λ_0*. Then for a complex function* $f(\lambda)$*,*

$$\frac{1}{2\pi j}\int_\Gamma \frac{f(\lambda)}{\lambda - \lambda_0} d\lambda = f(\lambda_0) \tag{A.39}$$

Suppose we want a function $f(\lambda)$ such that

$$\int_\Gamma \frac{R}{\lambda - \lambda_0} d\lambda = 2\pi j R \tag{A.40}$$

Then, consider some rational function, such as indicated in Figure A.17 and its expansion,

$$G(s) = (\text{polynomial terms}) + \frac{R_1}{s - p_1} + \frac{R_2}{s - p_2} + \cdots + \frac{R_n}{s - p_n} \tag{A.41}$$

$$\int_\Gamma G(s)ds = 2\pi j \sum_k R_k \tag{A.42}$$

This is the residue formula, true for more general functions.

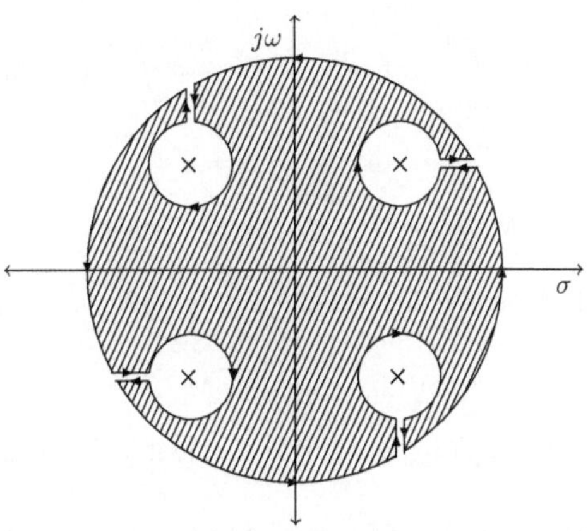

Figure A.17 G(s) contour.

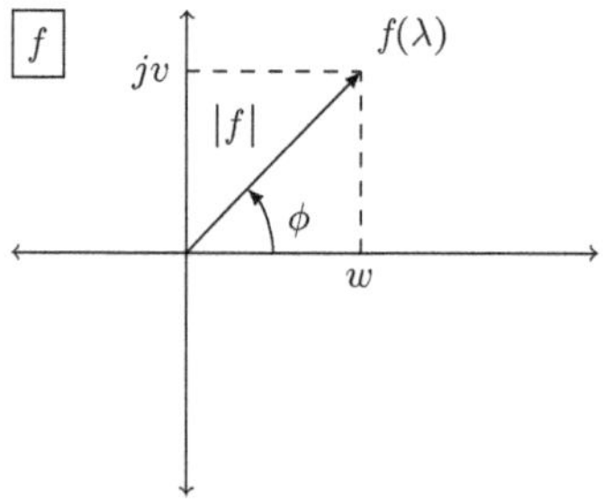

Figure A.18 Polar coordinates of contour C_0.

A.5.1 Terminology

The modulus, $|f|$ and argument, ϕ are defined as:

$$|f| = \sqrt{w^2 + v^2}, \quad \phi = \tan^{-1}\frac{v}{w} \tag{A.43}$$

as indicated in Figure A.18.

A.6 Maximum Modulus Theorem

In this section, we introduce the Maximum Modulus Principle and Maximum Modulus Theorem. With this, we will be able to start developing principles for control system properties.

Consider an application of the Cauchy integral theorem to a complex function $f(\lambda)$ and along a contour C_0, as shown in (Figure A.19):

$$\frac{1}{2\pi j}\int_{C_0}\frac{f(\lambda)}{\lambda-\lambda_0}\mathrm{d}\lambda = f(\lambda_0) \tag{A.44}$$

The modulus:

$$|f(\lambda_0)| \leq \frac{1}{|2\pi j|}\int_{C_0}\frac{|f(\lambda)|}{|\lambda-\lambda_0|}|\mathrm{d}\lambda| = \frac{1}{2\pi r_0}\int_{C_0}|f(\lambda)|r_0\mathrm{d}\lambda \tag{A.45}$$

$$|f(\lambda_0)| \leq \frac{1}{2\pi}\int_0^{2\pi}|f(\lambda_0 + r_0e^{j\theta})| = F_0 \tag{A.46}$$

where F_0 is the average value of $F(\lambda)$ on C_0. Let M_0 be the maximum modulus of $f(\lambda)$.

$$|f(\lambda_0)| \leq M_0 \tag{A.47}$$

Also,

$$|f(\lambda_0 + r_0e^{j\theta})| \leq M_0 \tag{A.48}$$

Therefore,

$$F_0 = \frac{1}{2\pi}\int_0^{2\pi}|f(\lambda_0 + r_0e^{j\theta})| \leq \frac{M_0}{2\pi}\int_0^{2\pi}\mathrm{d}\theta = M_0 \tag{A.49}$$

What is the possibility that $|f(\lambda_0)| = M_0$?

$$M_0 \leq \frac{1}{2\pi}\int_0^{2\pi}|f(\lambda_0 + r_0e^{j\theta})| = F_0 \tag{A.50}$$

If $|f(\lambda_0)| < M_0$ for some θ, then it must be for some interval of θ, implying $M_0 < M_0$, a contradiction. We can then conclude that

$$|f(\lambda_0 + r_0e^{j\theta})| = M_0 \tag{A.51}$$

A similar analysis applies for $C_1(r_1 < r_0)$, thus, $|f(\lambda_0)| = M_0$ if and only if $|f(\lambda_0)| \equiv M_0$ throughout the entire region.

Lemma A.1 (Maximum Modulus Principle) *If f is analytic at λ_0, then every neighborhood of λ_0 in which f is not a constant contains points λ such that $|f(\lambda)| > |f(\lambda_0)|$.*

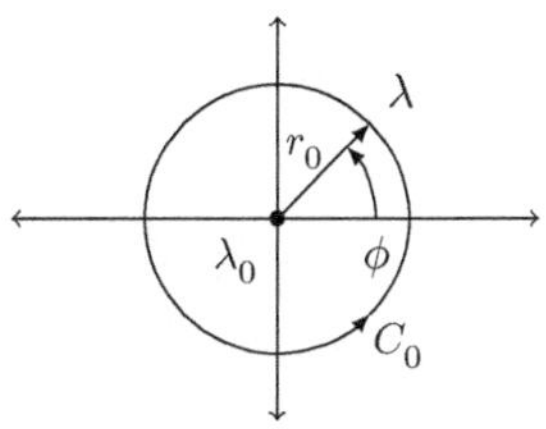

Figure A.19 Polar coordinates for an application of the Cauchy Integral Theorem.

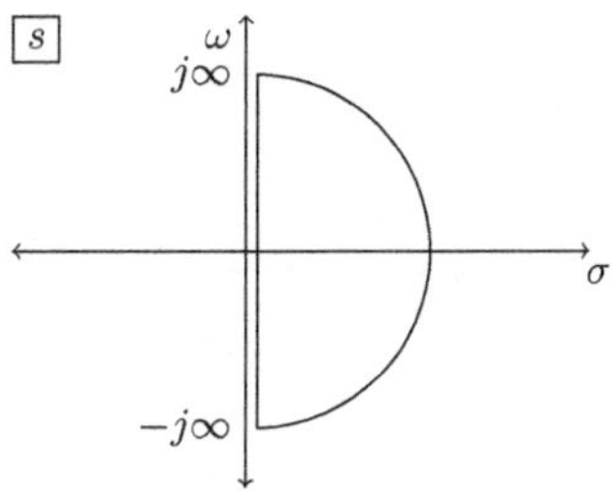

Figure A.20 Contour encompassing the RHP.

Theorem A.5 (Maximum Modulus Theorem) *If $f(\lambda)$ is analytic throughout Ω and on its boundary Γ, and if f is not identically a constant, then the maximum modulus of $f(\lambda)$ is attained on the boundary.*

As an application of this theorem, let us suppose that we have a well-designed feedback: $T(s)$ and $S(s)$ are stable (no poles on the $j\omega$-axis or in the RHP). Consider the contour that extends along the $j\omega$-axis and encompasses the entire right-hand plane, as in Figure A.20. Here, $T(s)$ and $S(s)$ are analytical inside and on the contour and max $|T(s)|$ and max $|S(s)|$ occur on the boundary.

If the feedback is strictly proper, then $|S| = |T| = 0$ on the curved portion of the contour, so max $|T(s)|$ and max $|S(s)|$ must occur on the $j\omega$-axis. If the feedback is proper, max $|T(s)|$ and max $|S(s)|$ occur on the $j\omega$-axis as well.

$$\max |S(s)| = \max_{\omega} |S(j\omega)| \tag{A.52}$$

Suppose $L(s)$ has a pole $s = p$ in the RHP and a non-minimum phase zero $s = z$. $L(s)$ can be written with this unstable pole pulled out, such that

$$L(s) = \frac{1}{(s-p)} L'(s) \tag{A.53}$$

The pole $s = p$ will therefore be in the numerator of $S(s)$, making $S(s)$ non-minimum phase:

$$S(s) = (s-p)S' \tag{A.54}$$

$S(s)$ can be factored into a minimum phase transfer function and a non-minimum phase all-pass filter:

$$S(s) = S_{ap}(s)S_{mp}(s) \tag{A.55}$$

$S_{ap}(s)$ is the all-pass filter portion of $S(s)$, while S_{mp} is the minimum phase part.

Example A.3 Consider the following non-minimum phase transfer function:

$$S(s) = \frac{1-s}{(2-s)(s+1)} \tag{A.56}$$

$$S(s) = \underbrace{\frac{1-s}{1+s} \cdot \frac{2+s}{2-s}}_{\text{all-pass}} \cdot \underbrace{\frac{1}{s+2}}_{\text{min-phase}} \tag{A.57}$$

$$|S(j\omega)| = |S_{mp}(s)| \tag{A.58}$$

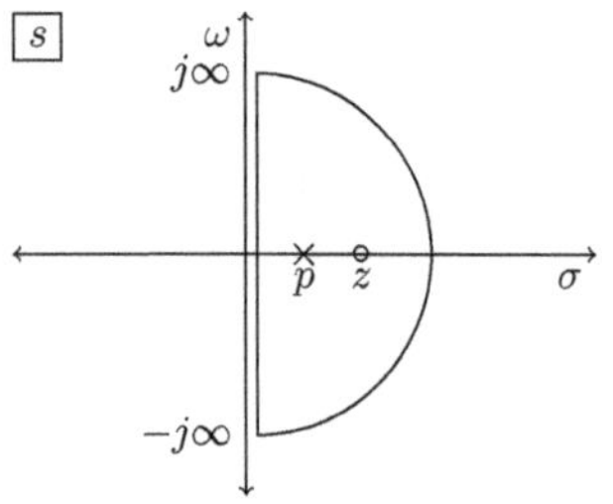

Figure A.21 Poles and zeros of L within the RHP contour.

Example A.4

$$S = \frac{(1-s)(1+s)}{s(s+2)} = \underbrace{\frac{1-s}{1+s}}_{S_{ap}} \cdot \underbrace{\frac{(1+s)^2}{s(s+2)}}_{S_{mp}} \tag{A.59}$$

Example A.5

$$S = \frac{(1+s)}{(s+3)(s+2)} = \underbrace{1}_{S_{ap}} \cdot \underbrace{\frac{(1+s)}{(s+3)(s+2)}}_{S_{mp}} \tag{A.60}$$

Now, consider the interpolation conditions for the aforementioned system:

$$S(z) = 1 \tag{A.61}$$

The all-pass factor of $S(s)$ is

$$S_{ap} = \frac{s-p}{s+p} \tag{A.62}$$

Therefore, the interpolation condition becomes

$$S(z) = S_{ap}(z)S_{mp}(z) = 1 \tag{A.63}$$

$$S_{mp}(z) = \frac{1}{S_{ap}(z)} = \frac{z+p}{z-p} \tag{A.64}$$

$$\max_{\omega} |S(j\omega)| = \max_{\omega} |S_{mp}(j\omega)| \geq |S_{mp}(z)| \tag{A.65}$$

$$\max_{\omega} |S(j\omega)| \geq \left|\frac{z+p}{z-p}\right| \tag{A.66}$$

By the maximum modulus theorem, we have arrived at an analytical constraint. Non-minimum phase zeros and unstable poles, such as in Figure A.21 create a lower bound on the M_2 margin.

A.7 Poisson Integral Formula

Consider an $f(\lambda)$ that is analytical on and within the contour D, which has a radius R, as in Figure A.22.

$$\int_D \frac{f(\lambda)}{\lambda - \lambda_0} \mathrm{d}\lambda = 2\pi j f(\lambda_0) \tag{A.67}$$

$$\int_D \frac{f(\lambda)}{\lambda + \bar{\lambda}_0} \mathrm{d}\lambda = 0 \cdot 2\pi j f(\lambda_0) \tag{A.68}$$

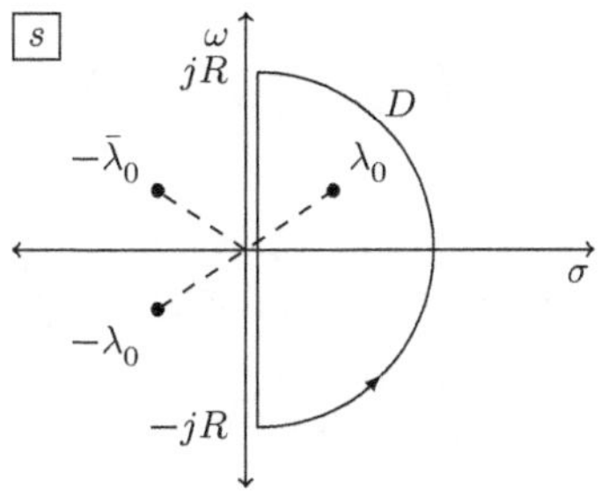

Figure A.22 A point λ_0 within the RHP contour.

By subtracting the second equation from the first,

$$f(\lambda_0) = \frac{1}{2\pi j}\int_D f(\lambda)\left(\frac{\lambda_0 + \bar{\lambda}_0}{(\lambda - \lambda_0)(\lambda + \bar{\lambda}_0)}\right)\mathrm{d}\lambda \tag{A.69}$$

$$f(\lambda_0) = \frac{-1}{2\pi j}\int_R^{-R} f(j\omega)\left(\frac{2\sigma_0}{\sigma_0^2 + (\omega - \omega_0)^2}\right)j\mathrm{d}\omega$$

$$+\frac{1}{2\pi j}\int_{-\frac{\pi}{2}}^{\frac{\pi}{2}} f(Re^{j\theta})\left(\frac{2\sigma_0}{(Re^{j\theta} - \lambda_0)(Re^{j\theta} + \bar{\lambda}_0)}\right)j\mathrm{d}\theta \tag{A.70}$$

$$f(\lambda_0) = \frac{1}{\pi}\int_{-R}^{R}\frac{f(j\omega)\sigma_0}{\sigma_0^2 + (\omega - \omega_0)^2}\mathrm{d}\omega + I_z \tag{A.71}$$

$$I_z \leq \frac{\sigma_0}{\pi R}|f|_{max}\underbrace{\int_{-\frac{\pi}{2}}^{\frac{\pi}{2}}\frac{\mathrm{d}\theta}{\left(e^{j\theta} - \frac{\lambda_0}{R}\right)\left(e^{j\theta} + \frac{\bar{\lambda}_0}{R}\right)}}_{\text{integral exists as } R\to\infty} \tag{A.72}$$

Therefore, $|I_z| \to 0$ if $f(\lambda)$ is bounded. This results gives us the Poisson integral formula.

Theorem A.6 (Poisson Integral Theorem) *If $f(\lambda)$ is analytical and of bounded magnitude on the imaginary axis, and is analytical in the RHP, and if*

$$\lambda_0 = \sigma_0 + j\omega_0, \quad \sigma > 0 \tag{A.73}$$

Then,

$$f(\lambda_0) = \frac{1}{\pi}\int_{-\infty}^{\infty} f(j\omega)\frac{\sigma_0}{\sigma_0^2 + (\omega - \omega_0)^2}\mathrm{d}\omega \tag{A.74}$$

As an application of the Poisson integral formula, consider a transfer function:

$$G(s) = G_{ap}(s)G_{mp}(s) \tag{A.75}$$

For $s_0 = \sigma_0 + j\omega_0$ in the RHP, let

$$f(s) = \ln G_{mp}(s) \tag{A.76}$$

The Poisson integral formula therefore means

$$f(s_0) = \frac{1}{\pi}\int_{-\infty}^{\infty}\frac{f(j\omega)\sigma_0}{\sigma_0^2 + (\omega - \omega_0)^2}\mathrm{d}\omega \tag{A.77}$$

By inverting the definition of $f(s)$,

$$G_{mp} = e^{f(s)} = e^{\mathrm{Re}(f) + j\mathrm{Im}(f)} = e^{\mathrm{Re}(f)}e^{\mathrm{I}(f)} \tag{A.78}$$

Therefore,

$$|G_{mp}| = e^{\mathrm{Re}(f)}, \quad \ln|G_{mp}| = \mathrm{Re}(f) \tag{A.79}$$

The real part of $f(s_0)$ is

$$\mathrm{Re}(f(s_0)) = \frac{1}{\pi}\int_{-\infty}^{\infty} \mathrm{Re}(f(j\omega))\frac{\sigma_0}{\sigma_0^2 + (\omega - \omega_0)^2}\mathrm{d}\omega \tag{A.80}$$

So,

$$\ln|G_{mp}(s_0)| = \frac{1}{\pi}\int_{-\infty}^{\infty} \ln|G_{mp}(j\omega)|\frac{\sigma_0}{\sigma_0^2 + (\omega - \omega_0)^2}\mathrm{d}\omega \tag{A.81}$$

$$\ln|G_{mp}(s_0)| = \frac{1}{\pi}\int_{-\infty}^{\infty} \ln|G(j\omega)|\frac{\sigma_0}{\sigma_0^2 + (\omega - \omega_0)^2}\mathrm{d}\omega \tag{A.82}$$

Now consider that $G(s)$ is the sensitivity transfer function of some feedback system, $S(s) = G(s)$:

$$M_1 = \max_{\omega_1 \le \omega \le \omega_2} |S(j\omega)| \tag{A.83}$$

For example, suppose the disturbance d_y is the in frequency range $\omega_1 \le \omega \le \omega_2$:

$$M_2 = \max_{\omega} |S(j\omega)| \tag{A.84}$$

Using the above,

$$\ln|S_{mp}(s_0)| \le \ln M_1 \int_{\omega_1}^{\omega_2} \frac{\sigma_0}{\sigma_0^2 + (\omega - \omega_0)^2}\mathrm{d}\omega + \ln M_2\left(\int_{-\infty}^{\omega_1} \frac{\sigma_0}{\sigma_0^2 + (\omega - \omega_0)^2}\mathrm{d}\omega + \int_{\omega_2}^{\infty} \frac{\sigma_0}{\sigma_0^2 + (\omega - \omega_0)^2}\mathrm{d}\omega\right) \tag{A.85}$$

$$\implies \ln|S_{mp}(s_0)| \le c_1 \ln M_1 + c_2 \ln M_2 \tag{A.86}$$

Suppose that $G_p(s)$ has a non-minimum phase zero, $z > 0$:

$$S(z) = 1 = S_{mp}(z)S_{ap}(z) \tag{A.87}$$

$$S_{mp}(z) = \frac{1}{S_{ap}(z)} \tag{A.88}$$

Therefore,

$$\ln\left|\frac{1}{S_{ap}(z)}\right| \le c_1 \ln M_1 + c_2 \ln M_2 \tag{A.89}$$

A.8 Cauchy's Argument Principle

Suppose that $f(\lambda)$ is analytical inside and on Γ except for at most a finite number of poles, N_p. $f(\lambda)$ has no zeros on Γ and at most a finite number of zeros interior to Γ, N_0. This is illustrated in Figure A.23.

Consider, first of all, a zero repeated N_0 times:

$$f(\lambda) = (\lambda - \lambda_0)^{N_0} g(\lambda) \tag{A.90}$$

$g(\lambda)$ is the remainder of $f(\lambda)$ that is analytical and has no zeros:

$$\frac{\mathrm{d}f}{\mathrm{d}\lambda} = N_0(\lambda - \lambda_0)^{N_0 - 1} g(\lambda) + (\lambda - \lambda_0)^{N_0}\frac{\mathrm{d}g}{\mathrm{d}\lambda} \tag{A.91}$$

$$\frac{1}{f(\lambda)} \cdot \frac{\mathrm{d}f}{\mathrm{d}\lambda} = \frac{N_0}{\lambda - \lambda_0} + \frac{1}{g(\lambda)} \cdot \frac{\mathrm{d}g}{\mathrm{d}\lambda} \tag{A.92}$$

It can be shown that where $g(\lambda)$ is analytic, $\frac{\mathrm{d}g}{\mathrm{d}\lambda}$ is analytic. Therefore, $\frac{1}{g(\lambda)} \cdot \frac{\mathrm{d}g}{\mathrm{d}\lambda}$ is analytic.

Then,

$$\frac{1}{2\pi j}\int_{\Gamma} \frac{1}{f(\lambda)} \cdot \frac{\mathrm{d}f}{\mathrm{d}\lambda}\mathrm{d}\lambda = \frac{1}{2\pi j}\int_{\Gamma} \frac{N_0}{\lambda - \lambda_0}\mathrm{d}\lambda + \frac{1}{2\pi j}\int_{\Gamma} \frac{1}{g(\lambda)} \cdot \frac{\mathrm{d}g}{\mathrm{d}\lambda}\mathrm{d}\lambda \tag{A.93}$$

$$\frac{1}{2\pi j}\int_{\Gamma} \frac{1}{f(\lambda)} \cdot \frac{\mathrm{d}f}{\mathrm{d}\lambda}\mathrm{d}\lambda = N_0 + 0 \tag{A.94}$$

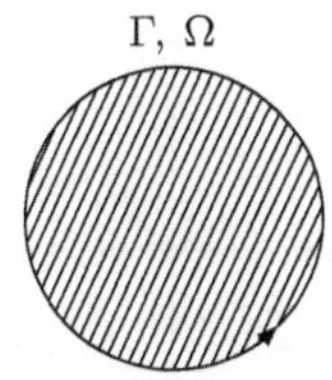

Figure A.23 An arbitrary line and surface integral in the complex plane.

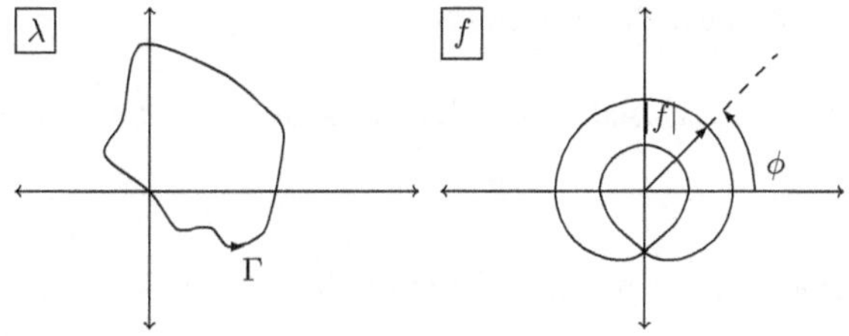

Figure A.24 Mapping of contour Γ.

Next, consider a pole repeated N_p times:

$$f(\lambda) = (\lambda - \lambda_0)^{-N_p} g(\lambda) \tag{A.95}$$

$$\frac{df}{d\lambda} = -N_p(\lambda - \lambda_0)^{-N_p-1} g(\lambda) + (\lambda - \lambda_0)^{-N_p} \frac{dg}{d\lambda} \tag{A.96}$$

$$\frac{1}{f(\lambda)} \cdot \frac{df}{d\lambda} = \frac{-N_p}{\lambda - \lambda_0} + \frac{1}{g(\lambda)} \cdot \frac{dg}{d\lambda} \tag{A.97}$$

$$\frac{1}{2\pi j} \int_\Gamma \frac{1}{f(\lambda)} \cdot \frac{df}{d\lambda} d\lambda = -N_p \tag{A.98}$$

With both poles and zeros,

$$\frac{1}{2\pi j} \int_\Gamma \frac{1}{f(\lambda)} \cdot \frac{df}{d\lambda} d\lambda = N_0 - N_p \tag{A.99}$$

This results in the mapping shown in Figure A.24.

We can rewrite $\frac{1}{f(\lambda)} \cdot \frac{df}{d\lambda}$:

$$\frac{1}{f(\lambda)} \cdot \frac{df}{d\lambda} = \frac{d \ln f}{d\lambda}, \quad f = |f| e^{j\phi} \tag{A.100}$$

$$\frac{1}{f(\lambda)} \cdot \frac{df}{d\lambda} = \frac{d}{d\lambda}(\ln |f| + j\phi) \tag{A.101}$$

$$\frac{1}{2\pi j} \int_\Gamma \frac{1}{f(\lambda)} \cdot \frac{df}{d\lambda} d\lambda = \frac{1}{2\pi j} \int_\Gamma \frac{d}{d\lambda}(\ln |f| + j\phi) d\lambda = \frac{1}{2\pi j} j\Delta\phi \tag{A.102}$$

Theorem A.7 (Principle of Argument)

$$\frac{\Delta\phi}{2\pi} = N_0 - N_p \tag{A.103}$$

For an application of the argument principle, consider a closed-loop transfer function:

$$\frac{G_D(s)}{1 + L(s)} \tag{A.104}$$

The denominator is some rational function:

$$1 + L(s) = \frac{\overbrace{\delta_L(s) + \eta_L(s)}^{\text{closed-loop poles}}}{\underbrace{\delta_L(s)}_{\text{open-loop poles}}} \tag{A.105}$$

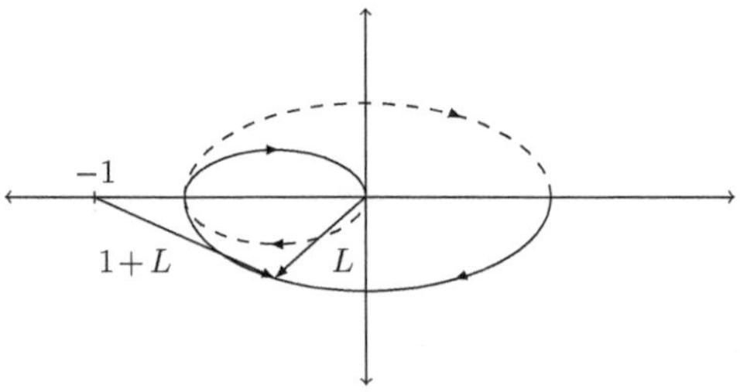

Figure A.25 A Nyquist locus.

For $f(s) = 1 + L(s)$, N_0 equals the number of closed-loop poles, N_c, while N_p is the number of open-loop poles. Then,

$$\frac{\Delta\phi}{2\pi} = N_c - N_p \tag{A.106}$$

Rotation of $1 + L$ creates encirclements of -1 by $L(j\omega)$. Suppose that $N_p = 0$. We want $N_c = 0$, so our criterion is $\Delta\phi = 0$. Now, suppose that $N_p = 1$. For $N_c = 0$, the criterion is $\frac{\Delta\phi}{2\pi} = -1$. In other words, if there is one open-loop pole in the right-hand plane, $L(j\omega)$ must encircle -1 in the counterclockwise direction one time for the system to be stable (or more accurately, $1 + L(j\omega)$ must encircle the origin in the counterclockwise direction one time). Figure A.25, for instance, shows zero clockwise encirclements and would indicate stability for $N_p = 0$ and instability for $N_p = 1$.

A.9 Nyquist Stability Criterion

Let N be the number of clockwise encirclements, N_p be the number of open-loop poles in the RHP, and N_c be the number of closed-loop poles in the RHP. Then,

$$N = N_c - N_p \tag{A.107}$$

$$N_c = N + N_p \tag{A.108}$$

The stability of the closed-loop system can be determined from the open-loop poles and from counting the number of clockwise encirclements!

Example A.6 Consider the open-loop transfer function:

$$L(s) = \frac{K}{\tau s + 1} \tag{A.109}$$

$$N_p = 0 \Longrightarrow N = 0 \text{ (desired)} \tag{A.110}$$

This produces the Nyquist contour shown in Figure A.26.

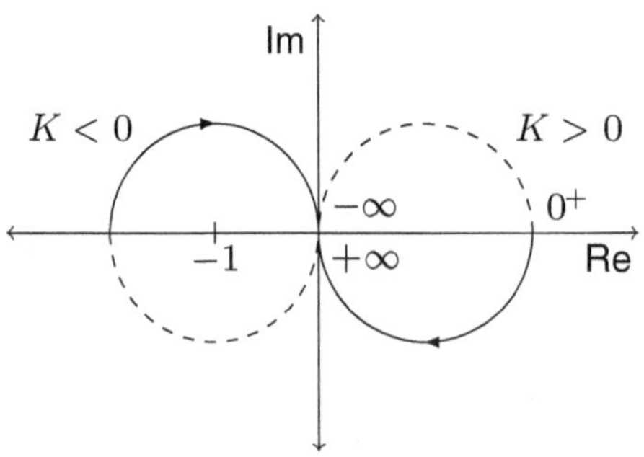

Figure A.26 Nyquist locus for a first-order transfer function. Negative frequencies have dashed lines.

Example A.7 Consider the open-loop transfer function:

$$L(s) = \frac{K}{(\tau_1 s + 1)(\tau_2 s + 1)(\tau_3 s + 1)} \tag{A.111}$$

$N_p = 0$, and for small K, $N = 0$, so $N_c = 0$ and the closed-loop response is stable. Increasing K, however, causes encirclements of -1 and therefore closed-loop instability. Figures A.27 and A.28 illustrate this phenomena – note that the negative frequencies are drawn as dashed lines.

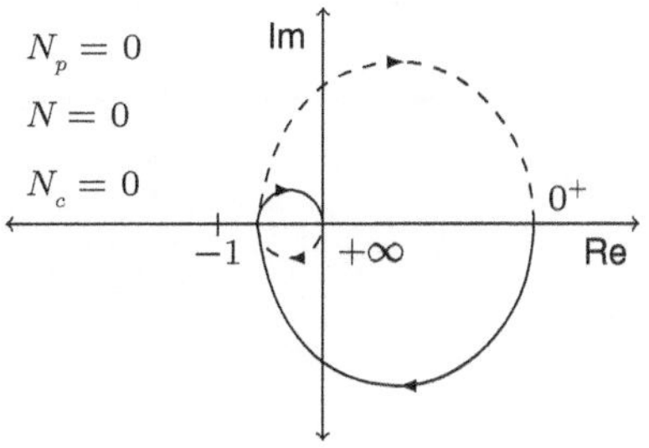

Figure A.27 Nyquist locus for small K.

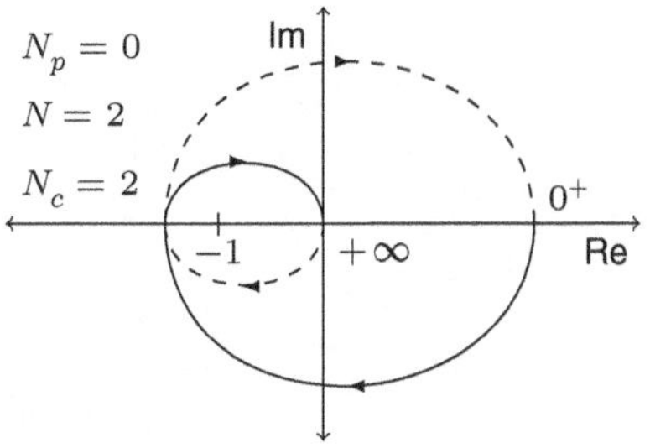

Figure A.28 Nyquist locus for large K.

Example A.8 Consider the open-loop transfer function with zeros on the imaginary axis:

$$L(s) = \frac{(\tau_1 s + 1)}{s^2(\tau_2 s + 1)} \tag{A.112}$$

$N_p = 2$, so we must select a K so that $N = -2$, giving $N_c = 0$. Figure A.29 illustrates the Nyquist contour for this system, while Figure A.30 shows how the D-shaped contour must encompass the singularities on the $j\omega$-axis.

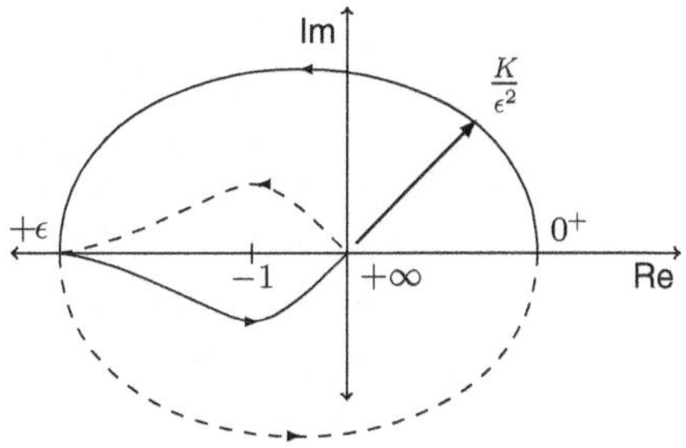

Figure A.29 Nyquist locus with poles on the $j\omega$ axis.

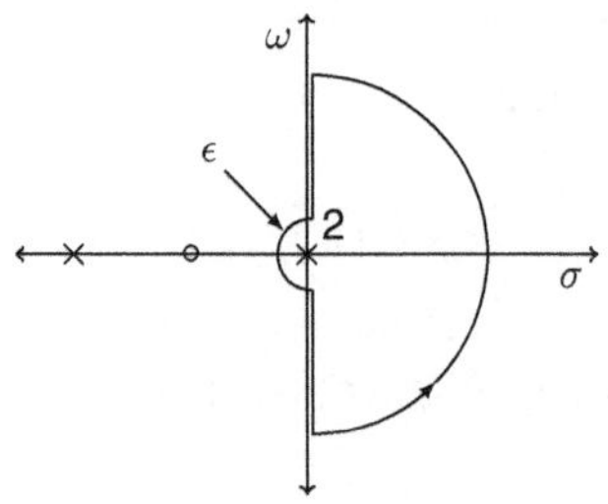

Figure A.30 Nyquist contour with poles on the $j\omega$ axis.

B

Singular Value Properties

This appendix expands upon and develops proofs for the singular value properties discussed in Chapter 11.

B.1 Spectral Norm Proof

Consider a process with input $\underline{x}$ and output $M\underline{x}$. Then,

$$\frac{\|M\underline{x}\|_E^2}{\|\underline{x}\|_E^2} = \frac{(M\underline{x})^* M\underline{x}}{\underline{x}^*\underline{x}} = \frac{\underline{x}^* M^* M\underline{x}}{\underline{x}^*\underline{x}} \tag{B.1}$$

Next, we will consider the transformation:

$$\underline{x} = \Omega\underline{z} \tag{B.2}$$

where Ω is the right singular vectors of M:

$$\Omega = \left[\ \underline{r}_1 \ \vdots\ \underline{r}_2 \ \vdots\ \cdots \ \vdots\ \underline{r}_m\ \right] \tag{B.3}$$

Then,

$$\underline{x}^* M^* M\underline{x} = \underline{z}^*\Omega^* M^* M\Omega\underline{z} \tag{B.4}$$

$$= \underline{z}^* \begin{bmatrix} \sigma_1 \underline{l}_1^* \\ \sigma_2 \underline{l}_2^* \\ \vdots \\ \sigma_m \underline{l}_m^* \end{bmatrix} \begin{bmatrix} \sigma_1 \underline{l}_1^* & \sigma_2 \underline{l}_2^* & \cdots & \sigma_m \underline{l}_m^* \end{bmatrix} \underline{z}$$

$$= \underline{z}^* \begin{bmatrix} \sigma_1^2 & 0 & \cdots & 0 \\ 0 & \sigma_2^2 & & \\ \vdots & & \ddots & \\ 0 & & & \sigma_m^2 \end{bmatrix} \underline{z}$$

$$= \sigma_1^2 |z_1|^2 + \sigma_2^2 |z_2|^2 + \cdots + \sigma_m^2 |z_m|^2 \tag{B.5}$$

Also,

$$\underline{x}^*\underline{x} = \underline{z}^*\Omega^*\Omega\underline{z} \tag{B.6}$$

$$= \underline{z}^*\underline{z}$$

$$= |z_1|^2 + |z_2|^2 + \cdots + |z_m|^2 \tag{B.7}$$

So,

$$\frac{\|M\underline{x}\|_E^2}{\|\underline{x}\|_E^2} = \frac{\sigma_1^2 |z_1|^2 + \sigma_2^2 |z_2|^2 + \cdots + \sigma_m^2 |z_m|^2}{|z_1|^2 + |z_2|^2 + \cdots + |z_m|^2} \tag{B.8}$$

Then, given $\sigma_k^2 \leq \sigma_{max}^2$,

$$\frac{\|M\underline{x}\|_E^2}{\|\underline{x}\|_E^2} \leq \sigma_{max}^2 \frac{|z_1|^2 + |z_2|^2 + \cdots + |z_m|^2}{|z_1|^2 + |z_2|^2 + \cdots + |z_m|^2}$$

$$\leq \sigma_{max}^2$$

Robust Control: Youla Parameterization Approach, First Edition. Farhad Assadian and Kevin R. Mallon.

Companion Website: www.wiley.com/go/assadian/robust

Next, we let $\underline{x} = \underline{r}_{max}$. Then,

$$M\underline{x} = \sigma_{max}\underline{l}_{max} \tag{B.9}$$

So,

$$\begin{aligned}\|M\underline{x}\|_E^2 &= \sigma_{max}^2 \underline{l}_{max}^* \underline{l}_{max} \\ &= \sigma_{max}^2\end{aligned} \tag{B.10}$$

This implies

$$\frac{\|M\underline{x}\|_E^2}{\|\underline{x}\|_E^2} = \max_{\text{all } \underline{x}} \frac{\|M\underline{x}\|_E^2}{\|\underline{x}\|_E^2} = \sigma_{max}^2 \tag{B.11}$$

Thus,

$$\|M\|_S = \sigma_{max} \tag{B.12}$$

B.2 Proof of Bounded Eigenvalues

Consider a square matrix M. Let λ be an arbitrary eigenvalue of M corresponding to the eigenvector $\underline{\epsilon}$. Then,

$$M\underline{\epsilon} = \lambda\underline{\epsilon} \tag{B.13}$$

So,

$$\|M\underline{\epsilon}\|_E = \|\lambda\underline{\epsilon}\|_E \tag{B.14}$$

$$\begin{aligned}&= \sqrt{(\lambda\underline{\epsilon})^* \lambda\underline{\epsilon}} \\ &= |\lambda| \sqrt{\underline{\epsilon}^*\underline{\epsilon}} \\ &= |\lambda| \cdots \|\underline{\epsilon}\|_E\end{aligned} \tag{B.15}$$

We can divide both sides by $\|\underline{\epsilon}\|_E$. Then,

$$|\lambda| = \frac{\|M\underline{\epsilon}\|_E}{\|\underline{\epsilon}\|_E} \leq \sigma_{max} \tag{B.16}$$

So,

$$|\lambda| \leq \sigma_{max}(M) \tag{B.17}$$

Similarly, we can start with

$$\frac{1}{\lambda}\underline{\epsilon} = M^{-1}\underline{\epsilon} \tag{B.18}$$

Thus,

$$\frac{1}{|\lambda|} \leq \sigma_{max}(M^{-1}) = \frac{1}{\sigma_{min}(M)} \tag{B.19}$$

So,

$$|\lambda| \geq \sigma_{min}(M) \tag{B.20}$$

Therefore,

$$\sigma_{min}(M) \leq |\lambda| \leq \sigma_{max}(M) \tag{B.21}$$

B.3 Proof of Matrix Inequality

$$\sigma_{min} \leq \frac{\|M\underline{x}\|_E}{\|\underline{x}\|_E} \leq \sigma_{max} \tag{B.22}$$

$$\sigma_{min}\|\underline{x}\|_E \leq \|M\underline{x}\|_E \leq \sigma_{max}\|\underline{x}\|_E \tag{B.23}$$

What if $\underline{x}$ multiplies into MN? We will look at the upper and lower bounds separately.

B.3.1 Upper Bound

$$\|MN\underline{x}\|_E = \|M(N\underline{x})\|_E \le \sigma_{max}(M)\|N\underline{x}\|_E \tag{B.24}$$

$$\|MN\underline{x}\|_E \le \sigma_{max}(M)\sigma_{max}(N)\|\underline{x}\|_E \tag{B.25}$$

So,

$$\frac{\|MN\underline{x}\|_E}{\|\underline{x}\|_E} \le \sigma_{max}(M)\sigma_{max}(N) \tag{B.26}$$

For any $\underline{x}$ in particular,

$$\max_{\underline{x}} \frac{\|MN\underline{x}\|_E}{\|\underline{x}\|_E} = \sigma_{max}(MN) \tag{B.27}$$

So,

$$\sigma_{max}(MN) \le \sigma_{max}(M)\sigma_{max}(N) \tag{B.28}$$

B.3.2 Lower Bound

Take $\underline{x} = N\underline{r}_{max}(N)$. Then,

$$\sigma_{min}(M) \le \frac{\left\|M(N\underline{r}_{max})\right\|_E}{\left\|N\underline{r}_{max}\right\|_E} \tag{B.29}$$

$$\le \frac{\left\|\underline{r}_{max}\right\|_E}{\left\|\underline{r}_{max}\right\|_E} \cdot \frac{\left\|MN\underline{r}_{max}\right\|_E}{\left\|N\underline{r}_{max}\right\|_E}$$

$$\le \frac{\left\|\underline{r}_{max}\right\|_E}{\left\|\underline{r}_{max}\right\|_E} \cdot \frac{\left\|MN\underline{r}_{max}\right\|_E}{\left\|N\underline{r}_{max}\right\|_E} = \frac{\left\|\underline{r}_{max}\right\|_E}{\left\|N\underline{r}_{max}\right\|_E} \cdot \frac{\left\|MN\underline{r}_{max}\right\|_E}{\left\|\underline{r}_{max}\right\|_E}$$

$$\le \frac{1}{\sigma_{min}(N)} \cdot \frac{\left\|MN\underline{r}_{max}\right\|_E}{\left\|\underline{r}_{max}\right\|_E} \tag{B.30}$$

So,

$$\sigma_{min}(M)\sigma_{max}(N) \le \frac{\left\|MN\underline{r}_{max}(N)\right\|_E}{\left\|\underline{r}_{max}(N)\right\|_E} \le \sigma_{max}(MN) \tag{B.31}$$

It can similarly be shown that

$$\sigma_{max}(M)\sigma_{min}(N) \le \frac{\left\|MN\underline{r}_{max}(MN)\right\|_E}{\left\|\underline{r}_{max}(MN)\right\|_E} = \sigma_{max}(MN) \tag{B.32}$$

Aside: Recall that

$$\sigma(M^{-1}) = \frac{1}{\sigma(M)} \tag{B.33}$$

So,

$$\sigma_{max}(M) = \frac{1}{\sigma_{min}(M^{-1})} \tag{B.34}$$

and

$$\sigma_{min}(M) = \frac{1}{\sigma_{max}(M^{-1})} \tag{B.35}$$

B.3.3 Combined Inequality

When put together, we then have

$$\max \left\{ \begin{array}{cc} \sigma_{min}(M) & \sigma_{max}(N) \\ \sigma_{max}(M) & \sigma_{min}(N) \end{array} \right\} \leq \sigma_{max}(MN) \leq \sigma_{max}(M)\sigma_{max}(N) \tag{B.36}$$

B.4 Triangle Inequality

What about if $\underline{x}$ multiplies into $(M + N)$?

$$\|(M + N)\underline{x}\|_E \leq \|M\underline{x}\|_E + \|N\underline{x}\|_E \tag{B.37}$$

B.4.1 Upper Bound

$$\max_{\underline{x}} \frac{\|(M + N)\underline{x}\|_E}{\|\underline{x}\|_E} \leq \max_{\underline{x}} \left(\frac{\|M\underline{x}\|_E}{\|\underline{x}\|_E} + \frac{\|N\underline{x}\|_E}{\|\underline{x}\|_E} \right) \tag{B.38}$$

$$\leq \max_{\underline{x}} \frac{\|M\underline{x}\|_E}{\|\underline{x}\|_E} + \max_{\underline{x}} \frac{\|N\underline{x}\|_E}{\|\underline{x}\|_E}$$

$$\sigma_{max}(M + N) \leq \sigma_{max}(M) + \sigma_{max}(N) \tag{B.39}$$

B.4.2 Lower Bound

$$\sigma_{max}(M) = \sigma_{max}(M + N - N) \leq \sigma_{max}(M + N) + \sigma_{max}(N) \tag{B.40}$$

So,

$$\sigma_{max}(M) - \sigma_{max}(N) \leq \sigma_{max}(M + N) \tag{B.41}$$

Similarly,

$$\sigma_{max}(N) - \sigma_{max}(M) \leq \sigma_{max}(M + N) \tag{B.42}$$

B.4.3 Combined Inequality

$$\left|\sigma_{max}(M) - \sigma_{max}(N)\right| \leq \sigma_{max}(M + N) \leq \sigma_{max}(M) + \sigma_{max}(N) \tag{B.43}$$

C

Bandwidth

In the past, there have been many inconsistent attempts to describe bandwidth of dynamic systems. Similar definitions of bandwidth have been extended to not only describe the closed-loop bandwidth of feedback control systems but the frequency at which usefulness of feedback is lost. In this appendix, we discuss a measure of bandwidth that describes the cutoff frequency at which the feedback control system's usefulness is no longer advantageous.

C.1 Introduction

A somewhat slippery term in the control literature is that of bandwidth: "Slippery" because the definition is not given in a consistent and precise manner (if it is given at all) and the meaning and use of the term is somewhat vague.

Still, the term is very useful and many attempts have been made to describe it. The idea is to convey an estimate of the length of a frequency range that is important on the Bode-magnitude curve of some dynamic system. This length is, roughly speaking, proportional to the speed with which the system responds to, say, the unit step input.

Of course, in the past, experts tried to provide a precise mathematical meaning to bandwidth where the following fundamental theorem is proved [57]:

$$\Delta\omega \times \Delta t > 1/4 \tag{C.1}$$

where $\Delta\omega$ denotes the bandwidth and the effective duration of transient by Δt. This theorem clearly shows that the duration of a transient is reduced by increasing the bandwidth. However, as is obvious, this definition only provides a lower bound to the bandwidth.

Other descriptions rather than mathematical definitions for bandwidth include the *bandwidth* is that range of positive frequency bounded by those frequencies where the Bode-magnitude falls to 3 dB below the maximum magnitude. Sometimes, 6 dB or even 12 dB are used instead and so it was referred to the −3 dB or to the −6 dB bandwidth. Additionally, only positive frequencies are used in this convention so that, as is often the case, where the Bode diagram is monotonically decreasing, the bandwidth is taken to lie from the zero frequency to the −3 dB (or the −6 dB) frequency.

The relationship of bandwidth to speed of response is also established for both the first- and second-order dynamics systems. For the first-order system, the Bode modulus is described by:

$$M(\omega) = \frac{1}{\sqrt{1 + (\tau\omega)^2}} \tag{C.2}$$

where τ is the system time constant. Clearly, the −3 dB bandwidth (the $\frac{1}{\sqrt{2}}$ frequency) is

$$\omega_B = \frac{1}{\tau} \tag{C.3}$$

The higher the bandwidth, the smaller the τ and the faster such a system responds to, say, step inputs.

Next, consider the quadratic factor whose Bode modulus is

$$M(\omega) = \frac{\omega_n^2}{\sqrt{(\omega_n^2 - \omega^2) + 4\zeta^2\omega_n^2\omega^2}} \tag{C.4}$$

where ω_n is the undamped natural frequency. A formula for bandwidth, as a function of undamped natural frequency and damping ratio, can be derived [20] but it turns out to be overly complex for most purposes, so it is easier to consider the

Robust Control: Youla Parameterization Approach, First Edition. Farhad Assadian and Kevin R. Mallon.

Companion Website: www.wiley.com/go/assadian/robust

high- frequency asymptote instead. This asymptote is obtained by setting the condition $\omega \gg \omega_n$ to obtain

$$\left[\frac{\omega_n}{\omega}\right]^2 \tag{C.5}$$

This approximation drops to the −6 dB value at

$$\omega_B = \sqrt{2} \times \omega_n \tag{C.6}$$

Given the relationships between rise and settling times with the damping ratio, it follows that, at a fixed value of damping ratio, the rise and settling times diminish as bandwidth is increased.

In all this development, the bandwidth is related to the system response time and not given in a consistent and precise manner. In addition, use of a bandwidth to characterize the speed of response of dynamic systems to step inputs could be at times misleading, e.g. where the dynamic system is an integrator. In addition, when higher than second order dynamic systems are considered, all these definitions of bandwidth become impractical unless these higher-order systems could be approximated by a second-order system.

These similar definitions of bandwidth have been extended to characterize the speed of response of feedback control systems. A very useful measure for feedback control designers would be: at what frequency the feedback control system loses its usefulness? In this work, we will show that this measure has a precise definition and this measure could either be used as a new way we look at the feedback control bandwidth or a precise way to characterize a cutoff frequency at which a transition from feedback to no feedback is made. Next, we use the machinery of neoclassical feedback control development to define this measure in terms of the closed-loop feedback system transfer function.

C.2 Information as a Precise Measure of Bandwidth

C.2.1 Neoclassical Feedback Control

Recall the three closed-loop transfer functions that are shaped in the frequency domain to meet both the performance and robustness criteria: The closed-loop transfer functions consist of T, complementary sensitivity, S, sensitivity, and Y, Youla.

Given the feedback control loop above, these closed-loop transfer functions can be written as:

$$T = \frac{G_c G_p}{1 + G_c G_p} \tag{C.7}$$

$$S = \frac{1}{1 + G_c G_p} \tag{C.8}$$

$$Y = \frac{G_c}{1 + G_c G_p} \tag{C.9}$$

Where, from above, we can write $T = YG_p$ and $S = 1 - YG_p$.

C.2.2 Defining a Measure to Characterize the Usefulness of Feedback

In Figure C.1, signal y is measured and fed back and it is known that the act of measurement introduces noise (sensor noise) to the feedback loop. It turns out that sensor noise, due to a typical sensor construction, is a high-frequency phenomenon and is considered as high-frequency uncertainty. In addition to sensor noise, dynamic system models cannot be modeled perfectly, meaning that in the frequency domain, dynamic models have a limited frequency range in which they can accurately represent reality. The model accuracy is in the low-frequency range and hence, these models introduce high-frequency uncertainty (known as unstructured model uncertainty) into the feedback loop. Due to these high-frequency uncertainties (sensor and unstructured model uncertainties), we are forced to abandon the feedback control loop at particular frequency. *It is very interesting to notice that the act of measurement is what gives meaning to the feedback loop and the same act results in abandoning this control loop!*

The closed-loop control system bandwidth (ω_B) is utilized to specify the frequency transition from feedback to no feedback, in addition to describing the closed-loop system speed of response. Interestingly enough, the definition of this closed-loop bandwidth is exactly the same as what is provided in Section C.1. We argue in this work that the use of ω_B to specify the frequency transition from feedback to no feedback could be very conservative. In fact, we specify a new

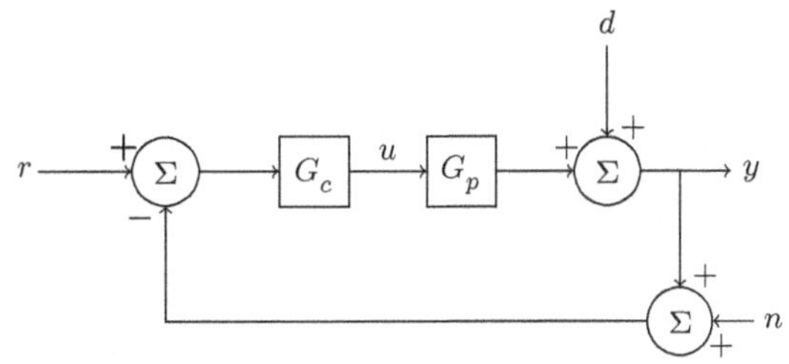

Figure C.1 Feedback Control Loop.

frequency called (ω_{BC}) to differentiate this new frequency from the traditional closed-loop system bandwidth, ω_B. We will provide an exact computation of this frequency to illustrate the transition from feedback to no feedback when designing feedback control systems.

C.2.3 Computation of New Bandwidth

The frequency at which a transition from feedback to no feedback (ω_{BC}) is computed by the use of following theorem:

Theorem C.1 *The frequency, ω_{BC}, at which a transition from feedback to no feedback is made, is computed using the following three closed-loop transfer functions:*

$$T = \frac{G_c G_p}{1 + \alpha G_c G_p} \tag{C.10}$$

$$S = \frac{1}{1 + \alpha G_c G_p} \tag{C.11}$$

$$Y = \frac{G_c}{1 + \alpha G_c G_p} \tag{C.12}$$

where α is selected to approach from 1 (usual feedback) to 0 (no feedback).

Proof: Let's start by the following feedback control loop given in Figure C.2.

When $\alpha \to 1$, we have our usual feedback control system, but when $\alpha \to 0$ then obviously, there is no information that is fed back. When $\alpha \to 0$ both the magnitudes of complementary sensitivity and sensitivity transfer functions are given by:

$$|T| \to |L| \tag{C.13}$$

$$|S| \to 1 \tag{C.14}$$

where L is the open-loop return ratio and is equal to $G_c G_p$. For computation of Youla, we use Figure C.3.

Given Figure C.3, when $\alpha \to 0$ then,

$$|Y| \to |G_c| \tag{C.15}$$

Therefore, as $\alpha \to 0$,

$$|T| \to |L|$$
$$|S| \to 1$$
$$|Y| \to |G_c|$$

To show when $\alpha \to 0$ then $\omega \to \omega_{BC}$ it suffices to look at the usual feedback when $\alpha \to 1$ and knowing at some high frequency, $|G_c G_p| \ll 1$ then this result in three closed-loop transfer functions computed in Eqs. (C.13)–(C.15). ■

It is interesting to see that these transitions, $|T| \to |L|$, $|S| \to 1$, and $Y \to |G_c|$, which are known to happen at some high frequency due to forcing the magnitude of the open-loop ratio, $|L|$, to be small, is directly related to feedback information constant α. With the use of this constant, then one can compute the frequency, ω_{BC}, at which the benefits of feedback are lost.

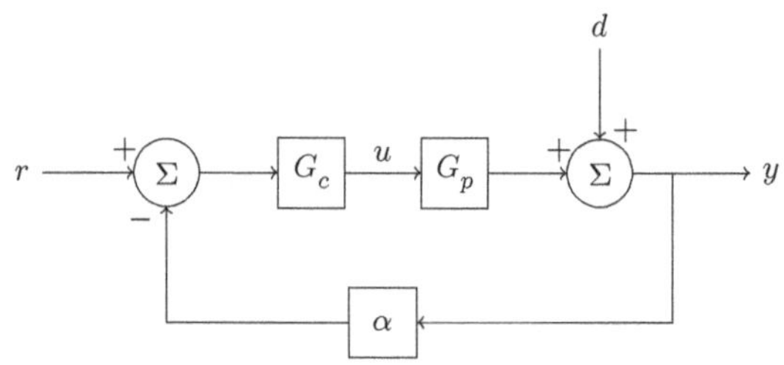

Figure C.2 Feedback control loop with α block.

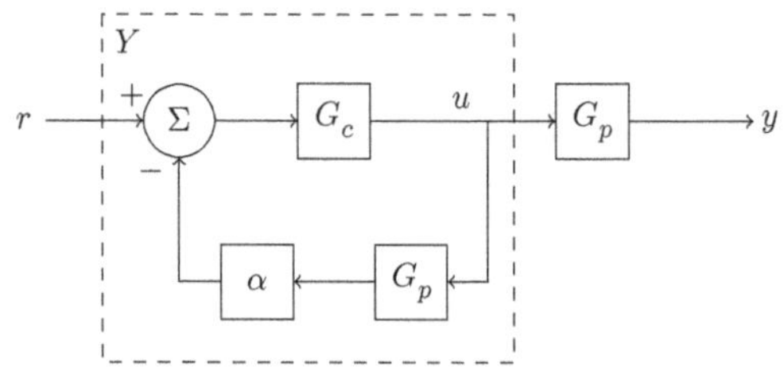

Figure C.3 Youla in feedback control loop with α block.

C.3 Examples

In this section, we present two simple examples to illustrate the application of the proposed measure or so-called new bandwidth. It is important to note that this measure equivalently applies to more complex single input single output (SISO) and multiple inputs multiple outputs (MIMO) cases.

We can write a simple MATLAB function to compute this bandwidth for a specified value of α that utilizes MATLAB's existing ability to evaluate frequency responses. In other words, it computes the frequency at which the Youla frequency response reaches a specified percent of its final value. Effectively, this is finding when the sensitivity transfer function S settles to within a specified error bound around 0 dB. This function is given as follows:

```
function w_BC = NeoClassicalBW(S,alpha)
%NEOCLASSICALBW returns the neoclassical measure of bandwidth.
% w_BC = NEOCLASSICALBW(S, alpha) returns the neoclassical
% bandwidth for the given sensitivity transfer function S
% and the specified cutoff value of alpha, where 0 < alpha < 1.

upper = 1+alpha;
lower = 1-alpha;

[mag,~,w] = bode(S);

isBounded = or(mag>upper, mag<lower);
index = find(isBounded,1,'last');
w_BC = w(index);

end
```

Greater care could be taken to obtain a more precise measurement for the neoclassical bandwidth, but for the sake of these examples, the above function will suffice. We can now consider two example applications.

Example C.1 In this example, we examine the "Interpolation Constraints" and the consequences of violating these constraints in feedback design. Consider a force $F(t)$ applied to a mass to control velocity as shown in Figure C.4.

$$\frac{\hat{v}}{\hat{v}} = \frac{1}{ms} \implies \text{BIBO unstable where} \quad \hat{v} = F(t) \tag{C.16}$$

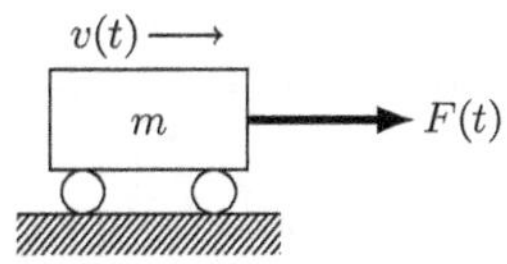

Figure C.4 Single mass system with force input and velocity output.

We will examine a Youla parameter that cancels the dynamics of the mass but does *not* meet the interpolation condition:

$$Y = \frac{ks}{\tau s + 1} \tag{C.17}$$

Then,

$$T = \frac{k}{m(\tau s + 1)} \tag{C.18}$$

Suppose that $k = m$. Then at $s = 0$, $T = 1$, so the interpolation condition is met [14]. We will find the corresponding controller.

$$S = 1 - T = \frac{\tau s}{\tau s + 1} \quad \text{where} \quad S = (1 - G_p Y) \tag{C.19}$$

$$G_c = \frac{Y}{S} = \frac{ms}{\tau s + 1}\frac{\tau s + 1}{\tau s} \tag{C.20}$$

$$\implies \quad G_c = \frac{m}{\tau} \tag{C.21}$$

where S is the sensitivity transfer function and G_c is a pure gain, in this case proportionally equivalent to the mass m and inversely proportional to the time constant τ. The return ratio transfer function is given by

$$L = \frac{m}{\tau} \cdot \frac{1}{ms} = \frac{1}{\tau s} \tag{C.22}$$

For the sake of this simple example, let's assume both m and τ are equal to one. In this case, $L = \frac{1}{s}$, $G_c = 1$, $T = \frac{1}{s+1}$, and $S = \frac{s}{s+1}$.

Figure C.5 illustrates the closed-loop complementary sensitivity (T in dark gray and starting at 0dB), sensitivity (S in light gray), and the return ratio (L in gray and starting at 40dB) transfer functions. The frequency ω_{BC} is around 5 rad/s. This is the frequency that these transfer functions are within 2% of their final values.

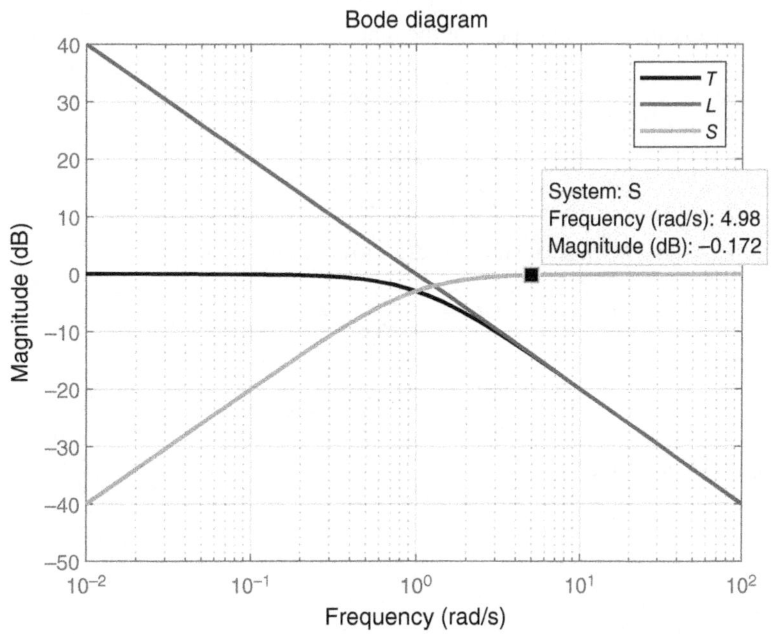

Figure C.5 Closed-Loop and Return Ratio Transfer Functions for Example C.1.

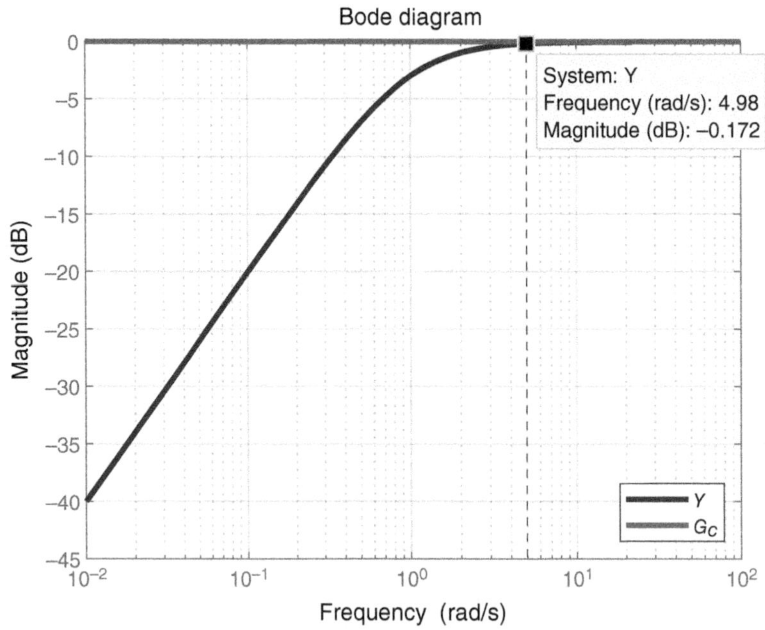

Figure C.6 Controller and Youla transfer functions for Example C.1.

The Youla (Y in dark gray and starting at −40dB) and the controller (G_c in gray and starting at 0dB) transfer functions are shown in Figure C.6. It is again around 5 rad/s, where Youla reaches 2% of its final value where $G_c = 1$.

Example C.2 Consider again a force $F(t)$ applied to a mass again, but this time, to control position rather than velocity, as shown in Figure C.7.

$$\frac{\hat{y}}{\hat{v}} = \frac{1}{ms^2} \tag{C.23}$$

We choose a Youla parameter to cancel the dynamics of the mass:

$$Y = \frac{mks^2}{(\tau s + 1)^2} \tag{C.24}$$

Then,

$$T = \frac{k}{(\tau s + 1)^2} \tag{C.25}$$

The interpolation conditions, to ensure internal stability, are $T(0) = 1$ and $\frac{dT(0)}{ds} = 0$.

$$T(0) = \frac{k}{(\tau \cdot 0 + 1)^2} = 1 \tag{C.26}$$

Therefore $k = 1$. So,

$$\left.\frac{dT}{ds}\right|_{s=0} = \frac{-2\tau}{(\tau \cdot 0 + 1)^3} = 0 \tag{C.27}$$

Therefore, $\tau = 0$. This means we now have

$$Y = ms^2 \tag{C.28}$$

Y is now improper, which is unacceptable.

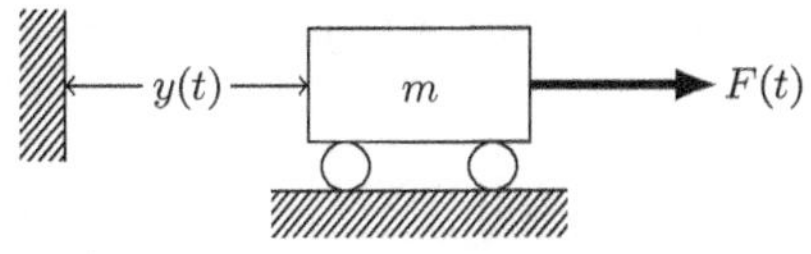

Figure C.7 Single mass system with force input and position output.

We have shown previously in Section 3.5 that adding a zero will allow the interpolation conditions to be met. So, the next $Y(s)$ we consider is

$$Y(s) = \frac{ms^2(\tau_1 s + 1)}{(\tau s + 1)^3} \tag{C.29}$$

We added an additional pole and zero to Y with an extra tuning parameter τ_1. Thus,

$$T(s) = \frac{(\tau_1 s + 1)}{(\tau s + 1)^3} \tag{C.30}$$

The first interpolation condition, $T(0) = 1$, is satisfied and it is straightforward to show that $\frac{dT(0)}{ds}$ can be written as:

$$\frac{dT(0)}{ds} = \left[\frac{\tau_1}{(\tau s + 1)^3} - \frac{3\tau(\tau_1 s + 1)}{(\tau s + 1)^4}\right] \Big|_{s=0} \tag{C.31}$$

$$= \tau_1 - 3\tau = 0 \tag{C.32}$$

so that an internally stabilizing value of the Youla transfer function is

$$Y(s) = \frac{ms^2(3\tau s + 1)}{(\tau s + 1)^3} \tag{C.33}$$

An internally stabilizing compensator is calculated similar to Example C.1:

$$G_c(s) = \frac{ms^2(3\tau s + 1)/(\tau s + 1)^3}{1 - (3\tau s + 1)/(\tau s + 1)^3} \tag{C.34}$$

$$= \frac{ms^2(3\tau s + 1)}{\tau^2 s^2(\tau s + 3)} \tag{C.35}$$

$$= \frac{m}{3\tau^2} \frac{(3\tau s + 1)}{(\tau/3)s + 1} \tag{C.36}$$

This turns out to be a lead controller. As a check for internal stability, we consider the fact that $G_c G_p$ does not contain the wrong kind of cancellation (cancellation in the RHP or on $j\omega$-axis). Also, $T(s) = (3\tau s + 1)/(\tau s + 1)^3$ so that the closed loop is stable; as a matter of fact, we have placed all three closed-loop poles at $s = -1/\tau$ in this example. It should be noted, using this approach, we have the flexibility to select Youla and shape the closed-loop transfer function T any way we desire but we have to meet interpolation conditions and as result internal stability irrespective of these choices.

Figure C.8 illustrates the closed-loop transfer functions T (in dark gray and starting at 0dB) and S (in gray and ending at 0dB) and the return ratio L (in light gray). Again the frequency ω_{BC} is marked on this figure, where according to the aforementioned theorem, it is the frequency where $|T| \rightarrow |L|$, $|S| \rightarrow 1$, and $Y \rightarrow |G_c|$.

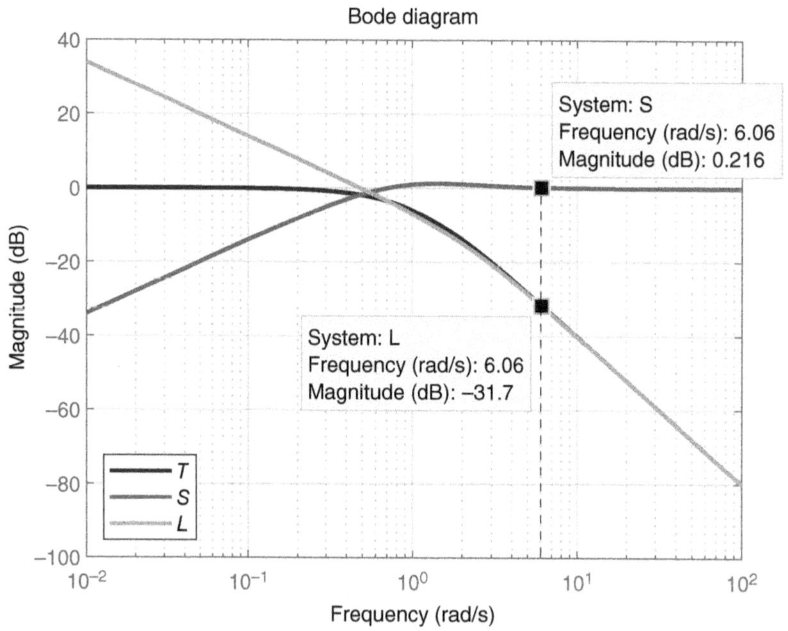

Figure C.8 Closed-Loop and Return Ratio Transfer Functions for Example C.2.

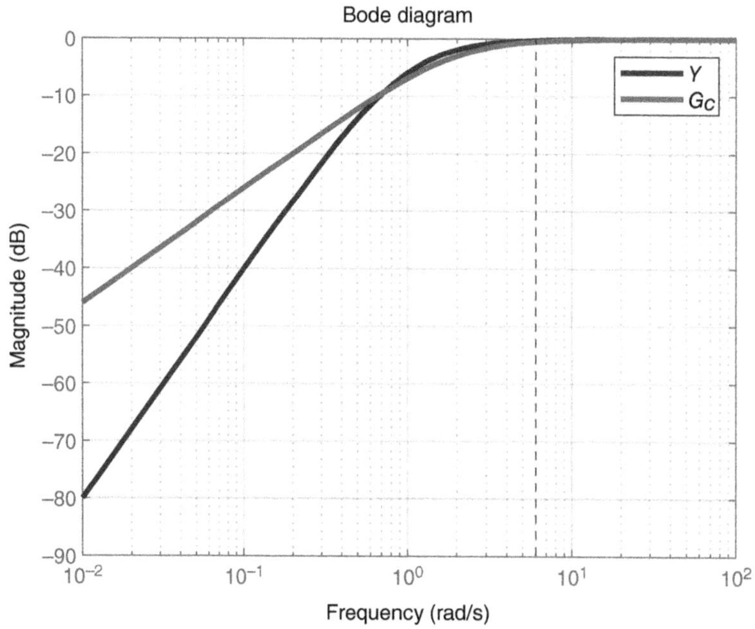

Figure C.9 Controller and Youla transfer functions for Example C.2.

The frequency where $Y \to |G_c|$ is illustrated in Figure C.9, this frequency corresponds to the same frequency ω_{BC} in Figure C.8.

C.4 Summary

The concept of closed-loop control system bandwidth has been discussed in many control texts and publications in the past. This concept is not well defined when one tries to use this frequency to relate the speed of closed-loop response in time domain due to inputs such as the unit step to a corresponding frequency known as closed-loop bandwidth in the frequency domain. It turns out that when one tries to extend the concept of bandwidth to the frequency at which benefits of feedback is lost, then this bandwidth does not simply play a role. In this work, we proposed a new and precise measure which is directly related to a particular frequency, when designing closed-loop feedback systems, after which the benefits of feedback are lost. We showed that high-frequency asymptotes of three closed-loop transfer functions, T, S, and Y are exactly the same asymptotes as when feedback information is lost. We modeled this information through a binary number α.

D

Example Matlab Code

This appendix provides three example m-files for designing and simulating Youla and H_∞ controllers in Matlab using Matlab's Control Systems Toolbox and Robust Control Toolbox.

D.1 Example 1

Consider a plant

$$G_p = 20\begin{bmatrix} \dfrac{1}{(s+2)^2} & \dfrac{s+1}{s+2} \\ \dfrac{-(s+1)}{s+2} & \dfrac{s+1}{s+2} \end{bmatrix} \tag{D.1}$$

Use Matlab to design and simulate a controller for this system.

```
%% defining plant tranfer function matrix
s = tf('s');

Gp = 20*[1/(s+2)^2 (s+1)/(s+2);-(s+1)/(s+2) (s+1)/(s+2)];

UL = [1 0;((s+1)*(s+2)) 1];
UR = [1 -((s+1)*(s+2));0 1];

%% Selecting Youla
Y1=1/5;
Y2=(1/5)*1/((s+1)*(s+2)*(s^2+3*s+3));

My=[Y1 0;0 Y2];

%% Transform into coupled system
Y = zpk(minreal(UR*My*UL,1e-3)); % bodemag(Y)

Ty = zpk(minreal(Gp*Y,1e-3)); % bodemag(T)
Sy = zpk(minreal(eye(2)-Ty,1e-3)); % bodemag(S)
Gc = zpk(minreal(Y*Sy^-1,1e-3)); % bodemag(Gc)
Ly = zpk(minreal(Gp*Gc));

Tu = zpk(minreal(Y*Gp,1e-3));
Su = zpk(minreal(eye(2)-Tu,1e-3));

%% Plot Results
figure(1)
sigma(Gp)
hold on
sigma(Gc)
sigma(Y)
hold off
legend('Gp','Gc','Y')
grid
```

Robust Control: Youla Parameterization Approach, First Edition. Farhad Assadian and Kevin R. Mallon.

Companion Website: www.wiley.com/go/assadian/robust

```
figure(2)
sigma(Ly)
hold on
sigma(Ty)
sigma(Sy)
hold off
legend('Ly','Ty','Sy')
grid

figure(3)
sigma(Sy)
hold on
sigma(Su)
hold off
legend('Sy','Su')
grid

figure(4)
t=0:0.01:10;
y =step(Ty,t);
k=1:length(t);
plot(t,y(k,1,1),'-');
title('Plot of step reponses to the first input')
hold on
plot(t,y(k,2,1),'--')
legend('y1(t)','y2(t)');
grid

figure(5)
t=0:0.01:10;
y =step(Ty,t);
k=1:length(t);
plot(t,y(k,1,2),'-');
title('Plot of step reponses to the second input')
hold on
plot(t,y(k,2,2),'--')
legend('y1(t)','y2(t)');
grid

figure(6)
t=0:0.01:10;
y =step(Y,t);
k=1:length(t);
plot(t,y(k,1,1),'-');
title('Plot of actuator reponses to the first input')
hold on
plot(t,y(k,2,1),'*')
legend('u1(t)','u2(t)');
grid

figure(7)
t=0:0.01:10;
y =step(Y,t);
k=1:length(t);
plot(t,y(k,1,2),'-');
title('Plot of actuator reponses to the first input')
hold on
plot(t,y(k,2,2),'*')
legend('u1(t)','u2(t)');
grid
```

D.2 Example 2

Consider a plant

$$G_p = \begin{bmatrix} 2 & \dfrac{-1}{s+1} \\ \dfrac{-1}{s^2} & \dfrac{1}{s^2(s+1)} \end{bmatrix} \tag{D.2}$$

Use Matlab to design and simulate a controller for this system.

```
%% Defining plant tranfer function matrix
s = tf('s');

Gp= [2 -1/(s+1);-1/s^2 1/(s^2*(s+1))];

ULinv=[-s^2 1;1 0];
URinv = [-(s+1) 1;1 0];
UL=ULinv^-1;
UR=URinv^-1;

%% Selecting Youla
Y1=(s^2*(4*s+1))/(s+1)^3;
Y2=1/(s+1)^3;

My=[Y1 0;0 Y2];

%% Transform into coupled system
Y = zpk(minreal(UR*My*UL,1e-3)); % bodemag(Y)

Ty = zpk(minreal(Gp*Y,1e-3)); % bodemag(T)
Sy = zpk(minreal(eye(2)-Ty,1e-3)); % bodemag(S)
Gc = zpk(minreal(Y*Sy^-1,1e-3)); % bodemag(Gc)
Ly=zpk(minreal(Gp*Gc));

Tu = zpk(minreal(Y*Gp,1e-3));
Su = zpk(minreal(eye(2)-Tu,1e-3));

%% Plot results
w=10^-2:0.01:10^3;
figure(1)
sigma(Gp,w)
hold on
sigma(Gc,w)
sigma(Y,w)
hold off
legend('Gp','Gc','Y')
grid

figure(2)
sigma(Ly,w)
hold on
sigma(Ty,w)
sigma(Sy,w)
hold off
legend('Ly','Ty','Sy')
grid

figure(3)
sigma(Sy,w)
hold on
sigma(Su,w)
hold off
legend('Sy','Su')
grid
```

```
figure(4)
t=0:0.01:10;
y =step(Ty,t);
k=1:length(t);
plot(t,y(k,1,1),'-');
title('Plot of step reponses to the first input')
hold on
plot(t,y(k,2,1),'--')
legend('y1(t)','y2(t)');
grid

figure(5)
t=0:0.01:10;
y =step(Ty,t);
k=1:length(t);
plot(t,y(k,1,2),'-');
title('Plot of step reponses to the second input')
hold on
plot(t,y(k,2,2),'--')
legend('y1(t)','y2(t)');
grid

figure(6)
t=0:0.01:10;
y =step(Y,t);
k=1:length(t);
plot(t,y(k,1,1),'-');
title('Plot of actuator reponses to the first input')
hold on
plot(t,y(k,2,1),'*')
legend('u1(t)','u2(t)');
grid

figure(7)
t=0:0.01:10;
y =step(Y,t);
k=1:length(t);
plot(t,y(k,1,2),'-');
title('Plot of actuator reponses to the first input')
hold on
plot(t,y(k,2,2),'*')
legend('u1(t)','u2(t)');
grid
```

D.3 Example 3

Consider a plant

$$G_p = 100\begin{bmatrix} \frac{s}{s+1} & \frac{s}{s+1} & \frac{1}{s+1} \\ \frac{1}{s+1} & \frac{1}{s+1} & \frac{1}{s+1} \end{bmatrix} \tag{D.3}$$

Use Matlab to design and simulate a controller for this system.

```
%% Defining plant tranfer function matrix
s = tf('s');

Gp= 100*[s/(s+1) s/(s+1) 1/(s+1);1/(s+1) 1/(s+1) 1/(s+1)];

ULinv=[1 -1;1 0];
URinv = [1 1 1;1 1 0;1 0 0];
```

```
UL=ULinv^-1;
UR=URinv^-1;

%% Selecting Youla
zeta=0.707;
wn1=1;
wn2=1;

Y1=(0.01*(wn2^2*(s+1)))/(s^2+2*zeta*wn2*s+wn2^2);
Y2=(0.01*(wn1^2*(s+1)))/(s^2+2*zeta*wn1*s+wn1^2);

My=[Y1 0;0 Y2;0 0];

%% Transform into coupled system
Y = zpk(minreal(UR*My*UL,1e-3)); % bodemag(Y)

Ty = zpk(minreal(Gp*Y,1e-3)); % bodemag(T)
Sy = zpk(minreal(eye(2)-Ty,1e-3)); % bodemag(S)
Gc = zpk(minreal(Y*Sy^-1,1e-3)); % bodemag(Gc)
Ly=zpk(minreal(Gp*Gc));

Tu = zpk(minreal(Y*Gp,1e-3));
Su = zpk(minreal(eye(3)-Tu,1e-3));

%% Plot Results
w=10^-2:0.01:10^3;
figure(1)
sigma(Gp,w)
hold on
sigma(Gc,w)
sigma(Y,w)
hold off
legend('Gp','Gc','Y')
grid

figure(2)
sigma(Ly,w)
hold on
sigma(Ty,w)
sigma(Sy,w)
hold off
legend('Ly','Ty','Sy')
grid

figure(3)
sigma(Sy,w)
hold on
sigma(Su,w)
hold off
legend('Sy','Su')
grid

figure(4)
t=0:0.01:10;
y =step(Ty,t);
k=1:length(t);
plot(t,y(k,1,1),'-');
title('Plot of step reponses to the first input')
hold on
plot(t,y(k,2,1),'--')
legend('y1(t)','y2(t)');
grid

figure(5)
```

```
t=0:0.01:10;
y =step(Ty,t);
k=1:length(t);
plot(t,y(k,1,2),'-');
title('Plot of step reponses to the second input')
hold on
plot(t,y(k,2,2),'--')
legend('y1(t)','y2(t)');
grid

figure(6)
t=0:0.01:10;
y =step(Y,t);
k=1:length(t);
plot(t,y(k,1,1),'-');
title('Plot of actuator reponses to the first input')
hold on
plot(t,y(k,2,1),'*')
legend('u1(t)','u2(t)');
grid

figure(7)
t=0:0.01:10;
y =step(Y,t);
k=1:length(t);
plot(t,y(k,1,2),'-');
title('Plot of actuator reponses to the first input')
hold on
plot(t,y(k,2,2),'*')
legend('u1(t)','u2(t)');
grid
```

D.4 Example 4

Consider a plant with damping ration $\zeta = 1/20$, natural frequency $\omega_n = 20$ rad/s, and static gain $K = 1/10$.

$$G_p = \frac{40}{s^2 + 2\left(\frac{1}{20}\right) 20s + 20^2} \tag{D.4}$$

Design an H_∞ controller for this system using Matlab.

```
%% Define Plant
numg=(1/10)*20^2;
deng=[1, 2*(1/20)*20, 20^2];

sys1=tf(numg,deng);
Gp=tf(sys1);

%% Hinf shaping filter
numw1=1.5*[1 350];
denw1=[1 1];
w1=[numw1;denw1];
W1=tf(numw1,denw1);

% numw2=100*[1 10]; % try different things here. TFs,
% constants, etc...
% denw2=[1 1000];
% w2=[numw2;denw2];
w2=[];

numw3=1.5*8000*[1 350]/350;
denw3=[1 8000];
w3=[numw3;denw3];
```

```
W3=tf(numw3,denw3);

% plot
figure(1)
sigma(inv(W1));hold on;
sigma(Gp/(1+Gp))
sigma(1/(1+Gp))
sigma(inv(W3));hold off;

%% Hinf Controller
ssga_=augtf(sys1,w1,w2,w3);
[sys3,sscl,gamma,info]=hinfsyn(ssga_); % Hinf controller

sysc = sys1*sys3;

sysclosed = feedback(sysc,1);

T=minreal(tf(sysclosed));
S=minreal(tf(1-sysclosed));
Gp=minreal(tf(sys1));
Y=minreal(T/Gp);
Gc=minreal(Y/S);

%% Plot Results
figure(2)
step(sysclosed,linspace(0,0.2,0.0001)')
title('closed-loop step response')

figure(3)
bode(1 - sysclosed)
hold on
bode(sysclosed)
bode(denw1,numw1)
bode(denw3,numw3)
hold off
legend('sensitivity','complementary sensitivity','invW1', 'invW3')

figure(4)
bode(denw1,numw1)
hold on
bode(denw3,numw3)
hold off
legend('1/W1', '1/W3')

figure(5)
bode(sysclosed)
title('frequency response')

figure(6)
bode(T);
hold on;
bode(S);
bode(Y);
bode(Gc);
bode(Gp);
legend('T','S','Y','Gc','Gp')
hold off
```

References

1 Youla, D.C., Bongiorno, J.J., and Lu, C.N. (1974). Single-loop feedback-stabilization of linear multivariable dynamical plants. *Automatica* 10 (2): 159–173.

2 Maciejowski, J.M. (1989). *Multivariable Feedback Design*, 1e. Wokingham: Addison-Wesley.

3 Skogestad, S. and Postlethwaite, I. (2007). *Multivariable Feedback Control: Analysis and Design*, vol. 2. New York: Wiley.

4 Krstic, M., Kokotovic, P.V., and Kanellakopoulos, I. (1995). *Nonlinear and Adaptive Control Design*. Wiley.

5 Curtain, R.F. (1991). State-space approaches to H-infinity control for infinite-dimensional linear systems. *Transactions of the Institute of Measurement and Control* 13 (5): 253–261.

6 Protter, M.H. (1998). *Basic Elements of Real Analysis*. Springer Science & Business Media.

7 Churchill, R.V. (1972). *Operational Mathematics*. McGraw-Hill College: New York, NY.

8 Lathi, B.P. and Green, R.A. (2005). *Linear Systems and Signals*, vol. 2. New York: Oxford University Press.

9 Brewer, J.W. (1974). *Control Systems: Analysis, Design, and Simulation (Book)*, 560 p. Englewood Cliffs, NJ: Prentice-Hall, Inc.

10 Hörmander, L. (1976). *Linear Partial Differential Operators: 4th Printing*. Springer.

11 Hanta, V. and Procházka, A. (2009). Rational approximation of time delay. *Institute of Chemical Technology in Prague. Department of Computing and Control Engineering. Technická* 5 (166): 28.

12 Sun, T., Assadian, F., and Du, P.-p. (2017). An integrated mechanical/control design approach to optimisation of an electric power steering system under objective evaluation of vehicle handling and stability. *International Journal of Vehicle Design* 74 (3): 257–279.

13 Morari, M. and Zafiriou, E. (1989). *Robust Process Control* Prentice Hall: Englewood Cliffs, NJ.

14 Doyle, J.C., Francis, B.A., and Tannenbaum, A.R. (2013). *Feedback Control Theory*. Courier Corporation.

15 Ozkan, B., Margolis, D., and Pengov, M. (2008). The controller output observer: estimation of vehicle tire cornering and normal forces. *Journal of Dynamic Systems, Measurement, and Control* 130 (6): 061002.

16 Luenberger, D. (1971). An introduction to observers. *IEEE Transactions on Automatic Control* 16 (6): 596–602.

17 Welch, G. and Bishop, G. (1995). *An Introduction to the Kalman Filter* University of North Carolina at Chapel Hill: Chapel Hill, NC.

18 Vinnicombe, G. (2000). *Uncertainty and Feedback: H∞ Loop-Shaping and the ν-Gap Metric*. World Scientific.

19 Kuo, B.C. (1987). *Automatic Control Systems*. Prentice Hall PTR.

20 Nise, N.S. (2015). *Control Systems Engineering*, 7e. Wiley.

21 Garcia, C.E. and Morari, M. (1982). Internal model control. A unifying review and some new results. *Industrial & Engineering Chemistry Process Design and Development* 21 (2): 308–323.

22 Crosby, M.J., Harwood, R.A., and Karnopp, D. (1974). Vibration control using semi-active force generators. *Journal of Engineering for Industry* 96 (2): 619–626.

23 Kalman, R.E. (1960). A new approach to linear filtering and prediction problems. *Journal of Basic Engineering* 82 (1): 35–45.

24 Grip, H.F., Imsland, L., Johansen, T.A. et al. (2009). Vehicle sideslip estimation. *IEEE Control Systems* 29 (5): 36–52.

25 Pacejka, H. (2005). *Tire and Vehicle Dynamics*. Elsevier.

26 JVa (2006). *SCA830-D05 Single Axis Accelerometer with Digital SPI Interface*. VTI Technologies Oy.

27 Assadian, F., Beckerman, A.K., and Alcantar, J.V. (2018). Estimation design using Youla parametrization with automotive applications. *Journal of Dynamic Systems, Measurement, and Control* 140 (8): 081015–11.

Robust Control: Youla Parameterization Approach, First Edition. Farhad Assadian and Kevin R. Mallon.

Companion Website: www.wiley.com/go/assadian/robust

28 Brogan, W.L. (1991). *Modern Control Theory*. India: Pearson Education.
29 Lavretsky, E. and Wise, K. (2013). *Robust and Adaptive Control*. Springer-Verlag: London.
30 Doyle, J. and Stein, G. (1979). Robustness with observers. *IEEE Transactions on Automatic Control* 24 (4): 607–611.
31 Zames, G. (1981). Feedback and optimal sensitivity: model reference transformations, multiplicative seminorms, and approximate inverses. *IEEE Transactions on Automatic Control* 26 (2): 301–320.
32 Mackenroth, U. (2004). *Robust Control Systems: Theory and Case Studies*. Springer Science & Business Media.
33 Assadian, F. and Aneke, E. (2006). Two distinct methods for integration of active differential and active roll control systems. *International Journal of Vehicle Design* 42 (3–4): 348–369.
34 Stengel, R.F. (1994). *Optimal Control and Estimation*. Courier Corporation.
35 Glover, K. (1984). All optimal Hankel-norm approximations of linear multivariable systems and their l_∞-error bounds. *International Journal of Control* 39 (6): 1115–1193.
36 Doyle, J.C. (1985). Structured uncertainty in control system design. *1985 24th IEEE Conference on Decision and Control*, pp. 260–265. IEEE.
37 Boyd, S., Boyd, S.P., and Vandenberghe, L. (2004). *Convex Optimization*. Cambridge University Press.
38 Beckerman, A. and Assadian, F. (2017). A new estimation concept for automotive suspensions. *Proceedings of the 20th World Congress*. International Federation of Automatic Control.
39 Beckerman, A. and Assadian, F. (2016). Generalized template for suspension dynamic modeling with the use of bond graphs. *Proceedings of the 49th Annual Simulation Symposium*, p. 10. Society for Computer Simulation International.
40 Beckerman, A. and Assadian, F. (2016). Modeling ground vehicles with active suspension kinematics using bond graphs. *Proceedings of the International Conference on Bond Graph Modeling and Simulation*, pp. 54–63. Society for Computer Simulation International.
41 Gillespie, T.D. (1985). Heavy Truck Ride. *Technical report 850001*. SAE Technical Paper.
42 Tan, K.K., Lee, T.H., and Huang, S. (2007). *Precision Motion Control: Design and Implementation*. Springer Science & Business Media.
43 Karnopp, D. (2013). *Vehicle Dynamics, Stability, and Control*. CRC Press.
44 Lee, H.G., Sung, K.G., and Choi, S.B. (2009). Ride comfort characteristics with different tire pressure of passenger vehicle featuring MR damper. *Journal of Physics: Conference Series* 149: 012069.
45 Balas, G., Chiang, R., Packard, A., and Safonov, M. (2021). *Robust Control Toolbox User's Guide*. The Math Works, Inc.: Natick, MA.
46 Isermann, R. (2014). *Engine Modeling and Control*, 1017. Berlin, Heidelberg: Springer.
47 Petersen, D., Charvat, J., Somers, J. et al. (2020). Apollo to Artemis: Mining 50-Year Old Records to Inform Future Human Lunar Landing Systems.
48 Klumpp, A.R. (1974). Apollo lunar descent guidance. *Automatica* 10 (2): 133–146.
49 Quaide, W.L. and Oberbeck, V.R. (1969). Geology of the Apollo landing sites. *Earth-Science Reviews* 5 (4): 255–278.
50 Dueri, D., Açıkmeşe, B., Scharf, D.P., and Harris, M.W. (2017). Customized real-time interior-point methods for onboard powered-descent guidance. *Journal of Guidance, Control, and Dynamics* 40 (2): 197–212.
51 Scharf, D.P., Açıkmeşe, B., Dueri, D. et al. (2017). Implementation and experimental demonstration of onboard powered-descent guidance. *Journal of Guidance, Control, and Dynamics* 40 (2): 213–229.
52 Açıkmeşe, B., Carson, J.M., and Blackmore, L. (2013). Lossless convexification of nonconvex control bound and pointing constraints of the soft landing optimal control problem. *IEEE Transactions on Control Systems Technology* 21 (6): 2104–2113.
53 Açıkmeşe, B., Scharf, D., Blackmore, L., and Wolf, A. (2008). Enhancements on the convex programming based powered descent guidance algorithm for mars landing. *AIAA/AAS Astrodynamics Specialist Conference and Exhibit*, p. 6426.
54 Ge, Z.M. and Cheng, Y.H. (1982). Extended Kane's equations for nonholonomic variable mass system 49 (2): 429–431. https://doi.org/10.1115/1.3162105.
55 Wang, Y., Rajamani, R., and Zemouche, A. (2018). Sequential LMI apprach for the design of a BMI-based robust observer state feedback controller with nonlinear uncertanties. *International Journal of Nonlinear Control* 28 (1): 1246–1260.
56 Pacejka, H.B. and Bakker, E. (1992). The magic formula tyre model. *Vehicle System Dynamics* 21 (S1): 1–18.
57 Brewer, J. (1995). *MAE 272 - Real and Complex Analysis in Signal, Dynamic System, and Feedback Theories, Lecture Notes*. Davis, CA: University of California.

Index

Robust Control: Youla Parameterization Approach, First Edition. Farhad Assadian and Kevin R. Mallon.

Companion Website: www.wiley.com/go/assadian/robust